Next Generation Artificial Vision Systems

Reverse Engineering the Human Visual System

Artech House Series
Bioinformatics & Biomedical Imaging

Series Editors
Stephen T. C. Wong, The Methodist Hospital Research Institute
Guang-Zhong Yang, Imperial College

Advances in Diagnostic and Therapeutic Ultrasound Imaging, Jasjit S. Suri, Chirinjeev Kathuria, Ruey-Feng Chang, Filippo Molinari, and Aaron Fenster, editors

Biological Database Modeling, Jake Chen and Amandeep S. Sidhu, editors

Biomedical Informatics in Translational Research, Hai Hu, Michael Liebman, and Richard Mural

Genome Sequencing Technology and Algorithms, Sun Kim, Haixu Tang, and Elaine R. Mardis, editors

Life Science Automation Fundamentals and Applications, Mingjun Zhang, Bradley Nelson, and Robin Felder, editors

Microscopic Image Analysis for Life Science Applications, Jens Rittscher, Stephen T. C. Wong, and Raghu Machiraju, editors

Next Generation Artificial Vision Systems: Reverse Engineering the Human Visual System, Maria Petrou and Anil Bharath, editors

Systems Bioinformatics: An Engineering Case-Based Approach, Gil Alterovitz and Marco F. Ramoni, editors

Next Generation Artificial Vision Systems

Reverse Engineering the Human Visual System

Anil Bharath
Maria Petrou

Imperial College London

ARTECH HOUSE
BOSTON | LONDON
artechhouse.com

The catalog record for this book is available from the U.S. Library of Congress.
The catalogue record for this book is available from the British Library.

ISBN 1-59693-224-4
ISBN 13 978-1-59693-224-1

Cover design by Igor Valdman
© 2008 ARTECH HOUSE, INC.
685 Canton Street
Norwood, MA 02062

All rights reserved. Printed and bound in the United States of America. No part of this book may be reproduced or utilized in any form or by any means, electronic or mechanical, including photocopying, recording, or by any information storage and retrieval system, without permission in writing from the publisher.

All terms mentioned in this book that are known to be trademarks or service marks have been appropriately capitalized. Artech House cannot attest to the accuracy of this information. Use of a term in this book should not be regarded as affecting the validity of any trademark or service mark.

10 9 8 7 6 5 4 3 2 1

Contents

Preface — xiii

CHAPTER 1
The Human Visual System: An Engineering Challenge — 1

1.1 Introduction — 1
1.2 Overview of the Human Visual System — 2
    1.2.1 The Human Eye — 3
        1.2.1.1 Issues to Be Investigated — 8
    1.2.2 Lateral Geniculate Nucleus (LGN) — 10
    1.2.3 The V1 Region of the Visual Cortex — 12
        1.2.3.1 Issues to Be Investigated — 14
    1.2.4 Motion Analysis and V5 — 15
        1.2.4.1 Issues to Be Investigated — 15
1.3 Conclusions — 15
    References — 17

PART I
The Physiology and Psychology of Vision — 19

CHAPTER 2
Retinal Physiology and Neuronal Modeling — 21

2.1 Introduction — 21
2.2 Retinal Anatomy — 21
2.3 Retinal Physiology — 25
2.4 Mathematical Modeling—Single Cells of the Retina — 27
2.5 Mathematical Modeling—The Retina and Its Functions — 28
2.6 A Flexible, Dynamical Model of Retinal Function — 30
    2.6.1 Foveal Structure — 31
    2.6.2 Differential Equations — 32
    2.6.3 Color Mechanisms — 34
    2.6.4 Foveal Image Representation — 36
    2.6.5 Modeling Retinal Motion — 37
2.7 Numerical Simulation Examples — 38
    2.7.1 Parameters and Visual Stimuli — 38
    2.7.2 Temporal Characteristics — 39
    2.7.3 Spatial Characteristics — 41
    2.7.4 Color Characteristics — 43
2.8 Conclusions — 45
    References — 46

CHAPTER 3
A Review of V1 — 51

3.1 Introduction — 51
3.2 Two Aspects of Organization and Functions in V1 — 52
    3.2.1 Single-Neuron Responses — 52
    3.2.2 Organization of Individual Cells in V1 — 53
        3.2.2.1 Orientation Selectivity — 55
        3.2.2.2 Color Selectivity — 56
        3.2.2.3 Scale Selectivity — 57
        3.2.2.4 Phase Selectivity — 58
3.3 Computational Understanding of the Feed Forward V1 — 58
    3.3.1 V1 Cell Interactions and Global Computation — 59
    3.3.2 Theory and Model of Intracortical Interactions in V1 — 61
3.4 Conclusions — 62
    References — 63

CHAPTER 4
Testing the Hypothesis That V1 Creates a Bottom-Up Saliency Map — 69

4.1 Introduction — 69
4.2 Materials and Methods — 73
4.3 Results — 75
    4.3.1 Interference by Task-Irrelevant Features — 76
    4.3.2 The Color-Orientation Asymmetry in Interference — 81
    4.3.3 Advantage for Color-Orientation Double Feature but Not Orientation-Orientation Double Feature — 84
    4.3.4 Emergent Grouping of Orientation Features by Spatial Configurations — 87
4.4 Discussion — 92
4.5 Conclusions — 98
    References — 99

PART II
The Mathematics of Vision — 103

CHAPTER 5
V1 Wavelet Models and Visual Inference — 105

5.1 Introduction — 105
    5.1.1 Wavelets — 105
    5.1.2 Wavelets in Image Analysis and Vision — 107
    5.1.3 Wavelet Choices — 107
    5.1.4 Linear vs Nonlinear Mappings — 112
5.2 A Polar Separable Complex Wavelet Design — 113

	5.2.1	Design Overview	113
	5.2.2	Filter Designs: Radial Frequency	114
	5.2.3	Angular Frequency Response	116
	5.2.4	Filter Kernels	118
	5.2.5	Steering and Orientation Estimation	119
5.3	The Use of V1-Like Wavelet Models in Computer Vision		120
	5.3.1	Overview	120
	5.3.2	Generating Orientation Maps	121
	5.3.3	Corner Likelihood Response	123
	5.3.4	Phase Estimation	123
5.4	Inference from V1-Like Representations		124
	5.4.1	Vector Image Fields	125
	5.4.2	Formulation of Detection	126
	5.4.3	Sampling of $(\mathcal{B}, \mathcal{X})$	127
	5.4.4	The Notion of "Expected" Vector Fields	128
	5.4.5	An Analytic Example: Uniform Intensity Circle	129
	5.4.6	Vector Model Plausibility and Extension	129
	5.4.7	Vector Fields: A Variable Contrast Model	130
	5.4.8	Plausibility by Demonstration	131
	5.4.9	Plausibility from Real Image Data	132
	5.4.10	Divisive Normalization	133
5.5	Evaluating Shape Detection Algorithms		135
	5.5.1	Circle-and-Square Discrimination Test	135
5.6	Grouping Phase-Invariant Feature Maps		138
	5.6.1	Keypoint Detection Using DTCWT	138
5.7	Summary and Conclusions		140
	References		141

CHAPTER 6
Beyond the Representation of Images by Rectangular Grids — 145

6.1	Introduction	145
6.2	Linear Image Processing	145
	6.2.1 Interpolation of Irregularly Sampled Data	146
	6.2.1.1 Kriging	146
	6.2.1.2 Iterative Error Correction	151
	6.2.1.3 Normalized Convolution	153
	6.2.2 DFT from Irregularly Sampled Data	156
6.3	Nonlinear Image Processing	157
	6.3.1 V1-Inspired Edge Detection	158
	6.3.2 Beyond the Conventional Data Representations and Object Descriptors	162
	6.3.2.1 The Trace Transform	162
	6.3.2.2 Features from the Trace Transform	165

6.4	Reverse Engineering Some Aspect of the Human Visual System	167
6.5	Conclusions	168
	References	169

CHAPTER 7
Reverse Engineering of Human Vision: Hyperacuity and Super-Resolution 171

7.1	Introduction	171
7.2	Hyperacuity and Super-Resolution	172
7.3	Super-Resolution Image Reconstruction Methods	173
	7.3.1 Constrained Least Squares Approach	174
	7.3.2 Projection onto Convex Sets	177
	7.3.3 Maximum A Posteriori Formulation	180
	7.3.4 Markov Random Field Prior	180
	7.3.5 Comparison of the Super-Resolution Methods	183
	7.3.6 Image Registration	183
7.4	Applications of Super-Resolution	184
	7.4.1 Application in Minimally Invasive Surgery	184
	7.4.2 Other Applications	187
7.5	Conclusions and Further Challenges	188
	References	188

CHAPTER 8
Eye Tracking and Depth from Vergence 191

8.1	Introduction	191
8.2	Eye-Tracking Techniques	192
8.3	Applications of Eye Tracking	195
	8.3.1 Psychology/Psychiatry and Cognitive Sciences	195
	8.3.2 Behavior Analysis	196
	8.3.3 Medicine	197
	8.3.4 Human–Computer Interaction	199
8.4	Gaze-Contingent Control for Robotic Surgery	200
	8.4.1 Ocular Vergence for Depth Recovery	202
	8.4.2 Binocular Eye-Tracking Calibration	204
	8.4.3 Depth Recovery and Motion Stabilization	206
8.5	Discussion and Conclusions	209
	References	210

CHAPTER 9
Motion Detection and Tracking by Mimicking Neurological Dorsal/Ventral Pathways 217

9.1	Introduction	217
9.2	Motion Processing in the Human Visual System	218
9.3	Motion Detection	219

		9.3.1	Temporal Edge Detection	221
		9.3.2	Wavelet Decomposition	224
		9.3.3	The Spatiotemporal Haar Wavelet	225
		9.3.4	Computational Cost	230
	9.4	Dual-Channel Tracking Paradigm		230
		9.4.1	Appearance Model	231
		9.4.2	Early Approaches to Prediction	232
		9.4.3	Tracking by Blob Sorting	233
	9.5	Behavior Recognition and Understanding		237
	9.6	A Theory of Tracking		239
	9.7	Concluding Remarks		241
		References		242

PART III
Hardware Technologies for Vision — 249

CHAPTER 10
Organic and Inorganic Semiconductor Photoreceptors
Mimicking the Human Rods and Cones — 251

10.1	Introduction		251
10.2	Phototransduction in the Human Eye		253
	10.2.1	The Physiology of the Eye	253
	10.2.2	Phototransduction Cascade	255
		10.2.2.1 Light Activation of the Cascade	257
		10.2.2.2 Deactivation of the Cascade	258
	10.2.3	Light Adaptation of Photoreceptors: Weber-Fechner's Law	258
	10.2.4	Some Engineering Aspects of Photoreceptor Cells	259
10.3	Phototransduction in Silicon		260
	10.3.1	CCD Photodetector Arrays	262
	10.3.2	CMOS Photodetector Arrays	263
	10.3.3	Color Filtering	265
	10.3.4	Scaling Considerations	268
10.4	Phototransduction with Organic Semiconductor Devices		269
	10.4.1	Principles of Organic Semiconductors	270
	10.4.2	Organic Photodetection	271
	10.4.3	Organic Photodiode Structure	273
	10.4.4	Organic Photodiode Electronic Characteristics	274
		10.4.4.1 Photocurrent and Efficiency	274
		10.4.4.2 The Equivalent Circuit and Shunt Resistance	277
		10.4.4.3 Spectral Response Characteristics	281
	10.4.5	Fabrication	281
		10.4.5.1 Contact Printing	282
		10.4.5.2 Printing on CMOS	284
10.5	Conclusions		285
	References		286

CHAPTER 11
Analog Retinomorphic Circuitry to Perform Retinal
and Retinal-Inspired Processing — 289

- 11.1 Introduction — 289
- 11.2 Principles of Analog Processing — 290
  - 11.2.1 The Metal Oxide Semiconductor Field Effect Transistor — 292
    - 11.2.1.1 Transistor Operation — 293
    - 11.2.1.2 nMOS and pMOS Devices — 293
    - 11.2.1.3 Transconductance Characteristics — 293
    - 11.2.1.4 Inversion Characteristics — 294
    - 11.2.1.5 MOSFET Weak Inversion and Biological Gap Junctions — 295
  - 11.2.2 Analog vs Digital Methodologies — 296
- 11.3 Photo Electric Transduction — 296
  - 11.3.1 Logarithmic Sensors — 297
  - 11.3.2 Feedback Buffers — 298
  - 11.3.3 Integration-Based Photodetection Circuits — 298
  - 11.3.4 Photocurrent Current-Mode Readout — 300
- 11.4 Retinimorphic Circuit Processing — 300
  - 11.4.1 Voltage Mode Resistive Networks — 301
    - 11.4.1.1 Limitations with This Approach — 303
  - 11.4.2 Current Mode Approaches to Receptive Field Convolution — 303
    - 11.4.2.1 Improved Horizontal Cell Circuitry — 305
    - 11.4.2.2 Novel Bipolar Circuitry — 305
    - 11.4.2.3 Bidirectional Current Mode Processing — 306
    - 11.4.2.4 Dealing with Multiple High Impedance Processing Channels — 307
    - 11.4.2.5 The Current Comparator — 310
  - 11.4.3 Reconfigurable Fields — 312
  - 11.4.4 Intelligent Ganglion Cells — 314
    - 11.4.4.1 ON–OFF Ganglion Cells — 315
    - 11.4.4.2 Pulse Width Encoding — 316
- 11.5 Address Event Representation — 317
  - 11.5.1 The Arbitration Tree — 318
  - 11.5.2 Collisions — 322
  - 11.5.3 Sparse Coding — 322
  - 11.5.4 Collision Reduction — 323
- 11.6 Adaptive Foveation — 324
  - 11.6.1 System Algorithm — 325
  - 11.6.2 Circuit Implementation — 326
  - 11.6.3 The Future — 329
- 11.7 Conclusions — 330
  - References — 330

CHAPTER 12
Analog V1 Platforms — 335

12.1 Analog Processing: Obsolete? — 335
12.2 The Cellular Neural Network — 340
12.3 The Linear CNN — 340
12.4 CNNs and Mixed Domain Spatiotemporal Transfer Functions — 342
12.5 Networks with Temporal Derivative Diffusion — 345
    12.5.1 Stability — 348
12.6 A Signal Flow Graph–Based Implementation — 349
    12.6.1 Continuous Time Signal Flow Graphs — 349
    12.6.2 On SFG Relations with the MLCNN — 352
12.7 Examples — 355
    12.7.1 A Spatiotemporal Cone Filter — 355
    12.7.2 Visual Cortical Receptive Field Modelling — 360
12.8 Modeling of Complex Cell Receptive Fields — 362
12.9 Summary and Conclusions — 363
    References — 364

CHAPTER 13
From Algorithms to Hardware Implementation — 367

13.1 Introduction — 367
13.2 Field Programmable Gate Arrays — 367
    13.2.1 Circuit Design — 369
    13.2.2 Design Process — 369
13.3 Mapping Two-Dimensional Filters onto FPGAs — 369
13.4 Implementation of Complex Wavelet Pyramid on FPGA — 370
    13.4.1 FPGA Design — 370
    13.4.2 Host Control — 373
    13.4.3 Implementation Analysis — 374
    13.4.4 Performance Analysis — 375
        13.4.4.1 Corner Detection — 377
    13.4.5 Conclusions — 377
13.5 Hardware Implementation of the Trace Transform — 377
    13.5.1 Introduction to the Trace Transform — 377
    13.5.2 Computational Complexity — 381
    13.5.3 Full Trace Transform System — 382
        13.5.3.1 Acceleration Methods — 382
        13.5.3.2 Target Board — 383
        13.5.3.3 System Overview — 383
        13.5.3.4 Top-Level Control — 384
        13.5.3.5 Rotation Block — 384
        13.5.3.6 Functional Blocks — 386
        13.5.3.7 Initialization — 386
    13.5.4 Flexible Functionals for Exploration — 387
        13.5.4.1 Type A Functional Block — 388

	13.5.4.2 Type B Functional Block	388
	13.5.4.3 Type C Functional Block	389
13.5.5	Functional Coverage	389
13.5.6	Performance and Area Results	389
13.5.7	Conclusions	391
13.6 Summary		391
References		392

CHAPTER 14
Real-Time Spatiotemporal Saliency 395

14.1 Introduction		395
14.2 The Framework Overview		396
14.3 Realization of the Framework		398
14.3.1	Two-Dimensional Feature Detection	398
14.3.2	Feature Tracker	399
14.3.3	Prediction	404
14.3.4	Distribution Distance	406
14.3.5	Suppression	410
14.4 Performance Evaluation		411
14.4.1	Adaptive Saliency Responses	411
14.4.2	Complex Scene Saliency Analysis	412
14.5 Conclusions		413
References		413

Acronyms and Abbreviations	415
About the Editors	419
List of Contributors	420
Index	423

Preface

If you were trying to capture a photo by inserting the film back to front and by shaking the camera, justifiably, you might be considered mad. And yet, that is what the human visual system does! The human visual system is dynamic, multirate, adaptive, and bizarre from the engineering point of view. In fact, scientists have tried to develop artificial vision systems for years, and they go about the task at orthogonal directions to what Nature has done: the human visual system relies on analog rather than digital technology; it relies on organic rather than silicon sensors; it uses irregular sampling patterns to create an image rather than sensors arranged on a rectangular grid; and it works!

This book is the result of the collective work of several scientists in the past 5 years, working towards the development of technologies inspired by the human visual system. Such an endeavor is multidisciplinary and very broad in scope, requiring the expertise of neurologists, psychologists, mathematicians, physicists, computer scientists, and software as well as hardware engineers. No single group, yet alone a single person, can solve the problems associated with understanding the human visual system in a few years, and so inevitably, this book covers only some aspects of the issues involved. Generally speaking, there are three recognised perspectives to "reverse-engineering" the human visual system: approaching the problem from the viewpoint of the physiologist/cognitive psychologist; approaching the problem from the viewpoint of the software engineer, and approaching the problem from the hardware perspective. We organized the contents of this book to reflect these three perspectives.

The first chapter of the book presents a general introduction and gives a broad overview of the human visual system, containing pointers to the remaining chapters of the book. Part I of the book then follows, containing chapters on the physiology/psychology of vision. Chapter 2 provides an example of a modeling approach to the retina which is dynamical, and in which individual cells of the retina are explicitly modelled via a coupled set of differential equations that describe physiological cell-level functions. This is contrasted, in Chapter 3, which is an introduction to the broader functional organisation of receptive field characteristics of cells in V1, the primary visual cortex. Finally, Chapter 4 considers psychophysical experiments designed to probe models of V1 processing through psychophysical experiments that address visual attention. These three areas are further addressed in Parts II and III, from the mathematical and software, and from the hardware engineering perspectives, respectively.

Part II begins with Chapter 5, which models the workings of V1 as a spatial frequency analyser and discusses wavelet analysis from this point of view. Starting from the observation that the sensors in the human retina are placed on an irregular sampling pattern, Chapter 6 provides the mathematical background for performing image processing with irregularly sampled data. Chapter 7 examines

xiii

the relationship between super-resolution techniques and some characteristic movements of the human eye, known as tremor and microsaccades. Chapter 8 exploits the vergence of the eyes of the viewer to estimate the depth of the viewed surface, and thus compensate for its motion, with the specific application in mind of robotic assisted surgery. Finally, Chapter 9 considers motion detection algorithms inspired by the way the human brain detects motion.

Part III deals with hardware aspects of human vision-inspired technology. Starting with the development of polymer sensors that can imitate spectral response characteristics in the human retina, presented in Chapter 10, it proceeds to consider in Chapter 11 the development of hybrid chips that combine organic (polymer) sensors and analog circuitry. Chapter 12 deals with the design of models of very large scale integrated (VLSI) analog circuits, that may be used to implement classical simple cell V1 receptive fields. Chapter 13 shows how some of the algorithms discussed in Part II may be implemented in digital hardware, while Chapter 14 considers aspects of pre-attentive vision in terms of saliency and attention caused by spatial as well as temporal saliency created by motion.

In these 14 chapters, the subject is by no means exhausted and it will probably keep scientists busy for many years to come. This book is a small contribution towards the holy grail of cognitive vision: to reverse engineer the human visual system and to reproduce its functionality for robotic applications, like robotic assisted surgery and automatic driving, medical research, like drug development and impairment diagnosis, and finally the development of artificial low powered implantable retinas.

The authors of this book are very grateful to the Research Councils of the UK for their generous support of this research over a period of 4 years, under the scheme "Basic Technology."

For more information, please go to http://www.bg.ic.ac.uk/staff/aab01/book/

Anil Bharath
Maria Petrou
July 2008

CHAPTER 1
The Human Visual System: An Engineering Challenge

Maria Petrou and Anil Bharath

1.1 Introduction

The functionality of certain physical organs has fascinated and inspired humans ever since they developed self-consciousness. An example is the wings of birds. From the myth of Icarus to the aspiring fools of medieval times, countless humans lost their lives attempting to fly by wearing wings that *looked like* those they could see on the bodies of birds, and jumping off cliffs while flapping them. Flying was not achieved until people understood the laws of physics and engineered their own solutions to the problem of flying. Artificial vision research has often followed similar paths: a pair of cameras to play the role of eyes; a computer to analyze the digital images; a dead end! This was, for example, the fate of the line of research followed by early computer vision researchers in the 1960s, who developed vision systems for block world understanding[1]: most of the solutions offered were not transferable to vision in the natural world. In spite of this, that line of research opened the way for many interesting directions of research, all moving painfully slowly towards the ultimate aim of unrestricted human-like artificial vision. It is not because researchers have not appreciated the significance of the analogy with flying; it is mostly because just like flying, our understanding of the physics, the maths, and the engineering (and perhaps some other science with concepts not yet fully invented or developed) behind biological vision has not matured yet. And just like airplane flying, which depends on different mechanisms from the mechanism used by birds, perhaps after we have developed that understanding of biological vision, we might be able to invent solutions to the problems that are based on totally different principles from those of biological vision.

This book was compiled in order to contribute towards this direction. The topic of biological vision is very broad, and it is not possible to cover all aspects of it in a single volume. We have selected some parts of it that we feel are most relevant to current technologies, and for which our understanding is closer to our grasp, and examine them from an engineering point of view. But, we go beyond that: we use the human visual system as a springboard that inspires us to question aspects of our current technology that are often taken for granted, for example, the convention of sampling the continuous world using a discrete regular grid of sampling points. We also present methodologies that were developed trying to imitate the human visual system, like methods of constructing organic sensors with responses to the

---

1. Early computer vision systems were assuming that the world in which they had to operate consisted of geometric solids or blocks. This line of research is known as "block world understanding."

electromagnetic spectrum similar to those of the human eye, which have much broader applications in printing, broadcasting, and the entertaining industry.

This chapter presents an overview of the human visual system and highlights the aspects of it that inspired the research that is presented in the subsequent chapters. It also underlines the engineering questions behind these aspects and explains why an engineer should be interested in studying them.

1.2 Overview of the Human Visual System

The best way to present the human visual system is to follow the path of light as it enters the eye. This is depicted in Figure 1.1, where the human brain is shown as seen from above.

Light enters the eye and stimulates the sensors at the back of it. The signal that is created then travels through the optic nerve, crossing over the nerve coming from the other eye, and reaching an organ, deep inside the brain in the thalamus area, called the lateral geniculate nucleus (LGN). Outputs of the LGN are sent to the visual cortex at the back of the brain. The visual cortex is perhaps the most complex part of the human body. It is the place in the brain where most of the visual processing takes place. In order to study it, neuroscientists have divided it into regions labelled with the letter "V" followed by a number. The largest of these regions, where most of the preliminary processing is believed to take place, is the so-called region "V1." V1 is also known as striate cortex, because it lies within a region having a striated appearance in cross section, and as anatomical region 17. The other regions of the visual cortex, namely V2, V3, V4, and V5, lie outside the striated region of the brain. Recently, other retinotropically mapped areas with

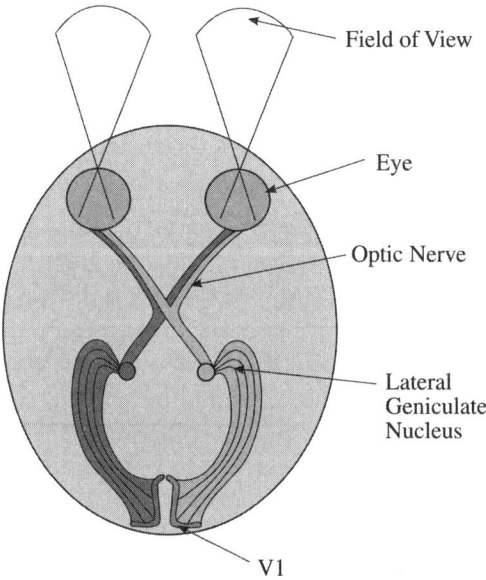

Figure 1.1 The path the optical signal follows inside the human head.

suggested designations V6, V7, and V8 have also been tentatively identified. In this book we mainly concentrate on technologies inspired from the eye, from V1, and from the regions of the visual cortex which are responsible for motion detection. Although several parts of the visual cortex contribute to motion analysis, the region most associated with it is V5, and that is why we shall review briefly in this chapter the structure of the eye, of V1, and of V5. For the sake of completeness, we shall also include some information on LGN, although no chapter in the book is related to its functionality.

Generally speaking, the further along the visual pathway we move away from the eye, the less we understand what is going on. An exception to this is an increasingly clear understanding of the role of V1, more so than our understanding of the LGN, which is anatomically less easily accessible. In the rest of this section, we shall follow the path of the signal from the eye to the visual cortex, and at each stage we shall consider which aspects of the organs involved are of interest to a vision engineer and why, and point out the relevant chapters of the rest of the book.

1.2.1 The Human Eye

Figure 1.2 is a photograph of what the eye looks like when viewed through the pupil. This visible layer is called the *retina* and it contains two important regions: the *fovea* and the place, known as the *optic disc*, where the optic nerve leaves the eye to transfer its signal to the brain. Figure 1.3(a) shows a schematic representation of a cross section of the human eye. At the back of it is the retina which contains the photoreceptors. Figure 1.3(b) shows in magnification a small cross section of the retina. It consists of five layers of distinct cells. The remarkable thing apparent from this cross section is that the photosensitive cells are those at the back of the layer, that is, those farthest away from the entrance of the eye. This means that the light has to travel through four layers of cells before it reaches the sensors!

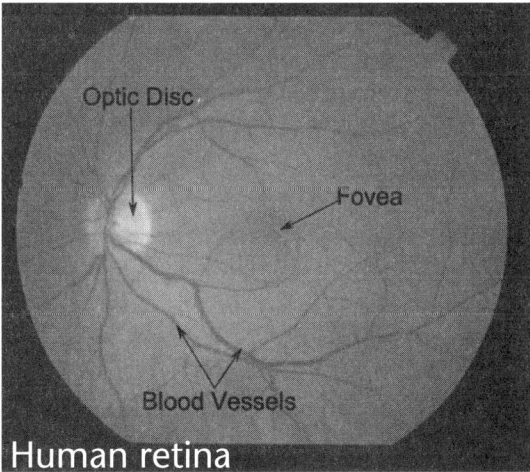

Figure 1.2 The inside of the human eye as viewed through the eye lens.

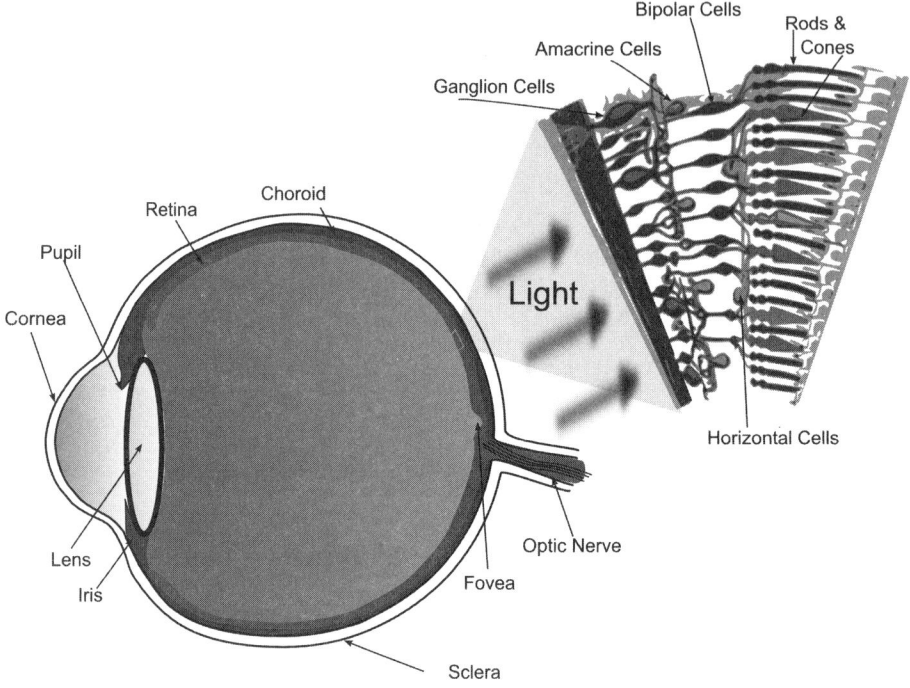

Figure 1.3 (a) A cross section of the human eye. (b) A cross section of the human retina.

These extra layers are better shown in Figure 1.4. The first cells met by the light are the *ganglion* cells. Then, we have the *amacrine* cells, that seem to connect several ganglion cells together, the *bipolar* cells, the *horizontal* cells, and finally the photoreceptors.

There has not been a definitive explanation of why the photoreceptors of the retina are not directly exposed to the incoming light. Some scientists suggest that this is an accident of evolution. Others say that the location of the sensors at the back shields the sensors from scattered light, that is, the layers of the cells in front of them act the same way as a collimator acts for a gamma camera. Yet another possible engineering explanation is that the photoreceptors have to be able to "empty" quickly, and they need a heat-sink with large capacity next to them to absorb their extra "cargo" so that they can quickly recover to respond to new signals. The highly perfused wall of the eye, being next to them, plays this role.

All photoreceptors in the human eye are designed to respond to all wavelengths of the electromagnetic spectrum in the range [∼400 nm, ∼700 nm], but not uniformly. The first type of sensor are called *rods*, due to the shape they have. They have a broad sensitivity curve peaking at about 499–505 nm. Then there are three types of photoreceptor with conic shape, called *cones*, that have broad sensitivity curves, peaking at the long wavelengths (red part of the electromagnetic spectrum), the medium wavelengths (green part of the electromagnetic spectrum), or the short wavelengths (blue part of the electromagnetic spectrum), called L or

1.2 Overview of the Human Visual System

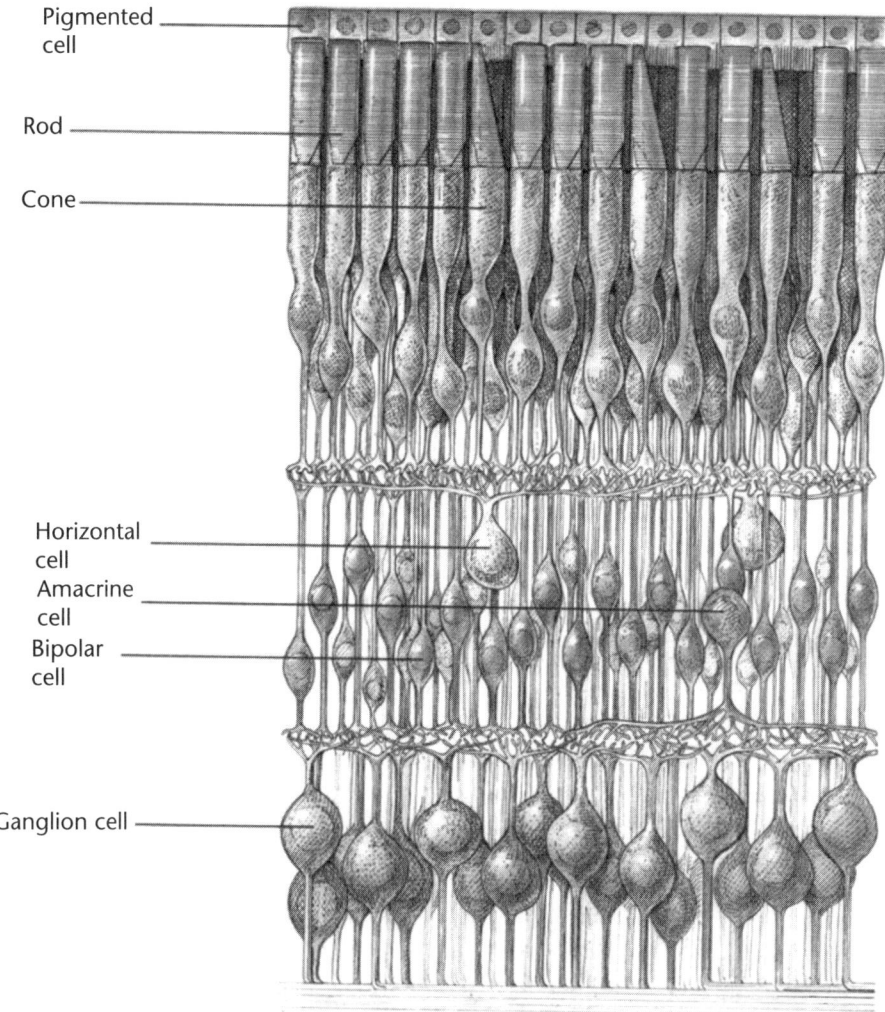

Figure 1.4 A cross section of the human retina. The path of the light is from the bottom of the page upwards. The photoreceptors are the cells that form the top layer (reprinted with permission from [8]).

red, M or *green*, and S or *blue* cones, respectively. The last three types of sensor are responsible for color vision, not because of their sensitivity curves (which are not much different from those of the rods), but due to the use of their recordings in combination with each other. So, color vision is the result of the circuitry of the sensors rather than the sensitivity curves of the sensors. Of course, if all sensors had identical sensitivity curves, no color vision would have been possible. The farther away we move from the fovea, the more sparse the cones become and the denser the rods become. This is shown schematically in Figure 1.5. Thus, peripheral vision is mostly due to rods. In the fovea itself we have virtually no rods. The red and

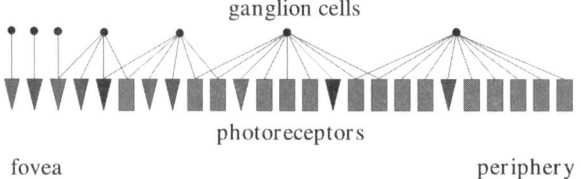

Figure 1.5 There are more cones in the fovea than in the periphery. The ganglion cells may accept input from several photoreceptors, which constitute their receptive fields (redrawn from [15]).

green cones are many more than the blue ones. Further, the sensors are irregularly placed, as shown in Figure 1.6.

Each ganglion cell, through the horizontal, bipolar, and amacrine cells, receives input from several photoreceptors as shown schematically in Figure 1.5. The photoreceptors that feed into a ganglion cell constitute its *receptive field*. There are two types of ganglion cell. Scientists performed experiments with electrodes attached to the ganglion cells of some animals [8]. They noticed that when a bright light was shown to some photoreceptors, the cell would be activated. When the spatial extent of the light was increased, contrary to expectations, the cell became quiet, as if receiving no stimulus, or as if its activation had been totally suppressed [8]. It turned out that such cells behaved as if they were spot detectors, responding only when a bright spot surrounded by a dark annulus was shown to them. These cells may be called *center-on-surrounding-off* cells. A complementary behavior was ob-

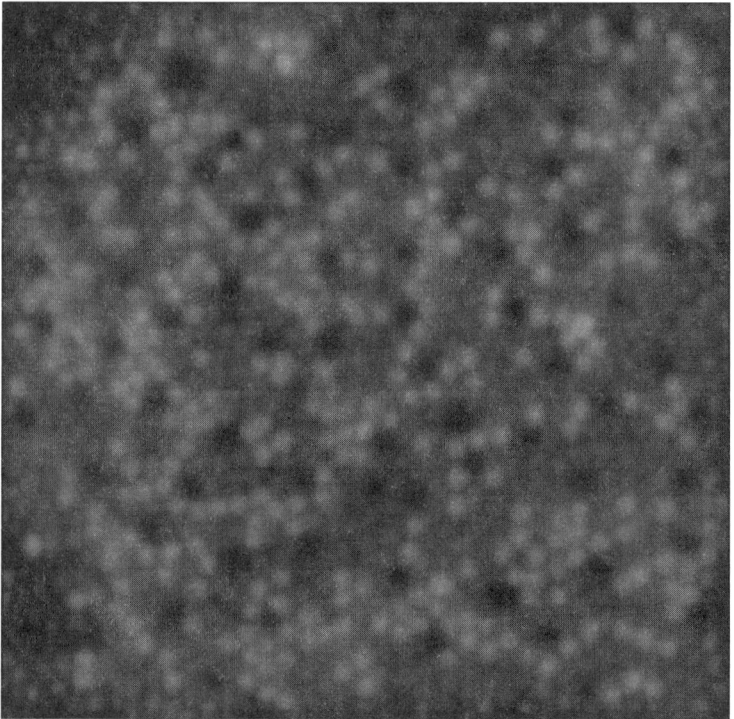

Figure 1.6 The sensors in the retina are not placed on a regular grid (reproduced from [18]).

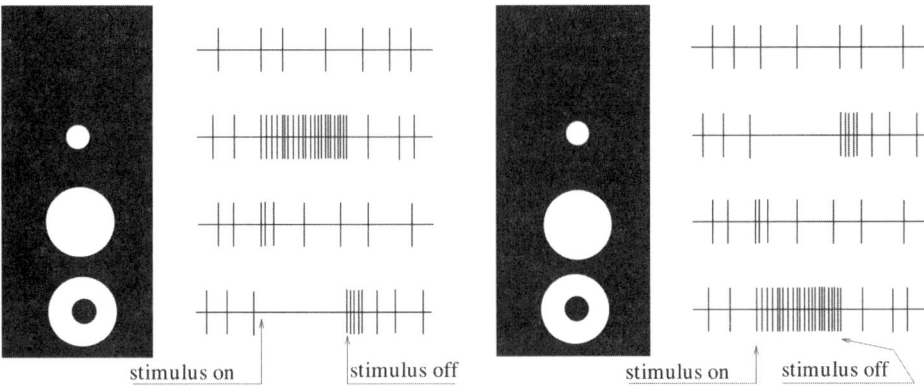

Figure 1.7 Each horizontal axis measures time. The stimulus is turned on at a particular instant and turned off a little later. The spikes plotted along each time axis indicate pulses produced by the monitored ganglion cell. In the absence of any stimulus, cells tend to fire at random and occasionally. When they are stimulated they fire very vigorously, and when they are suppressed they remain silent. The cell on the left is activated by bright spots surrounded by a dark annulus and that is why it may be called center-on-surrounding-off. The cell on the right is activated by dark spots surrounded by a bright annulus and that is why it may be called center-off-surrounding-on (redrawn from [8]).

served in a second type of ganglion cell, which was activated when it was shown a dark spot surrounded by a bright annulus. Such cells are dark spot detectors and may be called *center-off-surrounding-on*. These responses are summarized in Figure 1.7.

One other interesting fact about the eye is that it is not a static device. It vibrates and shakes at random with specific frequencies. These vibrations are known

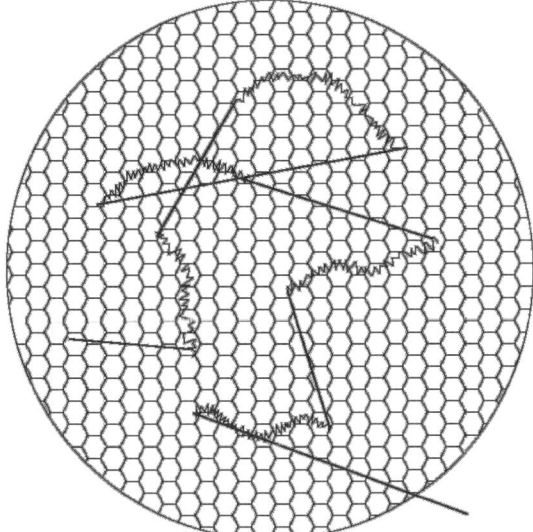

Figure 1.8 The hexagonal cells are supposed to be the photoreceptors in the retina. The straight line segments represent the microsaccades, while the curved lines represent the drifts, superimposed by the tremor vibration (redrawn from [12]).

as *tremor* (an aperiodic, wave-like motion of the eye with amplitude about the diameter of a cone in the fovea and frequency ~ 90 Hz), *drifts* (slow motions of the eye which may move the fixated object over a dozen photoreceptors with mean speed about 6 minutes of arc per second), and *microsaccades* (jerk-like movements that may move the fixated object over several dozen photoreceptors, lasting for about 25 ms) [12] (see Figure 1.8). These movements are different from the *saccades* which are associated with perception and are the result of gaze control and gaze direction. The role of tremor, drifts, and microsaccades is not fully understood. It is possible that it has to do with increasing the resolution of the viewed scene beyond that allowed by the placement of the sensors.

1.2.1.1 Issues to Be Investigated

The response curves of the photoreceptors to the various wavelengths are very different in the human eye and in the CCD sensors we currently use. This may not appear to be a crucial point. However, there are applications where the artificial vision systems we develop are designed to replace the human in a repetitive task, for example, batching ceramic tiles in color shades. This involves the categorization of color shades in sets of reflectance functions[2] which, under the same illumination conditions, will be perceived by the human visual system as identical. We say that such reflectance functions constitute the *metamers* of the human eye. Remember that a photosensor is an integrator that accumulates photons of different wavelengths modulated by its sensitivity function. Many different combinations of photons (corresponding to different reflectance functions of the viewed materials) may produce the same value recorded by the sensor. These different combinations of photon energies, making up a variety of reflectance functions, are the metamers. If the sensitivity function of the sensor changes, its metamers change too. So, different reflectance functions could be treated as identical by two sensors that have different sensitivity curves. The net result is that color shade categorization performed by a CCD sensor does not divide a set of colors in the same categories of color shades as the human eye does. Thus, automatic industrial color categorization aimed for the consumer, who is concerned about the color shade of the product, is very difficult with conventional cameras (see, however, [2,3] for finding ways around this problem).

The first question that arises, therefore, is whether we can construct photosensitive sensors that have sensitivity curves similar to those of the cones in the human eye. The answer is yes, and Chapter 10 of this book deals with this problem exactly, discussing methodologies for developing such organic photoreceptors.

The next issue is the interfacing of such sensors with electronic circuitry. In particular, cells act as analog devices between their limits of activation threshold and saturation. Analog devices have a major advantage over digital devices: they require extremely low levels of power, of the order to nWatt (10^{-9} Watt) (see Figure 1.9). Such power can easily be harnessed from the movement of the head,

---

2. The reflectance function that characterizes a material expresses the fraction of the incident light that is reflected by the material as a function of the wavelength of the incident light.

1.2 Overview of the Human Visual System

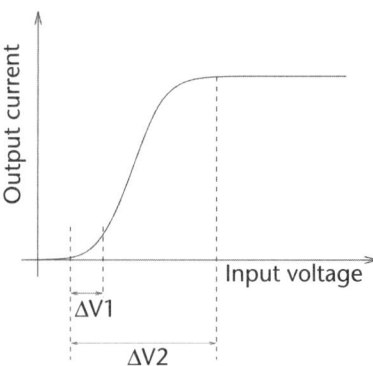

Figure 1.9 Typical curve of a diode. In order to operate as a digital device, the voltage has to change by $\Delta V2$. When operating as an analog device, the voltage may change only by $\Delta V1 \ll \Delta V2$.

for example. If we were to achieve one day the development of an implantable retina, we would need to combine the organic polymer photosensors with analog electronic control. This topic is dealt with in Chapter 11 of this book.

If we really want to work towards an implantable retina, we must have a tool to emulate the behavior of the retina in the most accurate way, down to the cellular level, so that we can perform experiments, make hypotheses and test them, and make predictions of treatments for various conditions. A physiologically accurate and versatile computational and mathematical model then of the retina is a valuable tool to have. Chapter 2 of this book deals with this issue.

Something we did not discuss in the previous section is the well-known adaptability of the eye. The eye exhibits an enormous range of brightness adaptability, but also adaptability in focusing on objects in various distances. The latter is achieved by the muscles that control it: they can change the focal length of the lens but they can also change the gazing direction so both eyes gaze at the same physical point. The closer the point, the more the optical axes of the two eyes have to move away from being parallel to each other. It is clear, therefore, that vergence of the optical axes of the eyes is related to the distance of the viewed object. This idea is exploited in Chapter 8, where an eye tracker is used to track the gaze of a surgeon when she performs an operation. From the vergence of her eyes, depth information may be extracted to stabilize the moving surface of the imaged organ.

Finally, one thing that strikes us as we see Figure 1.6 is the irregularity of the points with which the eye samples the continuous scene. Artificial vision relies on imaging sensors that sample the scene on a regular rectangular grid. How does the human eye allow us to see straight lines even though it samples the world at irregularly placed points? This is particularly striking with our peripheral vision, which relies on sparsely and irregularly placed sensors with mostly identical sensitivity curves, but allows us to perceive straight lines perfectly well without seeing, for example, loss or degradation of color or shape at the periphery of our field of view. Another example is our acuity in texture perception: texture is a spatial pattern, and as such its representation is expected to depend strongly on the way we sample space. How can we perceive and recognize textures that are irregularly sampled? Some possible explanation must lie in the presence of the vibrations executed by the eye. It has been known for some time that vibrating the reconstruction grid over a scene sampled with low resolution improves the quality of the constructed image significantly (e.g., [14]). However, this might explain the vision of straight lines perceived from irregularly placed samples, but it does not explain the fine detail perceived beyond the resolution of the available sensors. Such detail can only be recovered if one exploits the aliased part of the spectrum of the recorded signal. This naturally leads to methods that do exactly that, known collectively as *super-resolution*. Chapter 7 of this book expands upon these aspects of the human visual system and discusses super-resolution technology.

All the above raise the following questions: can we do image processing with irregularly sampled data? Which of our well-established image processing algorithms are immediately transferable to irregular grids and which are not? But is there really any reason to worry that our technology is based on rectangular grids? Maybe after all, we can stick with our rectangular grid and the eye with its own irregular one. The truth is that in many disciplines, the data we collect are collected at sampling points we *can* collect data from, as opposed to at sampling points we *desire* to collect data from. Such data could be data collected by geoscientists, astronomers, and so on. Images which have not been captured by a camera but computed from collected data, for example, seismic section images concerning the crust of the Earth, are simply resampled versions of irregularly sampled scenes. In addition, ordinary images may contain occlusions and faults resulting in samples that do not form a regular rectangular grid, or made up from the combination of several low-resolution images the samples of which do not constitute a regular grid either. So, motivated by studying the human visual system, we start wondering how we may process such irregularly sampled data. This issue is dealt with in Chapter 6 of this book.

1.2.2 Lateral Geniculate Nucleus (LGN)

The lateral geniculate nucleus is made up of six anatomical layers as shown in Figure 1.10. The layers alternately receive inputs from the left and right portions of the visual field (see Figure 1.11). More significantly, the first two layers primarily receive signals from a type of retinal ganglion cell called the *parasol cell*, while the last four layers mainly receive signals from another type of ganglion cell, called the *midget cell*. Parasol cells are the most common ganglion cells in the retina forming about 85% of the total population. They have large receptive fields and quickly saturate in response to changing visual stimuli. This allows them to respond well to motion in the visual field. By contrast, midget retinal ganglion cells have very small receptive fields and have lower contrast sensitivities. They only represent 10% of the ganglion cells and are concentrated in the high spatial acuity region of the retina, that is, the fovea. The parasol cells appear to be responsible for extracting information from the scene as *fast* as possible, while the midget cells appear to be responsible for extracting as *much* information as possible from the scene [10]. The parasol-connected first two layers of the LGN constitute the *magnocellular*

1.2 Overview of the Human Visual System 11

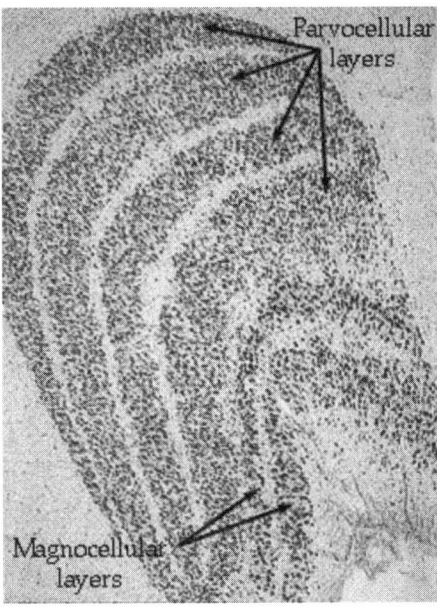

Figure 1.10 A cross section of the lateral geniculate nucleus showing its layered structure (reprinted with permission from [24]).

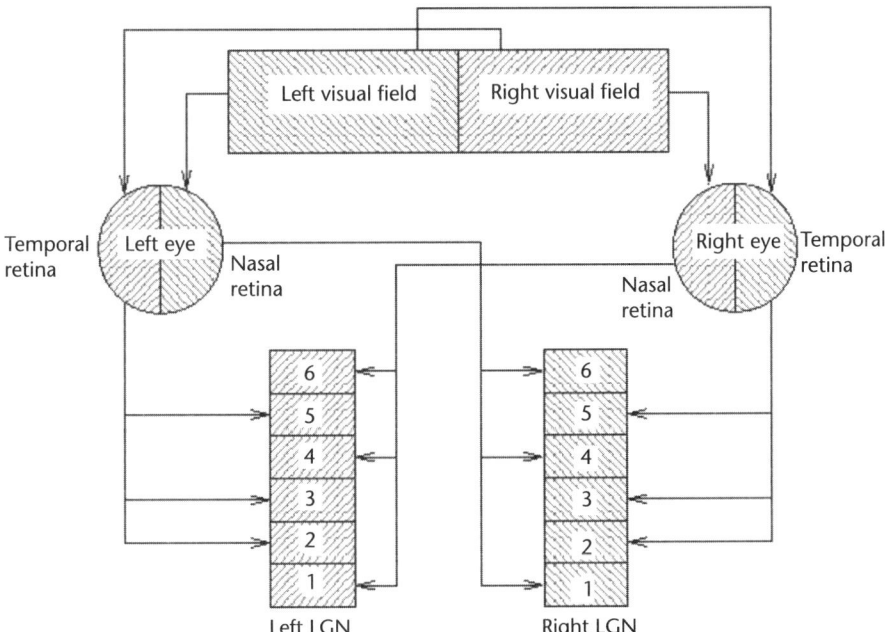

Figure 1.11 The layers of the lateral geniculate nucleus receive information from different eyes (redrawn from [26]).

(also called the *where* or *M*) pathway, which is concerned with computing and representing motion. The midget-connected last four layers of the LGN make up the *parvocellular* (also called the *what* or *P*) pathway, which is responsible for form, color, and texture [9]. Different layers of the LGN send information to different parts of V1. The middle "intercalated" two layers have recently been further identified as forming a separate pathway called *koniocellular* [7]. The neurons in the koniocellular pathway project mostly to the so-called layers 2/3 of V1, while the neurons of the remaining parvocellular pathways, called neoparvocellular neurons, send signals to the so-called $4C\beta$ layer of V1, and the magnocellular neurons send signals to the so-called $4C\alpha$ layer of V1. According to [11], the neoparvocellular layers mainly receive input from midget cells connected to L and M photoreceptors, while the koniocellular layers receive significant input from S photoreceptors.

As LGN relays information from ganglion cells in the retina to the visual cortex, the receptive fields of the LGN neurons[3] are similar to those of the retinal ganglion cells. LGN neurons in the parvocellular and magnocellular layers differ mainly in the number of input connections they have, which have been estimated to be of the order of 500 and 3,000, respectively [23]. Magnocellular cells also relay information faster than their parvocellular brethren. In terms of connections to other areas of the visual system, the LGN is the target of massive cortical feedback from V1 and V2 (ten times as many connections as feed forward to V1). This feedback pathway provides more connections to the LGN than any other source of input [5]. The feedback connections are thought to play an important role in changing the normally isotropic center-surround receptive fields of LGN neurons into orientation selective receptive fields [20]. The feedback pathway also appears to be fundamentally linked to the orientation preference of the cortical cells in V1 [13]. The role of the feedback pathway has not been clearly established, and the function of the LGN is poorly understood beyond that of relaying signals from the retina to V1.

As the role of LGN in vision is relatively poorly understood, we do not plan to investigate it in isolation much further in this book. The reader should note, however, that a consideration of V1 necessarily incorporates the effect of the LGN, and so, by mapping the receptive fields in V1 by stimuli induced in the retina, one is effectively probing a multistage system, which includes the processing of the LGN and, indeed, any feedback it might receive from V1.

1.2.3 The V1 Region of the Visual Cortex

In V1, the information from the two eyes is still kept separately, but it is fragmented and interlaced as shown in Figure 1.12. Just like LGN, V1 is also layered. Most of the information it receives comes from LGN. The layers of the LGN correspond alternatively to the two eyes and this layering is maintained in V1 in the form of *ocular dominance*, corresponding to left or right eye alternating columns.

---

3. The receptive field of an LGN neuron is defined as the effective transformation of the visual inputs from the retinal photoreceptors to the neuron output, including all processing by any intermediate neurons.

1.2 Overview of the Human Visual System   13

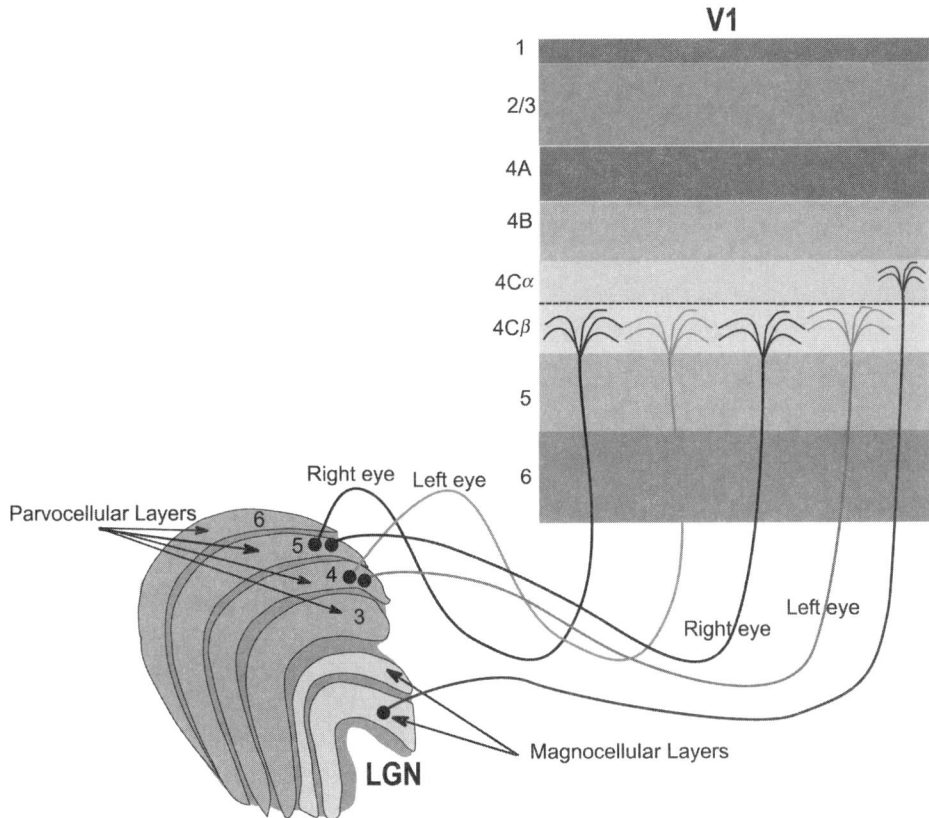

Figure 1.12 The columns of the V1 receive information from alternating layers of the lateral geniculate nucleus.

V1 contains simple and composite neurons. Figure 1.13 demonstrates schematically the role of a simple neuron which acts as if it were a rod or line detector. Such a neuron receives information from the ganglion cells in the eye, which we saw to behave as spot detectors. This particular type of V1 neuron fires only when all three ganglion cells it receives input from fire. Note that if these ganglion cells are of the type center-on-surrounding-off, and if their receptive fields are aligned, then if a bright bar is shown to them coinciding with their aligned receptive fields, all three will be activated, because their peripheries will receive relatively little signal. The V1 cell then will be activated too. If the bright line has a different direction, at most one of the ganglion cells will be activated and so the V1 cell will not be activated. V1 cells have also a temporal response which produces some motion selectivity. This connects the analysis performed by V1 with that performed by V5 and the other areas of the visual cortex responsible for motion analysis.

V1 is thought to be responsible for the first stage of vision, the part that is more akin to image processing, and acts as a gate to preattentive vision. V1 is often portrayed as either performing line and edge detection, or "gating awareness" to the rest of the visual system. A closer analogy might be to say that V1 provides a representation of the visual world to "higher" visual areas in the brain and appears

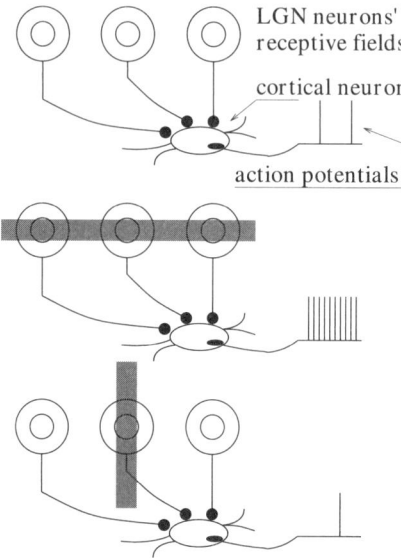

Figure 1.13 Some of the neurons in V1 act as line detectors by combining the input from several ganglion cells from the eye with aligned receptive fields.

to be effective at permitting object detection and recognition, as well as motion estimation. A recent successful demonstration of the ability to reproduce some physiologically observed aspects of V5 responses, using a model of V1 processing, is to be found in [19]. Moreover, V1 appears to receive much input from other visual areas, suggesting that its representation is adaptive and may even depend on context. A demonstration of disrupting V1-gated perception, using transcranial magnetic stimulation (TMS) applied to V5 and V1 in near simultaneity, is discussed in [17].

1.2.3.1 Issues to Be Investigated

Artificial vision systems are still very far from having acceptable performance in a generic environment. Understanding how our benchmark system works helps us improve them. Chapter 3 of this book deals in detail with our current understanding of the physiology of V1. Chapter 4 is dedicated to the functionality of V1 as a salienator. The model of V1 acting as a salienator is used in Chapter 6 as an edge detector, that detects edges on the basis of their saliency, as opposed to their contrast, like conventional edge detectors do. Chapter 5 considers the function of V1 as a frequency analyzer, that is, as a processor that takes the wavelet transform of the imaged scene. Issues of population encoding[4] are also discussed there.

Another important aspect of vision is the speed with which it is performed. Hardware implementation of some of the algorithms of Chapters 5 and 6 is discussed in Chapter 13, while Chapter 12 deals with the design of models of very

---

4. Population encoding refers to vision processes that rely on the synergy of many neurons.

large scale integrated (VLSI) analog circuits that may be used to implement classical simple-cell V1 receptive fields.

1.2.4 Motion Analysis and V5

V5 [otherwise known as *middle temporal* (*MT*)] is an area of the extrastriate visual region known to be heavily involved in motion processing. The nature of its functionality appears similar to that of optical flow processing, widely used in computer vision. V5 derives much of its input from layer IVb of V1, and from areas V2 and V3. Like V1, V5 also displays a retinotopic organization,[5] but its receptive fields appear to be larger than those of V1. A further notable difference is that, while there is selectivity to the direction of motion, there is much lower selectivity to the spatial orientation of a moving bar, and this analogy is made stronger by noting that the columnar organization present in V5 is in relation to different velocities of motion, rather than to different spatial orientations, which is the case with the columnar organization of V1. Models of processing in V5 have been explored for some time [21], with the most recent one being a V1-derived model that successfully demonstrates that orientation-independent V5 responses may be obtained by grouping responses emerging from V1-type computations [19]. Controversy continues, however, because some researchers have demonstrated that V5 may yield responses to motion even when V1 has been destroyed, and there is some evidence that a subset of projections from the retina to V5 bypass V1 entirely. Nevertheless, what seems certain is that V5 plays some role in guiding eye movements during object pursuit [1].

1.2.4.1 Issues to Be Investigated

Motion analysis is a very important visual cue. Not only does it act as a recognition cue for an individual, something that inspired gait-based biometrics research, but it also acts as an important cue for attracting our attention: motion detection with our peripheral vision is a survival skill which might offer the viewer his next meal or save him from his next predator. So, motion saliency is an important preattentive visual characteristic: a sudden change of direction or pace is something that might draw our attention. Chapter 14 presents a hardware-implemented algorithm that detects such saliency in real time. Finally, Chapter 9 presents motion detection algorithms that are inspired by V5.

1.3 Conclusions

In this chapter we reviewed some of the physiological characteristics of the human visual system, following the path the signal follows from the retina to the visual cortex. The literature on the physiology of the human vision system is vast, and we made no attempt to review it. Instead, we selected some aspects that inspired us as engineers. At each stage of the path of the signal we discussed, we considered

---

5. It preserves the topology of the image formed on the retina.

the engineering aspects that inspired the chapters of this book. In particular, we saw how Chapters 2, 7, 8, 10, and 11 are inspired by the structure of the human eye, while Chapters 3, 4, 5, and 12 from the physiology and functionality of V1. The rest of the chapters have been inspired by less localized aspects of the human visual system.

We discussed in somewhat greater detail the physiology of LGN because LGN is not discussed in any other chapter of this book due to the mystery that surrounds its role in the human visual system. It is possible that LGN is a feedback control center for gaze direction and foveation. It may also act as a pathway connecting visual input directly to the motor control center of the brain. For example, it is known that frogs have cells in their retinas that respond directly to flying insects, and these responses are directly transmitted to the muscles that control the tongue of the frog: when a fly passes by, the signals from the frog's eyes trigger the tongue to shoot out and catch the fly. The location of the fly must be relayed to the muscles that control the tongue in a hardwired way. If the information from the visual sensors first had to be transferred to the brain of the frog, where it would be analyzed and subsequently acted upon by the right part of the frog's brain and nervous system, the frog might have starved to death! It is possible, therefore, that the human LGN, being strategically located in the middle of the brain, may play the role of the organ where direct connections necessary for survival are hardwired, bypassing the visual cortex. The technology of in vivo imaging has not advanced enough yet to allow us to "see" in real time the signals as they are transmitted from the eye to the brain and back. Until this has been achieved, issues of feedback control mechanisms and direct connections cannot be tested and confirmed. The technologies that one day may resolve these issues may be electroencephalography (EEG) and magnetoencephalography (MEG), as these techniques have the necessary temporal resolution to allow us to "image thought." Techniques like fMRI have much better spatial resolution, but they currently lack the necessary temporal resolution. Perhaps an instrument, yet to be invented, that will combine both technologies may resolve the issues of "thought imaging." From the artificial vision point of view, technologies that are related to the possible direct connection between visual sensors and motor control are in the area of distributed

Figure 1.14 An image and the path of the gaze of a viewer.

sensor-actuator control systems. In particular, there have been artificial retinas that consist of the so-called "intelligent pixels," where some processing takes place at the pixels themselves [4, 6, 16].

However, apart from the possible direct connections between certain visual stimuli and motor responses, the most challenging aspect of the human visual system is visual cognition itself. This aspect of vision is well beyond the current engineering technologies. Capturing images is one thing; knowing what is there is another. Scientists for many years have been puzzled as to how we actually see. Experiments with eye trackers (see Figure 1.14) show the gaze of the viewer when seeing an image and trying to understand it [22].

Such gaze paths depend not on the preattentive saliency expressed by V1 models, but rather by higher-order modules of vision that feedback to the eyes and direct them to points of interest. This loop of looking-seeing-controlling is an engineering challenge tackled by robotic scientists, but it is still in its infancy. It will take considerable time before an autonomous robot with artificial vision as good as that of the human emerges.

References

[1] Born, R. T. and Bradley, D. C., "Structure and function of visual area MT." *Annual Review of Neuroscience*, Vol. 28, 2005, pp. 157–189.

[2] Boukouvalas, C. and Petrou, M., "Perceptual correction for colour grading using sensor transformations and metameric data." *Machine Vision and Applications*, Vol. 11, 1998, pp. 96–104.

[3] Boukouvalas, C. and Petrou, M., "Perceptual correction for colour grading of random textures." *Machine Vision and Applications*, Vol. 12, 2000, pp. 129–136.

[4] Dubois, J., Ginhac, D., and Paindavoine, M., "Design of a 10,000 frames/s CMOS sensor with in-situ image processing." In *Proceedings of the 2nd International Workshop on Reconfigurable Communication-centric Systems-on-Chip* (ReCoSoc 2006), 2006, pp. 177–182.

[5] Erisir, A., Horn, S. V., and Sherman, S., "Relative numbers of cortical and brainstem inputs to the LGN." *National Academy of Science, USA*, Vol. 94, 1997, pp. 1517–1520.

[6] Funatsu, E., Nitta, Y., and Kyuma, K., "A 128×128-pixel artificial retina LSI with two-dimensional filtering functions." *Japanese Journal of Applied Physics*, 38(8B), 1999, pp. L938–L940.

[7] Hendry, S. and Reid, R., "The koniocellular pathway in primate vision." *Annual Review of Neuroscience*, Vol. 23, 2000, pp. 127–153.

[8] Hubel, D. H., "Eye, brain and vision," *Scientific American Library*, ISBN 0-7167-5020-1. http://neuro.med.harvard.edu/site/dh/, 1988.

[9] Kaplan, E. and Benardete, E., "The dynamics of primate retinal ganglion cells." *Progress in Brain Research*, Vol. 134, 2001, pp. 1–18.

[10] Li, Z., "Different retinal ganglion cells have different functional goals." *International Journal of Neural Systems*, Vol. 3(3), 1992, pp. 237–248.

[11] Martin, P., "Colour through the thalamus." *Clinical and Experimental Optometry*, Vol. 87, 2004, pp. 249–257.

[12] Martinez-Conde, S., Machnik, S. L. and Hubel, D. H., "The role of fixational eye movements in visual perception," *Nature Reviews*, Vol. 5, pp. 229–240.

[13] Murphy, P., Duckett, S., and Sillito, A., "Feedback connections to the lateral geniculate nucleus and cortical response properties." *Science*, Vol. 19(286), 1999, pp. 1552-1554.

[14] Kadyrov, A. and Petrou, M., "Reverse engineering the human vision system: a possible explanation for the role of microsaccades." *Proceedings of the 17th International Conference on Pattern Recognition*, ICPR04, August 22-26, Cambridge, UK, Vol. 3, 2004, pp. 64-67.

[15] Olshausen, B., "General aspects of information processing in the visual system," http://redwood.berkeley.edu/bruno/npb261b/, 2005.

[16] Paillet, F., Mercier, D., and Bernard, T. M., "Second generation programmable artificial retina." In *Proceedings of the 12th Annual IEEE International ASIC/SOC Conference*, pp. 304-309, 1999.

[17] Pascual-Leone, A. and Walsh, V., "Fast backprojections from the motion to the primary visual area necessary for visual awareness." *Science*, Vol. 292, 2001, pp.510-512.

[18] Roorda, A. and Williams, D. R., "The arrangement of the three cone classes in the living human eye." *Nature*, Vol. 397, 1999, pp. 520-522.

[19] Rust, N. C., Mante, V., Simoncelli, E. P., and Movshon, J. A., "How MT cells analyse the motion of visual patterns." *Nature Neuroscience*, Vol. 9, 2006, pp. 1421-1431.

[20] Sillito, A., Cudeiro, J., and Murphy, P., "Orientation sensitive elements in the corticofugal influence on center-surround interactions in the dorsal lateral geniculate-nucleus." *Experimental Brain Research*, Vol. 93, 1993, pp. 6-16.

[21] Simoncelli, E. P. and Heeger, D. J., "A model of neuronal responses in visual area MT." *Vision Research*, Vol. 38, 1998, pp. 743-761.

[22] Yarbus, A. L. 1967. *Eye, movements and vision*. Plenum Press, New York, Library of Congress Catalogue number 66-19932.

[23] Wilson, J., "Synaptic organisation of individual neurons in the macaque lateral geniculate nucleus." *Journal of Neuroscience*, Vol. 9(8), 1989, pp. 2931-2953.

[24] http://thalamus.wustl.edu/course/cenvis.html

[25] http://webvision.med.utah.edu/sretina.html

[26] http://www.lifesci.sussex.ac.uk/home/George_Mather/Linked%20Pages/Physiol/LGN.html

PART I
The Physiology and Psychology of Vision

CHAPTER 2
Retinal Physiology and Neuronal Modeling

H. Momiji and A. A. Bharath

2.1 Introduction

The study of biological visual systems benefits from knowledge of several fields: anatomy, physiology, signal processing, and information theory, to name a few. Broad reviews of visual anatomy are available in [1–3], and an excellent overview of how various topics fit together are available from Wandell [4]. The anatomy and physiology of the retina are summarized concisely in [5], and publicized in detail on the Internet in [6]. Signal processing analogies and interpretations are described in detail in [4], where particular emphasis is placed on the significance of the assumptions of the properties of linearity and shift-invariance in modeling biological visual systems. More general approaches to the mathematical modeling of neurons are introduced in [7], with exercises using a multipurpose neural simulator called GENESIS, and modeling is further detailed in [8].

The retina receives visual information in the form of light, and transmits such information to the brain, more precisely to the LGN (lateral geniculate nucleus) in the form of an axonal spike train. The retina is also the first organ that processes visual information. Its functions include: photo-electric transduction; light-intensity adaptation; smoothing (noise reduction); spatial band-pass filtering; creation of parallel pathways; creation of color opponency channels, and analog-to-digital conversion (see, for example, Tovée [2]).

Mammalian retinae contain more than 50 distinct anatomical cell types, each with a different function [5]. This fact, together with the existence of some distinct, purpose-specific pathways (for example, P, M, red/green, and blue/yellow), leads to the suggestion that the retinal output signal is made up of many different purpose-specific parallel representations of the same visual information, as thought to be the case in the rest of the visual system.

2.2 Retinal Anatomy

Animals for which the retina has been studied in detail include the tiger salamander, fish, and turtle. In mammals, the species covered have included rabbits, rats, and primates. Although there is a common basic structure, the details of retinal structure and function differ significantly between species of vertebrates. Some of this variation is a reflection of the degree of local retinal processing. Primate retinal organization has been studied mainly in the macaque monkey, whose retina is known to have marked similarities with the human retina [9].

The primate retina consists of two anatomically distinct layers, the central fovea and the peripheral retina. A retina has a basic laminar structure comprising photoreceptors, horizontal cells, bipolar cells, amacrine cells, and ganglion cells which are stacked (roughly speaking) from the outer to inner layers (Figure 2.1). Electrical signals flow in both directions: from outer to inner layers (feed forward), and from inner to outer layers (feedback). They also flow horizontally. The following is a brief description of the function of each cell class.

- **Photoreceptors** work as photoelectric transducers. There are two distinct types: cones and rods.
    - **Cones** are color sensitive and divided further into three types: L, M, and S cones, each displaying peak sensitivity to long, medium, and short wavelengths of light, respectively (see Chapter 10 for response curves). This is known as *trichromacy* with $\lambda_{max} = 420$, 530, and 565 nm in humans. Some vertebrates are di- or tetrachromatic. Fish, birds, reptiles, and some mammals (mice) have a fourth photoreceptor type that is sensitive to ultraviolet light [10]. S-cones are morphologically distinct and can easily be distinguished. For primates, S-cones represent about 5% to 10% of the total number of cones. The ratio between the numbers of L- and M-cones

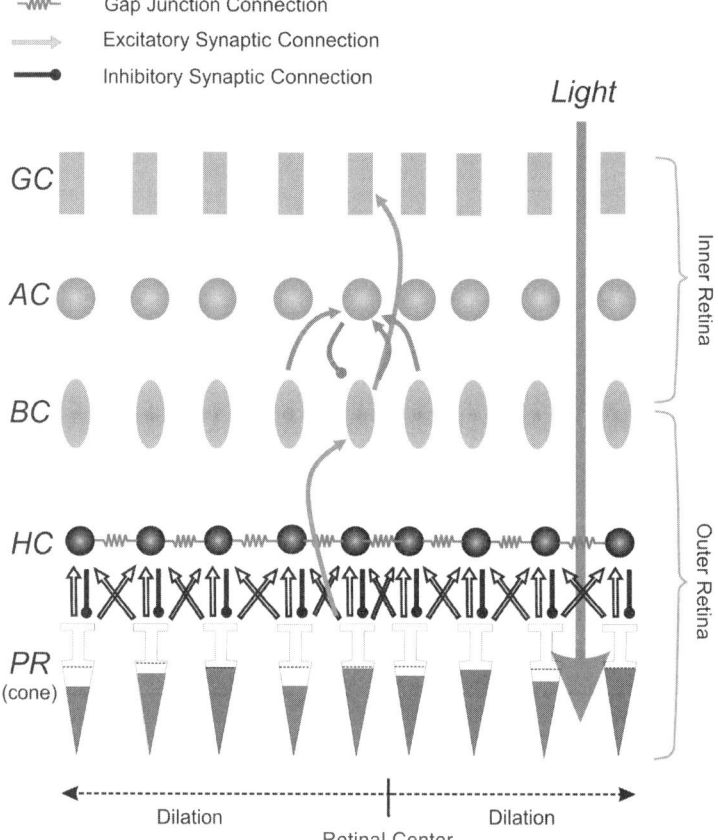

Figure 2.1 Illustration of retinal cell layers and dominant connection types in the central fovea.

varies across individuals. For example, L/M ratios in the fovea of 1.15 and 3.79 have been reported for two human subjects with normal color vision [11]. The fovea is dominated by cones, which are closely and hexagonally packed. Furthermore, the very center of the fovea is devoid of S-cones, which peak in number on the foveal slope. L- and M-cones are not arranged regularly in the two-dimensional foveal space, but may, for example, form random clusters.

- **Rods** are achromatic but very sensitive to dim light, such as starlight at night. Rods are dominant in the periphery of the retina, with a small number of sparsely spaced cones. Peak spectral sensitivity occurs at a wavelength, $\lambda_{\max}$, of 499 nm in humans.
- Another class of photoreceptive cells has been deduced for some time. This third type, for which there are several different candidates, and may be termed collectively nonrod, noncone photoreceptors (see, for example, [12]), is thought to be responsible chiefly for tuning circadian rhythm and for pupillary constriction.
- **Horizontal cells** receive their input from photoreceptors and send antagonistic signals to photoreceptors (lateral inhibition), and probably to bipolar cells, which make up the center-surround receptive field. Horizontal cells form a network throughout the whole retina. The horizontal cells of the primate retina are classified anatomically into three types: HI, HII, and HIII, but the physiological and functional differences between these types do not seem to be altogether clear.
- **Bipolar cells** are the first cells in the retinal network which express a center-surround receptive field [13], and this is thought to originate from antagonistic signals from photoreceptors and horizontal cells.[1] The output of a bipolar cell is sent to amacrine and ganglion cells. Two main classes of bipolar cell are the *midget* and *diffuse* bipolar cells, which are found in P- and M-pathways, respectively. The fovea has only midget bipolar cells. Taking another "functional axis" for classification, bipolar cells may be functionally categorized into two types: ON and OFF, which either flip or maintain the sign of the signal from the photoreceptors, respectively.
- **Amacrine cells** receive input from bipolar cells and yield a sign-inverted signal to ganglion cells. Their main function is thought to be fine-tuning of lateral inhibition [14]. Amacrine cells also form a network throughout the whole retina and represent the most numerous and diverse retinal anatomical cell type.
- **Ganglion cells** integrate "analog" information received from the bipolar and amacrine cells and convert it into the form of an action potential spike train, which is subsequently sent to the LGN.

---

1. Illustrations in some sources do not show the connection between horizontal and bipolar cells. Some retinal physiologists speculate that the feedback from a horizontal cell to cones is the origin of the center-surround receptive field. Such a feedback mechanism is mathematically sufficient to explain the center-surround receptive field.

Table 2.1 Some Anatomical Quantities and Their Typical Values in Humans. After [4].

Quantity	Typical Values	Units
Retinal area for 1 deg of visual angle	0.3	mm
Number in retina:		
cones	5×10^6	—
rods	10^8	—
optic nerve fibers	1.5×10^6	—
Fovea diameter:		
whole area	1.5	mm
rod-free area	0.5	mm
rod-free, capillary-free area	0.3	mm
Photoreceptor diameter:		
foveal cone	1–4	μm
extrafoveal cone	4–10	μm
rod near fovea	1	μm
Peak cone density	1.6×10^5	mm^{-2}
Photoreceptor-ganglion cell ratio:		
in fovea	1:3	—
in retina	125:1	—

- In addition to this classical type of ganglion cell, a type of *photosensitive* ganglion cell has also been identified. Compared to rods and cones, photosensitive ganglion cells are less sensitive and very slow in their light response. They are primarily connected to the suprachiasmatic nucleus (SCN, the master circadian pacemaker) and, therefore, thought to be associated with fine-tuning the circadian rhythm. Their role in vision remains under debate [12].

Some useful anatomical quantities are summarized in Table 2.1.

The basic circuitry that connects these types of cells has become clear over the past 60 years (Figure 2.1). Direct pathways, mediated by bipolar cells, are modified by horizontal and amacrine cells with sign-inverted signals. This antagonistic circuitry has a concentric two-dimensional structure, namely the center-surround receptive field, to which the center responds positively while the surround[2] does so negatively (see Chapter 1, Figure 1.7), or vice versa.

Anatomically, the structure of the retinal center-surround receptive field varies significantly in species. For example, unlike those of humans, the rabbit's center-surround system is strongly anisotropic.

There are two distinct functional pathways in primate vision, known as the parvocellular (P) and magnocellular (M) pathways. The P-pathway provides high spatial but low temporal resolution, while the resolution of the M-pathway is temporally high but spatially low. The P-pathway is sensitive also to color in higher mammals. These pathways may be traced back to anatomically distinct ganglion cells in the retina. *Midget* ganglion cells, which have relatively small cell bodies and dendritic trees, provide the predominant *afferent* inputs to the LGN P-layers. *Parasol* ganglion cells have relatively large sizes of dendritic tree structures and cell bodies, and these project (provide inputs to) the M-pathways.

---

2. The term "surround" is often used in the neuroscience literature to refer to the outer (excitatory or inhibitory) part of a receptive field.

The circuitry of the retina varies with eccentricity[3] from the center of the fovea. The human fovea is made up of predominantly midget ganglion cells that are further subdivided into ON- and OFF-pathways in a *private-line* manner: each cone connects to one ON- and one OFF-midget bipolar cell and these connect, respectively, to one ON- and one OFF-midget ganglion cell. These two pathways are modified by lateral inhibition with horizontal and amacrine cells.

2.3 Retinal Physiology

A cell's contents are separated from the outside by its membrane. A cell contains ions that are involved in the electrical activity through the ion channels embedded in the membrane. Because of the difference in the potassium ion, K^+, population on either side of the cell membrane, the cell is usually charged negatively. The potential when the cell is at equilibrium (resting potential) can be estimated by summing the potential associated with each ion type calculated from the Goldman equation (known, more expansively, as the Goldman-Hodgkin-Katz equation), which is a refinement of the famous Nernst equation, widely used in electrochemistry. The resting potential is typically -70 mV. Note that in the physiologists' convention, the sign of ionic currents is usually taken negative when positive ions flow into the cell (referred to as inward negative) [15].

Of the five main types of retinal cell, only ganglion cells have an excitable membrane that produces a spike train. All the other cells have graded potentials (Figure 2.1), although the synaptic terminal of a bipolar cell has recently been found to generate spontaneous spikes [16].

A ganglion cell may be thought of, to a first approximation, as an analog-to-digital converter. Many ganglion cells fire spontaneously even in darkness. The firing interval is neither regular, nor is it "random" [17]. The frequency of the action potential of a ganglion cell is positively correlated to its membrane potential. It is thought that the incoming information in the form of a graded potential is coded primarily into the form of firing rate (in other words, frequency). There is, however, a view that the timing of firing (in other words, phase) may also be important for later signal processing, due to the fact that ganglion cells have been observed to show concerted firing [18].

There are two types of interconnection between cells: synaptic input by means of a neurotransmitter, such as glutamate (excitation) and or γ-aminobutyric acid, or GABA, (inhibition); and direct electrical coupling between neighboring cells via the gap junctions. The coupling via gap junctions is thought to be ohmic, being proportional to the voltage difference between the cells [19].

Horizontal cells are coupled spatially via gap junctions, forming a network throughout the whole retina [20]. The strength of the connection between each

---

3. The *eccentricity* measurement comes from the field of *perimetry*, concerned with describing visual acuity. We may consider the center of the pupil of the eye as the geometric center of a sphere of radius equal to that of the pupil-foveal distance. The center of the fovea may then be taken as corresponding to an azimuthal angle of $0°$ in a spherical polar coordinate system, and off-axis angles, in any direction, describe loci of a particular eccentricity, measured in degrees.

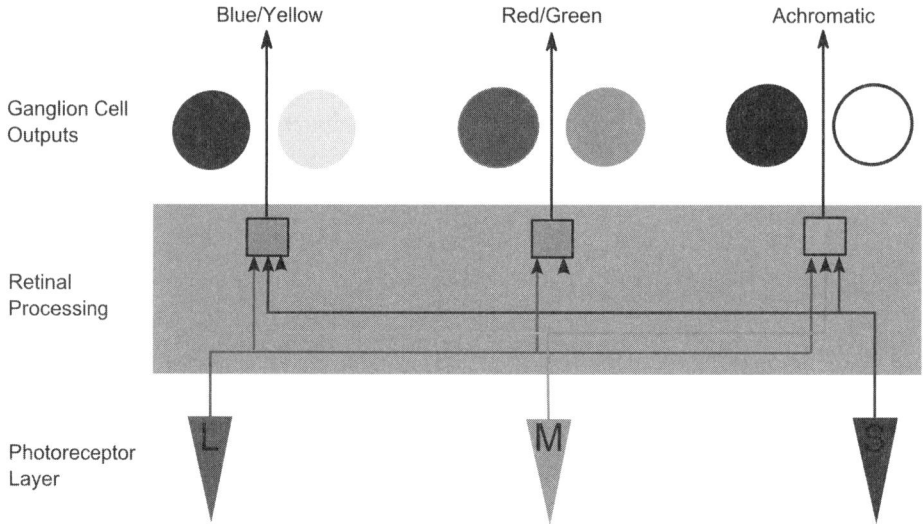

Figure 2.2 Luminance and color-opponent channels.

horizontal cell changes by adapting to the visual scene, in a somewhat Hebbian manner. Although neighboring cones are coupled in the same way, the strength of this intercone connection is thought to be negligible compared with the connections between horizontal cell pairs (HC–HC connections). Gap junctions have been modeled with a voltage-dependent resistor [20], but some nonlinear behavior is also apparent, and they may occasionally behave as rectifiers. Amacrine cells are also laterally connected but the connection strength is thought to be weaker.

Although a cone receives feedback signals from both horizontal and bipolar cells, the feedback from a bipolar cell is thought to be much smaller than that from a horizontal cell.

Functionally, a retina may be regarded as a mosaic of center-surround receptive fields (RFs). Generally, the receptive fields become larger with eccentricity and display greater overlap with each other [21]. Under the assumption of linearity, and the further (and strong) assumption that the receptive field pattern at one location might be approximately repeated at other nearby spatial locations (spatial invariance), the retina might be modeled as a spatial bandpass filter [4]. In fact, if one relaxes the assumption of spatial invariance to one of mere linearity, one may loosely regard the primate retina as achieving an isotropic, irregularly sampled wavelet representation, in which the visual field is sampled and projected onto a set of shifted and dilated functions.[4] See also Chapter 5, which describes the modeling of *cortical* visual processes with wavelets.

---

4. A set of "basis" functions for a wavelet transform can be generated by translations and dilations of one or more prototype wavelets; the former may be associated with the spatial preservation of visual input onto the retinal surface and foveation, and the latter with the retinal variation of the receptive field size with eccentricity.

Because S-cones make up about 5% to 10% of the total number of cones, the overall spectral sensitivity of the eye is well modeled by combining L- and M-cone spectra alone. The rate of physiological input between L- and M-cones reflects the anatomical cone ratio and is preserved from outer to inner layers [9].

There appear to be two main channels involved in encoding color (Figure 2.2). In the red–green opponency pathway, signals from L- and M-cones are differenced, whereas in the blue–yellow opponency pathway, signals from S-cones are opposed to an (L + M) cone signal [9].

Some center-surround receptive fields have color-opponent properties: red center–green surround; green center–red surround; blue center–yellow surround; and yellow center–blue surround.

With the "private-line" structure connecting single cones and bipolar cells in the fovea and the convergent connections in the peripheral retina, red–green opponency is stronger at the center of the fovea than in the peripheral retina [9].

2.4 Mathematical Modeling—Single Cells of the Retina

Neural activities can be considered in terms of membrane potentials and ionic currents through the ion channels embedded in the membrane. One approach to neuronal modeling is to derive an equivalent electrical circuit. General-purpose ionic current models have been developed at Caltech, [23]; [7], at Duke [24]; and at Pennsylvania [25]. These simulators are publicized on the Internet, and can be used freely (see the References for the Web addresses).

A single cell may be divided into compartments, which are, ideally, small enough to be at approximately the same potential [7]. Such a compartment is usually modeled as a parallel-conductance circuit, comprising a capacitor and sets made up of a battery and a voltage-variable resistor each. Generally, the number of such sets reflects the number of ion channels.

These so-called conductance-based models were catalyzed by the work of Hodgkin and Huxley [26], who derived their classic equations to describe the electrical properties of the squid giant axon and, among other membrane behavioral characteristics, predict the form, duration, amplitude, and conduction velocity of an axonal spike. Extensions to the Hodgkin-Huxley equations have also been used to describe spike generation in the soma of neurons by incorporating extra ionic channels. Under certain conditions, it can be shown [27] that, due to minimal coupling between some of the quantities in the set of four main equations comprising the Hodgkin-Huxley model, the complexity can be reduced to a *leaky integrate-and-fire* model, in which a neuron is viewed as a leaky integrator which resets its state to zero when firing occurs.

The ionic current models have also been developed specifically for photoreceptors [28], horizontal cells [20, 29, 30], and bipolar cells [31–33]. These models are in the same formulation as the Hodgkin-Huxley equations, and provide very precise descriptions of a single cell. In [28], for example, the dynamics of a single cone are modeled as a set of nonlinear differential equations involving over twenty variables. Numerically simulated results of these models were compared with physiological data, mostly those of the voltage-clamp and current-clamp experiments, with good agreement. These approaches could eventually provide the basis for a model of the whole retina.

Signal transmission in a dendrite can be modeled by considering the dendrite as a simple delay line comprising a set of resistors (R) and capacitors (C), with assumptions that the membrane is passive and that the input is a current (cable theory). Cable theory is also applicable to a branching tree of dendrites [34,35]. A more complete description, which permits solutions to a wider class of geometries and general branching structures, is again compartmental modeling, in which a neuron is treated as a branching train of connected equivalent circuits.

In this chapter, we do not consider further the modeling of single cells or of dendritic structures, because these are topics that are extremely well covered in other articles [27] and texts, such as [36] and [40], which contains both introductory and detailed material. In addition, for more details of photoreceptor models, the reader is referred to Chapter 10.

Now, we move onto models to describe the activities of spatially distributed groups of cells, showing how this leads to a description of ganglion-cell activity in the form of a spatial receptive field; this sets the scene for "abstracting" the concept of spatial receptive fields, a topic that we discuss in more detail, in the context of the primary visual cortex, in Chapter 5.

2.5 Mathematical Modeling—The Retina and Its Functions

In principle, any scene that is viewed by the retina could be decoded from the spike trains transmitted along the optic nerve fibers. However, without a full understanding of the neural code of the retina, such a task is technically rather difficult at present! Even the activity of a single ganglion cell may be difficult to interpret [17].

Perhaps the simplest model of retinal function is an array of center-surround receptive fields defined on a two-dimensional lattice. Retinal center-surround receptive fields can be represented by the difference of Gaussians (DoG):[5]

$$D(x,y) = k_c \exp\left(-\frac{x^2 + y^2}{2r_c^2}\right) - k_s \exp\left(-\frac{x^2 + y^2}{2r_s^2}\right) \tag{2.1}$$

Here k_c, r_c, k_s, and r_s are positive constants and (x,y) denotes the two-dimensional spatial location. This function, illustrated in Figure 2.3, captures the spatial bandpass nature of a retina [4].

With the DoG function, Rodieck [5] developed a linear spatiotemporal model for the firing rate (R) of a ganglion cell attributed to a center-surround receptive field. The brief flash of a small light spot on the center of the receptive field induces

---

5. Not to be confused with the spatial derivatives of a Gaussian function. Note that in place of the difference of two Gaussians, Laplacian of Gaussian (LoG) functions have also been used for retinal receptive field modeling. The LoG is the second order, isotropic spatial derivative of a Gaussian, and although it permits less control over the balance between center and surround responses, it requires fewer parameters to be specified.

2.5 Mathematical Modeling—The Retina and Its Functions

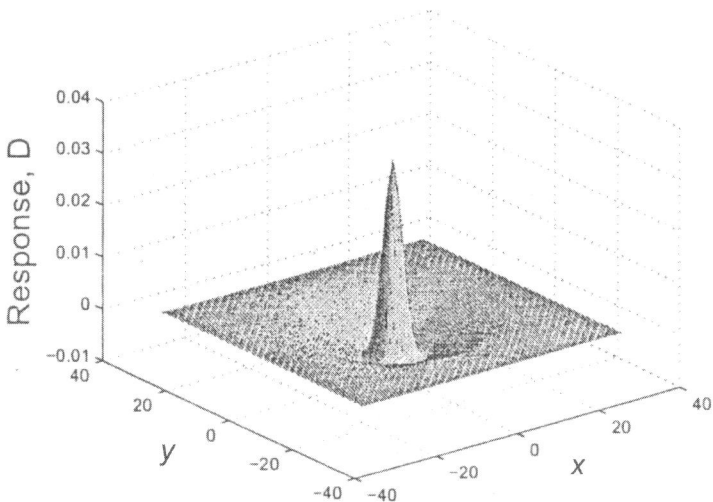

Figure 2.3 Illustration of a difference of Gaussians as a model for an ON-center retinal receptive field. The parameters used to generate this plot were $r_c = 2, k_c = 1/(2\pi r_c^2), r_s = 10, k_s = 5/(2\pi r_s^2)$. Note that the location of peak response of Equation (2.1) is at $(0,0)$.

a brief increase in firing rate followed by an undershoot and gradual recovery towards the baseline firing rate (R_0). This time course of firing rate, $A(t)$, may be represented as

$$A(t) = \delta(t) - h\exp\left(-\frac{t}{\tau}\right) \quad (2.2)$$

where $\delta(t)$ denotes the delta function, and h and τ are positive constants. If assuming linearity both in space and in time, the firing rate for the light illumination ($I(x,t)$) may be written:

$$R(t) = R_0 + \int\int I(x,t')B(x)A(t-t')dxdt' \quad (2.3)$$

Since the firing rate cannot be negative, rectification must be applied at the output of this, essentially linear, stage. This model agrees well with some experiments [39]. Variations of the Rodieck model, such as one containing a time delay in the surround signal due to the propagation delay induced by a horizontal cell, are reviewed briefly by Meister [17].

Another type of computational retinal model was developed by Shah and Levine [40, 41], aimed at robotic vision. The model comprises seven types of cells, namely cones, horizontal cells, midget and diffuse bipolar cells, amacrine cells, and P and M ganglion cells. These cells were stacked on a lattice, and each cell was modeled as having a Gaussian-spreading, time-decaying receptive field:

$$G(\mathbf{r};\sigma) = \frac{1}{2\pi\sigma^2}\exp\left(\frac{-|\mathbf{r}|^2}{2\sigma^2}\right) \quad (2.4)$$

$$K(t;\tau) = \frac{1}{\tau}\exp\left(\frac{-t}{\tau}\right) \text{ if } t \geq 0, \tau > 0$$
$$= 0 \qquad\qquad\quad \text{if } t < 0 \tag{2.5}$$

Here σ and τ are positive constants and **r** denotes the horizontal position. Parameter σ should be larger than the lattice constant (in other words, photoreceptor spacing). The model consists of two main streams of signal flow: cone–bipolar cell--M-type ganglion cell; and cone–midget bipolar cell–P-type ganglion cell, modified by horizontal and amacrine cells. Four other models of this type are listed and compared by Chae et al. [42], who concluded that Shah and Levine's [40,41] was the best model available at that time.

Roska and his colleagues have been developing a mammalian retinal model within the framework of Cellular Neural Networks (CNNs). Roska and colleagues tune the parameters of the network so that its responses match those of the rabbit retina [43–45]. See, also, the recent articles describing temporal dynamics of a simulated retina [46]. As described in detail in Chapter 12, a CNN consists of an array of similar analog processing cells with local interactions, which can perform parallel analog computations on corresponding pixel values of a given image. In [45], only nearest-neighbor connections were considered. An interesting point of their model(s) is that ON- and OFF-pathways are introduced in an asymmetric manner [44].

Both Shah and Levine's model [40, 41] and that of Roska and Werblin [43] and Bálya et al. [44, 45], use conceptually the same modeling framework: various types of retinal cell are defined on a two-dimensional lattice, and the state of each cell is represented only by its membrane potential. In short, Shah and Levine's model is concise and easy to understand, but lacks some biological detail, whereas the more recent models from Bálya, Roska, and Levine are much more detailed, and, due to the existence of a training phase, are tunable to match biological responses.

The third type of model contains more details of ionic current interactions. Based on their study of single cells, Fujii et al. developed a model of the rod network on a hexagonal grid connected via gap junctions [47]. Ishii et al. [19] developed a model, comprising a horizontal cell network connected via gap junctions, to estimate the spatiotemporal input signal to horizontal cells from the response of their membrane potentials. Fujii [48] further developed computational models for subsets of the retina consisting of (a) rods and horizontal cells, and (b) cones and horizontal cells.

2.6 A Flexible, Dynamical Model of Retinal Function

In the sections to follow, we describe a model for the representation of the central foveal region, describing the interactions between individual cells by using a series of coupled differential equations. The benefit of doing this is that quite precise behavior of both individual and collective responses of cells may be modeled at a level that is intermediate between detailed cellular-level physiology and macroscopic function. Furthermore, more complex networks of nonlinear cores or processor nodes should be able to accurately reproduce, in real time, the response of the static retina. To illustrate the flexibility of such an approach, we point out that the model has also been extended to describe the peripheral retina as well, where it has been applied to the study of color-opponent properties [49].

2.6.1 Foveal Structure

The model consists of five layers of photoreceptors (*pr*), horizontal cells (*hc*), bipolar cells (*bc*), amacrine cells (*ac*), and ganglion cells (*gc*), and considers only ON-pathways. For each layer, a two-dimensional continuous space is considered, whereby cells can be defined anywhere in that space, such that they do not occupy the same location within the same layer.

As described earlier, the primate fovea is dominated by cones, which are hexagonally packed. In a model description that captures this spatial structure, the cone packing should be denser at the retinal center ($\approx 160{,}000/\text{mm}^2$), becoming sparser with eccentricity [3]. Furthermore, the central fovea, whose diameter is about 100 cones ($\approx 1°$ in visual angle), should be devoid of S-cones, with the L- and M-cones forming random clusters [11]. Secondly, the central area should display a one-to-one connection between cones and ganglion cells via midget bipolar cells [50].

There are other aspects to consider. Foveal horizontal cells contact only 6 to 7 cones [3]. Because amacrine cells spatially fine-tune receptive fields [14], connections made via gap junctions between cones and between amacrine cells should be included, but should be weaker than those that occur between horizontal cells.

By considering these features, the central area of the fovea may be modeled as follows. First, a uniform triangular grid labeled by (p, q) is defined to mathematically describe the hexagonal close packing of cones. Then, each cell type and its inter-connections are defined with the following rules, which are schematically illustrated in Figure 2.1. In essence, (a) either an M- or L-cone (PR) is defined randomly on every lattice site (p, q) [51]; (b) a horizontal cell (HC) is defined on every site on the lattice (p, q); (c) a bipolar cell (BC) is defined on every site on the lattice (p, q); (d) an amacrine cell (AC) is defined on every site on the lattice (p, q); (e) a ganglion cell (GC) is defined on every site on the lattice (p, q); (f) each HC has feed forward connections from the seven nearest PRs [3]—each PR, say at (p, q), has a feedback connection from the HC at (p, q) [40,52]; (g) each BC, say at (p, q), has a feed forward connection from the PR at (p, q); (h) each GC, say at (p, q), has a feed forward connection from the BC at (p, q); (i) each AC has feed forward connections from the seven nearest BCs; (j) each BC, say at (p, q), has a feedback connection from the AC at (p, q); (k) each HC has connections to the nearest six HCs via gap junctions.

Finally, the triangular grid is dilated about the origin as a function of eccentricity. This is described by the mapping:

$$(x, y) = (x_0, y_0)(x_0^2 + y_0^2)^{0.1} \tag{2.6}$$

where (x_0, y_0) and (x, y) are the respective lattice site locations on Cartesian coordinates before and after the dilation, respectively. According to Equation (2.6),

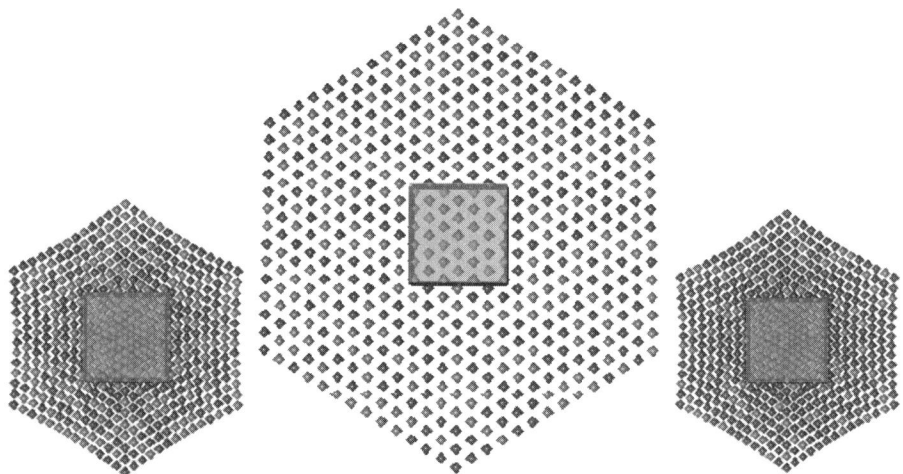

Figure 2.4 Cone mosaics at 0 (left, right) and 1 degree (center) eccentricities. All contain the same number of cones and are displayed at the same scale. Size of a 15 × 15 stimulus is indicated for reference. L/M ratios are 1.10 (left, center) and 4.03 (right).

cone density at an eccentricity of 10 degree is about 1/16 of the peak value at the foveal center [3]. Figure 2.4 shows the resulting model structures seen from the cone layer at eccentricities of 0 degree (left) and 1 degree (center). The ratio of cell type populations is PR:HC:BC:AC:GC = 1:1:1:1:1 as in the primate fovea [53]. Since cells are sequentially countable for each type, they are referenced by a single label i in the following. This notion is quite generic and applicable to any structure with defects and/or deformations.

2.6.2 Differential Equations

In the following, superscripts pr, hc, bc, ac, gc, and xx denote, respectively, each cell type, and all cell types, whereas subscripts i and n identify the specific cell within each type.

The state of the system driven by light, I, at time t is represented as $(V_i^{gc}(t), V_i^{ac}(t), V_i^{bc}(t), V_i^{hc}(t), V_i^{pr}(t), [Ca]_i, [H]_i, [cGMP]_i, [PDE]_i; I_i(t))$, which follows a set of ordinary differential equations, which may be understood more easily by referring to Figure 2.1:

$$\frac{dV_i^{gc}}{dt} = -\alpha^{gc}\widehat{V_i^{gc}}(t) + G^{gc-bc}\,\widehat{V_i^{bc}}(t)(+\text{spike generation}) \qquad (2.7)$$

$$\frac{dV_i^{ac}}{dt} = -\alpha^{ac}\widehat{V_i^{ac}}(t) + G^{ac-bc}\sum_{n}^{\text{links}}\widehat{V_n^{bc}}(t) \qquad (2.8)$$

2.6 A Flexible, Dynamical Model of Retinal Function

$$\frac{dV_i^{bc}}{dt} = -\alpha^{bc}\widehat{V_i^{bc}}(t) + G^{bc-ac}\sum_n^{links}\widehat{V_n^{ac}}(t) + G^{bc-pr}V_i^{pr}(t) \tag{2.9}$$

$$\frac{dV_i^{hc}}{dt} = -\alpha^{hc}\widehat{V_i^{hc}}(t) + G^{hc-pr}\sum_n^{links}V_n^{pr}(t) + G^{hc-hc}\sum_n^{links}\left(\widehat{V_n^{hc}}(t) - \widehat{V_i^{hc}}(t)\right) \tag{2.10}$$

$$\frac{dV_i^{pr}}{dt} = \frac{1}{C_p}\left(q_P\frac{d[Ca]_i}{dt} + q_I\frac{d[H]_i}{dt}\right) + G^{pr-hc}\sum_n^{links}\left(\widehat{V_n^{hc}}(t)\right)^3 \tag{2.11}$$

$$\frac{d[Ca]_i}{dt} = \gamma(1 + c([cGMP]_i(t) - 1)) - \alpha[Ca]_i(t) \tag{2.12}$$

$$\frac{d[H]_i}{dt} = \frac{\lambda_H(1 - [H]_i(t))}{\exp((V_i^{pr}(t) + A_H)S_H) + 1} - \delta_H[H]_i(t) \tag{2.13}$$

$$\frac{d[cGMP]_i}{dt} = -\beta([Ca]_i(t) - 1) - [PDE]_i(t)[cGMP]_i(t) \tag{2.14}$$

$$\frac{d[PDE]_i}{dt} = \frac{1}{\tau}\left(A^{pr}S_{k;i}I_i(t)\exp\left[-\frac{1}{I^{pr}}\int_{t_0}^t I(t)dt\right] - [PDE]_i(t)\right) \tag{2.15}$$

where

$$\widehat{V_i^{xx}}(t) = V^{xx-max}\tanh\left(\frac{V_i^{xx}(t)}{V^{xx-max}}\right) \tag{2.16}$$

and V^{xx-max}, α^{xx}, G^{xx-yy}, G^{xx-xx}, C_p, q_P, q_I, γ, c, α, λ_H, A_H, S_H, δ_H, β, τ, A^{pr}, and I^{pr} are constants. V^{xx-max} is introduced to represent the limited membrane potential ($|V_i^{xx}(t)| < V^{xx-max}$); α^{xx} represents the electrically leaky nature of cells; G^{xx-yy} and G^{xx-xx} are the respective synaptic and gap-junction constants.

The type of synaptic coupling, excitatory or inhibitory, is reflected in the sign of G^{xx-yy} (see Table 2.2 with Figure 2.1). $S_{k;i}$ is the sensitivity of a cone to light and is dependent on type (k = S, M, L). The full expression for $S_{k;i}$ is given in Section 2.6.3. Light intensity is normalized by A^{pr}. Other constants of C_p, q_P, q_I, γ, c, α, λ_H, A_H, S_H, δ_H, β, τ and I^{pr} are used to collectively describe the intracone dynamics in (2.11) to (2.15) [54].[6]

---

6. There are some typographic errors in their paper, which have been corrected here.

The term C_p describes the membrane capacity, and q_P and q_I are the amounts of unit charge transported by the Ca^{2+} and H^+ ion currents, respectively. Terms γ and c collectively denote the influx rate of cGMP ions, while α denotes the efflux rate of Ca^{2+} ions.

The terms λ_H and δ_H denote the rate of increase and decrease of H ion concentration, respectively, while A_H denotes the activation of the photoreceptor at which the current is half-activated, S_H denotes the slope of the activation function and β denotes the strength of the resynthesis reaction of the cGMP ions. The time constant for PDE ion concentration is represented by τ, while I^{pr} represents the photoreceptor fatigue due to pigment bleaching. The reader should also consult Chapter 10 for further models and details of the nature of phototransduction.

The intercellular dynamics are formulated such that each cell senses the states of connected cells, then changes its own state. The state of each cell is represented by its membrane potential, V. For example, a horizontal cell may change its V^{hc} depending on the V^{pr} of a connected cone (feed forward, (2.10), second right term). Likewise, a cone may change its V^{pr} depending on the V^{hc} of a connected horizontal cell (feedback, Equation (2.11), second right term). Furthermore, for cones, simplified ionic interactions formulated by [54] are also considered. They involve concentrations[7] of Ca, H, $cGMP$, PDE ions (see [52] for a simpler model).

Ganglion cells generate spikes (action potentials). The firing rate is known to be positively correlated to the subthreshold membrane potential of a ganglion cell, while the subthreshold membrane potential dips for a few milliseconds after a spike event, which occurs erratically [17]. Spike generation may be modeled by specifying some time-varying firing period as a function of membrane potential. The spikes are, however, ignored in the present study as in some earlier models [40,44,52,54].

Although the ionic mechanism for the feedback from a horizontal cell to a cone has not been fully identified among three possible candidates [55], it is likely to be nonlinearly dependent on the horizontal cell membrane potential (V^{hc}) [56]. By considering symmetry, this feedback is modeled as being proportional to the cube of horizontal cell membrane potential ($V_i^{hc}(t))^3$. All other synaptic couplings are linearly modeled. Because of the presence of (2.11) and (2.13) to (2.16), the complete system is nonlinear.

Cones are modeled to sense (sample) light intensities at their locations. These intensities are bilinearly interpolated values of the image defined on the two-dimensional square grid. Equations (2.11) and (2.15) are different from the corresponding equations in [54], with the feedback mechanism from horizontal cells to a cone in (2.11), and the cone pigment bleaching effect (retinal fatigue) in (2.15).

2.6.3 Color Mechanisms

Color mechanisms should be modeled such that the sensitivity of each type of cone (S_k; $k = S, M, L$) varies with the normalized power spectrum of the incident light ($P_i(\lambda)$), where λ is wavelength in nm and $\int P_i(\lambda) \, d\lambda = 1$. According to [57], the

---

7. According to chemical notation, concentrations are indicated by square brackets ([...]).

2.6 A Flexible, Dynamical Model of Retinal Function

relative absorbance (A_k) of A1 pigments, found in humans, can be modeled for each k as:

$$A_{k;\alpha}(\lambda) = \frac{1}{\exp[A(a - \lambda_k/\lambda)] + \exp[B(b - \lambda_k/\lambda)] + \exp[C(c - \lambda_k/\lambda)] + D} \quad (2.17)$$

$$a = 0.8795 + 0.0459\exp[-(\lambda_k - 300)^2/11{,}940] \quad (2.18)$$

$$A_{k;\beta}(\lambda) = \bar{A}_\beta \exp\{-[(\lambda - \lambda_{m\beta})/b_\beta]^2\} \quad (2.19)$$

$$\lambda_{m\beta} = 189 + 0.315\lambda_k \quad (2.20)$$

$$b_\beta = -40.5 + 0.195\lambda_k \quad (2.21)$$

$$A_k(\lambda) = A_{k;\alpha}(\lambda) + A_{k;\beta}(\lambda) \quad (2.22)$$

where A, B, b, C, c, D, $\bar{A}_\beta$ are constants ($A = 69.7$, $B = 28$, $b = 0.922$, $C = -14.9$, $c = 1.104$, $D = 0.674$, $\bar{A}_\beta = 0.26$). For humans, $\lambda_S = 420$, $\lambda_M = 530$, and $\lambda_L = 565$ [2] (Figure 2.5).

S_k is then modeled as:

$$S_{k;i} = \int A_k(\lambda)P_i(\lambda)d\lambda \quad (2.23)$$

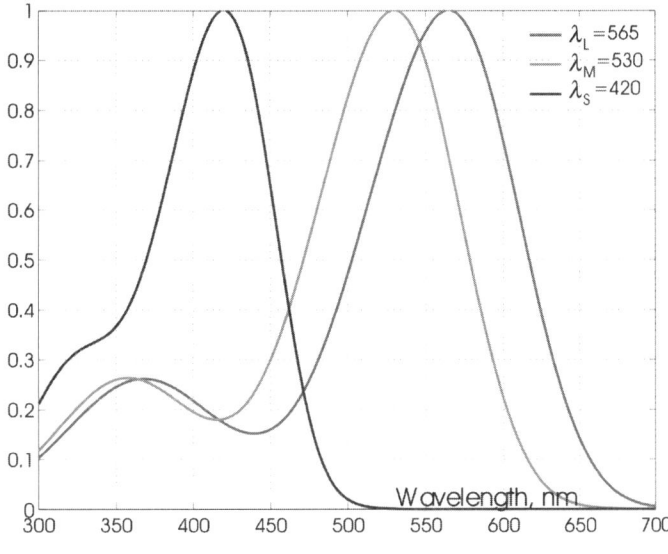

Figure 2.5 The relations between relative absorbance and wavelength calculated for S-, M-, and L-cones in humans [see (2.17–2.22)].

For a monochromatic light at waveletngth λ_k, $P(\lambda)$ is represented by the delta function $\delta(\lambda - \lambda_k)$, and therefore $S_k = A_k(\lambda_k) = 1$.

The model described here resembles two earlier cell-level models, Wilson's [52] and Hennig's [54]. Like the system described here, both these earlier models are formulated as a set of differential equations, and incorporate simplified ionic interactions only in the photoreceptors. However, both Wilson's [52] and Hennig et al.'s [54] models are described on uniform one-dimensional and two-dimensional hexagonal grids, respectively, while the model described in this section incorporates the hexagonal cone arrangement with continuous radial dilation. Many of the earlier models do not incorporate eye movements, although in their recent study on hyperacuity in the peripheral vision, Hennig et al. [58] did introduce eye movements. Negative feedback mechanisms between horizontal cells and cones, and between amacrine cells and bipolar cells, are present in Wilson's [52] model, but absent in Hennig's [54]. The intracone dynamics in Wilson's model are much simpler than those in Hennig's, which quantitatively reproduce physiological data [59]. Neither of these earlier cell-level models incorporates color mechanisms. Many of these points can be significant in describing particular aspects of a retinal representation.

2.6.4 Foveal Image Representation

As described in Section 2.5, a retinal representation can, with strong assumptions, be viewed as the convolution of an image and a set of DoG functions, thereby capturing the essence of a *static* spatial receptive field pattern. In the foveola, however, cone spacing increases at its fastest rate [3], so that the shift-invariance that is necessary for the justification of a convolution-type model cannot be realistically assumed. Also, the conventional time-dependent DoG model, mentioned earlier, fails to model spatiotemporally inseparable visual phenomena such as will be described later. For these reasons, it could be argued that a retinal representation is better modeled in terms of a population of ganglion cell activities constructed by cell-level models of the form described in this section.

For the simpler case of an achromatic image, a foveal representation is defined on a two-dimensional square grid as a map of ganglion cell membrane potentials ($\{V_i^{gc}\}$), and is calculated by bilinear interpolation from the dilating hexagonal arrangement (Figure 2.4). By "afterimages" we refer to the foveal representation after the termination of image projection.

Human color vision is considered to be shaped by luminance, red–green opponent, and blue–yellow opponent channels [2,4,60]. Based on this idea, the foveal representation of a color image may be modeled as follows (with reference to Figure 2.6). First, a set of ganglion cell membrane potentials ($\{V_i^{gc}\}$) is divided into two nonoverlapping subsets:

$$\{V_i^{gc}(t)\} = \{V_i^{gc}(t)\}^L \oplus \{V_i^{gc}(t)\}^M \qquad (2.24)$$

where, $\{V_i^{gc}\}^{L(M)}$ is a set of membrane potentials of a ganglion cell that has a direct link to an L- or M-cone.

2.6 A Flexible, Dynamical Model of Retinal Function

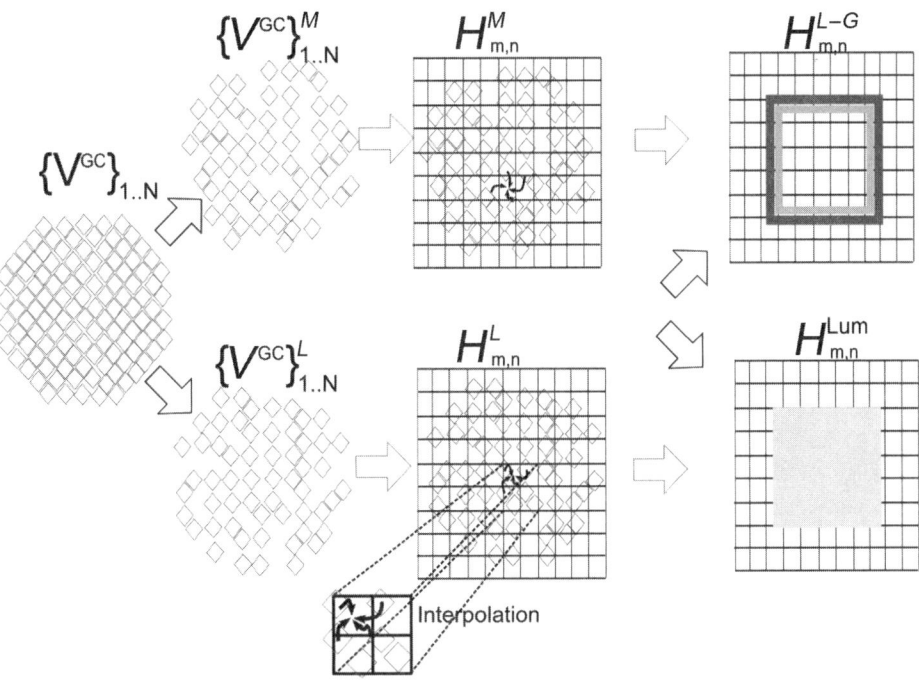

Figure 2.6 Schematic illustration of constructing foveal color representations.

Next, red and green representations (H^R, H^G) are defined on the same two-dimensional square grid as maps of $\{V_i^{gc}\}^L$ and $\{V_i^{gc}\}^M$, again employing bilinear interpolation:

$$H^R_{m,n}(t) \xleftarrow{\text{map}} \{V_i^{gc}(t)\}^L \qquad (2.25)$$

$$H^G_{m,n}(t) \xleftarrow{\text{map}} \{V_i^{gc}(t)\}^M \qquad (2.26)$$

where m and n are lattice indices. Finally, luminance (H^{Lum}) and R-G differential (H^{R-G}) representations are constructed:

$$H^{Lum}_{m,n}(t) = rH^R_{m,n}(t) + (1-r)H^G_{m,n}(t) \qquad (2.27)$$

$$H^{R-G}_{m,n}(t) = rH^R_{m,n}(t) - (1-r)H^G_{m,n}(t) \qquad (2.28)$$

where r is the rate of L-cone population. Examples of outputs of a retinal model to these stimuli are considered in Section 2.7.

2.6.5 Modeling Retinal Motion

In [61] tremor is modeled as the harmonic oscillation of the retina over a two-dimensional visual field with a period of 10 ms and amplitude of 1 cone distance, a motion that is consistent with that observed in the human retina [62]. Microsaccades are more difficult to introduce because the intermicrosaccadic interval is typically 1 second [62]. If included, a single microsaccadic event has the same effect as stimulus onset or cessation. Modeling drift requires simulation of motion (relative to the central fovea spacing) of roughly 2.5 cones within 250 ms, which corresponds to a mean speed of $0.1°/s$.

2.7 Numerical Simulation Examples

We now provide some examples of applying the model described above to simple scenes; this allows the demonstration of cell-level and macroscopic properties of a dynamical model as described in Section 2.6. These properties include fast and slow temporal behavior, and adaptation. Thus, both spatial and dynamical properties of a retina are nicely captured at several physical levels.

2.7.1 Parameters and Visual Stimuli

Table 2.2 lists the values of the parameters used in these examples. Time (t) and membrane potentials (V) are measured in ms and in mV, respectively. Light intensity (I) is measured in troland (td), and a normalization coefficient (A^{pr}) is determined by noting that a cone's response to light stimulus half saturates at about 10^3 td [40, 63]. When a grayscale image is used, sensitivities (S_k) may be set to 1 for all cones. For cells other than cones, time constants ($1/\alpha^{xx}$) are set to 10 ms, as they are of this order of magnitude [40, 45, 52, 54]. As noted in [52], it is difficult to physiologically determine values for the parameters that describe coupling strength between neurons. Values are, therefore, manually tuned so as to account for a wide variety of retinal properties.

The retina is assumed to be adapted to a dark uniform background until light onset at $t = t_0 = 0$ ms. Because cone pigment bleaching is thought to occur on the

Table 2.2 The Values of the Parameters Used in the Simulations

Symbol	Value	Unit	Symbol	Value	Unit
α^{gc}	0.1	ms^{-1}	α^{hc}	0.1	ms^{-1}
G^{gc-bc}	0.5	ms^{-1}	G^{hc-pr}	0.02	ms^{-1}
α^{ac}	0.1	ms^{-1}	G^{hc-hc}	1.5	ms^{-1}
G^{ac-bc}	0.02	ms^{-1}	G^{pr-hc}	-0.02	mV^{-2} ms^{-1}
α^{bc}	0.1	ms^{-1}			
G^{bc-ac}	-0.1	ms^{-1}			
G^{bc-pr}	-0.05	ms^{-1}			
V^{xx-max}	100	mV			
q_P/C_P	100.0	mV	λ_H	1	—
q_I/C_P	600.0	mV	A_H	400	mV
γ	0.4	ms^{-1}	S_H	0.01	mV^{-1}
c	0.14	—	δ_H	0.025	ms^{-1}
α	0.4	ms^{-1}	β	0.6	ms^{-1}
			τ	10	ms
			A^{pr}	10^{-4}	td^{-1}
dt	0.1	ms	I	$10^1, 10^2, 10^3, 10^4,$ or 10^5	td

order of tens of seconds under usual visual conditions [60], its effect is ignored here by assigning $I^{\mathrm{pr}} = +\infty$.

A 1-bit square image is used as light stimulus. Its size is chosen to be either a 15 × 15 grid of cones (0.15° × 0.15°) or a 3 × 3 grid. The 15 × 15 image is designed to allow the study of afterimage "patterns" while minimizing eccentricity effects. In contrast, the 3 × 3 image is chosen to elucidate properties in a ganglion cell receptive field, whose center size is comparable with this image size in the presence of tremor. An image is projected from $t = 0$ at an eccentricity of 0° (see [61] for off-axis studies). Each stimulation lasts indefinitely, or for 250 ms terminated either by complete darkness (dark poststimulus) or by the entire field being illuminated in uniform light whose intensity is the same as that during image projection (bright poststimulus) as in Figure 2.7. In addition, uniform (diffuse) light that covers the entire cone layer is used to investigate color properties.

The following model property is noted here. Despite no explicit delay term, for example, in the form of $V_i^{xx}(t - \tau)$, the model shows delays in simulated results between light stimulus (I_j) and ganglion cell membrane potentials (V_i^{gc}), because the terms involving $-\alpha^{xx}$ in (2.7) to (2.10), which are associated with the "leaky" nature of neurons, work as temporal lowpass filters. These delays are path-dependent, and hence are spatially varying. In the classical center-surround receptive field, the response to the surround illumination is known to be delayed relative to the response to the center illumination [17]. This feature is naturally reproduced in the model as described.

2.7.2 Temporal Characteristics

To illustrate the temporal dynamics of the network model described in Section 2.6, we may monitor the membrane potential of a ganglion cell in the center of the retina as a 3 × 3 light stimulus temporally modulated at 100% contrast with a square wave of 50% duty cycle.

One may "monitor" the response of individual cell components of the model, not only the ganglion cell outputs, under the effect of parameter changes to the model. Consider, first, the behavior of a cone during a 3 × 3 light stimulus.

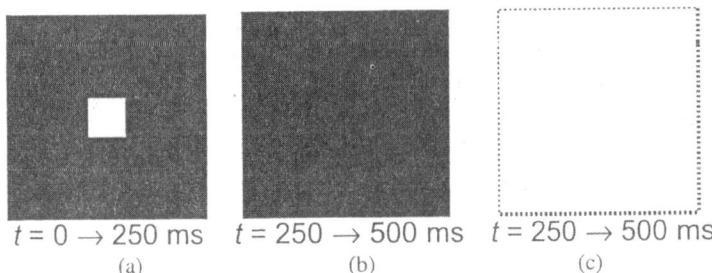

$t = 0 \to 250$ ms $t = 250 \to 500$ ms $t = 250 \to 500$ ms
(a)   (b)   (c)

Figure 2.7 (a) 1-bit square image used as a light stimulus in the study of afterimages. Size is 15 cone × 15 cone (0.15° × 0.15°) or 3 × 3. The stimulus lasts 250 ms followed by darkness, (b) dark poststimulus, or by (c) uniform light illumination (bright poststimulus).

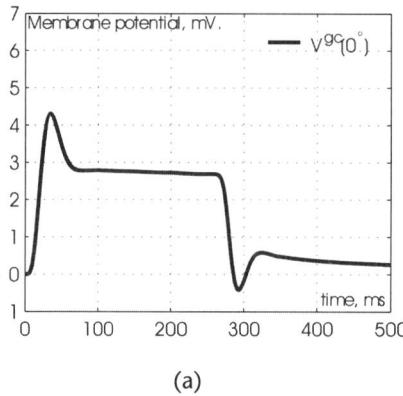
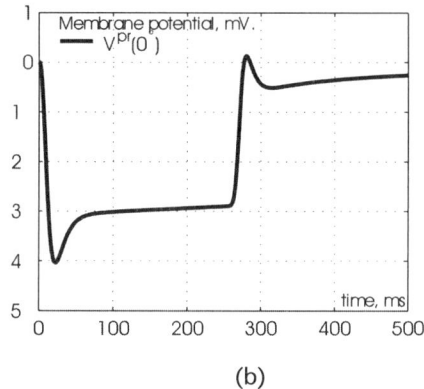

Figure 2.8 The modeled membrane potentials of a ganglion cell (left) and a cone (right) during and after a 3 × 3 stimulus. A bright poststimulus is used. The ganglion cell and the cone are directly interlinked (see Figure 2.1).

Figure 2.8 shows the membrane potential of a cone during and after the application of such a stimulus at the center of a retinal model. A bright poststimulus has been used. Note that the photoreceptor "phase" is inverted relative to that of the ganglion cell phase. This has significance in appreciating effects that are well known in human perception, such as the occurrence and nature of afterimages. In this study spikes are ignored.[8]

To illustrate the effects of tremor, consider the effect of illuminating the retina with light modulated at a 1 Hz stimulus frequency. We compare the results with and without the presence of tremor in Figure 2.9. The temporal interval of the simulation is 1 ms.

Without tremor, the receptive field center is small, and therefore the effect of lateral inhibition becomes more prominent. Compared with the control condition, which is the *presence* of tremor, the ganglion cell membrane potential (V^{gc}) is weaker. It is also seen that the delay in V^{gc} to the stimulus cessation is smaller, and that the following undershoot is larger. Tremor is, therefore, a factor to be considered in explaining certain fine aspects of visual perception.

Now, we consider the effect of varying the frequency of the square wave stimulus modulation through 10, 20, and 50 Hz. From psychophysical experiments, it is known that flickering light at about 10 Hz is perceived as being brighter than either steady illumination, or illumination that is being strobed at a faster rate. This is sometimes referred to as the Brücke-Bartley effect [60].

This feature (decreasing amplitude of response), is observed by comparing Figures 2.10(a–c), and, indeed, by comparing Figure 2.10(a) with Figure 2.9.

---

8. Ganglion cells induce spikes, but the rate of spike production is strongly correlated with the ganglion cell membrane potential. Thus, prediction of membrane potential by a physiological model is often considered a reasonable test for stimulus response. This assumption is challenged by spiking neuron models, which suggest that the phase (timing) of spikes is also significant and should be modeled too.

2.7 Numerical Simulation Examples

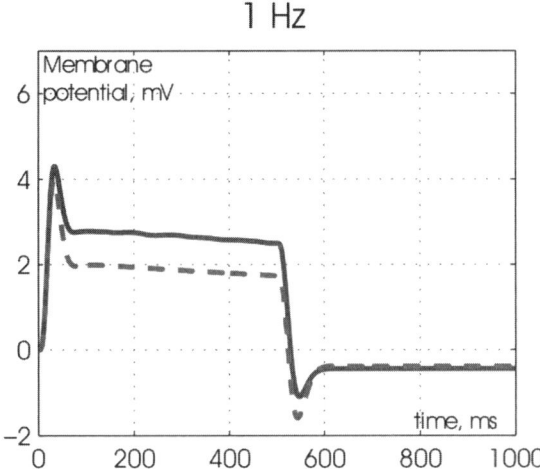

Figure 2.9 Simulation Example 1. The membrane potential of a ganglion cell is monitored under achromatic illumination at the retinal center. A 3 × 3 light stimulus is temporally modulated by an on–off square wave (100% contrast and 50% duty) with a frequency of 1 Hz. The dashed line is calculated without tremor. Note that spikes are ignored in the present study.

2.7.3 Spatial Characteristics

The receptive field (RF) size can be estimated at different eccentricities in order to assess whether the dilation of the network described in the preceding sections does indeed lead to the observed effect of a larger receptive field with increasing eccentricity. Momiji et al. [61] compared the receptive field sizes at 0°, 1°, and 2.5° off the retinal center for a model as described in Section 2.6. The ganglion cell membrane potential (V^{gc}) was monitored for 10^3-td light stimuli of different sizes. At each degree of selected eccentricity, the RF size is defined as the stimulus size that makes V^{gc} at 500 ms fall to 10% of its maximum value. The inverse of receptive field (RF) size at the retinal center was found to correspond to 8 cpd (cycles per degree), which mirrors the observation that the experimentally obtained contrast sensitivity curve peaks near 8 cpd for 900-td illumination [64]. The RF was indeed found to become larger with eccentricity, but the pace of increase became smaller as eccentricity increased. These features are consistent with experimental data on acuity [65].

The behavior of the spatial properties of Momiji's model [61] under different levels of light illumination was also tested. For example, it is known that in the retina [66], the receptive field size decreases with light intensity. This is somewhat reflected by the decrease in spatial acuity with lowering light levels. When testing the model at intensities ranging from 10^2 to 10^5 td at the retinal center, it was indeed confirmed that the RF size decreased with light intensity in a logarithmic fashion. The effect was found to saturate at about 10^4 td.

A final achromatic, spatial property illustrates the significance of a model that captures both spatial and temporal behavior: when a bright high-contrast square, of size 15 × 15, centered on a model retina, is switched on at $t = 0$ ms, then

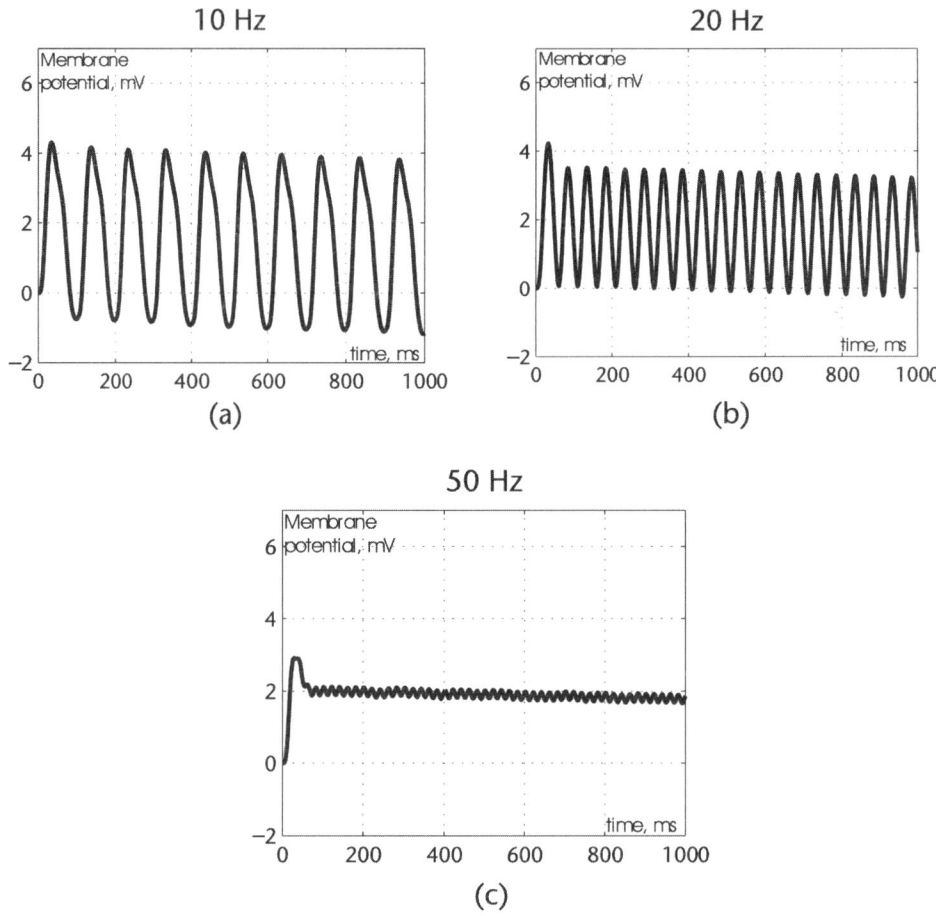

Figure 2.10 Simulation Example 2. The membrane potential of a ganglion cell is monitored under achromatic illumination at the retinal center. A 3 × 3 light stimulus is temporally modulated by an on–off square wave (100% contrast and 50% duty) with a frequency of (a) 10 Hz, (b) 20 Hz, and, (c) 50 Hz. Note the loss of temporal perception of the strobing (flicker fusion). Spikes are ignored in the present study.

switched off at $t = 250$ ms, with a dark poststimulus, an after-image appears in the spatial representation of ganglion cell outputs. The time course of the responses of ganglion cells (population response) encoded as an image is shown in Figure 2.11. For clarity, a profile of the firing rate, through a line of ganglion cells, is illustrated in Figure 2.12.

One may note three main features of retinal behavior. First, the clear presence of a negative afterimage in response to the switching of the stimulus. Second, one may note the bandpass filtering effect induced by the retinal receptive field characteristics. This is quite obvious from the profile at 250 ms, which shows clear enhancement of the borders of the square. Finally, one may note that the afterimage itself becomes a lowpass-filtered version of the stimulus. This is consistent with psychophysical observations [67].

2.7 Numerical Simulation Examples

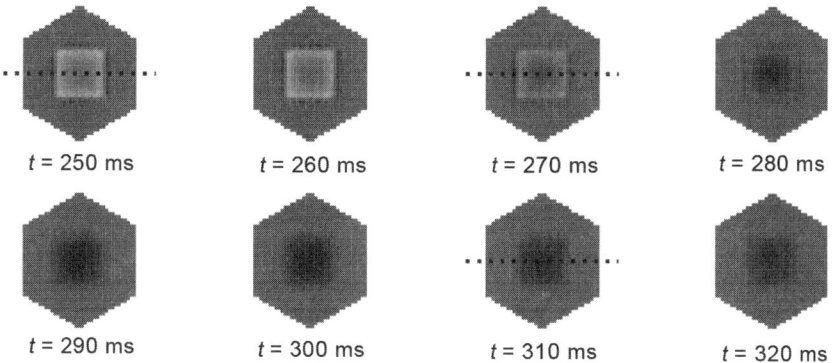

Figure 2.11 Snapshots of retinal ganglion cell outputs, encoded as an image, in response to a 15 × 15 square image being switched on, then off. The image projection begins at $t = 0$ ms and is terminated at 250 ms with a dark poststimulus. The light intensity is 10^3 td. A single grayscale is used to facilitate comparison.

2.7.4 Color Characteristics

It is difficult to separate cortical perceptual effects, or, indeed, effects that might be induced in the LGN, from those that emerge in retinal processing. Nevertheless, by a combination of psychophysical experiments, physiological experiments, and modeling approaches, the precise computational routes via which perception is achieved can be elucidated. Consider, now, questions of color perception: how do color opponent channels achieve an encoding of color, and what key aspects of color perception may be observed in this model? Second, how does color perception become invariant to the exact ratio of L/M-cones that might be present at any location in a random cluster, such as might be found in the retina?

A network-type model of the nature of that of Section 2.6 is useful in studying connection effects; one may monitor membrane potentials for individual cells in the model, may modify individual connection strength parameters, and also, because

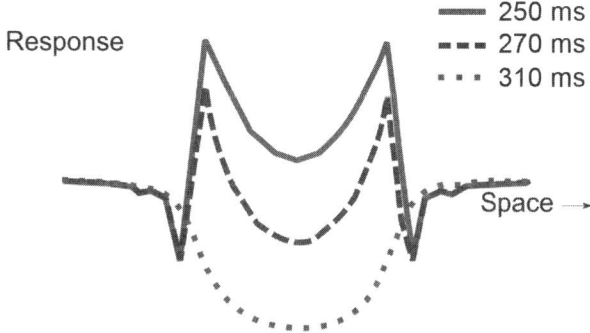

Figure 2.12 Afterimages induced by the switching of the 15 × 15 stimulus extracted at 250, 270, and 310 ms along the dotted lines of Figure 2.11, to illustrate better how the spatial structure of the afterimages is altered in time.

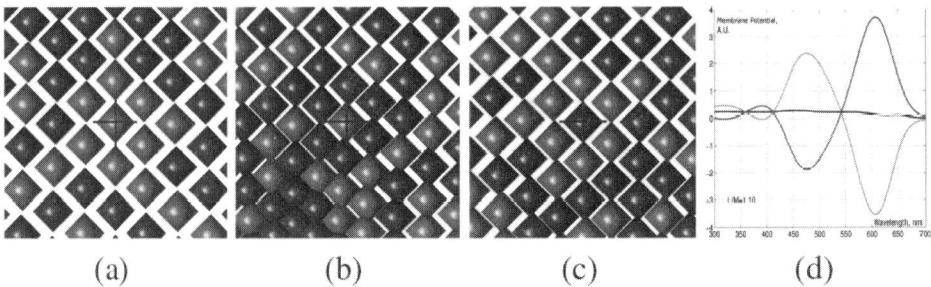

Figure 2.13 Experiment 10. Color sensitivity and L/M-cone arrangement. The entire cone layer shown in Figure 2.4 (left) is illuminated by monochromatic uniform light with various wavelengths. Local cone mosaics around three selected ganglion cells, each being labeled by a cross, are shown in (a), (b), and (c). Membrane potentials of the corresponding ganglion cells at 500 ms are shown (d) as a black curve, green curve, and red curve, respectively.

there is no constraint to a regular grid, it is possible to investigate the role of feedback in an instance of the model containing arbitrary clusters of "wiring."

Consider Figure 2.13(a–c). Three different center-surround arrangements are present, corresponding to green-only, red–green, and green–red. The wavelength of monochromatic, diffuse incident light is varied to yield the plots shown in Figure 2.13(d) of membrane potentials vs wavelength in nm at the three distinct ganglion cells.

To study the effect of varying L/M-cone populations, we may study the R–G opponent channel output, ϕ^{R-G}, as estimated from the model by summing the membrane potential of cells with links to L-cones, and subtracting that of ganglion cells with direct links to M-cones. This operation is performed as a function of wavelength, and at 500 ms after the monochromatic stimulus is presented. This quantity, ϕ^{R-G} is plotted for two different L/M-cone populations in Figure 2.14, on the left. Note that only relative values are significant on the vertical axis.

The slope of ϕ^{R-G} may be considered a measure of the ability of a subject to discriminate between two wavelengths. Thus, in a region of the curve where the slope is relatively small (around $\lambda = 475$ nm), we might expect to see the poorest color discrimination in this region. However, recall that the parameters of the model we have described here have been tuned in accordance with the foveola, and so we have omitted S-cones, which would, in a normal subject, contribute to color vision when the stimulus is projected even slightly outside the central foveola region.

There are some people that display the lack of an S-cone system entirely (overall, fewer than 1 in 3,000), and this does indeed lead to the poorest wavelength discrimination being observed at around 460 nm [68].

The simulation used to generate Figure 2.14 (left) is repeated in the right panel, but with the strengths of feedback connections from horizontal cells to photoreceptors and from amacrine cells to bipolar cells set to 0. Now, compare the curve pairs in the left- and right-hand panels. As the L/M-cone ratio moves from 1.10 to 4.03, the locations of the local maxima and minima of the ϕ^{R-G} vs

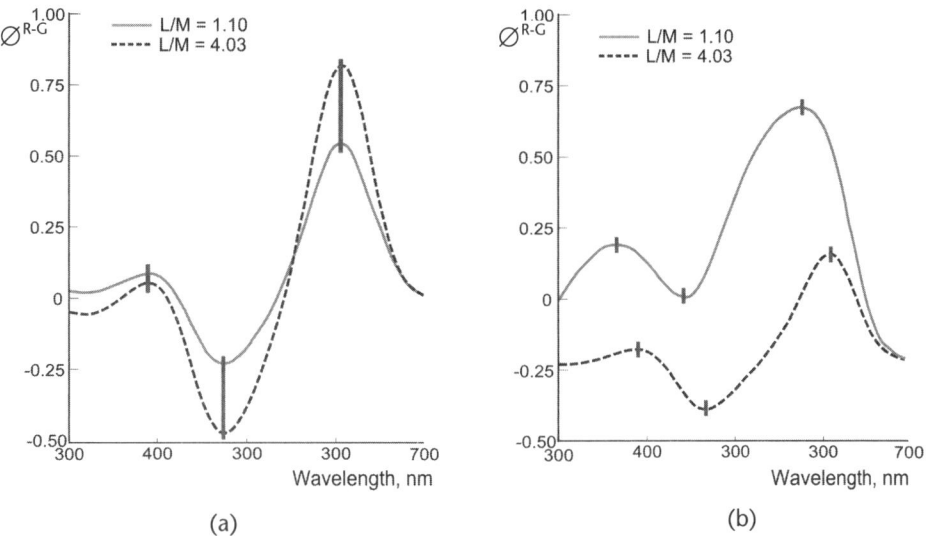

Figure 2.14 (Left) R–G channel outputs. Uniform monochromatic light is projected onto the retina, with various frequencies, on two cone mosaics with L/M ratios of 1.10 and 4.03. (Left) the control condition. (Right) the no-feedback condition (i.e., $G^{pr-hc} = G^{bc-ac} = 0.0$). Absolute values on the vertical axes are not meaningful.

λ characteristic remain in the same positions in the left-hand panel. This coincides with the observation that even widely differing L/M-cone ratios may be found in subjects with normal color vision [11]. On the other hand, in the absence (right panel) of the feedback "control" paths found in the retina, the curve of ϕ^{R-G} vs λ shifts its local maxima and minima towards the direction of decreasing wavelength, by about 30 nm. These observations are consistent with current thinking that retinal-level feedback mechanisms may compensate for diverse L/M ratios.

2.8 Conclusions

Mathematical models of retinal function have been applied successfully to understand and reproduce some of the macroscopic properties of retinal coding. Laplacian of Gaussian functions, or difference of Gaussians, both widely used in image processing, mirrors the function of the retina, but is often employed at a single scale, and in a static (nondynamical) manner.

In modeling psychophysical phenomena, more subtle perceptual effects, and effects that require either a temporally varying stimulus, or are themselves dynamic, are more difficult to reproduce with the simpler space–time retinal models. Moreover, linking such macroscopic models to cell physiology is, at best, restricted to the ganglion cell layer.

The dynamical systems approach described in this chapter leads to the ability to perform simulations that are physically detailed, and also permit the linkage of the macroscopic retinal properties to cell-level phenomena, creating a link between physiology and function.

Such detailed mathematical models may be useful for the development of medicines. For example, the side effect of *Zatebradine*, an antianginal agent, which is known in clinical trials to cause glaring or pronounced after-images, has been investigated by using ionic current models of a photoreceptor [69].

Caution is, of course, necessary in explaining perceptual effects without taking into account the role and complexity of cortical function. Furthermore, the model we have described here is computationally quite demanding, and does not necessarily lead to retinal representations that are of immediate practical use in image analysis or computer vision. In Chapter 3, we review some of the more important aspects of cortical function, with a view to the potential applications of developing artificial cortical representations, discussed in Chapter 5.

References

[1] Hubel, D. H., *Eye, Brain, and Vision*. Oxford, England: Scientific American Library, 1988.

[2] Tovée, M. J., *An Introduction to the Visual System*. Cambridge: Cambridge University Press, 1996.

[3] Wässle, H. and Boycott, B., "Functional architecture of the mammalian retina," *Physiological Reviews*, Vol. 71, 1991, pp. 447–480.

[4] Wandell, B. A., *Foundations of Vision*. Sunderland, Ma: Sinauer Associates, 1995.

[5] Masland, R. H., "The fundamental plan of the retina," *Nature Neuroscience*, Vol. 4, 2001, pp. 877–886.

[6] Kolb, H., Fernandez, E., and Nelson, R., Webvision, http://webvision.med.utah.edu/.

[7] Bower, J. M. and Beeman, D. eds., *The Book of GENESIS: Exploring Realistic Neural Models with the GEneral NEural SImulation System (2nd Ed.)*. New York: Springer-Verlag, 1998.

[8] Koch, C. and Segev, I. eds., *Methods in Neuronal Modeling*. Cambridge, MA: MIT Press, 1989.

[9] Dacey, D. M., "Parallel pathways for spectral coding in primate retina," *Annual Review of Neuroscience*, Vol. 23, 2000, pp. 743–775.

[10] Lyubarsky, A., Falsini, B., Pennisi, M. et al., "UV- and midwave-sensitive cone-driven retinal responses of the mouse: a possible phenotype for coexpression of cone photopigments," *Journal of Neuroscience*, no. 1, January 1999, pp. 442–455.

[11] Roorda, A. and Williams, D. R., "The arrangement of the three cone classes in the living human eye," *Nature*, Vol. 397, 1999, pp. 520–522.

[12] Foster, R. G. and Hankins, M. W., "Non-rod, non-cone photoreception in the vertebrates," *Progress in Retinal and Eye Research*, Vol. 21, 2002, pp. 507–527.

[13] Dacey, D., Packer, O. S., Diller, L., et al., "Center surround receptive field structure of cone bipolar cells in primate retina," *Vision Research*, Vol. 40, 2000, pp. 1801-1811.

[14] Cook, P. B. and McReynolds, J. S., "Lateral inhibition in the inner retina is important for spatial tuning of ganglion cells," *Nature Neuroscience*, Vol. 1, 1998, pp. 714-719.

References

[15] Nelson, M. and Rinzel, J., "The Hodgkin-Huxley model," in *The Book of GENESIS: Exploring Realistic Neural Models with the GEneral NEural SImulation System (2nd Ed.)*, J. M. Bower and D. Beeman, eds., New York: Springer-Verlag, 1998, pp. 29-49.

[16] Zenisek, D. and Matthews, G., "Calcium action potential in retinal bipolar neurons," *Visual Neuroscience*, Vol. 15, 1998, pp. 69-75.

[17] Meister, M. and Berry II, M. J., "The neural code of the retina," *Neuron*, Vol. 22, 1999, pp. 435-450.

[18] Brivanlou, I. H., Warland, D. K., and Meister, M., "Mechanisms of concerted firing among retinal ganglion cells," *Neuron*, Vol. 20, 1998, no. 3, pp. 527-539.

[19] Ishii, H., Yamamoto, M., Kamiyama, Y., and Usui, S., "Analysis of the spatial properties of synaptic input signals to the retinal horizontal cells," *The IEICE Transactions D-II*, Vol. J80-D-II, no. 7, pp. 1939-1946, 1997, in Japanese with English figures.

[20] Usui, S., Kamiyama, Y., Ishii, H., and Ikeno, H., "Reconstruction of retinal horizontal cell responses by the ionic current model," *Vision Research*, Vol. 36, 1996, no. 12, pp. 1711-1719.

[21] Bolduc, M. and Levine, M. D., "A review of biologically motivated space-variant data reduction models for robotic vision," *Computer Vision and Image Understanding*, Vol. 69, no. 2, 1998, pp. 170-184.

[22] Enroth-Cugell, C. and Robson, J. G., "The contrast sensitivity of retinal ganglion cells of the cat," *Journal of Physiology (London)*, Vol. 187, 1966, pp. 517-552.

[23] GENESIS, http://www.genesis-sim.org/GENESIS/, Caltech.

[24] NEURON, http://neuron.duke.edu/, Duke.

[25] Smith, R. G., NeuronC, http://retina.anatomy.upenn.edu/~rob/neuronc.html.

[26] Hodgkin, A. L. and Huxley, A. F., "A quantative description of membrane current and its application to conduction and excitation in nerve," *Journal of Physiology*, Vol. 117, 1952, pp. 500-544.

[27] Abbot, L. and Kepler, T., *Model Neurons: From Hodgkin-Huxley to Hopfield*, ser. Lecture Notes in Physics, L. Garrido, Ed. Springer-Verlag, 1990.

[28] Ogura, T., Kamiyama, Y. and Usui, S., "Ionic current model for the inner-segment of a retinal photoreceptor," *Systems and Computers in Japan*, Vol. 28, no. 13, 1997, pp. 55-66, Translated from The IEICE Transactions D-II, J78-D-II (10), 1501-1511 (1995) (in Japanese with English figures).

[29] Aoyama, T., Kamiyama, Y., Usui, S., Blanco, R., Vaquero, C. F., and de la Villa, P., "Ionic current model of rabit retinal horizontal cell," *Neuroscience Research*, Vol. 37, 2000, pp. 141-151.

[30] Aoyama, T., Kamiyama, Y., and Usui, S., "Simulation analysis of current-voltage relationship of retinal horizontal cells," *The IEICE Transactions D-II*, Vol. J85-D-II, no. 5, 2002, pp. 918-930, in Japanese with English figures.

[31] Ishihara, A., Kamiyama, Y., and Usui, S., "Ionic current model of the retinal bipolar cell," *The IEICE Transactions D-II*, Vol. J80-D-II, no. 12, 1997, pp. 3181-3190, in Japanese with English figures.

[32] ———, "The synaptic terminal model of retinal ON-center bipolar cell based on ionic mechanisms," *The IEICE Transactions D-II*, Vol. J83-D-II, no. 2, 2000, pp. 723-731, in Japanese with English figures.

[33] Usui, S., Ishihara, A., Kamiyama, Y., and Ishii, H., "Ionic current model of bipolar cells in the lower vertebrate retina," *Vision Research*, Vol. 36, no. 24, 1996, pp. 4069-4076.

[34] Rall, W., "Cable therory for dendritic neurons," in *Methods in Neuronal Modeling*, C. Koch and I. Segev, eds. Cambridge, MA: MIT Press, 1989, pp. 9–62.

[35] Segev, I., "Cable and compartmental models of dendritic trees," in *The Book of GENESIS: Exploring Realistic Neural Models with the GEneral NEural SImulation System (2nd Ed.)*, J. M. Bower and D. Beeman, eds. New York: Springer-Verlag, 1998, pp. 51–77.

[36] Fall, C. P., Marland, E. S., Wagner, J. M., and Tyson, J. J., *Computational Cell Biology*, ser. Interdisciplinary Applied Mathematics. Springer-Verlag.

[37] Dayan, P. and Abbott, L. F., *Theoretical Neuroscience*. MIT Press, 2001.

[38] Rodieck, R. W., "Quantitative analysis of cat retinal ganglion cell response to visual stimuli," *Vision Research*, Vol. 5, 1965, pp. 583–601.

[39] Rodieck, R. W. and Stone, J., "Response of cat retinal ganglion cells to moving visual patterns," *Journal of Physiology*, Vol. 28, 1965, pp. 819–832.

[40] Shah, S. and Levine, M. D., "Visual information processing in primate cone pathways-Part I: A model," *IEEE Transactions on Systems, Man, and Cybernetics—Part B*, Vol. 26, 1996, pp. 259–274.

[41] ———, "Visual information processing in primate cone pathways—part II: experiments," *IEEE Transactions on Systems, Man, and Cybernetics—Part B*, Vol. 26, 1996, pp. 275–289.

[42] Chae, S.-P., Lee, J.-W., Jang, W.-Y., et al., "Erg signal modeling based on retinal model," *IEICE Transactions on Fundamentals of Electronics Communications and Computer Sciences*, Vol. E84-A, no. 6, June 2001, pp. 1515–1524.

[43] Roska, B. and Werblin, F., "Vertical interactions across ten parallel, stacked representations in the mammalian retina," *Nature*, Vol. 410, 2001, pp. 583–587.

[44] Bálya, D., Roska, B., Roska, T., and Werblin, F. S., "A CNN framework for modeling parallel processing in a mammalian retina," *International Journal of Circuit Theory and Applications*, Vol. 30, 2002, pp. 363–393.

[45] Bálya, D., Rekeczky, C., and Roska, T., "A realistic mammalian retinal model implemented on complex cell CNN universal machine," in *ISCAS 2002: IEEE International Symposium on Circuits and Systems*, 2002, pp. IV-161–164.

[46] Werblin, F. S. and Roska, B., "The movies in our eyes," *Scientific American*, April 2007.

[47] Fujii, S., Kamiyama, Y., Takebe, S., and Usui, S., "A simulation analysis of retinal photoreceptor network on massively parallel system," in *BPES '98*, 1998, pp. 389–392, in Japanese with English abstract and figures.

[48] ———, "A simulation analysis of retinal outer plexiform layer model on massively parallel system," *The Institute of Electronics, Information and Communication Engineers, Technical Report of IEICE NC98-186*, 1999, in Japanese with English abstract.

[49] Momiji, H., Hankins, M., Bharath, A., and Kennard, K., "A numerical study of red-green colour opponent properties in the primate retina," *European Journal of Neuroscience*, Vol. 25, no. 4, 2007, pp. 1155–1165.

[50] Calkins, D., Schein, S., Tsukamoto, Y., and Stering, P., "M and L cones in macaque fovea connect to midget ganglion cells by different numbers of excitatory synapses," *Nature*, Vol. 371, 1994, pp. 70–72.

[51] Roorda, A., Metha, A., and Lennie, D., & Williams, P., "Packing arrangement of the three cone classes in primate retina," *Vision Research*, Vol. 41, 2001, pp. 1291–1306.

[52] Wilson, H., "A neural model of foveal light adaptation and afterimage formation," *Visual Neuroscience*, Vol. 14, 1997, pp. 403–423.

[53] Ahmad, K., Klug, K., Herr, S., Sterling, P., and Schein, S., "Cell density ratios in a foveal patch in macaque retina," *Visual Neuroscience*, Vol. 20, 2003, pp. 189–209.

References

[54] Hennig, M., Funke, K., and Wörgötter, F., "The influence of different retinal subcircuits on the nonlinearity of ganglion cell behavior," *Journal of Neuroscience*, Vol. 22, 2002, pp. 8726-8738.

[55] Kamermans, M. and Spekreijse, H., "The feedback pathway from horizontal cells to cones: A mini review with a look ahead," *Vision Research*, Vol. 39, 1999 pp. 2449-2468.

[56] Fahrenfort, I., Habets, R., Spekreijse, H., and Kamermans, M., "Intrinsic cone adaptation modulates feedback efficiency from horizontal cells to cones," *Journal of General Physiology*, Vol. 114, 1999, pp. 511-524.

[57] Govardovskii, V., Fyhrquist, N., Reuter, T., et al., "In search of the visual pigment template," *Visual Neuroscience*, Vol. 17, 2000, pp. 509-528.

[58] Hennig, M., Funke, K., and Wörgötter, F., "Eye micro-movements improve stimulus detection beyond the Nyquist limit in the peripheral retina," in *Advances in Neural Information Processing Systems*, Vol. 16., MIT Press, 2004, pp. 1475-1482.

[59] Schneeweis, D. and Schnapf, J., "The photovoltage of macaque cone photoreceptors: Adaptation, noise and kinetics," *Journal of Neuroscience*, Vol. 19, 1999, pp. 1203-1216.

[60] Schwartz, S. H., *Visual Perception: A Clinical Orientation*. McGraw-Hill Professional, 2004.

[61] Momiji, H., Bharath, A. A., Hankins, M., and Kennard, C., "Numerical study of short-term afterimages and associate properties in foveal vision," *Vision Research*, Vol. 46, no. 3, Feb 2006, pp. 365-381.

[62] Martinez-Conde, S., Macknik, S., and Hubel, D., "The role of fixational eye movements in visual perception," *Nature Reviews Neuroscience*, Vol. 5, 2004, pp. 229-240.

[63] Boynton, R. and Whitten, D., "Visual adaptation in monkey cones: Recordings of late receptor potentials," *Science*, Vol. 170, 1970, pp. 1423-1426.

[64] Lamming, D., *Vision and Visual Dysfunction*. London: Macmillan Press, 1991, vol. 5, ch. Contrast Sensitivity, pp. 35-43.

[65] Westheimer, G., *Adler's Physiology of the Eye: Clinical application (10th Ed.)*. London: Mosby, 2003, ch. Visual Acuity, pp. 453-469.

[66] Riggs, R., *Vision and Visual Perception*. John Wiley & Sons, 1965, ch. Visual Acuity, pp. 321-349.

[67] Kelly, D. and Martinez-Uriegas, E., "Measurements of chromatic and achromatic afterimages," *Journal of the Optical Society of America A*, Vol. 10, no. 1, 1993, pp. 29-37.

[68] Knoblauch, K., "Theory of wavelength discrimination in tritanopia," *Journal of the Optical Society of America A*, Vol. 10, no. 2, 1993, pp. 378-381.

[69] Ogura, T., Satoh, T.-O., Usui, S., and Yamada, M., "A simulation analysis on mechanisms of damped oscillation in retinal and rod photoreceptor cells," *Vision Research*, Vol. 43, 2003, pp. 2019-2028.

CHAPTER 3
A Review of V1
Jeffrey Ng and Anil Anthony Bharath

3.1 Introduction

In the visual system of primates, the primary visual cortex appears to play a vital role in extracting low-level visual features, such as orientation, scale, and local phase (or measure of symmetry), from the outputs of the retina, via the lateral geniculate nucleus (LGN). V1 performs a low-level "decomposition" of the visual input via the mapping defined by the receptive fields of many millions of neurons. The axons of these neurons subsequently project to the higher areas of the visual cortex for further processing. The importance of V1 in the human visual system is dramatically seen in cases of damage to V1, which usually results in total loss of vision [30]. In contrast, damage to the higher cortical areas only results in loss of particular visual abilities, such as perceiving faces and objects, and so on.

In V1, occasionally also known by its anatomical location as area 17, the neurons are arranged into a slightly folded two-dimensional sheet with six distinct layers. At an approximate surface size of 8 mm × 6 mm and a thickness of 1.7 mm [84], V1 is the largest of all the areas in the visual cortex. The primary visual cortex maintains a retinotopic map, such that nearby neurons in V1 receive inputs from nearby photoreceptors on the retina. The retinotopic map indicates that V1 uses the spatial location of stimuli in the field of view to "organize" its analysis. This may have important implications for its function. The number of neurons in primate V1 is thought to be at least 150 million [84], which puts the ratio of V1 neurons to retinal ganglion output cells at 100:1. V1 thus produces more outputs than its visual inputs, leading to an overcomplete representation of the visual field. Each neuron makes of the order of thousands of short-range interconnections with other neurons in V1 [84], an aspect that might be quite important in fully understanding the visual cortical code. The neuronal density, number, and destination of connections vary in the six layers of V1 [29].

The nature of V1 receptive fields was uncovered by the pioneering work of Hubel and Wiesel [33]. They discovered edge-selective neurons, topologically arranged into two-dimensional maps of the visual field, work that subsequently revealed the presence of a layered, "ice-cube" organization of the visual cortex, and also of strongly nonlinear receptive fields. Subsequent neurophysiological experiments have revealed a laminar specialization of both connections to other subcortical/cortical areas and anatomical/functional properties of the neurons themselves, organized three-dimensionally. The concept of local computations underlying edge-selectivity has however been challenged by discoveries of the effect of image context on individual neuron responses [1]. The role of these contextual effects has been recently proposed to serve preattentive segmentation, linking local neuroprocesses to global computational functions traditionally thought to be beyond the primary visual cortex, such as V2 and V4. For example, lateral interactions between neurons have been shown to lead to enhanced V1 responses to higher-level features, such as region boundaries (see Chapter 4). In this chapter, we first review the early work on V1, then provide an overview of more recent literature that challenges the view of V1 as a simple feature extractor.

V1 neurons were also found to be tuned to a variety of other stimuli, such as color, spatial frequency, eye of origin, motion, and visual disparity between the two eyes. Although the primary visual cortex is one of the most extensively examined cortical areas, much controversy still exists on the extent of its computational role. This review will, thus, include mainly established experimental facts about V1, and we only briefly mention areas of ongoing debate.

3.2 Two Aspects of Organization and Functions in V1

As discussed earlier, the primary visual cortex consists of a structured arrangement of neurons that retains a topographical map of the field of view across its surface [17, 80]. The physiology and patterns of connections among V1 neurons vary both across their six different layers and across functional subassemblies within individual layers that are called *columns*. Hubel and Wiesel [31] found the surprising fact, at the time, that individual neurons only responded to visual stimuli falling within their local spatial receptive fields on the retina, usually subtending a visual angle of no more than 1 or 2 degrees. The population responses of "simple" cells, one of the basic types of neurons in V1, can be approximated as a linear filtering of the visual signal by the weighting pattern of their particular receptive fields followed by a nonlinear pointwise transform, that is more or less monotonous with threshold and has a saturating behavior. That is, response $\propto f(\int K(\mathbf{x},\mathbf{y})I(\mathbf{y})d\mathbf{y})$ with nonlinearity $f(.)$, where $\mathbf{x}$ and $\mathbf{y}$ are spatial location vectors, with $\mathbf{y}$ being the dummy vector of weighted spatial integration and $I(\mathbf{y})$ is the input signal. $K(\mathbf{x},\mathbf{y})$ defines the weighting of the visual input over space in a vector coordinate $\mathbf{y}$ relative to location in the visual field $\mathbf{x}$. It, thus, also specifies a mapping that is permitted to vary with absolute spatial position (not necessarily spatially invariant). V1 contains a variety of cell types with different receptive field patterns and visual properties, that is different kernels $K(\mathbf{x},\mathbf{y})$.

3.2.1 Single-Neuron Responses

The seminal work of Hubel and Wiesel [31] led to Barlow's [5] formulation of the "cardinal cell theory," whereby the function and output of single neurons are important for visual perception. Information-theoretic methods have lent further support to this well-accepted view of V1 by showing that responses of pairs of neurons are only slightly redundant [25] or even independent [83].

The detection of edges or bars at "preferred" orientations by neurons in V1 is believed to arise from the selective arrangement of incoming connections carrying nonoriented visual input from the LGN to particular neurons in V1 [84]. The presence of extensive "lateral" connections between V1 cells and the recent discovery of a class of exclusively inhibitory cells in V1 [18], which do not directly receive visual input, prompted a re-examination of the predominantly feed forward role of neural circuits in V1 and in particular, the (linear) nature of the receptive field (RF). The classical RF was measured by Hubel and Wiesel [31] with point light sources and bars. Such simple visual stimuli did not elicit a response beyond a certain distance from the center of the RF. However, the lateral connections and inhibitory interneurons hint at significant interactions between the computation of nearby feed forward neurons. This interaction causes each cell's response to be significantly influenced by stimuli outside its classical receptive field in a region called its context. Consequently, a neuron's response can clearly signal global properties of input on a spatial scale much larger (e.g., that of a typical visual object) than the classical receptive field [46–48]. Such a global property clearly has exciting computational implications. One could also view the classical receptive field as arising from the feedforward processing of LGN inputs in V1, while the second contextual influence of stimuli outside the classical receptive field (CRF) arises from the recurrent or lateral processing in V1. For clarity, we shall first discuss the former before the latter.

A remark is in order here: some controversy still exists on the exact mechanisms that V1 employs for representing visual stimuli. More specifically, there is a discrepancy between the accuracy of the discrimination of visual properties, such as orientation by individual neurons and the perceptual performance of human subjects. Before Hubel and Wiesel's discovery of the local receptive field, Hebb [28] proposed that information is collectively encoded in the synchronous activity of subpopulations of neurons in the cortex. Recent evidence obtained by improved simultaneous recording of multiple neurons has shown increased information content in synchronous neuronal activity compared with the activity of single neurons [54, 59, 66].

3.2.2 Organization of Individual Cells in V1

The function of a V1 cell can be deduced from the optimal stimulus that triggers a significant jump in its spike output rate. The preferential response of neurons to a certain class of visual stimuli lies in what might really be thought of as tuning curves that describe response variations with respect to properties such as orientation, scale, spatial phase, direction of motion, and color. A distinction may also be made on the basis of ocular origin. The selectivity of the neurons to the first three properties can be explained by spatial weighting patterns applied to the photoreceptor signals. Selectivity to color and ocular origin can be achieved by biasing the weighting on the photoreceptor type (rod and class of cone) and eye of origin. However, directional motion selectivity involves the concept of time and nonseparable spatiotemporal receptive fields. Given the linear behavior of many of these neurons, Livingstone et al. [52] unsurprisingly found that they were usually simultaneously tuned to more than one of these properties, for instance, simultaneous tuning to both orientation and direction. However, V1 neurons usually have a stronger tuning to one specific property than to others.

The presence of the retinotopic map on the surface of V1 may be hard to reconcile with the simultaneous existence of other maps for properties such as orientation, ocular dominance, and scale or spatial frequency [13]. The cortical solution to this conundrum is the break-down of the retinotopic map at cortical scales smaller than 1.2 mm [8]. Under this threshold, neurons are organized by columns specializing in different stimuli. The result is a master retinotopic map with multiple overlapping feature maps for each property [8].

The initial discovery on V1's surface of the smoothly varying orientation map, subject to occasional breaks and reversals of direction, came together with the observation of constant preferred orientation, as one goes deeper into the layers [31]. Such functional columns have also been found for color and ocular dominance [41]. The presence of these columns has led to hypotheses of hypercolumns containing complete representations of orientation and ocular dominance [32], as well as spatial frequency [15]. Other multiparameter studies [15,16] led to the proposal of the complete integration region (CIR) covering an approximate surface area of 1 mm^2. It is estimated to contain about 100,000 cells, among which roughly 32,000 cells have narrow spatial frequency tuning over approximately a 3 octave range of peak spatial frequencies and 20 orientations. The discovery of columnar organizations prompted the emergence of the "ice-cube" model of V1 (illustrated in Figure 3.1).

Multiple parameter studies provide insight into the finer-scale organization of neurons within columns. In the cat, Freeman [23] ranked the similarity of tuning to preferred orientation, spatial frequency, and spatial phase in adjacent pairs of neurons. Preferred orientation turned out to be the most similar, partly explaining why orientation maps were the first to be discovered by Hubel and Wiesel [31]. Spatial frequency was the second most similar parameter while spatial phase was the most dissimilar. Freeman hypothesized that V1 may use a strategy of pooling spatial phase in local areas of the visual field to achieve phase-invariance, that is equivariant response to line and edge stimuli.

The neural circuitry of the primary visual cortex cannot be fully appreciated by the simple paradigm of two-dimensional feature maps. Laminar specialization shapes the sequence of transformations occurring on the incoming retinal signals. An illustration of the incoming connections from the LGN, the interlayer connections in V1, and projections to the upper areas of the cortex are provided in Figure 3.2. The anatomical layers are defined by the relative density of neurons, interconnections, and external connections to the LGN and other visual areas [84].

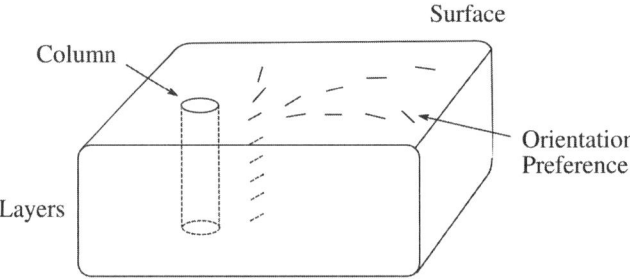

Figure 3.1 Ice-cube model of the V1 area. The function and selectivity of cells remain more or less similar in the laminar layers and vary gradually along the surface. Cytochrome oxidase staining reveals columns of color-sensitive cells in layer 2/3.

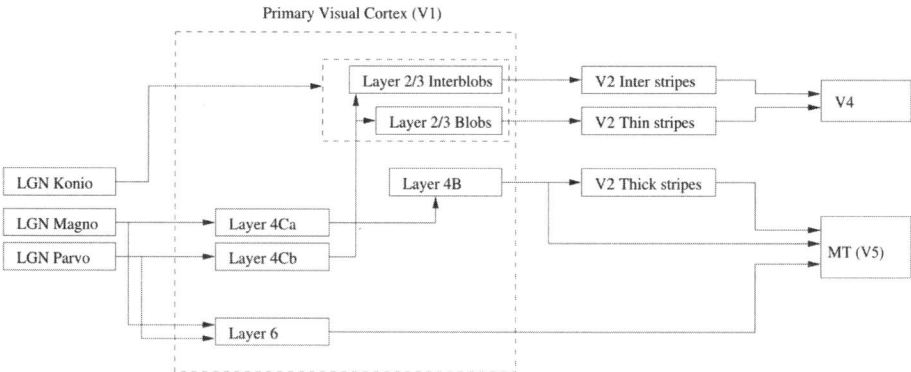

Figure 3.2 Connections from LGN, interlayer connections, and projections to other cortical areas. Note: layer 4C has been separated into 4Ca and 4Cb. Adapted from [20, 75].

Minor differences in the layers may cause further subdivisions. The first superficial layer contains relatively few neurons and does not perform any major processing. As in other cortical areas, the incoming connections from the LGN project to layer 4. The segregation of the magnocellular and parvocellular pathways is maintained, that is, the magnocellular and parvocellular layers of the LGN project to two neighboring but different subregions in sublayer 4C of V1, called layers 4Ca and 4Cb, respectively. The magnocellular pathway flows from layer 4Ca to 4B and then projects to the "thick stripes" of V2 and V5. The parvocellular pathway flows through layer 4Cb and then projects to the "blobs" and "interblobs" in layers 2 and 3. The "thick stripes" of area V2 and the "blobs" and "interblobs" of layer 2/3 in V1 take their names from the appearance of regions of cortical tissue after staining with the metabolic enzyme cytochrome oxidase. The latter participates in the electron transport chain in the brain.

The characteristics of the neurons in the blobs are tuning to color, low spatial frequencies, and a broad range of orientations [52]. In contrast, interblob neurons have no color selectivity, tight selectivity to particular orientations, and high spatial frequencies [52]. The presence of two types of regions in layer 2/3 and their separate projections to different regions of V2 (also revealed by cytochrome-oxidase staining) prompted the view that the parvo "what" pathway specializes into two subpathways in V1: parvo-B deals exclusively with color and parvo-I deals with orientation and high acuity perception [75]. New research, identifying neurons with joint tuning to color and orientation in the interblobs [41], has blurred the distinction between the two types of regions.

3.2.2.1 Orientation Selectivity

Being the first functional type of neurons to be identified by Hubel and Wiesel [31], orientation selective cells have a long history of studies. They come in two main flavors: simple and complex. The simple type prefers lines or edges at fixed orientations. Their response can be approximated by a linear filtering of their visual input with the spatial weighting pattern of their RF, as in two-dimensional signal processing. The two-dimensional Gabor function is the closest approximation to V1 RFs, for example, filter $\propto \exp[-x^2/(2\sigma_x^2) - y^2/(2\sigma_y^2)]\cos(2\pi f x + \phi)$, which is oriented along the y-axis, with width σ_x and length $\sigma_y > \sigma_x$, is tuned to optimal spatial frequency f, and has a receptive field phase ϕ. A phase of 0 or $\pi/2$ produces tuning to bars or edges respectively, thus controlling the symmetry (odd or even) of the preferred stimulus. More details are provided in Chapter 5. Smaller changes in phase induce a spatial offset in the location of the latter with respect to the center of the RF. In the cat, the degree of overlap of the RFs of simple cells is quite significant. The spacing of the centers of the RF of most simple cells is 1° which equates to less than a quarter of RF size [14].

Complex cells have an additional property of insensitivity to the location of the stimulus in their RF. Such behavior cannot be explained by linear mechanisms. A standard view [56] is that they receive inputs from simple cell subunits of different phases ϕ, such as a quadrature pair. Such units are abundant in layer 2/3 [75] and have also been reported in layer 6. A drifting sine grating stimulus would elicit a half-wave rectified modulation in the mean firing rate of a simple cell, and an approximately constant firing rate in a complex cell.

Recent studies of the impact of laminar specialization on orientation selectivity have shown variations in the bandwidth of orientation tuning [65]. These variations can be better understood by looking at the input–output response times of neurons across layers, which gives an indication of the sequence of the processing stages. For instance, the primary site of incoming projections from the LGN, layer 4C, appears to have response times of 45 ms and broad orientation tuning. The longest response times are found in layer 2 (70 ms), where orientation tuning width (defined as half-width at $\frac{1}{\sqrt{2}}$ height) is the narrowest at 20° on average, and in layer 6 (65 ms), which possess marginally sharper orientation tuning than layer 4C, with tuning widths at 25° on average. The authors propose that the increased delay in response time in layers 2 and 6 indicates significant lateral interactions between V1 cells to sharpen orientation selectivity. DeValois et al. [16] have earlier reported that orientation bandwidth in macaques at half-height of the maximum response ranges from 6° to 360° (unoriented) with a median near 40°. There are also cells untuned or poorly tuned to orientation; they tend to be tuned to color and are in the CO blobs [52]. The CO blobs are associated with cells tuned to lower spatial frequencies [19].

3.2.2.2 Color Selectivity

As described in Chapters 1 and 2, the human visual system senses the wavelength of incoming light through three classes of color photoreceptors in the retina. The photoreceptors have peak wavelength tuning to long (L), middle (M), and short (S) wavelengths in the visible spectrum, corresponding to red, green, and blue colors, respectively. Color-sensitive retinal ganglion cells combine spatial opponency (center-surround) with color opponency (red–green or blue–yellow), for example, red-center–green-surround [64]; see Chapter 2 for more details. This single opponent arrangement, which is also present in the LGN, produces selectivity to chromatic contrast. In the primary visual cortex, Livingstone et al. [52] found that orientation and color tuning are mutually exclusive in many neurons, while joint-tuning in other cells was reported by Landisman and Ts'o [41]. Interestingly, the opponency of color-tuned V1 neurons tends to be a double opponent arrangement, for example, the center and surround regions of the receptive field receive red–green and green–red opponent signals, respectively [52].

The distribution of color-selectivity in V1 cells operating inside and outside the central 2.5° of the visual field gives an indication on the use of color in central and peripheral vision. The population percentage of color-tuned orientation-indifferent neurons rises from 15% in the periphery to over 20% in central vision. Joint color and orientation tuning in central vision is at a similar percentage. This brings the percentage of the neural population that is tuned to color in central and peripheral vision to 47% and 23%, respectively.

The color-selectivity characteristics of neurons in the cytochrome oxidase blobs in layer 2/3 of V1 may provide more insight as to their function. The blobs contain predominantly one type of color opponency [41], either red–green or blue–yellow [79]. Blobs dedicated to red–green color opponency were 3 times more numerous than blue–yellow blobs [79], which is explained by the greater number of green and red photoreceptors than blue ones in the retina. However, areas of color selective neurons, called color patches, revealed by optical imaging[1] of neurons can extend beyond blobs and can even encompass two blobs. When two CO blobs of different color opponency are found in a color patch, the interblob region contains cells responding to both color opponency information in an additive manner. These additive dual-opponency responses are found in 44% of color patches [79]. Inside these patches, cells were also found to be more color selective and unoriented.

3.2.2.3 Scale Selectivity

One of the differences between the LGN and V1 is the change from broad spatial frequency tuning in the former to bandpass tuning in the latter [15]. From the top to the bottom of layer 4C (a to b) of V1, Hawken and Parker [27] have found a gradual decrease in RF size and contrast sensitivity. The findings agree with the common knowledge that the LGN mainly projects to layer 4C: LGN M- and P-cells project to sublayers 4Ca and 4Cb, respectively, and the difference in RF size of their corresponding parasol and midget retinal ganglion cells.

Bredfelt and Ringach [10] have observed that spatial frequency tuning varies dynamically over a limited range as stimuli are presented. More specifically, the bandpass selectivity of cells increases with time after initial response by about a fraction of an octave, although it is hard to ascertain from the format of the data. Then increasing low-frequency attenuation causes the peak of the tuning curve to shift towards higher frequencies by 0.62 ± 0.69 octaves. The authors found that feedback models agree better with their observations than feed forward models. More specifically, the delayed strengthening of a suppressive feedback loop could explain the delayed attenuation at the low frequencies of the spatial tuning curve.

---

1. High-resolution imaging of areal differences in tissue reflectance due to changes in blood oxygenation caused by neural activity.

DeValois et al. [15] proposed that the primary visual cortex performs a two-dimensional frequency filtering of the visual input signal with neurons jointly tuned to orientation and spatial frequency. Simple cells in the foveal region were found to have spatial frequency bandwidths at half amplitude from 0.4 octaves to greater than 2.6 octaves, with a peak in the number of cells at 1.2 to 1.4 octaves. The peak spatial frequency of the simple cells varied from 0.5 cycles/degree to greater than 16.0 cycles/degree. The frequency tuning bandwidth is around 1.5 octaves [2,40].

3.2.2.4 Phase Selectivity

The visual system uses spatial phase as a very important cue for feature detection. A study on the importance of image intensity vs phase showed that randomization of the latter destroyed pixel correlations and the ability of human subjects to recognize the image [61]. The randomization of image intensity produced no such effect. The tuning of neurons to specific spatial phases determines the symmetry of the features they detect: even-symmetry for lines and odd-symmetry for edges. Spatial phase can also be used to localize the position of maximum energy of features in the RF of a cell at a given scale or spatial frequency. Mechler et al. [57] describe lines and edges as examples of phase congruence whereby harmonic components across spatial scales share a common phase. This congruence of spatial phases across scales indicates significant structures in scale space [50]. Mechler et al. [57] investigated whether V1 neurons were sensitive to and encoded phase congruence in their responses. They found that simple cells primarily modulated phase congruence in their odd-response harmonics, but they were equally selective for even- and odd-symmetric spatial features. On the other hand, most complex cells modulated phase congruence in their even-response harmonics but were more selective in their odd-response harmonics. Very few cells were phase insensitive whereas most were tuned to phase congruence. Some cells were also tuned to intermediate symmetry between odd and even.

3.3 Computational Understanding of the Feed Forward V1

The vast amount of anatomical and physiological data on V1, captured under different experimental conditions and choice of stimuli, presents a serious obstacle to the compilation of the "big picture," that is, the main purpose and function of this piece of cortical tissue. Starting from an information coding perspective, Li and Atick [43,44,50,51] have shown that the receptive fields in V1 can be understood as part of an efficient code that removes the pairwise or second-order correlation between signals in two pixels. Both the coding representation in V1 and the center-surround receptive field of the retina are efficient in terms of removing such signal redundancy in their inputs. This framework explains and predicts how selectivity of V1 cells to different features should be correlated. For instance, the framework explains that cells tuned to color are also tuned to lower spatial frequencies and often not tuned to orientation, that cells tuned to higher spatial frequencies are often binocular, and that cells tuned to orientations are also tuned strongly or weakly to motion direction. The multiscale coding framework also explains that, a spatial frequency tuning bandwidth of about 1.6 octaves as those observed in the cortex [2,40], causes the neighboring cells tuned to the same spatial frequency tend to have 90 degree phase difference (phase quadrature) in their receptive fields, as in physiological data [62]. It also predicts the cell response properties to color, motion, and depth, and also how cells' receptive fields adapt to environmental changes.

Research into image representation and coding has independently produced filters qualitatively resembling their biological counterparts. In an attempt to find invertible wavelet transforms, the signal processing analogy of the computation of populations of cells with local receptive fields, for image reconstruction after wavelet decomposition, Kingsbury [37] obtained shift-invariant complex wavelets that were very similar in profile to V1 receptive fields; more detail is supplied on this construction in Chapter 5. Olshausen and Field [60] used independent component analysis (ICA) to find the minimum independent set of basis images that can be used to represent patches of 16×16 pixels from images of natural scenery. The basis images produced by ICA were again very similar to V1 receptive fields. These findings support the theory that V1 RFs evolved in the human brain to efficiently detect and represent visual features.

Simoncelli and Heeger [70] modelled the response of V1 complex cells by averaging over a pool of simple cells of the same orientation and RF spatial phase, but different and nearby RF centers. Hyvarinen and Hoyer [34] modified the ICA approach of Olshausen and Field to account for complex cells whose responses exhibit phase-invariant and limited shift-invariant properties. Miller and colleagues are one of the groups of researchers that have modelled and simulated how feed forward pathways from LGN and V1 intracortical circuitry contribute to the cell's response and selectivity properties such as the contrast-invariant tuning to orientation [21, 78].

Note also that Troyer et al. [78] have shown with computational simulations that contrast-invariant orientation tuning in V1 can be obtained from an LGN input consisting of two summed terms: a linear term encoding the temporal modulation of the sine grating used in the experiment and a non-linear term encoding the mean input.

3.3.1 V1 Cell Interactions and Global Computation

Cortical processing has traditionally been associated with feed forward integration of information over increasingly larger spatial areas as one travels along the processing pathways, for example, from V1 to V5, and feedback to enhance or suppress lower-level neuronal activity. Another route of information flow is lateral, that is through *recurrent* connections over long distances, at the early stages of visual processing; such lateral connections are found in V1. It has been found that a cell's response to stimulus within its receptive field can be influenced by stimuli outside the RF. In particular, an orientation tuned cell's response to an optimally oriented bar in its RF is suppressed by up to 80% when identically oriented bars surround the RF. This suppression is weaker when the surrounding bars are randomly oriented, and is weakest when the surrounding bars are orthogonally oriented [38]. When an optimally oriented low-contrast bar within the RF is flanked by collinear high-contrast bars outside the RF, such that the center and the surround bars could be part of a smooth line or contour, the response can be enhanced as much as a few times [36]. These contextual influences are fast, occur within 10 to 20 milliseconds after the initial cell response, and exist in cats and monkeys, in awake and anesthetized animals.

Classical receptive fields were mapped by measuring the response of individual cells to a moving high-contrast light source (point or bar) and plotting the responses with respect to the two-dimensional position of the light source. The more recent *reverse correlation technique* probes the phenomenological receptive field by correlating the cell's response with a grating of variable size centered upon the classical receptive field [4]. The region covered by the grating that elicits the highest response from the cell is called the "summation field" (SF), reflecting the fact that the response may be affected by lateral connections between multiple V1 neurons.

The resulting SF sizes are 2.3 times bigger than the average size of classical RFs found by Barlow et al. [6]. They depend on the contrast of stimuli, and are, on average, more than 100% larger in low-contrast than high-contrast conditions [67]. To gratings larger than the SF, the average firing rate of the neuron falls to an asymptotic level typically higher than the spontaneous firing rate.

Angelucci et al. [4] found that monosynaptic horizontal connections within area V1 are of an appropriate spatial scale to mediate interactions within the SF of V1 neurons and to underlie contrast-dependent changes in SF size. The spatial scale of feedback connections from higher visual areas was additionally suggested to be commensurate with the full spatial range of interactions, typically suppressive, between SF and the surrounding.

Recombinant adenoviruses containing a gene for a fluorescent protein can be used to image neural connections. With a green fluorescent adenovirus, Stettler et al. [73] found lateral connections in V1 to be slightly larger in diameter than previously believed and much denser than feedback connections from V2, contradicting results from [4]. Lateral connections were found to stretch almost 4 degree in diameter (on a retinotopic scale) from the center of the injection, which occurred over a 0.2-mm diameter; in comparison, the minimum RF size averaged 0.5 degree. In the central 1-mm diameter of the injection, the lateral connections were nonspecific whereas on the outside of the central 1-mm diameter, the connections had bell-curve-like densities with the centers located on neurons with similar orientation preference. Feedback connections from V2 stretched to 2.5 degree in diameter and originated from neurons with diverse orientation preferences. Bosking et al. [9] showed that the lateral connections projected farther along the axis of the receptive field than in the orthogonal direction. Kagan et al. [35] measured the classical RF and summation fields of macaque V1. Moving bright and dark (relative to background) orientation bars were used to measure the classical RF size of V1 cells. The cells were further subcategorized as simple, complex, and monocontrast. The latter responded either to bright or dark bars but not both. Expanding gratings were also used to measure SF size. However, they used a rectangular grating

3.3 Computational Understanding of the Feed Forward V1

Table 3.1 Dimensions of Summation Fields and Classical Receptive Fields (CRF) in Macaque V1

		SF Minarc	
Layer	Simple	Complex	Monocontrast
2/3	22 ± 3 (5)	23 ± 10 (49)	7 ± 2 (5)
4	12 ± 7 (18)	28 ± 11 (74)	18 ± 8 (7)
5/6	8 ± 5 (9)	49 ± 33 (29)	19 ± 12 (3)
All	13 ± 8 (33)	31 ± 18 (178)	14 ± 8 (17)
		CRF minarc	
Layer	Simple	Complex	Monocontrast
2/3	48 ± 11 (5)	26 ± 12 (49)	7 ± 2 (5) (74)
4	29 ± 16 (18)	32 ± 13 (74)	18 ± 8 (7)
5/6	19 ± 12 (9)	56 ± 338 (29)	17 ± 11 (3)
All	29 ± 16 (33)	35 ± 21 (178)	14 ± 8 (17)

Number of cells are in parentheses. ± values are standard deviations. Monocontrast cells respond either to light or dark bars but not to both.
Source: [35].
Note: $1° = 60$ minutes of arc (minarc).

whose length (in the direction of preferred orientation of the neuron) was restricted to the neuron's optimal length to a bar. This may not affect a neuron's classical RF but may limit the effect on SF by lateral interactions with other neurons as hypothesized by Angelucci et al. [4].

3.3.2 Theory and Model of Intracortical Interactions in V1

There have been various attempts to model the intracortical interactions and contextual influences in V1 [26, 71, 72, 86]. The dynamics of networks with recurrent interactions can easily induce stability problems. It is difficult to construct a well-behaved model that does not generate unrealistic or hallucinative responses to input stimuli, even when the model focuses only on a subset of the contextual influences such as collinear facilitation [86]. Because of this, many models, for example [71, 72], do not include explicit geometric or spatial relationships between visual stimuli, and thus have limited power to link physiology with visual behavior.

Recently, aided by a dynamically well-behaved model, including cells tuned to various orientations and different spatial locations and both suppressive and facilitative contextual influences as in physiology, Li [45–49] proposed that contextual influences enable V1 to generate a saliency map based on visual input defined mainly by contrast (see Chapter 4). With locally translation-invariant neural tuning properties and intracortical interactions, V1 responses may highlight the locations where translation invariance or homogeneity in input breaks down on a global scale. These locations can be at texture boundaries, at smooth contours, or at pop-out targets (such as red among greens or vertical among horizontals) in visual search behavior. This theory links V1 physiology/anatomy with visual behavior such as texture segmentation [7], contour integration [22], and pop-out and asymmetries in visual search [76, 77], which have often been thought of as complex global visual behavior dealt with beyond V1. The model proposes that V1 performs the initial stage of the visual segmentation task through the saliency map, and such segmentation has been termed preattentive segmentation in the psychological community. It is termed "segmentation without classification" (recognition) [47], which relates to the bottom-up segmentation in computer vision. This theory/model has generated testable predictions, some of which have been experimentally confirmed. An application of this model to edge postprocessing can be found in Chapter 6.

In particular, saliency maps are believed to provide a fast mechanism for directing the focus of attention of the visual system and eye-fixations without having to wait for slower feedback information from the higher-level cognitive areas of the visual cortex. Saliency maps may be useful for controlling pan-tilt zoom cameras and retinomorphic cameras capable of simultaneous image processing at different resolutions.

3.4 Conclusions

Neurophysiological and anatomical experiments have provided new insights into the functions of V1 neurons. The response of V1 cells has been shown to be more dynamic than previously believed. In particular, the spatial selectivity of V1 cells undergoes a temporal refinement process of the tuning curve from medium to high frequencies [10]. The underlying mechanism of V1 selectivity has also been shown to be more than a feed forward weighted-sum filtering (with or without static nonlinearities) of stimuli from the receptive fields [4]. Lateral connections among V1 neurons and feedback connections from higher cortical areas affect the response of V1 cells, and more importantly, determine the role of V1 in high-level visual functions beyond feature detection and representation. The following material could be useful for further readings on the primary visual cortex. Wandell [84] provides a textbook-style introduction to physiology and psychophysics of vision in general. Lennie [42] provides a review of V1 together with a proposal for its role in visual feature analysis in the context of functions of extrastriatal areas. Martinez and Alonso [55] review the properties and models regarding complex and simple cells. Schluppeck et al. [68] review and reconcile the conflicting evidence about color processing in V1 from optical imaging and single unit recording data. Angelucci et al. [3] review the anatomical origins of intracortical interactions in V1, while Lund et al. [53] review the anatomical substrate for the functional columns. Reid [63] reviews V1 in a textbook chapter for neuroscience. Fregnac [24] reviews the dynamics of functional connectivity in V1 at both molecular and network levels.

We do not cover in this chapter either the topic of development processes of V1, or plasticity and learning. Chapman and Stryker [11] review the development of orientation selectivity in V1, while Sengpiel and Kind [69] review the role of neural activities in development. The material in this review on the intracortical interactions and contextual influences is relatively recent, thus further developments and revised points of view on these aspects are expected. Furthermore, there is much interest regarding the possible roles of V1 in higher-level cognitive phenomena. These phenomena include alternative or unstable perceptions (e.g., binocular rivalry when subjects perceive one image of the two different and inconsistent images presented to the two eyes); attentional blindness, visual awareness, or consciousness [12, 74].

We shall return to considering V1 in Part 2 Chapter 5, where we focus on its role in attention and the concept of visual saliency. We shall also, in the next chapter, return to the implementation of practical models of V1 simple and complex cell processing and shall consider algorithms for feature and shape detection based on such models. In addition, we shall consider hardware mappings of V1 models in Chapters 12 and 13 of Part 3.

References

[1] Allman, J., Miezin, F., and McGuinness. E., "Stimulus specific responses from beyond the classical receptive field: Neurophysiological mechanisms for local-global comparisons in visual neurons." *Annual Review of Neuroscience*, Vol. 8, 1985, pp. 407-430.

[2] Andrews, B. W. and Pollen. D. A., "Relationship between spatial frequency selectivity and receptive field profile of simple cells." *Journal of Physiology (London)*, Vol. 287, 1979, pp. 163-176.

[3] Angelucci, A., Levitt, J. B., and Lund J. S., "Anatomical origins of the classical receptive field and modulatory surround field of single neurons in macaque visual cortical area V1." *Progress in Brain Research*, Vol. 136, 2002, pp. 373-388.

[4] Angelucci, A., Levitt, J. B., Walton, E. J. S., et al., "Circuits for local and global signal integration in visual cortex." *Journal of Neuroscience*, Vol. 22(19), 2002, pp. 8633-8646.

[5] Barlow, H. B., "Single units and sensation: A neuron doctrine for perceptual psychology?" *Perception*, Vol. 1(4), 1972, pp. 371-394.

[6] Barlow, H. B., Blakemore, C., and Pettigrew, J. D., "The neural mechanisms of binocular depth discrimination." *Journal of Physiology*, Vol. 193, 1967, pp. 327-342.

[7] Bergen, J. R. and Adelson, E. H., "Early vision and texture perception." *Nature*, Vol. 333, 1988, pp. 363-364.

[8] Blasdel, G. and Campbell, D., "Functional retinotopy of monkey visual cortex." *Journal of Neuroscience*, Vol. 21(20), October 2001, pp. 8286-8301.

[9] Bosking, W. H., Zhang, Y., Schofield, B., and Fitzpatrick, D., "Orientation selectivity and the arrangement of horizontal connections in tree shrew striate cortex." *Journal of Neuroscience*, Vol. 17(6), 1997, pp. 2112-2127.

[10] Bredfelt, C. E. and Ringach, D. L., "Dynamics of spatial frequency tuning in macaque V1." *Journal of Neuroscience*, Vol. 22(5), 2002, pp. 1976-1984.

[11] Chapman, B. and Stryker, M. P., "Origin of orientation tuning in the visual cortex." *Current Opinions in Neurobiology*, Vol. 2(4), 1992, pp. 498-501.

[12] Danckert, J. and Goodale, M. A., "A conscious route to unconscious vision." *Current Biology*, Vol. 10(2), 2000, pp. 64-67.

[13] Das, A., "Cortical maps: Where theory meets experiments." *Neuron*, Vol. 47(2), 2005, pp. 168-171.

[14] DeAngelis, G. C., Ghose, G. M., Ohzawa, I., and Freeman, R. D., "Functional microorganization of primary visual cortex: Receptive field analysis of nearby neurons." *Journal of Neuroscience*, Vol. 19(10), 1999, pp. 4046-4064.

[15] DeValois, R. L., Albrecht, D. G., and Thorell, L. G., "Spatial frequency selectivity of cells in macaque visual cortex." *Vision Research*, Vol. 22(5), 1982, pp. 545-559.

[16] DeValois, R. L., Yund, E. W., and Hepler, N., "The orientation and direction selectivity of cells in macaque visual cortex." *Vision Research*, Vol. 22(5), 1982, pp. 531-544.

[17] Dobelle, W. H., Turkel, J., Henderson, D. C., and Evans, J. R., "Mapping the representation of the visual field by electrical stimulation of human visual cortex." *American Journal of Ophthalmology*, Vol. 88(4), 1979, pp. 727–735.

[18] Dragoi, V. and Sur, M., "Dynamic properties of recurrent inhibition in primary visual cortex: Contrast and orientation dependence of contextual effects." *Journal of Neurophysiology*, Vol. 83(2), 2000, pp. 1019–1030.

[19] Edwards, D. P., Purpura, K. P., and Kaplan, E., "Contrast sensitivity and spatial frequency response of primate cortical neurons in and around the cytochrome oxidase blobs." *Vision Research*, Vol. 35(11), 1995, pp. 1501–1523.

[20] Felleman, D. J. and D. C. Van Essen. "Distributed hierarchical processing in the primate cerebral cortex." *Cerebral Cortex*, Vol. 1(1), 1991, pp. 1–47.

[21] Ferster, D. and Miller, K. D., "Neural mechanisms of orientation selectivity in the visual cortex." *Annual Review of Neuroscience*, Vol. 23, 2000, pp. 441–471.

[22] Field, D. J., Hayes, A., and Hess, R. F., "Contour integration by the human visual system: Evidence for a local 'association field.'" *Vision Research*, Vol. 33(2), 1993, pp. 173–193.

[23] Freeman, R. D., "Cortical columns: A multi-parameter examination." *Cerebral Cortex*, Vol. 13(1), 2003, 70–72.

[24] Fregnac, Y., "Dynamics of functional connectivity in visual cortical networks: An overview." *Journal of Physiology (Paris)*, Vol. 90(3-4), 1996, pp. 113–139.

[25] Gawne, T. J., Kjaer, T. W., Hertz, J. A., and Richmond, B. J., "Adjacent visual cortical complex cells share about 20% of their stimulus-related information." *Cerebral Cortex*, Vol. 6(3), 1996, pp. 482–489.

[26] Grossberg, S. and Mingolla, E., "Neural dynamics of perceptual grouping: Textures, boundaries, and emergent segmentations." *Perception and Psychophysics*, Vol. 38(2), 1985, pp. 141–171.

[27] Hawken, M. J. and Parker, A. J., "Contrast sensitivity and orientation selectivity in lamina-IV of the striate cortex of old-world monkeys." *Experimental Brain Research*, Vol. 54(2), 1984, 367–372.

[28] Hebb, D. O., "*The organization of behavior: A neuropsychological theory.*" Wiley, New York, 1949.

[29] Henry, G. H., "Afferent inputs, receptive field properties and morphological cell types in different laminae of the striate cortex." *Vision and Visual Dysfunction*, Vol. 4, 1989, pp. 223–245.

[30] Holmes, G., "Disturbances of vision by cerebral lesions." *British Journal of Ophthalmology*, Vol. 2, 1918, pp. 353–384.

[31] Hubel, D. and Wiesel, T., "Receptive fields, binocular interaction, and functional architecture in the cat's visual cortex." *Journal of Physiology*, Vol. 160(1), 1962, pp. 106–154.

[32] Hubel, D. H. and Wiesel, T. N., "Ferrier lecture. Functional architecture of macaque monkey visual cortex." In *Royal Society of London B Biological Science*, Vol. 198, 1977, pp. 1–59.

[33] Hubel, D. H. and Wiesel, T. N., "Early exploration of the visual cortex." *Neuron*, Vol. 20, 1998, pp. 401–412.

[34] Hyvarinen, A. and Hoyer, P., "Emergence of phase- and shift-invariant features by decomposition of natural images into independent feature subspaces." *Neural Computation*, Vol. 12(7), 2000, pp. 1705–1720.

[35] Kagan, I., Gur, M., and Snodderly, D. M., "Spatial organization of receptive fields of V1 neurons of alert monkeys: Comparison with responses to gratings." *Journal of Neurophysiology*, Vol. 88(5), 2001, pp. 2257–2574.

[36] Kapadia, M. K., Ito, M., Gilbert, C. D., and Westheimer, G., "Improvement in visual sensitivity by changes in local context: Parallel studies in human observers and in V1 of alert monkeys." *Neuron*, Vol. 15(4), 1995, pp. 843-856.

[37] Kingsbury, N. G., "Image processing with complex wavelets." *Philosophical Transactions of the Royal Society*, Vol. 357(1760), 1999, pp. 2543-2560.

[38] Knierim, J. J. and van Essen, D. C., "Neuronal responses to static texture patterns ion area V1 of the alert macaque monkeys." *Journal of Neurophysiology*, Vol. 67, 1992, pp. 961-980.

[39] Kovesi, P., "Phase congruency: A low-level image invariant." *Psychological Research*, Vol. 64(2), 2000, pp. 136-148.

[40] Kulikowski, J. J. and Bishop, P., "Linear analysis of the response of simple cells in the cat visual cortex." *Experimental Brain Research*, Vol. 44(4), 1981, pp. 386-400.

[41] Landisman, C. E. and Ts'o, D. Y., "Color processing in macaque striate cortex: Relationships to ocular dominance, cytochrome oxidase, and orientation." *Journal of Neurophysiology*, Vol. 87(6), 2002, pp. 3126-3137.

[42] Lennie, P., "Single units and visual cortical organization," *Perception*, Vol. 27(8), 1998, pp. 889-935.

[43] Li, Z., "Understanding ocular dominance development from binocular input statistics." In *The neurobiology of computation (Proceeding of computational neuroscience conference 1994), Editor: J. Bower*, pages 397-402. Kluwer Academic Publishers, 1995.

[44] Li, Z., "A theory of the visual motion coding in the primary visual cortex." *Neural Computation*, Vol. 8(4), 1996, pp. 705-730.

[45] Li, Z., "A neural model of contour integration in the primary visual cortex." *Neural Computation*, Vol. 10(4), 1998, 903-940.

[46] Li, Z., "Contextual influences in V1 as a basis for pop out and asymmetry in visual search." In *Proceedings of the National Academy of Science USA*, Vol. 96(18), 1999, pp. 10530-10535.

[47] Li, Z., "Visual segmentation by contextual influences via intracortical interactions in primary visual cortex." *Network: Computation in Neural Systems*, Vol. 10(2), 1999, 187-212.

[48] Li, Z., "Pre-attentive segmentation in the primary visual cortex." *Spatial Vision*, Vol. 13(1), 2000, pp. 25-50.

[49] Li, Z., "A saliency map in primary visual cortex." *Trends in Cognitive Sciences*, Vol. 6(1), 2002, pp. 9-16.

[50] Li, Z. and Atick, J. J., "Efficient stereo coding in the multiscale representation." *Network Computations in Neural Systems*, Vol. 5(2), 1994, pp. 157-174.

[51] Li, Z. and Atick, J. J., "Towards a theory of striate cortex." *Neural Computation*, Vol. 6(1), 1994, pp. 127-146.

[52] Livingstone, M. S. and Hubel, D. H., "Anatomy and physiology of a color system in primate visual cortex." *Journal of Neuroscience*, Vol. 4(1), 1984, pp. 309-356.

[53] Lund, J. S., Angelucci, A., and Bressloff, P. C., "Anatomical substrates for functional columns in macaque monkey primary visual cortex." *Cerebral Cortex*, Vol. 13(1), 2003, pp. 15-24.

[54] Maldonado, P. E. and Gerstein, G. L., "Neuronal assembly dynamics in the rat auditory cortex during reorganization induced by intracortical microstimulation." *Experimental Brain Research*, Vol. 112(3), 1996, pp. 431-441.

[55] Martinez, L. M. and Alonso, J. J., "Complex receptive fields in primary visual cortex." *Neuroscientist*, Vol. 9(5), 2003, pp. 317-331.

[56] Martinez, L. M. and Alonso, J. M., "Construction of complex receptive fields in cat primary visual cortex." *Neuron*, Vol. 32(3), 2001, pp. 515-525.

[57] Mechler, F., Reich, D. S., and Victor, J. D., "Detection and discrimination of relative spatial phase by V1 neurons." *Journal of Neuroscience*, Vol. 22(14), 2002, pp. 6129-6157.

[58] Merigan, W. H. and Maunsell, J. H. R., "How parallel are the primate visual pathways?" *Annual Review of Neuroscience*, Vol. 16, 1993, pp. 369-402.

[59] Nadasdy, Z., "Spike sequences and their consequences." *Journal of Physiology (Paris)*, Vol. 94(5), 2000, pp. 504-524.

[60] Olshausen, B. A. and Field, D. J., "Emergence of simple-cell receptive field properties by learning a sparse code." *Nature*, Vol. 381, 1996, pp. 607-609.

[61] Oppenheim, A. V. and Lim, J. S., "The importance of phase in signals." In *Proceedings of the IEEE*, Vol. 69, 1981, pp. 529-541.

[62] Pollen, D. A. and Ronner, S. F., "Phase relationships between adjacent simple cells in the cat." *Science*, Vol. 212(4501), 1981, pp. 1409-1411.

[63] Reid, R. C., "Vision." In *Fundamental Neuroscience*, pages 727-750. Academic Press, 2003.

[64] Reid, R. C. and Shapley, R. M., "Spatial structure of cone inputs to receptive fields in primate lateral geniculate nucleus." *Nature*, Vol. 356(6371), 1992, pp. 716-718.

[65] Ringach, D. L., Hawken, M. J., and Shapley, R., "Temporal dynamics and laminar organisation of orientation tuning in monkey primary visual cortex." In *European Conference on Visual Perception*, 1998.

[66] Samonds, J. M., Allison, J. D., Brown, H. A., and Bonds, A. B., "Cooperation between area 17 neuron pairs enhances fine discrimination of orientation." *Journal of Neuroscience*, Vol. 23(6), 2003, pp. 2416-2425.

[67] Sceniak, M. P., Ringach, D. L., Hawken, M. J., and Shapley, R. M., "Contrast's effect on spatial summation by macaque V1 neurons." *Nature Neuroscience*, Vol. 2(8), 1999, pp. 733-739.

[68] Schluppeck, D. and Engel, S. A., "Color opponent neurons in V1: A review and model reconciling results from imaging and single-unit recording." *Journal of Vision*, Vol. 2(6), 2002, pp. 480-492.

[69] Sengpiel, F. and Kind, P. C., "The role of activity in development of the visual system." *Current Biology*, Vol. 12(23), 1992, pp. 818-826.

[70] Simoncelli, E. P. and Heeger, D. J., "A model of neuronal responses in visual area MT." *Vision Research*, Vol. 38(5), 1998, pp. 743-761.

[71] Somers, D. C., Todorov, E. V., Siapas, A. G., et al., "A local circuit approach to understanding integration of long-range inputs in primary visual cortex." *Cerebral Cortex*, Vol. 8(3), 1998, pp. 204-217.

[72] Stemmler, M., Usher, M., and Niebur, E., "Lateral interactions in primary visual cortex: A model bridging physiology and psychophysics." *Science*, Vol. 269(5232), 1995, pp. 1877-1880.

[73] Stettler, D. D., Das, A., and Bennett, J., "Lateral connectivity and contextual interactions in macaque primary visual cortex." *Neuron*, Vol. 36(4), 2002, pp. 739-750.

[74] Tong, F., "Primary visual cortex and visual awareness." *Nature Review Neuroscience*, Vol. 4(3), 2003, pp. 219-229.

[75] Tovée, M. J., *An introduction to the human visual system*. Cambridge Press, 1996.

[76] Treisman, A. and Gelade, G., "A feature integration theory of attention." *Cognitive Psychology*, Vol. 12(1), 1980, pp. 97-136.

[77] Treisman, A. and Gormican, S., "Feature analysis in early vision: Evidence for search asymmetries." *Psychological Review*, Vol. 95(1), 1988, pp. 15-48.

[78] Troyer, T. W., Krukowski, A. E., and Miller, K. D., "LGN input to simple cells and contrast-invariant orientation tuning: An analysis." *Journal of Neurophysiology*, Vol. 87(6), 2001, pp. 2741-2752.

References

[79] Ts'o, D. Y. and Gilbert, C. D., "The organization of chromatic and spatial interactions in the primate striate cortex." *Journal of Neuroscience*, Vol. 8(5), 1988, pp. 1712-1727.

[80] Tusa, R. J., Palmer, L. A., and Rosenquist, A. C., "Retinotopic organization of area 17 (striate cortex) in the cat." *Journal of Comparative Neurology*, Vol. 177(2), 1978, pp. 213-235.

[81] Ungerleider, L. G. and Mishkin, M., "Two cortical visual systems." In D. J. Ingle, M. A. Goodale, and R. J. W. Mansfield, editors, *Analysis of visual behavior*. 1982.

[82] Ungerleider, L. G. and Mishkin, M., "Two cortical visual systems." In *Analysis of visual behavior*, pages 549-586. MIT Press, 1982.

[83] Victor, J. D., "How the brain uses time to represent and process visual information." *Brain Research*, Vol. 886(1), 2000, pp. 33-46.

[84] Wandell, B. A., *Foundations of vision*. Sinauer Associates Inc., 1995.

[85] Zeki, S., *A vision of the brain*. Blackwell Scientific Publications, 1993.

[86] Zucker, S. W., Dobbins, A., and Iverson, L., "Two stages of curve detection suggest two styles of visual computation." *Neural Computation*, Vol. 1(1), 1989, pp. 68-81.

CHAPTER 4
Testing the Hypothesis That V1 Creates a Bottom-Up Saliency Map

Li Zhaoping and Keith A. May[1]

4.1 Introduction

Due to information transmission and processing bottlenecks, such as the optic nerve and the attentional bottleneck, in the brain, only a limited amount of visual input information can be processed in detail. This necessitates the selection of the most appropriate information for such detailed or attentive processing somewhere along the visual pathway. Although we tend to notice our goal-directed or top-down selections, much of the selection occurs in a bottom-up or stimulus driven manner, particularly in selections immediately or very soon after visual stimulus onset [1–3]. For instance, a vertical bar among horizontal ones or a red dot among blue ones perceptually pops out automatically to attract attention [4, 5] and is said to be highly salient preattentively (see illustrative examples of pop-outs or otherwise in Figure 4.1). Physiologically, a neuron in the primary visual cortex (V1) gives a higher response to its preferred feature, for example, a specific orientation, color, or motion direction, within its receptive field (RF) when this feature is unique within the display, rather than when it is one of the elements in a homogeneous background [6–12]. This is the case even when the animal is under anesthesia [9], suggesting bottom-up mechanisms. This occurs because the neuron's response to its preferred feature is often suppressed when this stimulus is surrounded by stimuli of the same or similar features. Such contextual influences, termed isofeature suppression, and isoorientation suppression in particular, are mediated by intracortical connections between nearby V1 neurons [13–15]. The same mechanisms also make V1 cells respond more vigorously to an oriented bar when it is at the border, rather than the middle, of a homogeneous orientation texture, as physiologically observed [10], since the bar has fewer iso-orientation neighbors at the border. These observations have prompted suggestions that V1 mechanisms contribute to bottom-up saliency for pop-out features like the unique orientation singleton or the bar at an orientation texture border (e.g., [6–10]). This is consistent with observations that highly salient inputs can bias responses in extrastriate areas receiving inputs from V1 [16, 17].

Behavioral studies have examined bottom-up saliencies extensively in visual search and segmentation tasks [4, 18, 19], showing more complex, subtle, and general situations beyond basic feature pop-outs. For instance, a unique feature conjunction, for example, a red-vertical bar as a color-orientation conjunction,

---

1. This chapter is adapted from Zhaoping, L. and May, K. A. (2007). Psychophysical tests of the hypothesis of a bottom-up saliency map in primary visual cortex. *PLoS Computational Biology* 3(4):e62. doi:10.1371/journal.pcbi.0030062.

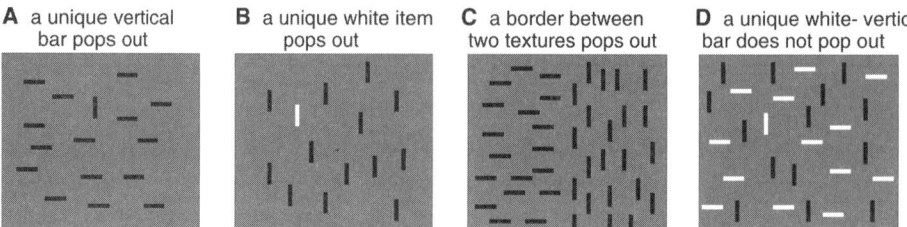

Figure 4.1 Visual input examples to demonstrate various bottom-up saliency effects. A: A unique vertical bar among horizontal bars pops out automatically, i.e., it attracts attention without top-down control. B: A white item also pops out among black ones. C: The segmentation between the two textures is easy because the border between the two textures pops out. D: Although the white-vertical bar is a unique conjunction of two features, white color and vertical orientation, in this display, it does not attract attention automatically. White color and vertical orientation are also abundant in the background items.

is typically less salient and requires longer search times (Figure 4.1D); ease of searches can change with target-distractor swaps; and target saliency decreases with background irregularities. However, few physiological recordings in V1 have used stimuli of comparable complexity, leaving it open as to how generally V1 mechanisms contribute to bottom-up saliency.

Meanwhile, a model of contextual influences in V1 [20–23], including iso-feature suppression and co-linear facilitation [24, 25], has demonstrated that V1 mechanisms can plausibly explain these complex behaviors mentioned above, assuming that the V1 cell with the highest response to a target determines its saliency and thus the ease of a task. Accordingly, V1 has been proposed to create a bottom-up saliency map, such that the receptive field (RF) location of the most active V1 cell is most likely selected for further detailed processing [20, 23]. We call this proposal the V1 saliency hypothesis. This hypothesis is consistent with the observation that microstimulation of a V1 cell can drive saccades, via superior colliculus, to the corresponding RF location [26], and that higher V1 responses correlate with shorter reaction times to saccades to the corresponding receptive fields [27]. It can be clearly expressed algebraically. Let $(O_1, O_2, \ldots, O_M)$ denote outputs or responses from V1 output cells indexed by $i = 1, 2, \ldots M$, and let the RFs of these cells cover locations $(x_1, x_2, \ldots, x_M)$, respectively. Then the location selected by bottom-up mechanisms is $\hat{x} = x_{\hat{i}}$ where $\hat{i}$ is the index of the most responsive V1 cell (mathematically, $\hat{i} = \mathrm{argmax}_i \{O_i\}$). It is then clear that (1) the saliency $\mathrm{SMAP}(x)$ at a visual location x increases with the response level of the most active V1 cell responding to it,

$$\mathrm{SMAP}(x) \text{ increases with } \max_{x_i = x} O_i, \text{ given an input scene} \qquad (4.1)$$

and the less activated cells responding to the same location do not contribute, regardless of the feature preferences of the cells; and (2) the highest response to a particular location is compared with the highest responses to other locations to determine the saliency of this location, since only the RF location of the most activated V1 cell is the most likely selected (mathematically, the selected

4.1 Introduction

location is $\hat{x} = \text{argmax}_x\{\text{SMAP}(x)\}$). As salience merely serves to order the priority of inputs to be selected for further processing, only the order of the salience is relevant [23]. However, for convenience we could write (4.1) as $\text{SMAP}(x) = [\max_{x_i=x} O_i]/[\max_j O_j]$, or simply $\text{SMAP}(x) = \max_{x_i=x} O_i$. Note that the interpretation of $x_i = x$ is that the receptive field of cell i covers location x or is centered near x.

In a recent physiological experiment, Hegde and Felleman [28] used visual stimuli composed of colored and oriented bars resembling those used in experiments on visual search. In some stimuli the target popped out easily (e.g., the target had a different color or orientation from all the background elements), whereas in others, the target was more difficult to detect, and did not pop out (e.g., a color-orientation conjunction search, where the target is defined by a specific combination of orientation and color). They found that the responses of the V1 cells, which are tuned to both orientation and color to some degree, to the pop-out targets were not necessarily higher than responses to non-pop-out targets, and thus raised doubts regarding whether bottom-up saliency is generated in V1. However, these doubts do not disprove the V1 saliency hypothesis since the hypothesis does not predict that the responses to pop-out targets in some particular input images would be higher than the responses to non-pop-out targets in other input images. For a target to pop out, the response to the target should be substantially higher than the responses to all the background elements. The absolute level of the response to the target is irrelevant: what matters is the relative activations evoked by the target and background. Since Hegde and Felleman [28] did not measure the responses to the background elements, their findings do not tell us whether V1 activities contribute to saliency. It is likely that the responses to the background elements were higher for the conjunction search stimuli, because each background element differed greatly from many of its neighbors and, as for the target, there would have been weak isofeature suppression on neurons responding to the background elements. On the other hand, each background element in the pop-out stimuli always had at least one feature (color or orientation) the same as all of its neighbors, so isofeature suppression would have reduced the responses to the background elements, making them substantially lower than the response to the target. Meanwhile, it remains difficult to test the V1 saliency hypothesis physiologically when the input stimuli are more complex than those of the singleton pop-out conditions.

Psychophysical experiments provide an alternative means to ascertain V1's role in bottom-up salience. While previous work [20–23] has shown that the V1 mechanisms can plausibly explain the commonly known behavioral data on visual search and segmentation, it is important to generate from the V1 saliency hypothesis behavioral predictions which are hitherto unknown experimentally so as to test the hypothesis behaviorally. This hypothesis testing is very feasible for the following reasons. There are few free parameters in the V1 saliency hypothesis since (1) most of the relevant physiological mechanisms in V1 are established experimental facts which can be modelled but not arbitrarily distorted, and (2) the only theoretical input is the hypothesis that the receptive field location of the most responsive V1 cell to a scene is the most likely selected. Consequently, the predictions from this hypothesis can be made precise, making the hypothesis falsifiable. One such psychophysical test confirming a prediction has been reported recently [29]. The current work aims to test the hypothesis more systematically, by providing nontrivial predictions that are more indicative of the particular nature of the V1 saliency hypothesis and the V1 mechanisms.

For our purpose, we first review the relevant V1 mechanisms in the rest of the introduction section. The Results section reports the derivations and tests of the predictions. The Discussion section will discuss related issues and implications of our findings, discuss possible alternative explanations for the data, and compare the V1 saliency hypothesis with traditional saliency models [18, 19, 30, 31] that were motivated more by the behavioral data [4, 5] than by their physiological basis.

The relevant V1 mechanisms for the saliency hypothesis are the receptive fields and contextual influences. Each V1 cell [32] responds only to a stimulus within its classical receptive field (CRF). Input at one location x evokes responses $(O_i, O_j, \ldots)$ from multiple V1 cells $i, j, \ldots$ having overlapping receptive fields covering x. Each cell is tuned to one or more particular features including orientation, color, motion direction, size, and depth, and increases its response monotonically with the input strength and resemblance of the stimulus to its preferred feature. We call cells tuned to more than one feature dimension *conjunctive cells* [23], for example, a vertical-rightward conjunctive cell is simultaneously tuned to rightward motion and vertical orientation [32], a red-horizontal cell to red color and horizontal orientation [33]. Hence, for instance, a red-vertical bar could evoke responses from a vertical-tuned cell, a red-tuned cell, a red-vertical conjunctive cell, and another cell preferring orientation two degrees from vertical but having an orientation tuning width of $15°$, and so on. The V1 saliency hypothesis states that the saliency of a visual location is dictated by the response of the most active cell responding to it [20, 23, 34], $\text{SMAP}(x) \propto \max_{x_i = x} O_i$, rather than the sum of the responses $\sum_{x_i = x} O_i$ to this location. This makes the selection easy and fast, since it can be done by a single operation to find the most active V1 cell ($\hat{i} = \text{argmax}_i\{O_i\}$) responding to any location and any feature(s). We will refer to saliency by the maximum response, $\text{SMAP}(x) \propto \max_{x_i = x} O_i$ as the MAX rule, to saliency by the summed response $\sum_{x_i = x} O_i$ as the SUM rule. It will be clear later that the SUM rule is not supported, or is less supported, by data, nor is it favored by computational considerations (see Discussion).

Meanwhile, intracortical interactions between neurons make a V1 cell's response context dependent, a necessary condition for signaling saliency, since, for example, a red item is salient in a blue but not in a red context. The dominant contextual influence is the isofeature suppression mentioned earlier, so that a cell responding to its preferred feature will be suppressed when there are surrounding inputs of the same or similar feature. Given that each input location will evoke responses from many V1 cells, and that responses are context dependent, the highest response to each location to determine saliency will also be context dependent. For example, the saliency of a red-vertical bar could be signaled by the vertical-tuned cell when it is surrounded by red horizontal bars, since the red-tuned cell is suppressed through isocolor suppression by other red-tuned cells responding to the context. However, when the context contains blue vertical bars, its saliency will be signaled by the red-tuned cells. In another context, the red-vertical conjunctive cell could be signaling the saliency. This is natural since saliency is meant to be context dependent.

Additional contextual influences, weaker than the isofeature suppression, are also induced by the intracortical interactions in V1. One is the colinear facilitation to a cell's response to an optimally oriented bar, when a contextual bar is aligned to this bar as if they are both segments of a smooth contour [24, 25]. Hence, iso-orientation interaction, including both iso-orientation suppression and colinear facilitation, is not isotropic. Another contextual influence is the general, feature-unspecific, surround suppression to a cell's response by activities in nearby cells regardless of their feature preferences [6, 7]. This causes reduced responses by contextual inputs of any features, and interactions between nearby V1 cells tuned to different features.

The most immediate and indicative prediction from the hypothesis is that task-irrelevant features can interfere in tasks that rely significantly on saliency. This is because at each location, only the response of the most activated V1 cell determines the saliency. In particular, if cells responding to task-irrelevant features dictate saliencies at some spatial locations, the task-relevant features become "invisible" for saliency at these locations. Consequently, visual attention is misled to task-irrelevant locations, causing delay in task completion. Second, different V1 processes for different feature dimensions are predicted to lead to asymmetric interactions between features for saliency. Third, the spatial or global phenomena often associated with visual grouping are predicted. This is because the intracortical interactions depend on the relative spatial relationship between input features, particularly in a non-isotropic manner for orientation features, making saliency sensitive to spatial configurations, in addition to the densities of inputs. These broad categories of predictions will be elaborated in the following sections in various specific predictions, together with their psychophysical tests.

This chapter is organized as follows. In Section 4.2 we present our methodology followed in our experiments. Section 4.3 presents the results of the experiments, Section 4.4 discussion, and Section 4.5 our conclusions.

4.2 Materials and Methods

Stimuli: In all our experiments, each stimulus pattern had 22 rows × 30 columns of items (of single or double bars) on a regular grid with unit distance 1.6° of visual angle. Each bar was a white (CIE illuminant C) 1.2 × 0.12 degree rectangle (for experiments in orientation feature dimensions only), or a colored 1.2 × 0.24 degree rectangle (for experiments involving color and orientation features). All bars had a luminance of 14 cd/m² unless otherwise stated, and the background was black. The colored bars were green or pink specified by their CIE 1976 coordinates (u', v'), with hue angles $h_{uv} = 130°$ or $310°$, respectively, where $\tan(h_{uv}) = (v' - v'_n)/(u' - u'_n)$, and (u'_n, v'_n) are the coordinates of CIE illuminant C (0.201, 0.461). All bars within a stimulus had the same saturation $s_{uv} = 13\sqrt{[(u' - u'_n)^2 + (v' - v'_n)^2]}$. For segmentation experiments, the vertical texture border between two texture regions was located randomly left or right, at 7, 9, or 11 interelement distances laterally from the display center. Stimuli in search tasks were made analogously to those in texture segmentation tasks, by reducing one of the two texture regions into a single target item. In each trial, the target was positioned randomly in one of the middle 14 rows; given the target's row number, its column number was such that the target was positioned randomly left or right, as close as possible to 16.8 degrees of visual angle from the display center. The noncolored bars were oriented either as specified in the captions of the figures and tables presented, or horizontally, vertically, and $\pm 45°$ from the vertical. The color and orientation of the target or left texture region in each trial were randomly green or pink (for colored stimuli) and left or right tilted (or horizontal or vertical) in the relevant orientations.

Subjects: The subjects were adults with normal or corrected to normal vision, and they are identified by letters, such as "LZ," in the figures and tables. Most subjects were naive to the purpose of the study, except for "LZ" (one of the authors), "LJ," and "ASL." Some subjects were more experienced at reaction time tasks than others. "AP," "FE," "LZ," "NG," and "ASL" participated in more experiments than others (such as "KC," "DY,' and "EW") who only participated in one or a few experiments.

Procedure and data analysis: The subjects were instructed to fixate centrally until the stimulus onset, to freely move their eyes afterwards, and to press a left or right key (located to their left- or right-hand side) using their left or right hand, respectively, quickly and accurately, to indicate whether the target or texture border (present in each trial) was in the left or right half of the display. The stimulus pattern stayed after the onset until the subject's response. There were 96 trials per subject per stimulus conditions shown. Average reaction times (RTs) were calculated (and shown in the figures and tables) excluding trials that were erroneous or had an RT outside 3 standard deviations from the mean. The number of such excluded trials was usually less than 5% of the total for each subject and condition, and our results did not change qualitatively even when we included all trials in calculating RTs or considered the speed-accuracy trade-off in performances. The error bars shown are standard errors. The experiments were carried out in a dark room. Within each figure plot, and each part (A, B, C, etc.) of Tables 4.1 and 4.2, all the stimulus conditions were randomly interleaved within an experimental session such that the subjects could not predict before each trial which stimulus condition would appear. For texture segmentation, the subjects were told to locate the border between two textures regardless of the difference (e.g., whether in color or orientation or both) between the two textures. For visual search, the subjects were told to locate the target which had a unique feature (such as orientation, color, or both, regardless of which orientation(s) and/or which color), for example the odd one out, within the display. The subjects were shown examples of the relevant stimulus conditions to understand the task before recording any data. Experiments (e.g., the one for Figure 4.6) requiring more than 300 to 400 trials in total were broken down into multiple sessions, such that each session typically took 10 to 20 minutes.

4.3 Results

For visual tasks in which saliency plays a dominant or significant role, the transform from visual input to behavioral response, particularly in terms of the reaction time in performing a task, via V1 and other neural mechanisms, can be simplistically and phenomenologically modelled as follows, for clarity of presentation.

$$\text{V1 responses } \mathbf{O} = (O_1, O_2, \ldots, O_M)$$
$$= f_{v1}(\text{visual input } \mathbf{I}; \boldsymbol{\alpha} = (\alpha_1, \alpha_2, \ldots))$$
$$\text{The saliency map SMAP}(x) \propto \max_{x_i = x} O_i \tag{4.2}$$

$$\text{The reaction time RT} = f_{\text{response}}(\text{SMAP}; \boldsymbol{\beta} = (\beta_1, \beta_2, \ldots)) \tag{4.3}$$

Here $f_{v1}(.)$ models the transform from visual input $\mathbf{I}$ to V1 responses $\mathbf{O}$ via neural mechanisms parameterized by α describing V1's receptive fields, see Chapter 5, and intracortical interactions, while $f_{\text{response}}(.)$ models the transform from the saliency map SMAP to RT via the processes parameterized by β modelling decision making, motor responses, and other factors beyond bottom-up saliency. Without quantitative knowledge of β, it is sufficient for our purpose to assume a monotonic transform $f_{\text{response}}(.)$ that gives a shorter RT to a higher saliency value at the task-relevant location, since more salient locations are more quickly selected. This is, of course, assuming that the reaction time is dominated by the time for visual selection by saliency, or that the additional time taken after visual selection and before the task response, say, indicated by button press, is a roughly constant quantity that does not vary sufficiently with the different stimuli being compared in any particular experiment. For our goal to test the saliency hypothesis, we will select stimuli such that this assumption is practically valid (see Discussion). Hence, all our predictions are qualitative, that is, we predict a longer reaction time (RT) in one visual search task than that in another rather than the quantitative differences in these RTs. This does not mean that our predictions will be vague or inadequate for testing the V1 saliency hypothesis, since the predictions will be very precise by explicitly stating which tasks should require longer RTs than which other tasks, making them indicative of V1 mechanisms. Meanwhile, the qualitativeness makes the predictions robust and insensitive to variations in quantitative details parameterized by α of the underlying V1 mechanisms, such as the quantitative strengths of the lateral connections, provided that the qualitative facts of the V1 neural mechanisms are fixed or determined. Therefore, as will be clear below, our predictions can be derived and comprehensible merely from our qualitative knowledge of a few facts about V1, for example, that neurons are tuned to their preferred features, that isofeature suppression is the dominant form of contextual influences, that V1 cells tuned to color have larger receptive fields than cells tuned to orientation, and so on, without resorting to quantitative model analysis or simulations which would only affect the quantitative but not the qualitative outcomes. Meanwhile, although one could quantitatively fit the model to behavioral RTs by tuning the parameters

α and β (within the qualitative range), it adds no value since model fitting is typically possible given enough parameters, nor is it within the scope of this chapter to construct a detailed simulation model that, for this purpose, would have to be more complex than the available V1 model for contextual influences [20–23]. Hence, we do not include quantitative model simulations in this study which is only aimed at deriving and testing our qualitative predictions.

4.3.1 Interference by Task-Irrelevant Features

Consider stimuli having two different features at each location, one task-relevant and the other task-irrelevant. For convenience, we call the V1 responses to the task-relevant and -irrelevant stimuli, relevant and irrelevant responses, respectively, and from the relevant and irrelevant neurons, respectively. If the irrelevant response(s) is stronger than the relevant response(s) at a particular location, this location's saliency is dictated by the irrelevant response(s) according to the V1 saliency hypothesis, and the task-relevant features become "invisible" for saliency. In visual search and segmentation tasks which rely significantly on saliency to attract attention to the target or texture border, the task-irrelevant features are predicted to interfere with the task by directing attention irrelevantly or ineffectively.

Figure 4.2 shows the texture patterns A, B, C to illustrate this prediction.[2] Pattern A has a salient border between two iso-orientation textures of left-oblique and right-oblique bars, respectively, activating two populations of neurons each for one of the two orientations. Pattern B is a uniform texture of alternating horizontal and vertical bars, evoking responses from another two groups of neurons for horizontal and vertical orientations, respectively. When all bars are of the same contrast, the neural response from the corresponding neurons to each bar would have been the same (ignoring neural noise), if there were no intracortical interactions giving rise to contextual influences. With iso-orientation suppression, neurons responding to the texture border bars in pattern A are more active than neurons responding to other bars in pattern A; this is because they receive iso-orientation suppression from fewer active neighboring neurons, since there are fewer neighboring bars of the same orientation. For ease of explanation, let us say, the highest neural responses to a border bar and a background bar are 10 and 5 spikes/second, respectively. This V1 response pattern makes the border more salient, so it pops out in a texture segmentation task. Each bar in pattern B has the same number of iso-orientation neighbors as a texture border bar in pattern A, so it evokes a comparable level of (highest) V1 response, that is, 10 spikes/second, to that evoked by a border bar in pattern A. If patterns A and B are superimposed, to give pattern C, the composite pattern will activate all neurons responding to patterns A and B,

---

2. To test the validity of a result, we often perform a so-called "significance test." A commonly used significance test is the t-test. In a significance test we calculate the probability p of making a mistake by rejecting the null hypothesis. The lower this probability is, the more safely we can reject the null hypothesis. The null hypothesis is usually chosen to be the opposite of what we wish to prove, so the lower the value of p, the more the support for our argument. In this case, the null hypothesis is "the reaction time for composite texture C is no higher than the reaction time for simple texture A."

4.3 Results

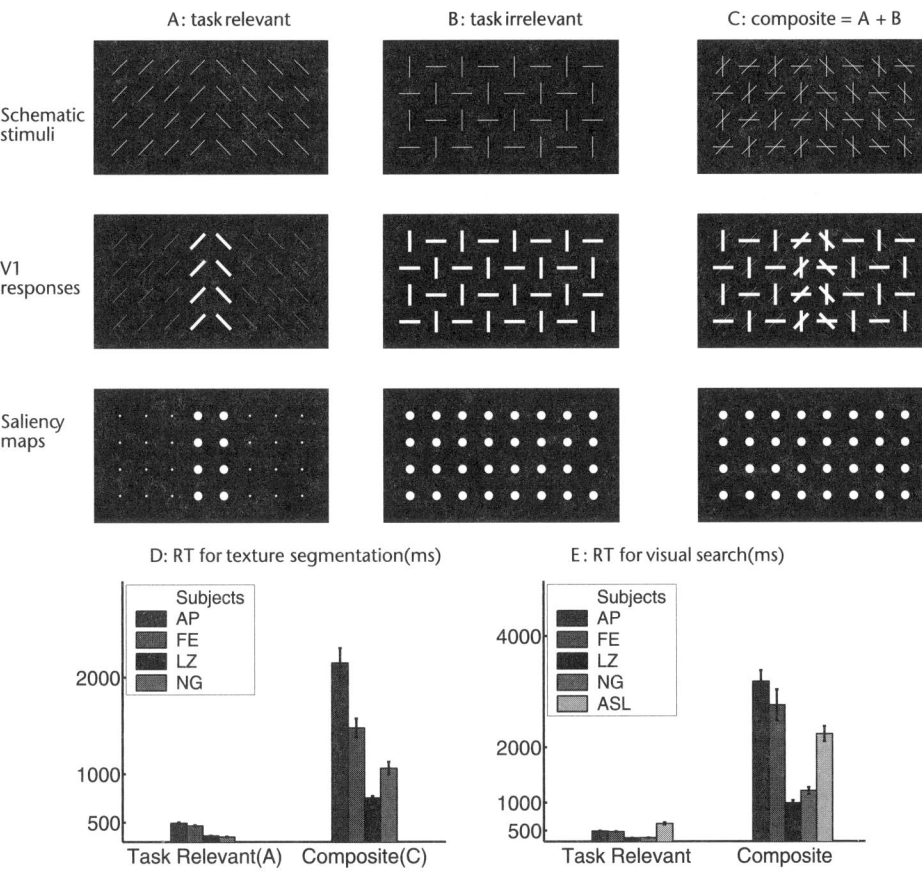

Figure 4.2 Prediction of interference by task-irrelevant features, and its psychophysical test. A, B, C are schematics of texture stimuli (extending continuously in all directions beyond the portions shown), each followed by schematic illustrations of its V1 responses, in which the orientation and thickness of a bar denote the preferred orientation and response level of the activated neuron, respectively. Each V1 response pattern is followed below by a saliency map, in which the size of a disk, denoting saliency, corresponds to the response of the most activated neuron at the texture element location. The orientation contrasts at the texture border in A and everywhere in B lead to less suppressed responses to the stimulus bars, since these bars have fewer iso-orientation neighbors to evoke iso-orientation suppression. The composite stimulus C, made by superposing A and B, is predicted to be difficult to segment, since the task-irrelevant features from B interfere with the task-relevant features from A, giving no saliency highlights to the texture border. D, E: reaction times (differently colored bars denote different subjects) for texture segmentation and visual search tasks testing the prediction. For each subject, RT for the composite condition is significantly higher than for the simple condition ($p < 0.001$). In all experiments in this chapter, stimuli consist of 22 rows × 30 columns of items (of single or double bars) on a regular grid with unit distance $1.6°$ of visual angle. http://www.bg.ic.ac.uk/aab01/book

each neuron responding approximately as it does to A or B alone [for simplicity, we omitted the general suppression between neurons tuned to different orientations, without changing our conclusion (see below)]. According to the V1 saliency hypothesis, the saliency at each texture element location is dictated by the most activated neuron there. Since the (relevant) response to each element of pattern A is lower than or equal to the (irrelevant) response to the corresponding element of pattern B, the saliency at each element location in pattern C is the same as for B, so there is no texture border highlight in such a composite stimulus, making texture segmentation difficult.

For simplicity in our explanation, our analysis above included only the dominant form of contextual influence, the isofeature suppression, but not the less dominant form of the contextual influence, the general surround suppression and colinear facilitation. Including the weaker forms of contextual influences, as in the real V1 or our model simulations [21–23], does not change our prediction here. So, for instance, general surround suppression between local neurons tuned to different orientations should reduce each neuron's response to pattern C from that to pattern A or B alone. Hence, the (highest) responses to the task-relevant bars in pattern C may be, say, 8 and 4 spikes/second at the border and the background, respectively. Meanwhile, the responses to the task-irrelevant bars in pattern C should be, say, roughly 8 spikes/second everywhere, leading to the same prediction of interference. In the rest of this chapter, for ease of explanation and without loss of generality or change of conclusions, we include only the dominant isofeature suppression in our description of the contextual influences, and ignore the weaker or less dominant colinear facilitation and general surround suppression unless their inclusion makes a qualitative or relevant difference (as we shall see in Section 4.3.4). For the same reason, our arguments do not detail the much weaker responses from cells not as responsive to the stimuli concerned, such as responses from motion direction selective cells to a nonmoving stimulus, or the response from a cell tuned to 22.5° to a texture element in pattern C composed of two intersecting bars oriented at 0° and 45°, respectively. (Jointly, the two bars resemble a single bar oriented at 22.5° only at a scale much larger or coarser than their own. Thus, the most activated cell tuned to 22.5° would have a larger RF, much of which would contain no contrast or luminance stimulus, leading to a response weaker than cells preferring *both the scale and the orientation* of the individual bars.) This is because these additional but nondominant responses at each location are "invisible" to saliency by the V1 saliency hypothesis and thus do not affect our conclusions.

Figure 4.2D shows that segmenting the composite texture C indeed takes much longer than segmenting the task-relevant component texture A, confirming the prediction. The reaction times were taken in a task when subjects had to report the location of the texture border, as to the left or right of display center, as quickly as possible.[3] In pattern C, the task-irrelevant horizontal and vertical features from component pattern B interfere with segmentation by relevant orientations from pattern A. Since pattern B has spatially uniform saliency values, the interference is not due to the noisy saliencies of the background [19, 35].

One may wonder whether each composite texture element in Figure 4.2C may be perceived by its average orientation at each location, see Figure 4.3F, thereby making the relevant orientation feature noisy to impair performance. Figure 4.3E demonstrates by our control experiment that this would not have caused as much impairment—RT for this stimulus is at least 37% shorter than that for the composite stimulus.

---

3. The actual stimuli used are larger, see Section 4.2.

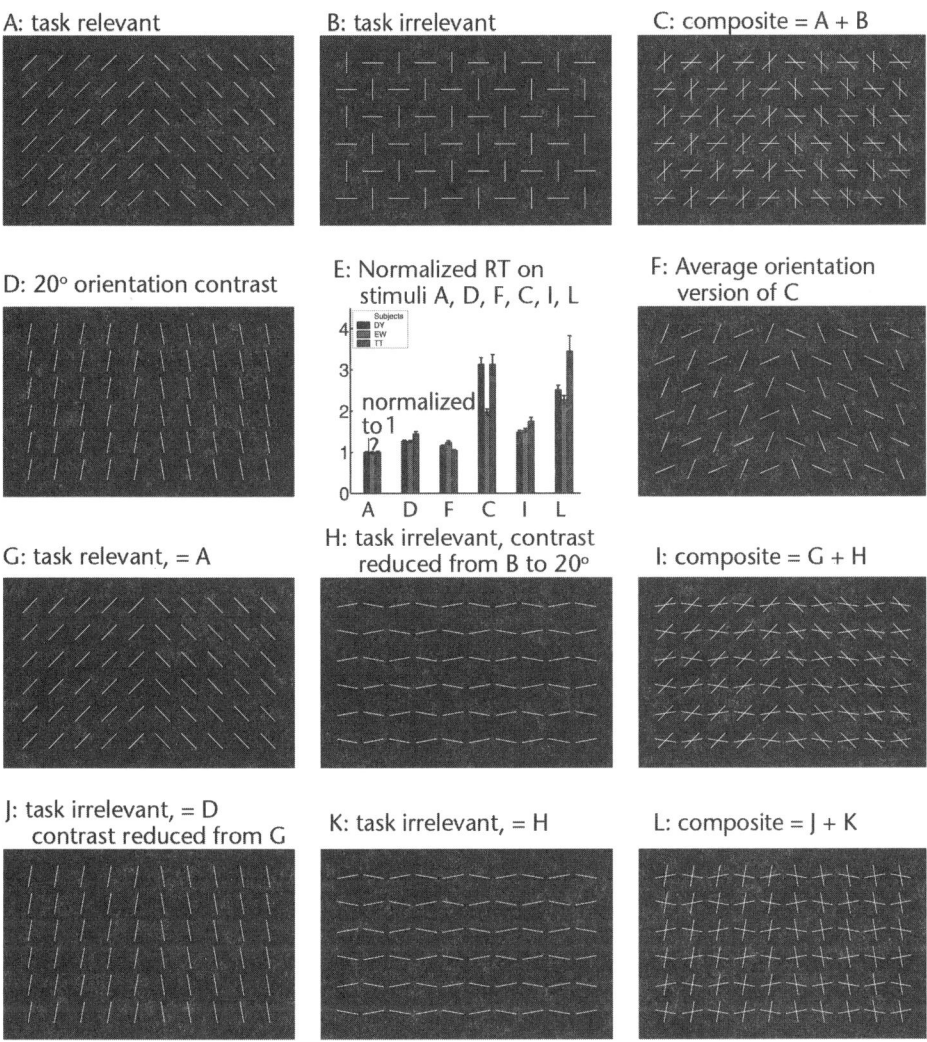

Figure 4.3 Further illustrations to understand interference by task-irrelevant features. A, B, and C are as in Figure 4.2 the schematics of texture stimuli of various feature contrasts in task-relevant and irrelevant features. D is like A, except that each bar is 10° from vertical, reducing orientation contrast to 20°. F is derived from C by replacing each texture element of two intersecting bars by one bar whose orientation is the average of the original two intersecting bars. G, H, and I are derived from A, B, and C by reducing the orientation contrast (to 20°) in the interfering bars, each is 10° from horizontal. J, K, and L are derived from G, H, and I by reducing the task-relevant contrast to 20°. E plots the normalized reaction times for three subjects, DY, EW, and TT, on stimuli A, D, F, C, I, and L randomly interleaved within a session. Each normalized RT is obtained by dividing the actual RT by the RT (which are 471, 490, and 528 ms, respectively, for subjects DY, EW, and TT) of the same subject for stimulus A. For each subject, RT for C is significantly ($p < 0.001$) higher than that for A, D, F, and I by, at least, 95%, 56%, 59%, and 29%, respectively. Matched sample t-test across subjects shows no significant difference ($p = 0.99$) between RTs for stimuli C and L.

If one makes the visual search analog of the texture segmentation tasks in Figure 4.2, by changing stimulus Figure 4.2A (and consequently stimulus Figure 4.2C) such that only one target of a left- (or right-) tilted bar is in a background of right- (or left-) tilted bars, qualitatively the same result (Figure 4.2E) is obtained. Note that the visual search task may be viewed as the extreme case of the texture segmentation task when one texture region has only one texture element.

Note that, if saliency were computed by the SUM rule $SMAP(x) \propto \sum_{x_i = x} O_i$ (rather than the MAX rule) to sum the responses O_i from cells preferring different orientations at a visual location x, interference would not be predicted since the summed responses at the border would be greater than those in the background, preserving the border highlight. Here, the texture border highlight H_{border} (for visual selection) is measured by the difference $H_{border} = R_{border} - R_{ground}$ between the (summed or maxed) response R_{border} to the texture border and the response R_{ground} to the background (where response R_x at location x means $R_x = \sum_{x_i = x} O_i$ or $R_x = \max_{x_i = x} O_i$, under the SUM or MAX rule, respectively). This is justified by the assumption that the visual selection is by the winner-take-all of the responses R_x in visual space x, hence the priority of selecting the texture border is measured by how much this response difference is compared with the levels of noise in the responses. Consequently, the SUM rule applied to our example of response values gives the same border highlight $H_{border} = 5$ spikes/second with or without the task-irrelevant bars, while the MAX rule gives $H_{border} = 0$ and 5 spikes/second, respectively. If the border highlight is measured more conservatively by the ratio $H_{border} = R_{border}/R_{ground}$ (when a ratio $H_{border} = 1$ means no border highlight), then the SUM rule predicts, in our particular example, $H_{border} = (10 + 10)/(5 + 10) = 4/3$ with the irrelevant bars, and $H_{border} = 10/5 = 2$ without, and thus some degree of interference. However, we argue below that even this measure of H_{border} by the response ratio makes the SUM rule less plausible. Behavioral and physiological data suggest that, as long as the saliency highlight is above the just-noticeable-difference (JND, [36]), a reduction in H_{border} should not increase RT as dramatically as observed in our data. In particular, previous findings [36, 37] and our data (in Figure 4.3E) suggests that the ease of detecting an orientation contrast (assessed using RT) does not reduce by more than a small fraction when the orientation contrast is reduced, say, from 90° to 20° as in Figure 4.3A and Figure 4.3D [36, 37], even though physiological V1 responses [38] to these orientation contrasts suggest that a 90° orientation contrast would give a highlight of $H_{90°} \sim 2.25$ and a 20° contrast would give $H_{20°} \sim 1.25$ using the ratio measurement for highlights. (Jones et al. [38] illustrated that the V1 response to a 90° and 20° orientation contrast, respectively, can be 45 and 25 spikes/second, respectively, over a background response of 20 spikes/second.) Hence, the very long RT in our texture segmentation with interference implies that the border should have a highlight $H_{border} \approx 1$ or below the JND, while a very easy segmentation without interference implies that the border should have $H_{border} \gg 1$. If O_{border} and O_{ground} are the relevant responses to the border and background bars, respectively, for our stimulus, and since O_{border} also approximates the irrelevant response, then applying the SUM rule gives border highlight $H_{border} = 2O_{border}/(O_{border} + O_{ground})$ and O_{border}/O_{ground}, with and without interference, respectively. Our RT data thus requires that $O_{border}/O_{ground} \gg 1$ and $2O_{border}/(O_{border} + O_{ground}) \approx 1$ should be satisfied simultaneously—this is difficult since $O_{border}/O_{ground} > 2$ means $2O_{border}/(O_{border} + O_{ground}) > 4/3$, and a larger O_{border}/O_{ground} would give a larger $2O_{border}/(O_{border} + O_{ground})$, making the SUM rule less plausible. Meanwhile, the MAX rule gives a border highlight $H_{border} = O_{border}/O_{border} = 1$ with interference and $H_{border} = O_{border}/O_{ground} > 1$ without. These observations strongly favor the MAX over the SUM rule, and we will show more data to differentiate the two rules later.

From our analysis above, we can see that the V1 saliency hypothesis also predicts a decrease of the interference if the irrelevant feature contrast is reduced, as demonstrated when comparing Figure 4.3GHI with Figure 4.3ABC, and confirmed in our data (Figure 4.3E). The neighboring irrelevant bars in Figure 4.3I are more similarly oriented, inducing stronger isofeature suppression between them, and decreasing their evoked responses, say, from 10 to 7 spikes/second. (Although colinear facilitation is increased by this stimulus change, since iso-orientation suppression dominates colinear facilitation physiologically, the net effect is decreased responses to all the task-irrelevant bars.) Consequently, the relevant texture border highlights are no longer submerged by the irrelevant responses. The degree of interference would be much weaker, though still nonzero since the irrelevant responses (of 7 spikes/second) still dominate the relevant responses (of 5 spikes/second) in the background, reducing the relative degree of border highlight from 5 to 3 spikes/second. Analogously, interference can be increased by decreasing task-relevant contrast, as demonstrated by comparing Figure 4.3J–L and Figure 4.3G–I, and confirmed in our data (Figure 4.3E). Reducing the relevant contrast makes the relevant responses to the texture border weaker, say from 10 to 7 spikes/second, making these responses more vulnerable to being submerged by the irrelevant responses. Consequently, interference is stronger in Figure 4.3L than Figure 4.3I. Essentially, the existence and strength of the interference depend on the relative response levels to the task-relevant and irrelevant features, and these response levels depend on the corresponding feature contrasts and direct input strengths. When the relevant responses dictate saliency everywhere and their response values or overall response pattern are little affected by the existence or absence of the irrelevant stimuli, there should be little interference. Conversely, when the irrelevant responses dictate saliency everywhere, interference for visual selection is strongest. When the relevant responses dictate the saliency value at the location of the texture border or visual search target but not in the background of our stimuli, the degree of interference is intermediate. In both Figure 4.3C and Figure 4.3L, the irrelevant responses (approximately) dictate the saliency everywhere, so the texture borders are predicted to be equally nonsalient. This is confirmed across subjects in our data (Figure 4.3E), although there is a large variation between subjects, perhaps because the bottom-up saliency is so weak in these two stimuli that subject-specific top-down factors contribute significantly to the RTs.

4.3.2 The Color-Orientation Asymmetry in Interference

Can task-irrelevant features from another feature dimension interfere? Figure 4.4A illustrates orientation segmentation with irrelevant color contrasts. As

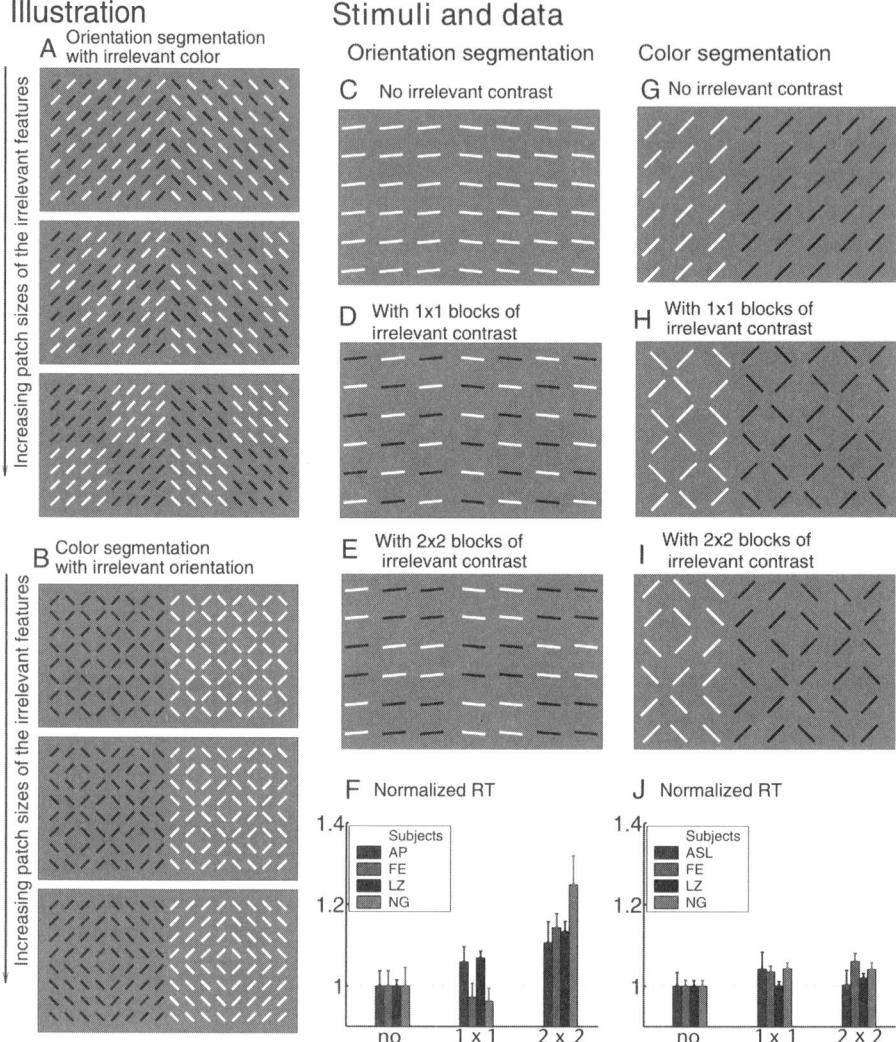

Figure 4.4 Interference between orientation and color, with schematic illustrations (left, A, B) and stimuli/data (right, C–J). A: Orientation segmentation with irrelevant color. B: Color segmentation with irrelevant orientation. Larger patch sizes of irrelevant color gives stronger interference, but larger patch sizes of irrelevant orientation do not make interference stronger. C, D, E: Schematics of the experimental stimuli for orientation segmentation, without color contrast (C) or with irrelevant color contrast in 1×1 (D) or 2×2 (E) blocks. All bars had color saturation $s_{uv} = 1$ and were $\pm 5°$ from horizontal. The actual colors used for the bars were green and pink against a black background, as described in Section 4.2. F: Normalized RTs for C, D, E for four subjects (different colors indicate different subjects). The "no," "1×1," "2×2" on the horizontal axis mark stimulus conditions for C, D, E, that is, with no or "$n \times n$" blocks of irrelevant features. The RT for condition "2×2" is significantly longer ($p < 0.05$) than that for "no" in all subjects, and than that of "1×1" in three out of the four subjects. By matched sample t-test across subjects, mean RTs are significantly longer in "2×2" than that in "no" ($p = 0.008$) and than that in "1×1" ($p = 0.042$). Each RT is normalized by dividing it with the subject's mean RT for the "no" condition, which for the four subjects AP, FE, LZ, and NG is 1,170, 975, 539, and 1,107 ms, respectively. G–J are for color segmentation, analogous to C–F, with stimulus bars oriented $\pm 45°$ and of color saturation $s_{uv} = 0.5$. Matched sample t-test across subjects showed no significant difference between RTs in different conditions. Only two out of the four subjects had their RTs significantly higher ($p < 0.05$) in interfering than no interfering conditions. The un-normalized mean RTs of the four subjects ASL, FE, LZ, and NG in the "no" condition are: 650, 432, 430, and 446 ms, respectively. http://www.bg.ic.ac.uk/staff/aab01/book

in Figure 4.2, the irrelevant color contrast increases the responses to the color features since the isocolor suppression is reduced. At each location, the response to color could then compete with the response to the relevant orientation feature to dictate the saliency. In Figure 4.2C, the task-irrelevant features interfere because they evoke higher responses than the relevant features, as made clear by demonstrations in Figure 4.3. Hence, whether color can interfere with orientation or vice versa depends on the relative levels of V1 responses to these two feature types. Color and orientation are processed differently by V1 in two aspects. First, cells tuned to color, more than cells tuned to orientation, are usually in V1's cytochrome oxidase stained blobs which are associated with higher metabolic and neural activities [39]. Second, cells tuned to color have larger receptive fields [33,40], hence they are activated more by larger patches of color. In contrast, larger texture patches of oriented bars can activate more orientation tuned cells, but do not make individual orientation tuned cells more active. Meanwhile, in the stimulus for color segmentation (e.g., Figure 4.4B), each color texture region is large so that color tuned cells are most effectively activated, making their responses easily the dominant ones. Consequently, the V1 saliency hypothesis predicts: (1) task-irrelevant colors are more likely to interfere with orientation than the reverse; (2) irrelevant color contrast from larger color patches can disrupt an orientation-based task more effectively than that from smaller color patches; and (3) the degree of interference by irrelevant orientation in a color-based task will not vary with the patch size of the orientation texture.

These predictions are apparent when viewing Figure 4.4A, B. They are confirmed by RT data for our texture segmentation task, shown in Figure 4.4C–J. Irrelevant color contrast can indeed raise RT in orientation segmentation, but is effective only for sufficiently large color patches. In contrast, irrelevant orientation contrast does not increase RT in color segmentation regardless of the sizes of the orientation patches. In Figure 4.4C–E, the irrelevant color patches are small, activating the color tuned cells less effectively. However, interference occurs under small orientation contrast which reduces responses to relevant features (as demonstrated in Figure 4.3). Larger color patches can enable interference even to a $90°$ orientation contrast at the texture border, as apparent in Figure 4.4A, and has been observed by Snowden [41]. In Snowden's design, the texture bars were randomly rather than regularly assigned one of two isoluminant, task-irrelevant colors, giving randomly small and larger sizes of the color patches. The larger color patches made task-irrelevant locations salient to interfere with the orientation segmentation task. Previously, the V1 saliency hypothesis predicted that Snowden's interference should become stronger when there are more irrelevant color categories, for example, each bar could assume one of three rather than two different colors. This is because more color categories further reduce the number of isocolor neighbors for each colored bar and thus the isocolor suppression, increasing responses to irrelevant color. This prediction was subsequently confirmed [29].

In Figure 4.4G–I, the relevant color contrast was made small to facilitate interference by irrelevant orientation, though unsuccessfully. Our additional data showed that orientation does not significantly interfere with color-based segmentation even when the color contrast was reduced further. The patch sizes, of 1×1

and 2 × 2, of the irrelevant orientation textures ensure that all bars in these patches evoke the same levels of responses, since each bar has the same number of iso-orientation neighbors (this would not hold when the patch size is 3 × 3 or larger). Such an irrelevant stimulus pattern evokes a spatially uniform level of irrelevant responses, thus ensuring that interference cannot possibly arise from nonuniform or noisy response levels to the background [19,35]. Patch sizes for irrelevant colors in Figure 4.4C–E were made to match those of irrelevant orientations in Figure 4.4G–I, so as to compare saliency effects by color and orientation features. Note that, as discussed in Section 4.3.1, the SUM rule would predict the same interference only if saliency highlight H_{border} is measured by the ratio between responses to the border and background. With this measure of H_{border}, our data in this subsection, showing that the interference only increases RT by a small fraction, cannot sufficiently differentiate the MAX from the SUM rule.

4.3.3 Advantage for Color-Orientation Double Feature but Not Orientation-Orientation Double Feature

A visual location can be salient due to two simultaneous feature contrasts. For instance, at the texture border between a texture of green, right-tilted bars and another texture of pink, left-tilted bars, in Figure 4.5C, both the color *and* orientation contrast could make the border salient. We say that the texture border has a color-orientation double feature contrast. Analogously, a texture border of an orientation-orientation double contrast, and the corresponding borders of single-orientation contrasts, can be made as in Figure 4.5E–G. We can ask whether the saliency of a texture border with a double feature contrast can be higher than both of those of the corresponding single feature–contrast texture borders. We show below that the V1 saliency hypothesis predicts a likely "yes" for color-orientation double feature but a definite "no" for orientation-orientation double feature.

V1 has color-orientation conjunctive cells which are tuned to both color and orientation, though their tuning to either feature is typically not as sharp as that of the single feature tuned cells [33]. Hence, a colored bar can activate a color tuned cell, an orientation tuned cell, and a color-orientation conjunctive cell, with cell outputs O_c, O_o, and O_{co}, respectively. The highest response $\max(O_c, O_o, O_{co})$ from these cells should dictate the saliency of the bar's location. Let the triplet of response be $[O_c^o, O_o^o, O_{co}^o]$ at an orientation texture border, $[O_c^c, O_o^c, O_{co}^c]$ at a color border, and $[O_c^{co}, O_o^{co}, O_{co}^{co}]$ at a color-orientation double feature border. Due to isofeature suppression, the response of a single feature cell is higher with than without its feature contrast, that is, $O_c^o < O_c^c$ and $O_o^c < O_o^o$. The single feature cells also have comparable responses with or without feature contrasts in other dimensions, that is, $O_c^c \approx O_c^{co}$ and $O_o^o \approx O_o^{co}$. Meanwhile, the conjunctive cell should have a higher response at a double than single feature border, that is, $O_{co}^{co} > O_{co}^o$ and $O_{co}^{co} > O_{co}^c$, since it has fewer neighboring conjunctive cells responding to the same color *and* same orientation. The maximum $\max(O_c^{co}, O_o^{co}, O_{co}^{co})$ could be O_c^{co}, O_o^{co}, or O_{co}^{co} to dictate the saliency of the double feature border. Without detailed knowledge, we expect that it is likely that, in at least some nonzero percentage of many trials, O_{co}^{co} is the dictating response, and when this happens, O_{co}^{co} is larger

4.3 Results

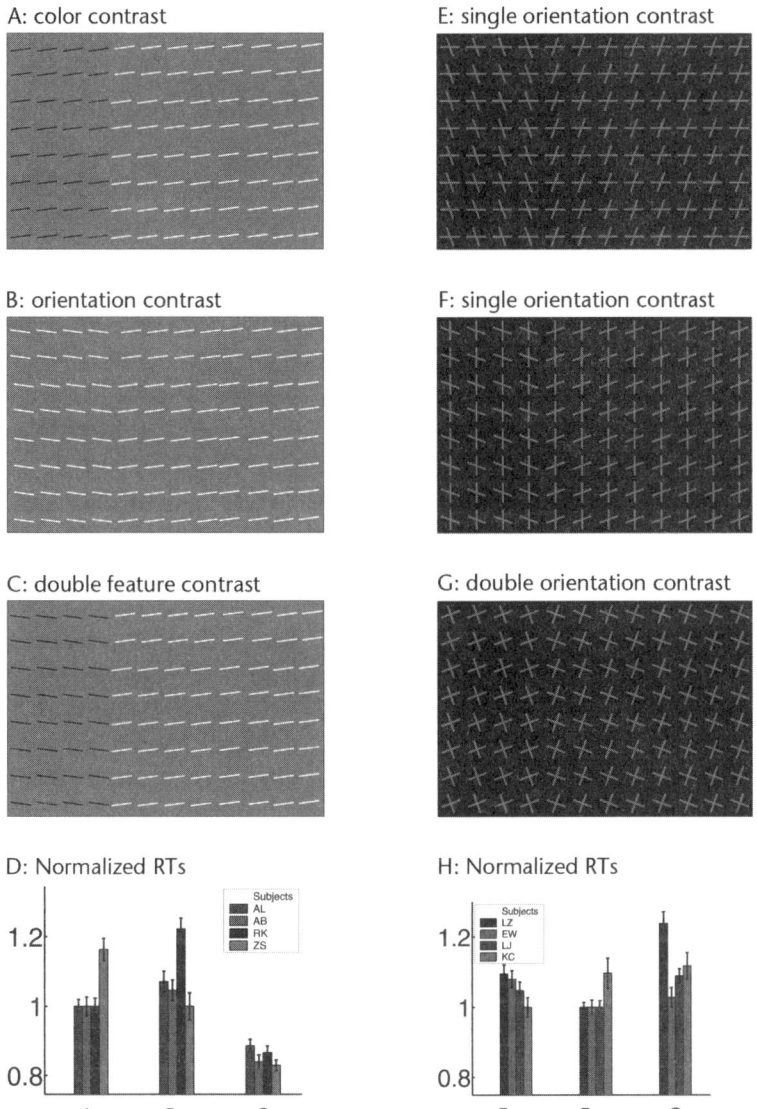

Figure 4.5 Testing the predictions of saliency advantage in color-orientation double feature, A–D, and the lack of it in orientation-orientation double feature, E–H. A–C: Schematics of the texture segmentation stimuli by color contrast, or orientation contrast, or by double color-orientation contrast. In the actual experiments, green and pink bars were used against a black background, as described in Section 4.2. D: Normalized RTs for the stimulus conditions A–C. Normalization for each subject is by the shortest mean RT of the two single feature contrast conditions (which for subjects AL, AB, RK, and ZS is 651, 888, 821, and 634 ms, respectively). All stimulus bars had color saturation $s_{uv} = 0.2$, and were at $\pm 7.5°$ from the horizontal direction. All subjects had their RTs for the double feature condition significantly shorter ($p < 0.001$) than those of both single feature conditions. E–G: The texture segmentation stimuli by single or double orientation contrast. Each oblique bar is $\pm 20°$ from the vertical direction in E and $\pm 20°$ from the horizontal direction in F. G is made by superposing the task-relevant bars in E and F. H: Normalized RTs for the stimulus conditions E–G (analogous to D). The shortest mean RT between the two single feature conditions, for the four subjects LZ, EW, LJ, and KC, are 493, 688, 549, and 998 ms, respectively. None of the subjects had RT for G lower than the minimum of the RTs for E and F. Averaged over the subjects, the mean normalized RT for the double orientation feature in G is significantly longer ($p < 0.01$) than that for the color orientation double feature in C. http://www.bg.ic.ac.uk/staff/aab01/book

than all responses from all cells to both single feature contrasts. Consequently, averaged over trials, the double feature border is likely more salient than both of the single feature borders and thus should require a shorter RT to detect. In contrast, there are no V1 cells tuned conjunctively to two different orientations, hence, a double orientation-orientation border definitely cannot be more salient than both of the two single orientation borders.

The above considerations have omitted the general suppression between cells tuned to different features. When this is taken into account, the single feature tuned cells should respond less vigorously to a double feature than to the corresponding effective single feature contrast. This means, for instance, $O_o^{co} \lesssim O_o^o$ and $O_c^{co} \lesssim O_c^c$. This is because general suppression grows with the overall level of local neural activities. This level is higher with double feature stimuli which activate some neurons more, for example, when $O_c^{co} > O_c^o$, and $O_o^{co} > O_o^c$ (at the texture border). In the color-orientation double feature case, $O_o^{co} \lesssim O_o^o$ and $O_c^{co} \lesssim O_c^c$ mean that $O_{co}^{co} > \max(O_c^{co}, O_o^{co})$ could not guarantee that O_{co}^{co} must be larger than all neural responses to both of the single feature borders. This consideration could somewhat weaken or compromise the double feature advantage for the color-orientation case, and should make the double orientation contrast less salient than the more salient one of the two single orientation contrast conditions. In any case, the double feature advantage in the color-orientation condition should be stronger than that of the orientation-orientation condition.

These predictions are indeed confirmed in the RT data. As shown in Figure 4.5D, H, the RT to locate a color-orientation double contrast border Figure 4.5C is shorter than both RTs to locate the two single feature borders Figure 4.5A and Figure 4.5B. Meanwhile, the RT to locate a double orientation contrast of Figure 4.5G is no shorter than the shorter one of the two RTs to locate the two single orientation contrast borders Figure 4.5E and Figure 4.5F. The same conclusion is reached (data not shown) if the irrelevant bars in Figure 4.5E or Figure 4.5F, respectively, have the same orientation as one of the relevant bars in Figure 4.5F or Figure 4.5E, respectively. Note that, to manifest the double feature advantage, the RTs for the single feature tasks should not be too short, since RT cannot be shorter than a certain limit for each subject. To avoid this RT floor effect, we have chosen sufficiently small feature contrasts to make RTs for the single feature conditions longer than 450 ms for experienced subjects and even longer for inexperienced subjects.

Nothdurft [42] also showed saliency advantage of the double feature contrast in color-orientation. The shortening of RT by feature doubling can be viewed phenomenologically as a violation of a race model which models the task's RT as the outcome of a race between two response decision-making processes by color and orientation features, respectively. This violation has been used to account for the double feature advantage in RT also observed in visual search tasks when the search target differs in both color and orientation from uniform distractors observed previously [43] and in our own data (Table 4.1A). In our framework, we could interpret the RT for color-orientation double feature as a result from a race between three neural groups—the color tuned, the orientation tuned, and the conjunctive cells.

It is notable that the findings in Figure 4.5H cannot be predicted from the SUM rule. With single or double orientation contrast, the (summed) responses to the background bars are approximately unchanged, since the iso-orientation suppression between various bars is roughly unchanged. Meanwhile, the total (summed) response to the border is larger when the border has double orientation contrast (even considering the general, feature unspecific, suppression between neurons). Hence, the SUM rule would predict that the double orientation contrast border is more salient than the single contrast one, regardless of whether one measures the border highlight H_{border} by the difference or ratio between the summed response to the texture border and that to the background.

4.3.4 Emergent Grouping of Orientation Features by Spatial Configurations

Combining iso-orientation suppression and colinear facilitation, contextual influences between oriented bars depend nonisotropically on spatial relationships between the bars. Thus, spatial configurations of the bars can influence saliency in ways that cannot be simply determined by densities of the bars, and properties often associated with grouping can emerge. Patterns A–G in Figure 4.6 are examples of these, and the RT to segment each texture will be denoted as RT_A, RT_B,..., RT_G. Patterns A and B both have a 90° orientation contrast between two orientation textures. However, the texture border in B seems more salient. Patterns C and D are both made by adding, to A and B, respectively, task-irrelevant bars ±45° relative to the task-relevant bars and containing a 90° irrelevant orientation contrast. However, the interference is stronger in C than in D. Patterns E and G differ from C by having zero orientation contrast among the irrelevant bars, pattern F differs from D analogously. As demonstrated in Figure 4.3, the interference in E and G should thus be much weaker than that in C, and that in F much weaker than that in D. The irrelevant bars are horizontal in E and vertical in G, on the same original pattern A containing only the ±45° oblique bars. Nevertheless, segmentation seems easier in E than in G. These peculiar observations all seem to relate to what is often called visual "grouping" of elements by their spatial configurations, and can in fact be predicted from the V1 saliency hypothesis when considering that the contextual influences between oriented bars are nonisotropic. To see this, we need to abandon the simplification used so far to approximate contextual influences by only the dominant component—isofeature suppression. Specifically, we now include in the contextual influences the subtler components: (1) facilitation between neurons responding to colinear neighboring bars and (2) general feature-unspecific surround suppression between nearby neurons tuned to any features.

Due to colinear facilitation, a vertical border bar in pattern B is salient not only because a neuron responding to it experiences weaker iso-orientation suppression, but also because it additionally enjoys full colinear facilitation due to the colinear contextual bars, whereas a horizontal border bar in B, or an oblique border bar in A, has only half as many colinear neighbors. Hence, in an orientation texture, the vertical border bars in B, and in general colinear border bars parallel to a texture border, are more salient than border bars not parallel to the border given the same orientation contrast at the border. Hence, if the highest response to each border bar in A is 10 spikes/second, then the highest response to each border bar in B

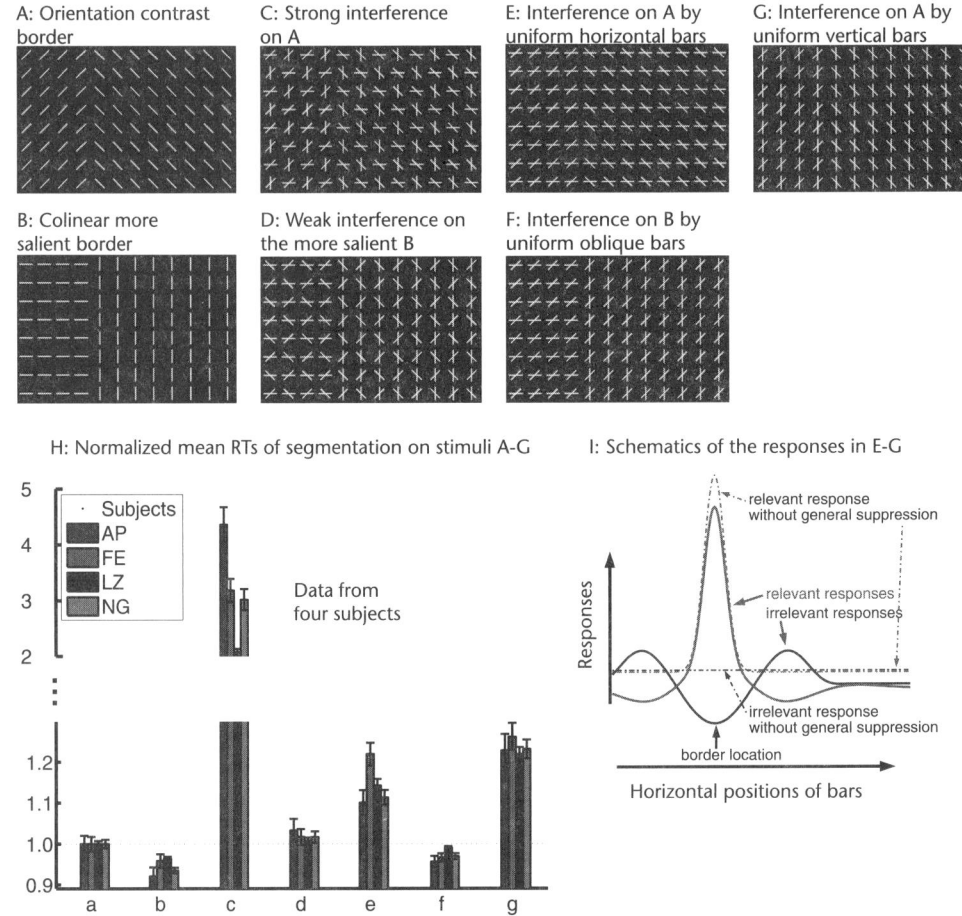

Figure 4.6 Demonstrating and testing the predictions on spatial grouping. A–G: Portions of different stimulus patterns used in the segmentation experiments. Each row starts with an original stimulus without task-irrelevant bars, followed by stimuli when various task-irrelevant bars are superposed on the original. H: RT data when different stimulus conditions are randomly interleaved in experimental sessions. The un-normalized mean RT for the four subjects AP, FE, LZ, and NG in condition A are 493, 465, 363, and 351 ms, respectively. For each subject, it is statistically significant that $RT_C > RT_A$ ($p < 0.0005$), $RT_D > RT_B$ ($p < 0.02$), $RT_A > RT_B$ ($p < 0.05$), $RT_A < RT_E$, RT_G ($p < 0.0005$), $RT_D > RT_F$, $RT_C > RT_E$, and RT_G ($p < 0.02$). In three out of the four subjects, $RT_E < RT_G$ ($p < 0.01$), and two out of the four subjects, $RT_B < RT_F$ ($p < 0.0005$). Meanwhile, by matched sample t-tests across subjects, the mean RT values between any two conditions are significantly different (p smaller than values ranging from 0.0001 to 0.04). I: Schematic representation of responses from relevant (gray) and irrelevant (black) neurons, with (solid curves) and without (dot-dashed curves) considering general suppressions, for situations in E–G. Interference from the irrelevant features arises from the spatial peaks in their responses away from the texture border.

could be, say, 15 spikes/second. Indeed, $RT_B < RT_A$, as shown in Figure 4.6H. (Wolfson and Landy [44] observed a related phenomenon, more details in Li [22]. Furthermore, the highly salient vertical border bars make segmentation less susceptible to interference by task-irrelevant features, since their evoked responses are more likely dominating to dictate salience. Hence, interference in D is much weaker than in C, even though the task-irrelevant orientation contrast is 90° in both C and D. Indeed, $RT_D < RT_C$ (Figure 4.6H), although RT_D is still significantly longer than RT_B without interference. All these are not due to any special status of the vertical orientation of the border bars in B and D, for rotating the whole stimulus patterns would not eliminate the effects. Similarly, when the task-irrelevant bars are uniformly oriented, as in patterns E and G (for A) and F (for B), the border in F is more salient than those in E and G, as confirmed by $RT_F < RT_E$ and RT_G.

The "protruding through" of the vertical border bars in D likely triggers the sensation of the (task-irrelevant) oblique bars as grouped or belonging to a separate (transparent) surface. This sensation arises more readily when viewing the stimulus in a leisurely manner rather than in the hurried manner of an RT task. Based on the arguments that one usually perceives the "what" after perceiving the "where" of visual inputs [45, 46], we believe that this grouping arises from processes subsequent to the V1 saliency processing. Specifically, the highly salient vertical border bars are likely to define a boundary of a surface. Since the oblique bars are neither confined within the boundary nor occluded by the surface, they have to be inferred as belonging to another, overlaying (transparent), surface.

Given no orientation contrast between the task-irrelevant bars in E–G, the iso-orientation suppression among the irrelevant bars is much stronger than that in C and D, and is in fact comparable in strength with that among the task-relevant bars sufficiently away from the texture border. Hence, the responses to the task-relevant and irrelevant bars are comparable in the background, and no interference would be predicted if we ignored general surround suppression between the relevant and irrelevant bars (detailed below). Indeed, $RT_E, RT_G \ll RT_C$, and $RT_F < RT_D$.

However, the existence of general surround suppression introduces a small degree of interference, making $RT_E, RT_G > RT_A$, and $RT_F > RT_B$. Consider E for example. Let us say that, without considering the general surround suppression, the relevant responses are 10 spikes/second and 5 spikes/second at the border and background, respectively, and the irrelevant responses are 5 spikes/second everywhere. The general surround suppression enables nearby neurons to suppress each other regardless of their feature preferences. Hence, spatial variations in the relevant responses cause complementary spatial variations in the irrelevant responses (even though the irrelevant inputs are spatially homogeneous), see Figure 4.6I for a schematic illustration. For convenience, denote the relevant and irrelevant responses at the border as $O_{border}(r)$ and $O_{border}(ir)$, respectively, and as $O_{near}(r)$ and $O_{near}(ir)$, respectively, at locations near but somewhat away from the border. The strongest general suppression is from $O_{border}(r)$ to $O_{border}(ir)$, reducing $O_{border}(ir)$ to, say, 4 spikes/second. This reduction in turn causes a reduction of iso-orientation suppression on the irrelevant responses $O_{near}(ir)$, thus increasing $O_{near}(ir)$ to, say, 6 spikes/second. The increase in $O_{near}(ir)$ is also partly due to a weaker general suppression from $O_{near}(r)$ (which is weaker than the relevant responses sufficiently away from the border because of the extra strong iso-orientation suppression from the very strong border responses $O_{border}(r)$ [47]). Mutual (iso-orientation) suppression between the irrelevant neurons is a positive feedback process that amplifies any response difference. Hence, the difference between $O_{border}(ir)$ and $O_{near}(ir)$ is amplified so that, say, $O_{border}(ir) = 3$ and $O_{near}(ir) = 7$ spikes/seconds, respectively. Therefore, $O_{near}(ir)$ dominates $O_{near}(r)$ somewhat away from the border, dictating and increasing the local saliency. As a result, the relative saliency of the border is reduced and some degree of interference arises, causing $RT_E > RT_A$. The same argument leads similarly to conclusions $RT_G > RT_A$ and $RT_F > RT_B$, as seen in our data (Figure 4.6H). If colinear facilitation is not considered, the degree of interference in E and G should be identical, predicting $RT_E = RT_G$. As explained below, considering colinear facilitation additionally will predict $RT_E < RT_G$, as seen in our data for three out of four subjects (Figure 4.6H). Stimuli E and G differ in the direction of the colinear facilitation between the irrelevant bars. The direction is *across* the border in E but *along* the border in G, and, unlike iso-orientation suppression, facilitation tends to equalize responses $O_{near}(ir)$ and

Table 4.1 RTs(ms) in Visual Search for Unique Color and/or Orientation, Corresponding to Those in Figures 4.4 and 4.5

	A: Single or Double Color/Orientation Contrast Search (Figure 4.5 A–D)		
Subjects	Color	Orientation	Color & Orientation
AP	512 ± 8 (1)	1378 ± 71 (1)	496 ± 7 (1)
FE	529 ± 12 (1)	1,509 ± 103 (3)	497 ± 12 (0)
LZ	494 ± 11 (3)	846 ± 37 (4)	471 ± 7 (0)
NG	592 ± 29 (2)	808 ± 34 (4)	540 ± 19 (0)
	B: Single or Double Orientation Contrast Search (Figure 4.5E–H)		
Subjects	Single Contrast 1 (Figure 4.5E)	Single Contrast 2 (Fig. 4.5F)	Double Contrast (Figure 4.5G)
LZ	732 ± 23 (1)	689 ± 18 (3)	731 ± 22 (1)
EW	688 ± 15 (0)	786 ± 20 (1)	671 ± 18 (2)
	C: Irrelevant Orientation in Color Search (Figure 4.4G–J)		
Subjects	No Irrelevant Contrast	1 × 1 Orientation Blocks	
AP	804 ± 30 (0)	771 ± 29 (0)	
FE	506 ± 12 (5)	526 ± 12 (0)	
LZ	805 ± 26 (1)	893 ± 35 (5)	
NG	644 ± 33 (1)	677 ± 34 (3)	
	D: Irrelevant Color in Orientation Search (Figure 4.4C–F)		
Subjects	No Irrelevant Contrast	1 × 1 Color Blocks	2 × 2 Color Blocks
AP	811 ± 30 (0)	854 ± 38 (0)	872 ± 29 (0)
FE	1,048 ± 37 (0)	1,111 ± 34 (0)	1,249 ± 45 (2)
LZ	557 ± 13 (1)	625 ± 22 (1)	632 ± 21 (1)
NG	681 ± 22 (1)	746 ± 27 (3)	734 ± 31 (1)

Each data entry is: RT ± its standard error (percentage error rate). In A, orientation of background bars: ±45° from vertical, orientation contrast: ±18°, $s_{uv} = 1.5$. In B, stimuli are the visual search versions of Figure 4.5E-G. In A and B, the normalized RT (normalized as in Fig. 4.5) for the double feature contrast is significantly ($p < 0.05$) longer in A than that in B. In C, luminance of bars = $1 cd/m^2$, $s_{uv} = 1.5$, bar orientation: ±20° from vertical or horizontal, irrelevant orientation contrast is 90°. No significant difference ($p = 0.36$) between RTs with and without irrelevant feature contrasts. In D, orientation of background/target bars: ±/∓81° from vertical, $s_{uv} = 1.5$, RTs for stimuli with irrelevant color contrast (of either condition) are significantly longer ($p < 0.034$) than those for stimuli without irrelevant color contrasts.

$O_{border}(ir)$ to the co-linear bars. This reduces the spatial variation of the irrelevant responses across the border in E such that, say, $O_{border}(ir) = 4$ and $O_{near}(ir) = 6$ spikes/second, thus reducing the interference.

The SUM rule (over V1's neural responses) would predict qualitatively the same directions of RT variations between conditions in this section only when the texture border highlight H_{border} is measured by the ratio rather than the difference between the (summed) response to the border and that to the background. However, using the same argument as in Section 4.3.1, our quantitative data would make the SUM rule even more implausible than it is in Section 4.3.1 (since, using the notation of Section 4.3.1, we note that O_{ground} approximates the irrelevant responses in E and G, whose weak interference would require a constraint of $H_{border} = (O_{border} + O_{ground})/2O_{ground} > 1 + \delta$ with $\delta \gg 0$, in addition to the other stringent constraints in Section 4.3.1 that made the SUM rule less plausible).

We also carried out experiments in visual search tasks analogous to those in Figures 4.4 to 4.6, as we did in Figure 4.2E analogous to Figure 4.2D. Qualitatively the same results as those in Figures 4.4 to 4.5 were found, see Table 4.1. For visual search conditions corresponding to those in Figure 4.6, however, since there were no elongated texture borders in the stimuli, grouping effects arising from the colinear border, or as the result of the elongated texture border, are not predicted, and indeed, not reflected in the data, see Table 4.2. This confirmed additionally that saliency is sensitive to spatial configurations of input items in the manner prescribed by V1 mechanisms.

Table 4.2 RTs (ms) for Visual Search for Unique Orientation, Corresponding to Data in Figure 4.6H

Conditions	AP	FE	LZ
	Subjects		
A	485 ± 8 (0.00)	478 ± 6 (0.00)	363 ± 2 (0.00)
B	479 ± 9 (0.00)	462 ± 6 (0.00)	360 ± 2 (0.00)
C	3,179 ± 199 (6.25)	2,755 ± 280 (5.21)	988 ± 50 (3.12)
D	1,295 ± 71 (1.04)	1,090 ± 53 (5.21)	889 ± 31 (3.12)
E	623 ± 20 (0.00)	707 ± 19 (0.00)	437 ± 9 (1.04)
F	642 ± 20 (0.00)	743 ± 21 (0.00)	481 ± 12 (3.12)
G	610 ± 21 (0.00)	680 ± 23 (0.00)	443 ± 10 (2.08)

Conditions	NG	ASL
	Subjects	
A	366 ± 3 (1.04)	621 ± 19 (0.00)
B	364 ± 3 (0.00)	592 ± 16 (1.04)
C	1,209 ± 62 (2.08)	2,238 ± 136 (11.46)
D	665 ± 22 (2.08)	1,410 ± 74 (4.17)
E	432 ± 7 (1.04)	838 ± 35 (0.00)
F	456 ± 9 (2.08)	959 ± 40 (1.04)
G	459 ± 12 (2.08)	1,042 ± 48 (3.12)

Stimulus conditions A–G are respectively the visual search versions of the stimulus conditions A–G in Figure 4.6. For each subject, no significant difference between RT_A and RT_B ($p > 0.05$). Irrelevant bars in C–G increase RT significantly ($p < 0.01$). All subjects as a group, no significant difference between RT_E and RT_G ($p = 0.38$); $RT_C > RT_D$ significantly ($p < 0.02$); RT_C, $RT_D > RT_E$, RT_F, RT_G significantly ($p < 0.01$). Each data entry is: RT ± its standard error (percentage error rate).

4.4 Discussion

In summary, we tested and confirmed several predictions from the hypothesis of a bottom-up saliency map in V1. All these predictions are explicit since they rely on the known V1 mechanisms and an explicit assumption of a MAX rule, SMAP$(x) \propto \max_{x_i=x} O_i$, for example, among all responses O_i to a location x, only the most active V1 cell responding to this location determines its saliency. In particular, the predicted interference by task-irrelevant features and the lack of saliency advantage for orientation-orientation double features are specific to this hypothesis since they arise from the MAX rule. The predictions of color-orientation asymmetry in interference, the violation (in the RT for color-orientation double feature) of a race model between color and orientation features, the increased interference by larger color patches, and the grouping by spatial configurations, stem one way or another from specific V1 mechanisms. Hence, our experiments provided direct behavioral test and support of the hypothesis.

As mentioned in Section 4.3.1, the predicted and observed interference by irrelevant features, particularly those in Figures 4.2 and 4.3, cannot be explained by any background "noise" introduced by the irrelevant features [19,35], since the irrelevant features in our stimuli have a spatially regular configuration and thus would by themselves evoke a spatially uniform or non-noisy response.

The V1 saliency hypothesis does not specify which cortical areas read out the saliency map. A likely candidate is the superior colliculus, which receives input from V1 and directs eye movements [48]. Indeed, microstimulation of V1 makes monkeys saccade to the receptive field location of the stimulated cell [26] and such saccades are believed to be mediated by the superior colliculus.

While our experiments support the V1 saliency hypothesis, the hypothesis itself does not exclude the possibility that other visual areas contribute additionally to the computation of bottom-up saliency. Indeed, the superior colliculus receives inputs also from other visual areas [48]. For instance, Lee et al. [49] showed that pop-out of an item due to its unique lighting direction is associated more with higher neural activities in V2 than those in V1. It is not inconceivable that V1's contribution to bottom-up saliency is mainly for the time duration immediately after exposure to the visual inputs. With a longer latency, especially for inputs when V1 signals alone are too equivocal to select the salient winner within that time duration, it is likely that the contribution from higher visual areas will increase. This is a question that can be answered empirically through additional experiments (e.g., [50]) beyond the scope of this chapter. These contributions from higher visual areas to bottom-up saliency are in addition to the top-down selection mechanisms that further involve mostly higher visual areas [51–53]. The feature-blind nature of the bottom-up V1 selection also does not prevent top-down selection and attentional processing from being feature selective [18,54,55], so that, for example, the texture border in Figure 4.2C could be located through feature scrutiny or recognition rather than saliency.

It is notable that while we assume that our RT data is adequate to test bottom-up saliency mechanisms, our stimuli remained displayed until the subjects responded by button press, for example, for a duration longer than the time nec-

4.4 Discussion

essary for neural signals to propagate to higher-level brain areas and feedback to V1. Although physiological observations [56] indicate that preparation for motor responses contribute a long latency and variations in RTs, our work needs to be followed up in the future to further validate our hopeful assumption that our RT data sufficiently manifests bottom-up saliency to be adequate for our purpose. We argue that to probe the bottom-up processing behaviorally, requiring subjects to respond to a visual stimulus (which stays on until the response) as soon as possible is one of the most suitable methods. We believe that this method would be more suitable than an alternative method to present stimulus briefly, with or, especially, without requiring the subjects to respond as soon as possible. After all, turning off the visual display does not prevent the neural signals evoked by the turned-off display from being propagated to and processed by higher visual areas [57], and, if anything, it reduces the weight of stimulus-driven or bottom-up activities relative to the internal brain activities. Indeed, it is not uncommon for subjects to experience in reaction time tasks that they could not cancel their erroneous responses in time even though the error was realized way before the response completion and at the initiation of the response according to EEG data [58], suggesting that the commands for the responses were issued considerably before the completion of the responses.

Traditionally, there have been other frameworks for visual saliency [18,19,30], mainly motivated by and developed from behavioral data [4, 5] when there was less knowledge of their physiological basis. Focusing on their bottom-up aspect, these frameworks can be paraphrased as follows. Visual inputs are analyzed by separate feature maps, for example, red feature map, green feature map, vertical, horizontal, left tilt, and right tilt feature maps, and so on, in several basic feature dimensions like orientation, color, and motion direction. The activation of each input feature in its feature map decreases roughly with the number of the neighboring input items sharing the same feature. Hence, in an image of a vertical bar among horizontal bars, the vertical bar evokes a higher activation in the vertical feature map than that by each of the many horizontal bars in the horizontal map. The activations in separate feature maps are summed to produce a master saliency map. Accordingly, the vertical bar produces the highest activation at its location in this master map and attracts visual selection. The traditional theories have been subsequently made more explicit and implemented by computer algorithms [31]. When applied to the stimulus in Figure 4.2C, it becomes clear that the traditional theories correspond to the SUM rule $\sum_{x_i=x} O_i$ for saliency determination when different responses O_i to different orientations at the same location x represent responses from different feature maps. As argued, our data in Sections 4.3.1, 4.3.2, and 4.3.4 on interference by task-irrelevant features is incompatible with or unfavorable for the SUM rule, and our data in Section 4.3.3 on the lack of advantage for the double orientation contrast is contrary to the SUM rule. Many of our predictions from the V1 saliency hypothesis, such as the color-orientation asymmetry in Sections 4.3.2 to 4.3.3 and the emergent grouping phenomenon in Section 4.3.4, arise specifically from V1 mechanisms, and could not be predicted by traditional frameworks without adding additional mechanisms or parameters. The traditional framework also contrasted with the V1 saliency hypothesis by implying that the saliency map should be in higher-level cortical areas where neurons are untuned to features, motivating physiological experiments searching for saliency correlates in areas like the lateral intraparietal area [59, 60], which, downstream from V1, could reflect bottom-up saliencies in its neural activities. Nevertheless, the traditional frameworks have provided an overall characterization of previous behavioral data on bottom-up saliency. This behavioral data provided part of the basis on which the V1 theory of saliency was previously developed and tested by computational modeling [20–23].

One may seek alternative explanations for our observations predicted by the V1 saliency hypothesis. For instance, to explain interference in Figure 4.2C, one may assign a new feature type to "two bars crossing each other at 45°," so that each texture element has a feature value (orientation) of this new feature type. Then, each texture region in Figure 4.2C is a checkerboard pattern of two different feature values of this feature type. So the segmentation could be more difficult in Figure 4.2C, just like it could be more difficult to segment a texture of "ABABAB" from another of "CDCDCD" in a stimulus pattern "ABABABABABCDCDCDCDCD" than to segment "AAA" from "CCC" in "AAAAAACCCCCC." This approach of creating new feature types to explain hitherto unexplained data could of course be extended to accommodate other new data. So for instance, new stimuli can easily be made such that new feature types may have to include other double feature conjunctions (e.g., color-orientation conjunction), triple, quadruple, and other multiple feature conjunctions, or even complex stimuli-like faces, and it is not clear how long this list of new feature types needs to be. Meanwhile, the V1 saliency hypothesis is a more parsimonious account since it is sufficient to explain all the data in our experiments without evoking additional free parameters or mechanisms. It was also used to explain visual searches for, for example, a cross among bars or an ellipse among circles without any detectors for crosses or circles/ellipses [20, 23]. Hence, we aim to explain the most data by the fewest necessary assumptions or parameters. Additionally, the V1 saliency hypothesis is a neurally based account. When additional data reveals the limitation of V1 for bottom-up saliency, searches for additional mechanisms for bottom-up saliency can be guided by following the neural basis suggested by the visual pathways and the cortical circuit in the brain [48].

Computationally, bottom-up visual saliency serves to guide visual selection or attention to a spatial location to give further processing of the input at that location. Therefore, by nature of its definition, bottom-up visual saliency is computed before the input objects are identified, recognized, or decoded from the population of V1 neural responses to various primitive features and their combinations. More explicitly, recognition or decoding from V1 responses requires knowing *both* the response levels *and* the preferred features of the responding neurons, while saliency computation requires only the former. Hence, saliency computation is less sophisticated than object identification and, therefore, it can be achieved more quickly. This is consistent with previous observations and arguments that segmenting or knowing "where is the input" precedes or is faster than identifying "what is the input" [45, 46], as well as being more easily impaired and susceptible to noise. On the one hand, the noise susceptibility can be seen as a weakness or a price paid for a faster computation; on the other hand, a more complete computation at the bottom-up selection level would render the subsequent, attentive, processing more redundant. This is particularly relevant when considering whether the MAX rule or the SUM rule, or some other rule (such as a response power summation rule) in between these two extremes, is more suitable for saliency computation. The MAX rule to guide selection can be easily implemented in a fast and feature blind manner, in which a saliency map read-out area (e.g., the superior colliculus) can simply treat the neural responses in V1 as values in a universal currency bidding for visual selection, to select (stochastically or deterministically) the receptive field location of the highest bidding neuron [34]. The SUM rule, or for the same reason the intermediate rule, is much more complicated to implement. The receptive fields of many (V1) neurons covering a given location are typically nonidentically shaped and/or sized, and many are only partially overlapping. It would be nontrivial to compute how to sum the responses from these neurons, whether to sum them linearly or nonlinearly, and whether to sum them with equal or nonequal weights of which values. More importantly, we should realize that these responses should not be assumed as evoked by the same visual object—imagine an image location around a green leaf floating on a golden pond above an underlying dark fish—deciding whether and how to sum the response of a green tuned cell and that of a vertical tuned cell (which could be responding to the water ripple, the leaf, or the fish) would likely require assigning the green feature and the vertical feature to their respective owner objects, for example, to solve the feature binding problem. A good solution to this assignment or summation problem would be close to solving the object identification problem, making the subsequent attentive processing, after selection by saliency, redundant. These computational considerations against the SUM rule are also in line with the finding that statistical properties of natural scenes also favor the MAX rule [61]. While our psychophysical data also favors the MAX over the SUM rule, it is currently difficult to test conclusively whether our data could be better explained by an intermediate rule. This is because, with the saliency map SMAP, reaction times RT = f(SMAP, β) [see (4.3)] depend on decision-making and motor response processes parameterized by β. Let us say that, given V1 responses O, the saliency map is, generalizing from (4.2), SMAP = SMAP(O, γ), where γ is a parameter indicating whether SMAP is made by the MAX rule or its softer version as an intermediate between MAX and SUM. Then, without precise (quantitative) details of O and β, γ cannot be quantitatively determined. Nevertheless, our data in Figure 4.5H favors a MAX rather than an intermediate rule for the following reasons. The response level to each background texture bar in Figure 4.5E–G is roughly the same among the three stimulus conditions, regardless of whether the bar is relevant or irrelevant, since each bar experiences roughly the same level of iso-orientation suppression. Meanwhile, let the relevant and irrelevant responses to the border bars be $O_E(r)$ and $O_E(ir)$, respectively, for Figure 4.5E, and $O_F(r)$ and $O_F(ir)$, respectively, for Figure 4.5F. Then the responses to the two sets of border bars in Figure 4.5G are approximately $O_E(r)$ and $O_F(r)$, ignoring, as an approximation, the effect of increased level of general surround suppression due to an increased level of local neural activities. Since both $O_E(r)$ and $O_F(r)$ are larger than both $O_E(ir)$ and $O_F(ir)$, an intermediate rule (unlike the MAX rule) combining the responses to two border bars would yield a higher saliency for the border in Figure 4.5G than for those in Figure 4.5E and Figure 4.5F, contrary to our data. This argument, however, cannot conclusively reject the intermediate rule, especially one that closely resembles the MAX rule, since our approximation to omit the effect of the change in general surround suppression may not hold.

Due to the difference between the computation for saliency and that for discrimination, it is not possible to predict discrimination performance from visual saliency. In particular, visual saliency computation could not predict subjects' sensitivities in discriminating between two texture regions (or in discriminating the texture border from the background). In our stimuli, the differences between texture elements in different texture regions are far above the discrimination threshold with or without task-irrelevant features. Thus, if instead of a reaction time task, subjects performed texture discrimination without time pressure in their responses, their performance will not be sensitive to the presence of the irrelevant features (even for briefly presented stimuli) since the task essentially probes the visual process for discrimination rather than saliency. Therefore, our experiments to measure reaction time in a visual segmentation or search task, requiring subjects to respond quickly regarding "where" rather than "what" about the visual input by pressing a button located congruently with "where," using trivially discriminable stimuli, are designed to probe bottom-up saliency rather than the subsequent object recognition (identification) or discrimination performance. This design assumes that a higher saliency of the texture border or the search target makes its selection easier and thus faster, manifesting in a shorter RT. This is why our findings in RTs cannot be explained by models of texture discrimination (e.g., [62]), which are based on *discriminating or identifying* texture features, that is, based on visual processing after visual selection by saliency. While our subjects gave different RTs to different stimuli, their response error rates were typically very small ($<5\%$) to all stimuli—as their reaction times were not used to measure discrimination sensitivities. For the same reason, if one were to explain the interference in Figure 4.2C by the noise added by the task-irrelevant features, this feature noise would not be strong enough to affect sufficiently the error rate, since the feature differences (between those of the irrelevant and relevant features) are many times larger than the just-noticeable feature difference for feature discrimination. Of course, some visual search tasks, especially those using hardly discriminable stimuli, rely more on recognition and/or less on bottom-up saliency computation. These tasks, while interesting to study for other purposes, would not be suitable for testing hypotheses on the bottom-up saliency, and we expect that cortical areas beyond V1 would be more involved for them and would have to read out from V1 the preferred features (labeled lines) *and* activities of more *and* less active neurons (i.e., beyond reading out the SMAP).

We also note that, since bottom-up saliency serves mainly to attract attention before detailed attentive processing, its purpose is transient and thus its transient effect is an adequate design for this purpose rather than a weakness. For the same reason, our hypothesis that V1's outputs represent bottom-up saliencies should be viewed as valid mainly in time windows very soon or transiently after new stimulus input. This hypothesis is also consistent with the theoretical framework that early stages along the human visual pathway serve to compress and select visual input data [63].

Our observations are related to Gestalt principles of perceptual organization and many previous observations of visual grouping and emergent properties [64, 65]. This suggests that V1 mechanisms could be the neural basis for many grouping phenomena, as has been shown in some examples [47, 66]. For instance, the main Gestalt principle of grouping by similarity is related to isofeature suppression in V1, since isofeature suppression, responsible for feature singleton pop-out, also makes a region of items of similar features less salient apart from the region border, which bounds, and induces the perception of, the region as a whole. Similarly, the principle of grouping by proximity is related to the finite length of the intracortical connections in V1 for contextual influences, and the principle of grouping by good continuation is related to the colinear facilitation in V1. Pomerantz [64] showed that certain features, particularly ones involving spatial properties such as orientation, interact in complex ways to produce emergent perceptual configurations that are not simply the sum of parts. One of his notable examples of what is termed "configuration superiority effect" is shown in Figure 4.7. One stimulus of a left-tilted bar among three right-tilted bars becomes a composite stimulus of a triangle among three arrows, when a noninformative stimulus of four identical "L" shaped items is added. As a result, the triangle is easier to detect among the arrows than the left-tilted bar among right-tilted ones in the original stimulus, as if the triangle is an emergent new feature. This superiority effect by spatial configurations of bars, the opposite of interference by irrelevant features in our data, could be accounted for by the following mechanism beyond V1. The added irrelevant "Ls" made the target triangle shape unique, while the original target bar was a rotated version of the bar distractors. It was recently shown [67] that, when the bottom-up saliency is not sufficiently high (as manifested in the longer than 1,000 ms RTs in Pomerantz's data, likely due to a small set size), object rotational invariance between target and distractors could introduce object-to-feature interference to drastically prolong RT. This interference is because the original target, identically shaped as the distractors, is confused as a distractor object. Whereas Gestalt principles and many psychological studies of

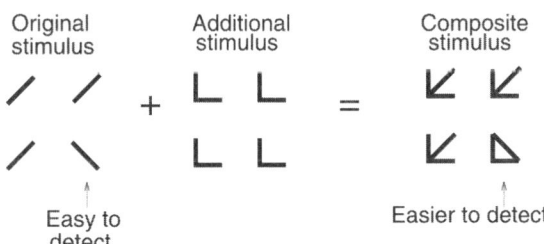

Figure 4.7 Illustration of Pomerantz's configuration superiority effect. The triangle is easier to detect among the three arrow shapes in the composite stimulus, than the left tilted bar among the right tilted bars in the original stimulus. Identical shape of the target and distractor bars in the original stimulus could lead to confusion and longer RTs.

emergent phenomena have provided excellent *summaries and descriptions* of a wealth of data, the V1 mechanisms provide *explanations* behind at least some of this data.

Meanwhile, the psychological data in the literature, including the vast wealth of data on visual grouping, can in turn predict the physiology and anatomy of V1 through the V1 saliency hypothesis, thus providing opportunities to further test the hypothesis through physiological/anatomical experiments. Such tests should help to explore the potentials and the limitations of the V1 mechanisms to explain the bottom-up selection factors. For example, knowing that color-orientation conjunctive search is difficult (e.g., [37]), searching for a red-vertical target among red-horizontal and blue-vertical distractors) and that color-orientation double feature is advantageous allow us to predict that, in V1, intracortical (disynaptic) suppressive connections should link conjunctive cells with other cells preferring *either* the same color *and/or* the same orientation. Data by Hegde and Felleman [28] is consistent with this prediction, although more direct and systematic tests of the prediction are desirable.

The V1 mechanisms for bottom-up saliency also have implications for mechanisms of top-down attention. First, if V1 creates a bottom-up saliency map for visual selection, then it would not be surprising that subsequent cortical areas/stages receiving input from V1 should manifest much interaction between bottom-up and top-down selectional and attentional factors. Second, by the V1 saliency hypothesis, the most active V1 cell attracts attention automatically to its receptive field location. This cell may be tuned to one or a few feature dimensions. Its response does not provide information about other feature dimensions to which it is untuned. Thus, such a bottom-up selection does not bind different features at the same location, and the top-down attention may have to bind the features subsequently [4]. Meanwhile, the conjunctive cells in V1 bind two (or more) features at the same location into a single cell by default (which may or may not be veridical). This suggests that top-down attentional mechanisms are required to determine, from the responses of the conjunctive and nonconjunctive cells, not only the relative strengths of the two features, but also whether the two features belong to the same objects or whether the two features need to be unbound. Our findings reported here should motivate new directions for research into the mechanisms and frameworks of bottom-up and top-down attentional selection, and post-selectional processes for problems including feature binding.

4.5 Conclusions

In conclusion, our psychophysical experiments tested and confirmed the predictions from the theoretical hypothesis that the primary visual cortex creates a bottom-up saliency map. Our findings reported here have since been followed up by more recent identifications of the fingerprints of V1 in visual saliency behavior, pointing to V1 rather than V2, for instance, as responsible for the early or fast component of the bottom up saliency ([68, 69]). Since the V1 hypothesis is a fundamental departure from the traditional framework of visual attention, our findings should motivate new ideas on bottom-up and top-down processes in vision and how they relate to various levels of visual perceptions and visually guided actions.

Acknowledgments

This work was supported in part by the Gatsby Charitable Foundation and by a grant GR/R87642/01 from the UK Research Council. We thank colleagues Neil Burgess, Peter Dayan, Michael Eisele, Nathalie Guyader, Michael Herzog, Alex Lewis, JingLing Li, Christoph Nothdurft, and Jeremy Wolfe for reading the draft versions of the manuscript and/or for very helpful comments. We also thank Stewart Shipp for help with the references.

References

[1] Jonides, J., "Voluntary versus automatic control over the mind's eye's movement," In J. B. Long & A. D. Baddeley (Eds.) *Attention and Performance IX*, 1981, pp. 187-203. Hillsdale, NJ: Lawrence Erlbaum Associates Inc.

[2] Nakayama, K. and Mackeben, M., "Sustained and transient components of focal visual attention." *Visual Research*, Vol. 29, 1989, pp. 1631-1647.

[3] Yantis, S., "Control of visual attention. In *Attention*, p. 223-256. Ed. H. Pashler, *Psychology Press*, 1998.

[4] Treisman, A. M. and Gelade, G., "A feature-integration theory of attention." *Cognitive Psychology*, Vol. 12(1), 1980, pp. 97-136.

[5] Julesz, B., "Textons, the elements of texture perception, and their interactions." *Nature*, Vol. 290, 1981, pp. 91-97.

[6] Allman, J., Miezin, F., and McGuinness, E., "Stimulus specific responses from beyond the classical receptive field: neurophysiological mechanisms for local-global comparisons in visual neurons." *Annual Review of Neuroscience*, Vol. 8, 1985, pp. 407-430.

[7] Knierim, J. J. and Van Essen, D. C., "Neuronal responses to static texture patterns in area V1 of the alert macaque monkey." *Journal of Neurophysiology*, Vol. 67(4), 1992, pp. 961-980.

[8] Sillito, A. M., et al., "Visual cortical mechanisms detecting focal orientation discontinuities." *Nature*, Vol. 378, 1995, pp. 492-496.

[9] Nothdurft, H. C., Gallant, J. L., and Van Essen, D. C., "Response modulation by texture surround in primate area V1: correlates of "popout" under anesthesia." *Visual Neuroscience*, Vol. 16, 1999, pp. 15-34.

[10] Nothdurft, H. C., Gallant, J. L., and Van Essen, D. C., "Response profiles to texture border patterns in area V1." *Visual Neuroscience*, Vol. 17(3), 2000, pp. 421-436.

[11] Jones, H. E., et al., "Surround suppression in primate V1." *Journal Neurophysiology*, Vol. 86(4), 2001, 2011-2028.

[12] Wachtler, T., Sejnowski, T. J., and Albright, T. D., "Representation of color stimuli in awake macaque primary visual cortex." *Neuron*, Vol. 37(4), 2003, pp. 681-691.

[13] Gilbert, C. D. and Wiesel, T. N., "Clustered intrinsic connections in cat visual cortex." *Journal Neuroscience*, Vol. 3(5), 1983, pp. 1116-1133.

[14] Rockland, K. S. and Lund, J. S., "Intrinsic laminar lattice connections in primate visual cortex." *Journal of Comparative Neurology*, Vol. 216(3), 1983, pp. 303-318.

[15] Hirsch, J. A. and Gilbert, C. D., "Synaptic physiology of horizontal connections in the cat's visual cortex." *Journal of Neuroscience*, Vol. 11(6), 1991, pp. 1800-1809.

[16] Reynolds, J. H. and Desimone, R., "Interacting roles of attention and visual salience in V4." *Neuron*, Vol. 37(5), 2003, pp. 853-863.

[17] Beck, D. M. and Kastner, S., "Stimulus context modulates competition in human extrastriate cortex." *Nature Neuroscience*, Vol. 8(8), 2005, pp. 1110-1116.

[18] Wolfe, J. M., Cave, K. R., and Franzel, S. L., "Guided search: an alternative to the feature integration model for visual search." *Journal of Experimental Psychology*, Vol. 15, 1989, pp. 419-433.

[19] Duncan, J. and Humphreys, G. W., "Visual search and stimulus similarity." *Psychological Review*, Vol. 96, 1989, pp. 1-26.

[20] Li, Z., "Contextual influences in V1 as a basis for pop out and asymmetry in visual search." *Proceedings of the National Academy of Science USA*, 1999a, 96(18):10530-10535.

[21] Li, Z., "Visual segmentation by contextual influences via intracortical interactions in primary visual cortex." *Network: Computation in Neural Systems*, Vol. 10(2), 1999b, pp. 187-212.

[22] Li, Z., "Pre-attentive segmentation in the primary visual cortex." *Spatial Vision*, Vol. 13(1), 2000, pp. 25-50.

[23] Li, Z., "A saliency map in primary visual cortex." *Trends in Cognitive Science*, Vol. 6(1), 2002, pp. 9-16.

[24] Nelson, J. I. and Frost, B. J., "Intracortical facilitation among co-oriented, co-axially aligned simple cells in cat striate cortex." *Experimental Brain Research*, Vol. 61(1), 1985, pp. 54-61.

[25] Kapadia, M. K., et al. "Improvement in visual sensitivity by changes in local context: parallel studies in human observers and in V1 of alert monkeys." *Neuron*, 15(4):843-856 (1995).

[26] Tehovnik, E. J., Slocum, W. M., and Schiller, P. H. "Saccadic eye movements evoked by microstimulation of striate cortex." *European Journal of Neuroscience*, Vol. 17(4), 2003, pp. 870-878.

[27] Super, H., Spekreijse, H., and Lamme, V.A., "Figure-ground activity in primary visual cortex (V1) of the monkey matches the speed of behavioral response." *Neuroscience Letters*, Vol. 344(2), 2003, pp. 75-78.

[28] Hegde, J. and Felleman, D. J., "How selective are V1 cells for pop-out stimuli?" *Journal of Neuroscience*, Vol. 23(31), 2003, pp. 9968-9980.

[29] Zhaoping, L. and Snowden, R. J., "A theory of a saliency map in primary visual cortex (V1) tested by psychophysics of color-orientation interference in texture segmentation." *Visual Cognition*, Vol. 14(4/5/6/7/8), 2006, pp. 911-933.

[30] Koch, C. and Ullman, S., "Shifts in selective visual attention: towards the underlying neural circuitry." *Human Neurobiology*, Vol. 4(4), 1985, pp. 219-227.

[31] Itti, L. and Koch, C., "A saliency-based search mechanism for overt and covert shifts of visual attention." *Vision Research*, Vol. 40(10-12), 2000, pp. 1489-1506.

[32] Hubel, D. H. and Wiesel, T. N. "Receptive fields and functional architecture of monkey striate cortex." *Journal of Physiology*, Vol. 195(1), 1968, pp. 215-243.

[33] Livingstone, M. S. and Hubel, D. H., "Anatomy and physiology of a color system in the primate visual cortex." *Journal of Neuroscience*, Vol. 4(1), 1984, pp. 309-356.

[34] Zhaoping, L., "The primary visual cortex creates a bottom-up saliency map." In *Neurobiology of Attention*, p. 570-575. Eds. L. Itti, G. Rees, and J. K. Tsotsos, Elsevier 2005.

[35] Rubenstein, B. S. and Sagi, D. "Spatial variability as a limiting factor in texture-discrimination tasks: implications for performance asymmetries." *Journal of the Optical Society of America A*, Vol. 7(9), 1990, pp. 1632-1643.

[36] Foster, D. H. and Ward, P. A., "Asymmetries in oriented-line detection indicate two orthogonal in early vision." *Proceedings of Royal Society: Biological Sciences*, Vol. 243(1306), 1991, pp. 75-81.

[37] Wolfe, J. M., "Visual search, a review." In *Attention*, p. 13-74. Ed. H. Pashler. Hove, East Sussex, UK, Psychology Press Ltd. (1998).

[38] Jones, H. E., Wang, W., and Sillito, A. M., "Spatial organization and magnitude of orientation contrast interactions in primate V1." *Journal of Neurophysiology*, Vol. 88(5), 2002, pp. 2796-2808.

[39] Deyoe, E. A., Trusk, T. C., and Wong-Riley, M. T., "Activity correlates of cytochrome oxidase-defined compartments in granular and supergranular layers of primary visual cortex of the macaque monkey." *Visual Neuroscience*, Vol. 12(4), 1995, pp. 629-639.

[40] Li, Z. and Atick J., "Towards a theory of striate cortex." *Neural Computation*, Vol. 6, 1994, pp. 127-146.

[41] Snowden, R. J., "Texture segregation and visual search: a comparison of the effects of random variations along irrelevant dimensions." *Journal Experimental Psychology: Human Perception and Performance*, Vol. 24, 1998, 1354-1367.

[42] Nothdurft, H. C., "Salience from feature contrast: additivity across dimensions." *Vision Research*, Vol. 40, 2000, pp. 1183-1201.

[43] Krummenacher, J., Muller, H. J., and Heller, D., "Visual search for dimensionally redundant pop-out targets: evidence for parallel-coactive processing of dimensions." *Perception and Psychophysphics*, Vol. 63(5), 2001, pp. 901-917.

[44] Wolfson, S. S. and Landy, M. S., "Discrimination of orientation-defined texture edges." *Vision Research*, Vol. 35(20), 1995, pp. 2863-2877.

[45] Sagi, D. and Julesz, B., " 'Where' and 'what' in vision." *Science*, Vol. 228(4704), 1985, pp. 1217-1219.

[46] Li, Z., "Visual segmentation without classification: a proposed function for primary visual cortex." *Perception*, 27, ECVP Abstract Supplement. (In Proceedings of the European Conference on Visual Perception; 24-28 August 1998; Oxford, United Kingdom.) http://www.perceptionweb.com/abstract.cgi?id=v980337

[47] Zhaoping, L., "V1 mechanisms and some figure-ground and border effects." *Journal of Physiology, Paris*, Vol. 97, 2003, pp. 503-515.

[48] Shipp, S., "The brain circuitry of attention." *Trends in Cognitive Science*, Vol. 8(5), 2004, pp. 223-230.

[49] Lee, T. S., et al., "Neural activity in early visual cortex reflects behavioral experience and higher-order perceptual saliency." *Nature Neuroscience*, Vol. 5(6), 2002, pp. 589-597.

[50] Guyader, N., et al., "Investigation of the relative contribution of 3-D and 2-D image cues in texture segmentation." (ECVP 2005 Abstract.) *Perception*, Vol. 34, 2005, pp. 55-55 Suppl. S.

[51] Tsotsos, J. K., "Analyzing vision at the complexity level." *Behavioral and Brain Sciences*, Vol. 13-3, 1990, pp. 423-445.

[52] Desimone, R. and Duncan J., "Neural mechanisms of selective visual attention." *Annual Review of Neuroscience*, Vol. 18, 1995, pp, 193-222.

[53] Yantis, S. and Serences, J. T., "Cortical mechanisms of space-based and object-based attentional control." *Current Opinions in Neurobiology*, Vol. 13(2), 2003, pp. 187-193.

[54] Treue, S. and Martinez-Trujillo, J. C., "Feature-based attention influences motion processing gain in macaque visual cortex." *Nature*, Vol. 399, 1999, pp. 575-579.

[55] Chelazzi, L., et al., "A neural basis for visual search in inferior temporal cortex." *Nature*, Vol. 363(6427), 1993, pp. 345-347.

[56] Thompson, K. G., et al., "Perceptual and motor processing stages identified in the activity

[57] Smithson, H. E. and Mollon, J. D. "Do masks terminate the icon?" *Quarterly Journal of Experimental Psychology*, Vol. 59(1), 2006, pp. 150-160.

[58] Gehring, W. J., et al., "A neural system for error detection and compensation." *Psychological Science*, Vol. 4(6), 1993, pp. 385-390.

[59] Gottlieb, J. P., Kusunoki, M., and Goldberg, M. E., "The representation of visual salience in monkey parietal cortex." *Nature*, Vol. 391(6666), 1998, pp. 481-484.

[60] Bisley, J. W. and Goldberg, M. E., "Neuronal activity in the lateral intraparietal area and spatial attention." *Science*, Vol. 299(5603), 2003, pp. 81-86.

[61] Lewis, A. and Zhaoping, L., "Saliency from natural scene statistics." Program No. 821.11, 2005 Abstract Viewer/Itineary planner. Washington, DC, Society for Neuroscience 2005. Online.

[62] Landy, M. S. and Bergen, J. R., "Texture segregation and orientation gradient." *Vision Research*, Vol. 31(4), 1991, pp. 679-691.

[63] Zhaoping, L., "Theoretical understanding of the early visual processes by data compression and data selection." *Network: Computation in Neural Systems*, Vol. 17(4), 2006, pp. 301-334.

[64] Pomerantz, J. R., "Perceptual organization in information processing." In *Perceptual Organization*, pp. 141-180. Eds. M. Kubovy and J. Pomerantz Hillsdale NJ, Erlbaum (1981).

[65] Herzog, M. H. and Fahle, M., "Effects of grouping in contextual modulation." *Nature*, Vol. 415, 2002, pp. 433-436.

[66] Herzog, M. H., et al., "Local interactions in neural networks explain global effects in the masking of visual stimuli." *Neural Computation*, Vol. 15(9), 2003, pp. 2091-2113.

[67] Zhaoping, L. and Guyader, N., "Interference with bottom-up feature detection by higher-level object recognition." *Current Biology*, Vol. 17, 2007, pp. 26-31.

[68] Koene, A. R. and Zhaoping, L., "Feature-specific interactions in salience from combined feature contrasts: evidence for a bottom-up saliency map in V1." *Journal of Vision*, 7(7):6, 1-14, http://journalofvision.org/7/7/6/, doi:10.1167/7.7.6.

[69] Zhaoping, L., "Popout by unique eye of origin: a fingerprint of the role of primary visual cortex in bottom-up saliency." Abstract 717.8, *Annual Meeting of Society for Neuroscience*, San Diego, California, Nov. 3-7, 2007.

PART II
The Mathematics of Vision

CHAPTER 5
V1 Wavelet Models and Visual Inference

Anil Anthony Bharath and Jeffrey Ng

5.1 Introduction

The receptive fields of the cells of the visual system show distinct spatial patterns of excitatory and inhibitory responses. Mathematical functions used to describe the spatial components of receptive fields may vary considerably. However, the class of functions termed wavelets might be argued as being quite well suited to model, for practical purposes, spatial receptive fields: first, wavelets constitute a sufficiently rich class of functions for modelling a variety of spatial receptive fields. Second, there is a class of linear coordinate transformations associated with wavelets that is well suited to both analytical studies and practical implementations. Third, there are well-defined numeric algorithms for implementing representations based on wavelets that are known to provide complete image decompositions of an image (visual data) and which may be readily decoded to recover the original image field. For the study of receptive fields as they relate to encoding visual data, this seems an ideal combination of desirable properties.

5.1.1 Wavelets

A *wavelet* is a type of mathematical function with certain properties. These properties are often quite intuitive. For example, for practical applications, a wavelet should be constrained to have a finite amount of energy. Thus, if we represent a one-dimensional wavelet by $\psi(x)$, we can say that

$$\int_{-\infty}^{\infty} |\psi(x)|^2 dx < \infty \tag{5.1}$$

where x might represent space, or time. It implies that while the function on the left of Figure 5.1 is a wavelet, the function on the the right is not! It is common to also specify another condition on the function ψ, which is that $\psi(x)$ must also be absolutely integrable:

$$\int_{-\infty}^{\infty} |\psi(x)| dx < \infty \tag{5.2}$$

A further requirement is also in terms of an integral, and it is expressed as follows:

$$\int_{-\infty}^{\infty} \psi(x) dx = 0 \tag{5.3}$$

This requirement is more subtle: it comes into play when the wavelet is used as part of a continuous wavelet transform, in which case it allows the specification of

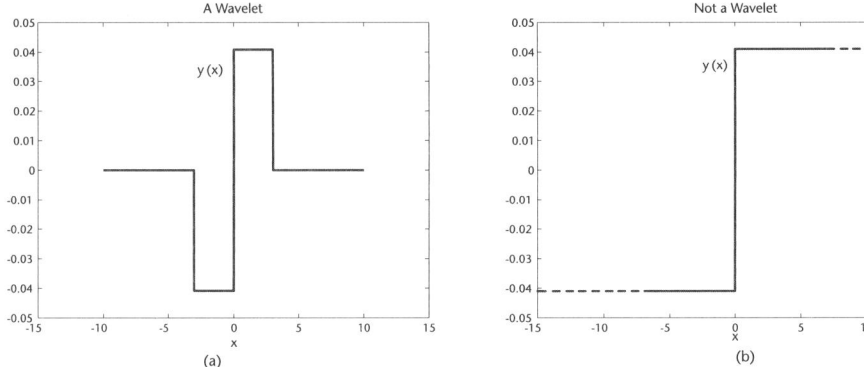

Figure 5.1 Properties of a wavelet: the figure on the right is not a wavelet, because despite having an apparent average of zero, it does not have finite energy. It also does not satisfy the admissibility condition of (5.2).

an *admissibility condition* [1], whereby the transform might be exactly invertible. An equivalent statement may be made in the Fourier domain, which is that

$$\Psi(\omega)|_{\omega=0} = 0 \tag{5.4}$$

where

$$\Psi(\omega) = \int_{-\infty}^{\infty} \psi(x) e^{-j\omega x} dx \tag{5.5}$$

is the Fourier transform of $\psi(x)$, and where $j = \sqrt{-1}$. This implies that while the function on the left of Figure 5.2 is a wavelet, that shown on the right is not. The finite-energy property (5.1) and "zero mean" (5.3) admissibility condition together suggest that wavelets have a localization ability in space. Furthermore, if a linear and spatially invariant system has an impulse response meeting these two requirements, the output of the system contains information only about changes

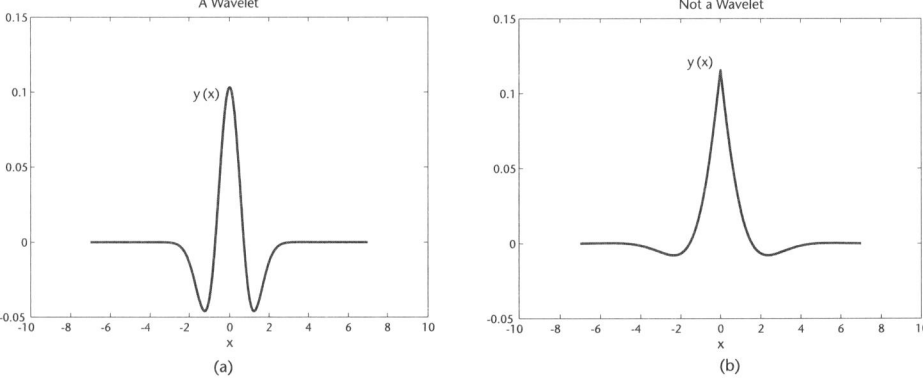

Figure 5.2 Properties of a wavelet: the figure on the right is not a wavelet, because despite having quickly decaying amplitude, its positive area strongly outweighs its negative area.

in the system's input, that is, the system yields no response to an input that does not change with x, and any such responses to change are localized in x.

5.1.2 Wavelets in Image Analysis and Vision

There are several methodologies of "classical" image processing that bear close parallels to the mechanisms of visual processing found in biology. For example, the scheme of Burt and Adelson [2], which triggered much progress in wavelet-based image compression, bears a striking resemblance both to Laplacian of Gaussian [3] and, more obviously, difference of Gaussian functions [4,5], used as models of retinal spatial receptive fields. Indeed, scale-space representations may be traced back to the biologically inspired work of Marr and Hildreth [6] on edge detection, and hence to the work of Witkin [7], Koenderink [8], and Lindeberg [9], culminating in the relatively recent, very successful scale-invariant feature transform (SIFT) method of keypoint detection [10].

Accepted methodologies that are in current use, and which bear close resemblance to the spatial computations performed by V1, include multirate filter-banks [11], which evolved primarily to meet the needs of image compression [12], and of event detection and characterization [13]. Requirements of scale- and orientation-invariant measures in computer vision also lend justification to scale-space approaches.

As mentioned in Chapter 3, Gabor functions remain the most widely used mathematical descriptions of spatial receptive field patterns. Gabor functions (see Figure 5.3 and 5.4 for examples of 2D Gabor functions) have the distinct advantage, from a modelling point of view, of requiring only a small number of intuitive parameters. Initially proposed as a phase-invariant way of representing, and simultaneously localizing, a signal in time-frequency space [14], Gabor functions have found wide applications. They continue to be used not only in speech and image processing, but also in visual neuroscience, where two-dimensional versions were constructed as very successful approximations of the cortical simple-cell receptive fields by Marcelja [15]. Subsequently, Daugman popularized their complex form [16] and very successfully applied them to characterizing the variegated patterns of the human iris for identity recognition [17]. Spatial Gabor functions, often with phases intermediate to cosine and sine oscillatory forms, are widely used in visual psychophysics, where they are termed Gabor *patches*. Note that alternative two-dimensional complex wavelet definitions that are arguably functionally better suited to visual processing [18] were first suggested by Granlund [19]. Granlund's work also included the earliest applications to texture analysis [20] of such V1-inspired methods.

5.1.3 Wavelet Choices

Despite the optimization of space, frequency, and orientation localization of two-dimensional Gabor functions [16], there are some computational issues in their efficient application to areas such as image compression. One of these is lack of easily specified orthogonality. Another issue is the potential instability in reconstructing an image perfectly from projections onto Gabor functions sampled on

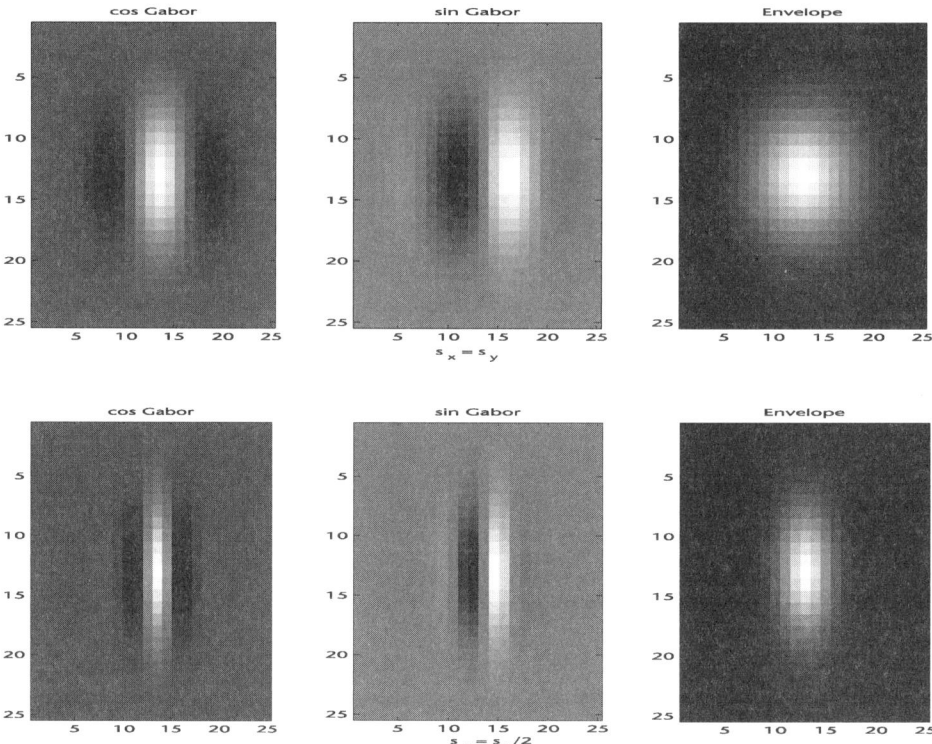

Figure 5.3 Gabor patches. (Top) A Gabor patch is a two-dimensional Gaussian function (*envelope*) modulated by a sine or cosine function of space. (Bottom) Illustrating the effect of changing the ratio between the σ parameters of the two-dimensional Gaussian in each Cartesian direction.

regular grids [21]. Finally, despite optimizing a space-frequency localization product, one-dimensional Gabor envelopes do not have strictly compact support in either the time or frequency space.[1] Daubechies, motivated by these problems [22] and inspired by a seminal paper by Mallat [23], which linked wavelet decompositions to both multiresolution spaces and to efficient filtering structures based on successive decimation, succeeded in constructing orthogonal wavelets of quite compact support [21]. Daubechies' initial wavelet constructions, together with spline wavelets, Symmlets, and so on, have been very successful in the domain of image compression.

Multirate filter bank structures are very efficient systems for performing orthogonal image decompositions, because of the strong underlying theoretical framework inherited from the field of digital filtering. However, despite early and promising successes in singularity detection and characterization [13], a problem arises when attempting to use such decompositions for more general visual pattern recognition, where the shifting of coefficient power between sub-bands presents practical complications. A property that is "missing" in maximally decimated filter

---

1. Indeed, because the Fourier transform, $\Psi(\omega_x, \omega_y)$, of a spatial Gabor function has a Gaussian envelope, it cannot be a wavelet, as $\Psi(0,0) \neq 0$.

5.1 Introduction

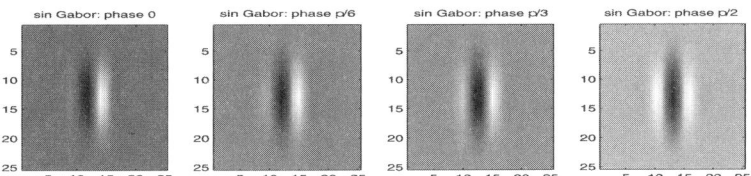

Figure 5.4 Gabor patches. Illustrating the effect of changing the *phase* of a Gabor function.

bank structures, as typified by dyadic decompositions employing a single mother wavelet, is therefore *shift invariance*.

Following Granlund's approach of employing locally steered one-dimensional Hilbert transforms, Simoncelli [24] addressed these shortcomings by suggesting *shiftable* multiscale transforms, whereby the aim of maintaining low redundancy is relaxed in favor of achieving greater coefficient stability in the sub-bands. Freeman and Adelson [25] also formalized and popularized the notion of steerability, closely tied to the principle of overcompleteness, as a means of synthesizing intermediate orientation channels from a set of fixed two- and three-dimensional filters. These ideas were further refined by Perona [26] to include more general kernels for early image processing, although many of these latter kernels are not known to have direct analogies among the receptive fields found in biological vision.

An important distinction between the multirate filtering approaches placed on a firm theoretical footing by Daubechies and Mallat, and their extension to two-dimensional, is that the number of highpass sub-bands is usually quite low, often 3 for each scale of decomposition, and has a correspondingly low angular selectivity. This is despite biological evidence, as described in Chapter 3, on the angular tuning curves of V1, and even, as pointed out by Perona [26], despite the growing body of evidence suggesting that many of the most promising methods for edge detection [27], stereo algorithms, and even compression at that time were using increased numbers of angular channels at each scale. For V1-like responses, Freeman's construction paradigm [25] partly alleviated this, allowing for a general

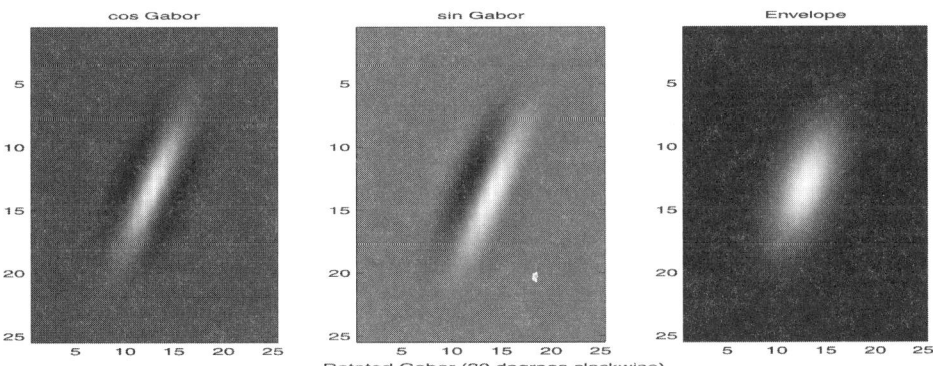

Figure 5.5 Gabor patches. Components of a complex, rotated Gabor pair. The components remain in spatial *quadrature* (see Section 5.1.4).

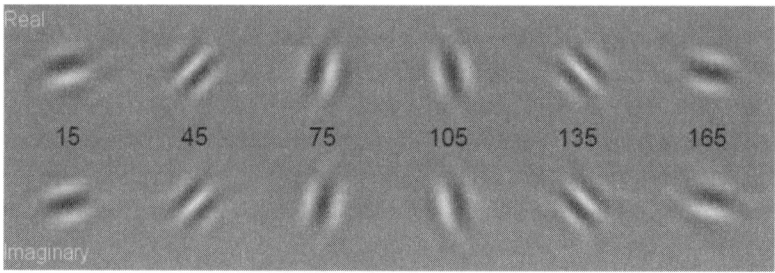

Figure 5.6 Example impulse responses from the original construction of Kingsbury's two-dimensional complex dual-tree wavelets.

nonseparable design where angular selectivity (bandwidth) may be specified easily and explicitly, because of the use of a Fourier-domain, polar-separable design methodology.

An alternative construction that achieves low overcompleteness (and therefore is efficient), yet also displays approximate shift-invariance and relatively high angular selectivity is the dual-tree complex wavelet transform (DTCWT) developed by Kingsbury [28]. Examples of the impulse responses of this decomposition are illustrated in Figure 5.6. The complex wavelet transform resulting from these responses has found applications in motion estimation, interpolation, denoising, and keypoint detection [29].

One practical issue with the original DTCWT transform, of particular relevance to detecting and characterizing visual structure, is that the high-pass channels do not lend themselves to easy orientation steering (i.e., altering only the direction tuning of the filtered response using outputs of filters at only a single scale). A recent modification of the dual-tree kernels [30] yields near rotation invariance and, consequently, orientation steerability within any one scale (see Figure 5.7). This makes the patterns of coefficients very stable under rotations, yet keeps the redundancy less than that of the fully shiftable, rotationally steerable wavelet decomposition containing undecimated band-pass channels, such as Freeman and Adelson's. Alternative constructions and the importance of the half-sample prin-

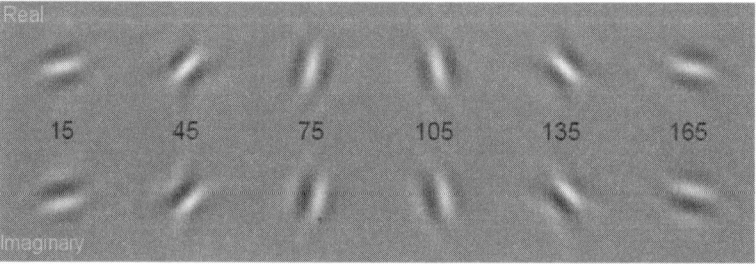

Figure 5.7 Example impulse responses of Kingsbury's modified complex dual-tree wavelets. Note that as direction is altered, the number of cycles per envelope remains approximately constant, which is directly associated with the property of approximate rotation invariance.

ciple introduced by dual-tree filtering structures may be found in the work of Selesnick; see, for example, [31]. Yet another construction is described by Bharath and Ng [32], and included in Section 5.2 as a design that has been successfully implemented on an FPGA platform (see Chapter 13) and on a programmable, high-performance graphics card. This design uses a radial frequency function that is known to yield good event characterization in discretized signals [33].

High-performance image denoising algorithms have been designed with the use of wavelet bandpass channels of very high angular selectivity, employing as many as 12 orientation channels in order to maintain steerability [34].

Another interesting demonstration of denoising through the use of the adaptive properties of a steerable wavelet decomposition was suggested by Bharath and Ng [32]. These authors proposed a structural-based denoising approach, in which the feature extraction properties of a complex wavelet pyramid were applied to adapt the local image "bases" through changing the gain of different directional filters as a function of position, a technique that is quite biologically plausible, and may be the computational analogy of dendritic gain control.

While it is debatable whether image denoising per se has many useful applications, it is certainly the case that the use of an appropriate wavelet domain for simultaneous denoising and feature detection does have certain advantages: it is more likely that the main application of denoising will be to lend robustness to feature detection in computer vision, where relatively small amounts of sensor noise might otherwise lead to corruption of image primitives. Indeed, Fauqueur and Kingsbury proposed precisely this approach in performing keypoint detection [29].

Ng and Bharath have also shown that color opponent wavelet decompositions based on complex, steerable wavelets permit the detection of perceptual boundaries [35], that is, both edges and line-line structures, even at single scales.

Parallels to classical V1 computation are also to be found in scale-space approaches, in which the original input image is embedded at the finest scale of a representation space. First and pure second-order partial derivatives in a Cartesian coordinate system at one scale are similar to symmetric and antisymmetric receptive fields of V1 neurons. However, it is quite common in scale-space approaches to use isotropic Gaussian kernels as the scale-space generator. This leads to low-order derivative kernels that lack strong directional selectivity. Higher-order kernels may fix this problem, but require a careful choice of scales and preferably the use of anisotropic envelopes to avoid excessive, biologically unrealistic spatial oscillations.

The scale-space image representation also lends itself elegantly to defining partial derivatives in *scale*, which can be used to, for example, determine optimal scales for processing an image, by using the general principle that the scale and spatial locations at which one finds a maximum magnitude of spatial derivative response provide an indication of the "optimal" scale for processing: see Seo and Yoo's [36] superb illustration of this principle. See also [37] and [38] for a practical application in blood vessel segmentation in retinal images.

5.1.4 Linear vs Nonlinear Mappings

The primary spatial operators of both the scale-space and filter-bank paradigms are *linear* and these bear close relevance indeed to the spatial properties of many classical V1 receptive fields. However, the majority of V1 receptive fields is actually of a complex nature, by which is meant they usually display spatially phase-invariant properties [39], or other, possibly unspecified nonlinearities. Interestingly, there is a generic model for how such properties arise in biological vision, which is efficiently mirrored by signal processing. For example, in order to "achieve" phase-invariant responses at one angle of orientation, one may use the model shown in Figure 5.8 (after [40]).

In order to generate a phase and orientation invariant measure of the edge likelihood, one may, thus, "pool" responses over all phases and orientations. An interesting generalization might be suggested, which is that in order to generate responses that are invariant with respect to some parameter, one combines the responses of receptive fields that have variations in that parameter via nonlinearities. This, indeed, has analogies to, say, gradient magnitude estimation used in image processing, whereby the responses of filters in different directions are combined to yield an orientation *invariant* response.

Returning to the specific example of receptive field phase, note that if two phases of receptive fields of the same orientation are in exact quadrature, the number of linear receptive fields feeding into the nonlinearity that is required to generate a phase-invariant response reduces to two, which are the so-called in-phase and

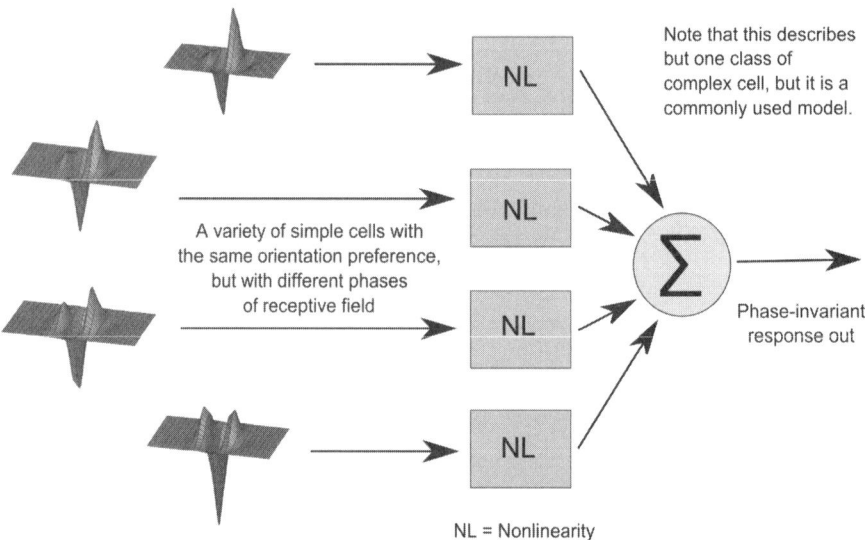

Figure 5.8 Generalized use of nonlinear combination of spatially coincident receptive fields which vary in their parameters in order to construct invariant feature measures. Note that the number of linear filters depends on parameter characteristics and the sharpness of their tuning. For example, to achieve phase-invariant responses from phase-quadrature receptive fields, this reduces to two linear operators per direction.

quadrature components associated with quadrature detections, and with complex Gabor functions (see Figure 5.5). In Section 5.4, we shall present a statistical analysis which suggests not only the concept of pooling filter responses over directions, but also a nonlinear normalization that is known in neuroscience as "divisive" normalization. This normalization suggests an alternative to standard feature-vector normalization, used in pattern recognition and computer vision for matching algorithms, and in Section 5.3, we show this to be effective even when applied to simulations of simple-cell responses for discriminating between planar shapes in noise.

5.2 A Polar Separable Complex Wavelet Design

The retinotopic mapping of V1 suggests that similar receptive field properties are present at different spatial locations in the visual cortex. Furthermore, eye movement will constantly alter the spatial mapping of the visual field onto cortical location. Thus, filtering with Gabor-like patches of a variety of scales, orientations and phases appear a reasonable computational analogy to V1 processing. Because this is computationally demanding, some simplifications must currently be pursued for real-time performance. *Multirate* schemes, employing pyramidal filtering structures, provide efficiency in both computation and storage requirements of any overcomplete representation that might be used to simulate the processes of V1. Furthermore, creating a wavelet design which is steerable, such that, for example, intermediate orientations of linear spatial receptive fields may be synthesized by weighted combination of the responses of fixed directional filters/receptive fields, lends further efficiencies.

For concreteness, the next section discusses the design of a particular steerable wavelet system [32] which has been applied to a number of problems in image analysis and computer vision.

5.2.1 Design Overview

To simulate several receptive field scales, the design of the system uses a dual channel, hierarchical approach, employing *decimation* in the lowpass channel in order to achieve a scaling of filter responses through successive levels of a pyramidal structure. The nature of the scheme is illustrated in Figure 5.9. For convenience in tuning the angular and radial characteristics of the filters, Fourier domain polar separability may be imposed, so that an analysis filter $G_{0,k}(\omega,\phi)$ in the kth direction in a filter set can be specified as the product of a radial frequency function $\Omega_0(\omega)$, and an angular frequency function $\Phi_{0,k}(\phi)$, that is, $G_{0,k}(\omega,\phi) = \Omega_0(\omega)\Phi_{0,k}(\phi)$. The matching synthesis filter, $G_{1,k}(\omega,\phi)$, can be specified as the product of the same angular frequency function $\Phi_{0,k}(\phi)$, and a synthesis radial frequency function $\Omega_1(\omega)$. For more details on the specification of wavelet filter-bank design, including extensive descriptions of the terminologies used in filter-bank design, the reader is directed to the excellent book by Vaidyanathan [11].

5.2.2 Filter Designs: Radial Frequency

For the isotropic lowpass radial frequency response, several options are available. We have used the following function in the radial frequency domain on the interval $-\pi < \omega \leq \pi$:

$$H_0(\omega, \phi) = H_0(\omega) = \frac{1}{1 + (\omega/\omega_c)^6} \quad (5.6)$$

where $\omega_c = 3\pi/8$. This is equivalent to a cascade of two 3rd order Butterworth filters. It was chosen to provide a reasonably flat power response, when used in combination with the bandpass radial frequency response, defined later, for radial frequency components in the range $[0, \omega_{\max}]$. The value $\omega_{\max}$ represents the peak frequency of the bandpass radial frequency response.

To address the construction of the bandpass analysis channels, the angular characteristics of the filter in the Fourier domain are assumed to satisfy the following:

$$\sum_{k=0}^{K-1} \Phi_{0,k}(\phi) \Phi_{1,k}(\phi) = C_\phi \quad (5.7)$$

where $k = 0, 1, ..., K - 1$ indexes a set of directions for the desired filter set. One can replace the K bandpass filters of Figure 5.9 with a single isotropic bandpass filter of radial frequency response $\Omega_0(\omega)$, which incorporates the angular gain factor, C_ϕ. This constraint allows the radial frequency characteristics of the filters to be tuned to address the requirements of a multirate system, since the structure of Figure 5.9 may be reduced to that shown in Figure 5.10.

With reference to the filtering and decimation scheme illustrated in Figure 5.10, the output, $Y(\omega)$, of an isotropic spatial filter is given by:

$$Y(\omega) = \frac{H_1(\omega)}{2} \{X(\omega)H_0(\omega) + X(\omega + \pi)H_0(\omega + \pi)\}$$
$$+ X(\omega)\Omega_0(\omega)\Omega_1(\omega) \quad (5.8)$$

To avoid aliasing, the constraints on design are different from a standard two-channel decomposition specifying full decimation. Practically, we require that the energy of the term $X(\omega + \pi)H_0(\omega + \pi)H_1(\omega)$ should be as small as possible. This can be ensured by employing the analysis (H_0) and synthesis (H_1) lowpass filters to have a cutoff near $\pi/2$.

For most natural scenery, the magnitude of the frequency components drops off rapidly with increasing radial spatial frequency, ω (see, for example, [41]). With this choice of lowpass characteristic, therefore, the aliasing is minimal, except in images containing an unusually high proportion of high-frequency components. One-dimensional filters with radial frequency responses, $\Omega_0(\omega)$, given by

$$\Omega_0(\omega) = \left(\frac{e}{14}\right)^7 \omega^7 e^{-\omega/2} U(\omega) \quad (5.9)$$

where $U(\omega)$ is the unit step function, were previously suggested for feature detection [33], and series approximations to associated two-dimensional filters were

5.2 A Polar Separable Complex Wavelet Design

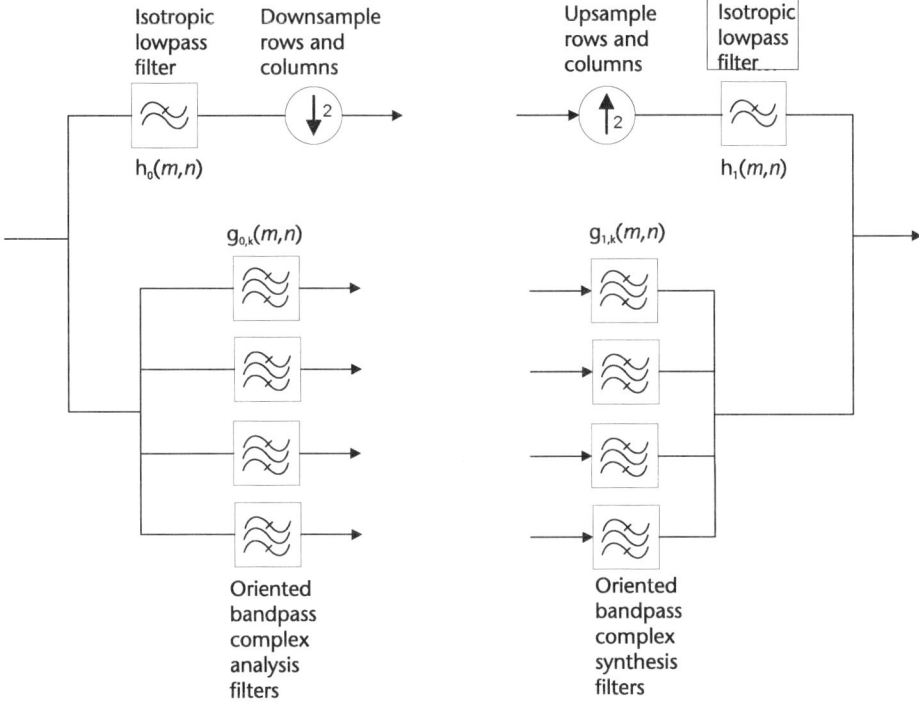

Figure 5.9 Basic pyramidal unit, illustrating a single stage of decomposition and reconstruction. After [32].

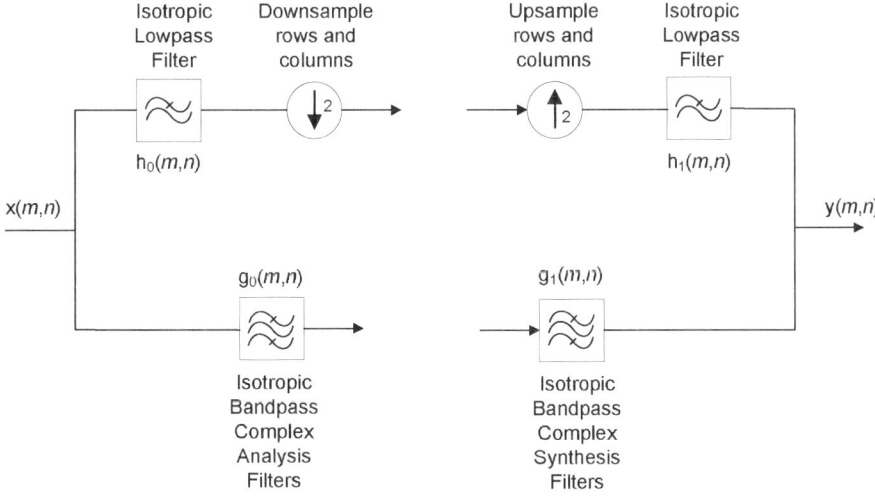

Figure 5.10 Representation of isotropic nature of pyramid by combining orientation selective channels, illustrating a single stage of decomposition and reconstruction. After [32].

investigated in [42]. These functions, known as *Erlang* functions have also approximately the shape of the warped functions suggested for scale steering in [24]. Recently, functions with similar radial frequency characteristics have been employed to generate the bandpass kernels of so-called Poisson α-scale spaces [43], for exploring structures in "deep" scale-space. Analyzing features in scale-space involves the comparison of filter responses across scales, which requires a stability of filter responses across the scale parameter. The family of α-scale spaces provides a parameterization of this scale stability and is based on lowpass smoothing kernels with a continuous transition from the identity operator, for $\alpha = 0$, to the Gaussian kernel, for $\alpha = 1$. The joint localization of Gaussian kernels in both the spatial and frequency domain causes transform coefficients to fall off more rapidly in magnitude as scale is increased [44]. Using an α value such that $\alpha < 1$ (for example, Poisson kernels, with $\alpha = 0.5$) biases the localization toward the frequency domain and provides an increased stability of transform coefficients across scales.

5.2.3 Angular Frequency Response

The angular characteristics of steerable wavelets are determined by several requirements. First, achieving near perfect reconstruction (PR condition) in the angular decomposition is desirable. Equation (5.7) specifies the condition that is sought if using real bandpass filters. To create quadrature complex filters, one may, instead, use a function that is zero over one-half of two-dimensional Fourier space. In the spatial domain, this yields complex filter kernels consisting of symmetric and antisymmetric spatial impulse responses, similar to the sine and cosine Gabor functions mentioned in Section 5.1.

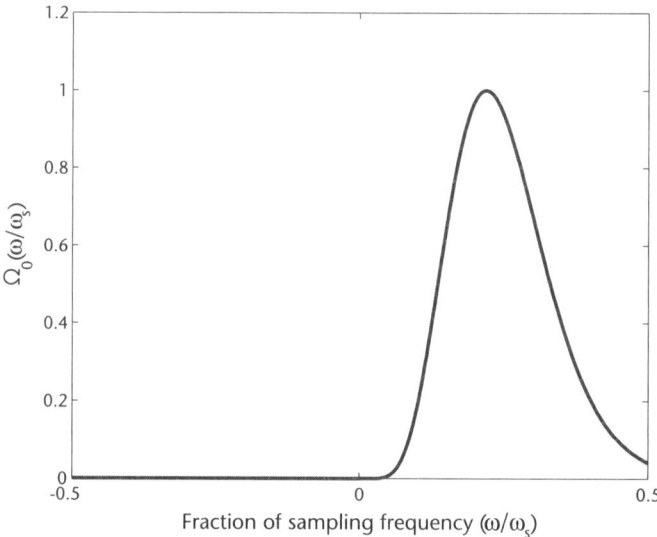

Figure 5.11 The radial frequency specification: the Erlang function $\Omega_0(\omega)$, retraced from [32].

5.2 A Polar Separable Complex Wavelet Design

The angular frequency characteristic may be selected to be a function such as

$$\Phi_0(\phi) = \cos^3(\phi)\text{rect}(\phi/\pi) \tag{5.10}$$

where $\text{rect}(\phi) = U(\phi + \frac{1}{2})U(\frac{1}{2} - \phi)$ and $U(.)$ is, again, the unit step function.

The prototype angular frequency response is rotated to generate the angular characteristics of oriented filters for a full set of directional filters by the following:

$$\Phi_{0,k}(\phi) = \Phi_0(\phi - \phi_k) \tag{5.11}$$

An illustration of the design for $K = 8$ complex filters evenly distributed on $[-\pi, \pi]$, with orientations at $0, \pi/4, \pi/2, 3\pi/4, \pm\pi, -3\pi/4, -\pi/2$, and $-\pi/4$, is shown in Figure 5.12. With this choice, filters specified by $\Phi_{0,k}$ that are π radians apart in orientation in the Fourier domain are complex conjugates of each other in the spatial domain. The filters for $k = K/2, K/2 + 1, \ldots, K - 1$ can be obtained by taking the complex conjugates of filters with $k = 0, 1, \ldots, K/2 - 1$. Therefore, the number of unique filter kernels necessary to implement this choice of a $K = 8$ filter set is only 4. Accordingly, Equation (5.7) should be modified to the following:

$$\sum_{k=0}^{K/2-1} \Phi_{0,k}(\phi)\Phi_{1,k}(\phi) + \Phi_{0,k}(\phi + \pi)\Phi_{1,k}(\phi + \pi) = C_\phi \tag{5.12}$$

To obtain near-perfect reconstruction across the angular selective bandpass filter channels, it is useful to have C_ϕ as close as possible to a constant value. This

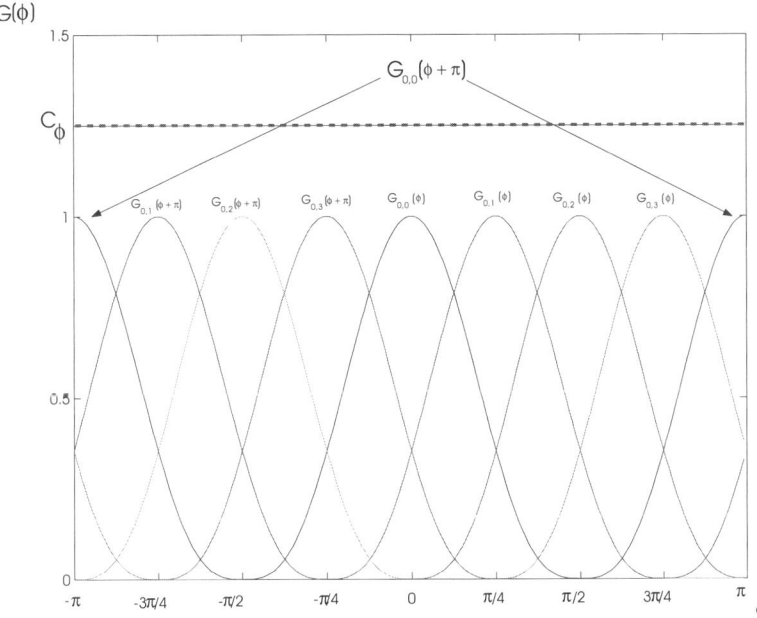

Figure 5.12 Angular frequency response of an 8 directional filter set using $\cos^3(\theta)$ angular selectivity. The sum of the squares across all 8 characteristic functions is very flat (dotted line). See text for details. Adapted from [32].

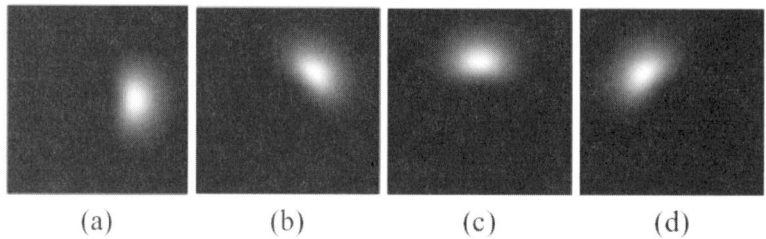

Figure 5.13 (a) Oriented bandpass frequency response $G_{0,0}(\omega,\phi)$ at 0 radians; (b) oriented bandpass frequency response $G_{0,1}(\omega,\phi)$ at $\pi/4$ radians; (c) oriented bandpass frequency response $G_{0,2}(\omega,\phi)$ at $\pi/2$ radians; (d) oriented bandpass frequency response $G_{0,3}(\omega,\phi)$ at $3\pi/4$ radians. Modified from [32].

is similar to the frame reconstruction requirements of decompositions addressed in [22]. Correction can be made for nonuniform $C_\phi(\phi)$ in a post-filtering step.

5.2.4 Filter Kernels

For each of the filter prototypes, $G_{0,k}(\omega,\phi)$, in the Fourier domain, a sampling on the two-dimensional interval $[-\pi,\pi] \times [-\pi,\pi]$ was performed, with a grid spacing of $\pi/64$ in each Cartesian direction. The choice of an odd matrix size for constructing the Fourier domain representation is tied to the symmetry of the filter kernels, which we have observed to be better on odd-sized grids. The sampled Fourier spaces for each of four oriented kernels is illustrated in Figure 5.13.

The inverse two-dimensional discrete Fourier transform was computed to extract 65×65 spatial impulse responses. These responses were each truncated to fit into a set of four 15×15 complex arrays. The kernels thus extracted are illustrated in Figure 5.14. The patterns of spatial symmetry of the filter kernels fall into

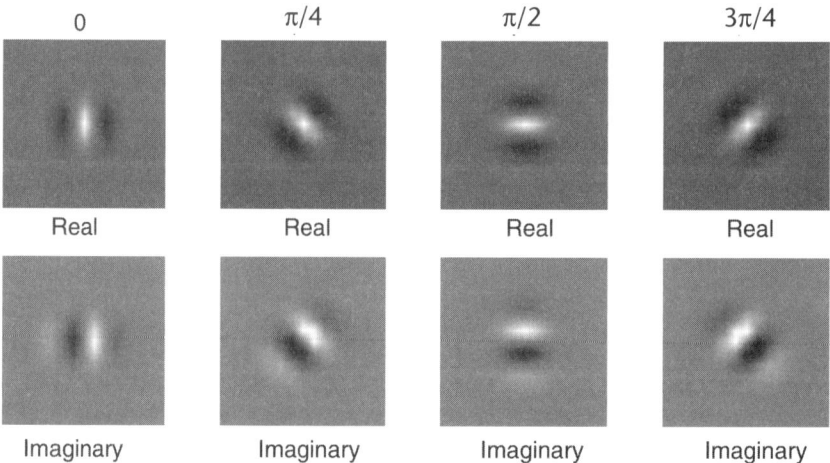

Figure 5.14 Real and imaginary parts of complex bandpass filter kernels $g_{0,k}(m,n)$ in the spatial domain.

five classes; this reduces the number of unique filter coefficient values necessary to specify a filter, and may also be used to reduce computational load.

5.2.5 Steering and Orientation Estimation

The steerable filtering framework allows both orientation adaptive filtering and orientation estimation to be performed in one framework. We first address the issue of orientation steering. From the angular filter prototypes of $\Phi_{0,k}(\phi)$, $k = 0, 1, 2, \ldots, K - 1$, we wish to construct $\Phi_0(\phi - \phi_s)$ by

$$\Phi_0(\phi - \phi_s) = \sum_{k=0}^{K-1} W_k(\phi_s) \Phi_{0,k}(\phi) \tag{5.13}$$

As described in [42], this may be treated as a one-dimensional regression problem against parameter ϕ_s where the K weighting functions, $W_k(.)$, are "learned" using a Moore-Penrose pseudoinverse. For example, Bharath and Ng [32] perform a dense, uniformly spaced sampling of ϕ_s over the range of allowed steering angles, $[-\pi/2, \pi/2]$. For each of these K_e values of ϕ_s, denoted by ϕ_k, function $\Phi_0(\phi - \phi_k)$ was sampled over ϕ at J locations in the range $[-\pi, \pi]$ to yield a $J \times 1$ column vector, Φ_{ϕ_k}. All column vectors may be collected together to form a $J \times K_e$ matrix, $\mathbf{F}_e$:

$$\mathbf{F}_e = \begin{pmatrix} | & | & | & | \\ \mathbf{g}_{\phi_1} & \mathbf{g}_{\phi_2} & \cdots & \mathbf{g}_{\phi_{K_e}} \\ | & | & | & | \end{pmatrix} \tag{5.14}$$

A $J \times K$ matrix $\mathbf{B}$ may be constructed by sampling the angular frequency response $\Phi_{0,k}(\phi)$ of the K fixed, complex bandpass filters in the variable ϕ at J locations spanning the range $[-\pi, \pi]$. One may then operate on the "example" matrix $\mathbf{F}_e$ with the pseudoinverse of $\mathbf{B}$ to obtain an interpolation matrix $\mathbf{W}$:

$$\mathbf{W} = \left(\mathbf{B}^T \mathbf{B}\right)^{-1} \mathbf{B}^T \mathbf{F}_e \tag{5.15}$$

Each row of $\mathbf{W}$ will then contain a series of weights, one in each column. The components of these vectors can be used to weight the outputs of the K fixed spatial domain filters in order to synthesize a response at the corresponding angle ϕ_s in the sampled space of possible kernel rotations. The linearity of the inverse Fourier transform and the polar separability of the kernels allow the weights to be applied to the filter responses in the *spatial* domain. In order to steer the filters to arbitrary angles, one may use least-squares fitting of polynomials to each of the weight vector *components*, with steering angle as the regressor. This approach has the advantage of reducing the weighting vectors to a relatively small number of coefficients, but relies on the weight vector tracing out a reasonably continuous manifold in the weight space as a function of steered angle. This is only achievable for a subset of all possible angular filter characteristics. We do not address here the conditions for good steering approximation, but see [45] for detailed studies of this issue from a Lie-algebra perspective. Weighting with trigonometric functions is also easily achievable.

At the ℓth level of pyramidal decomposition, we wish to combine the outputs of K fixed bandpass filters at K orientations, in order to synthesize an output from a bandpass filter that is oriented at some angle ϕ_s. Let $f_k^{(\ell)}(m,n)$, $k = 0, 1, 2, \ldots, K$, denote the output of the kth bandpass filter at level ℓ. First, note that, due to the nature of two-dimensional Fourier space, the outputs of the filters corresponding to $k = K/2, K/2 + 1, \ldots, K - 1$ are provided by the complex conjugates of the outputs of filters in the range $k = 0, \ldots, K/2 - 1$. Specifically, we have

$$f_{k+K/2}^{(\ell)}(m,n) = (f_k^{(\ell)}(m,n))^*, \quad k = 0, 1, 2, \ldots, K/2 - 1 \tag{5.16}$$

If the angle to which one wishes to steer the filters to is denoted by ϕ_s, then one may generate the steered output $f_s^{(\ell)}(m, n, \phi_s)$ of the bandpass filters at level ℓ of the decomposition by using

$$f_s^{(\ell)}(m, n, \phi_s) = \sum_{k=0}^{K/2-1} s_p(\phi, k) f_k^{(\ell)}(m, n) + \\ \sum_{k=0}^{K/2-1} s_q(\phi, k) f_k^{(\ell)*}(m, n) \tag{5.17}$$

Functions $s_p(\phi, k)$ and $s_q(\phi, k)$ are the polynomial steering functions, which have been obtained by regressing the weights in **W** on the parameter ϕ_s, and they control the weights given to the K fixed filters by exploiting the relation of Equation (5.16). The steering functions are compactly represented by using the coefficients p_k and q_k of the polynomial fits in steered angle, ϕ_s, to the steering weights as determined by:

$$s_p(\phi, k) = \sum_{n=1}^{P} p_k(n) \phi^n \tag{5.18}$$

$$s_q(\phi, k) = \sum_{n=1}^{Q} q_k(n) \phi^n \tag{5.19}$$

5.3 The Use of V1-Like Wavelet Models in Computer Vision

5.3.1 Overview

The complex wavelet decomposition may be treated as a linear operator, mapping the input image into an overcomplete wavelet representation. Due to the design of the wavelet pyramid discussed in Section 5.2, the transform has near-perfect reconstruction properties [32], making it suitable as a platform for a wide range of tasks in computer vision. Furthermore, due to its adaptability and ease of implementation on FPGA and other hardware, it constitutes quite a general platform for vision. Figure 5.15 illustrates how such a platform might be used in practice.

5.3 The Use of V1-Like Wavelet Models in Computer Vision

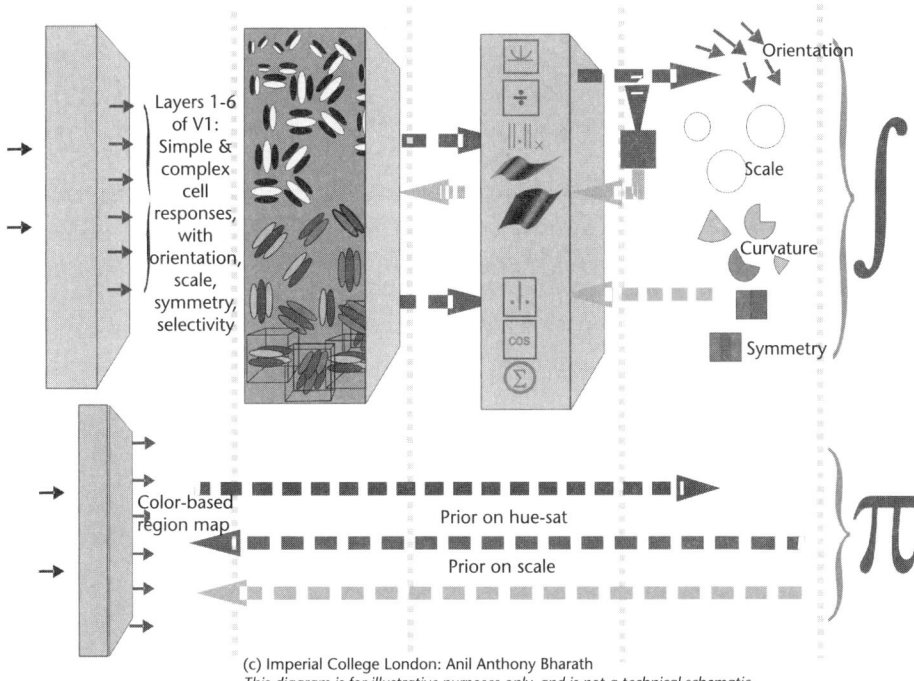

(c) Imperial College London: Anil Anthony Bharath
This diagram is for illustrative purposes only, and is not a technical schematic.

Figure 5.15 The figure above illustrates the principle of using wavelet representations to mimic the activity of V1. Wavelet decompositions applied to either intensity or to color-opponent spaces are mapped through nonlinear functions in order to generate local estimates of image properties, such as orientation, symmetry, and scale. These may in turn be used to adapt the wavelet representation, depending on the final application. Spatially mapped color information can also be incorporated to weight the measurements, for example, toward measurements more likely to correspond to skin tone, or other known prior information about color. http://www.bg.ic.ac.uk/staff/aab01/book

5.3.2 Generating Orientation Maps

A primary aim of the filter design is the construction of measures that are similar to the proposed saliency maps for modelling visual attention [46]. At the same time, it is useful to seek inspiration from specific models of visual processing that have been successfully developed in the pattern recognition field. Morrone and Burr [47], for example, proposed a model for feature detection that requires the estimation of orientation of local image energy in prescribed radial frequency bands. Their model for feature detection suggests that estimation of local structure orientation, followed by the use of the phase information along the normal to the orientation direction, provides an illumination invariant measure of the symmetry of the neighborhood along a locally defined axis of alignment.

A measure of *orientation dominance* has been found to be useful. This notion, employed, for example, in [48] for characterizing the angular tuning of orientation-selective cells in the primary visual cortex of mammals, provides a measure quite similar to the principal eigenvector of the Hessian, used in scale-space techniques [49]. Based on this measure, we construct an orientation dominance complex field as follows:

$$\mathbf{O}^{(\ell)}(m,n) = \frac{\sum_{k=0}^{K/2-1} |f_k^{(\ell)}(m,n)| e^{j2\phi_k}}{p + \left(\sum_{k=0}^{K/2-1} |f_k^{(\ell)}(m,n)|^2\right)^{\frac{1}{2}}} \qquad (5.20)$$

The conditioning constant, p, is set between 1% and 2% of the maximum pixel value in the image being analyzed.

This complex field is constructed using arrangements of fixed filters, in which a vectorial addition of the magnitude outputs of complex filter units is conveniently performed using complex number algebra; see [18] for similar techniques, but without the divisive normalization. In the vector formulation, the vector sum of the filter outputs, weighted by the components of the double-angle unit vectors defining filter directions, should be zero in image regions which yield exactly the same output magnitudes on all filters. Furthermore, because of the "divisive normalization," the magnitude of the scaled vector response ranges from 0 to $1 - \varepsilon(p)$, and it can be used as an indication of anisotropy: strongly isotropic neighborhoods will produce values close to 0 while strongly anisotropic neighborhoods will produce values close to 1. The justification for divisive normalization is presented in Section 5.4.

An illustration of this field, using the single-angle representation, is shown in Figure 5.16. This has been extracted from the second level of decomposition. From

Figure 5.16 Complex orientation dominance field; this consists of double-angled vectors [18] aligned along directions of dominant filter energy output. http://www.bg.ic.ac.uk/staff/aab01

this, a real scalar field, indicative of strength of dominance, may be constructed by taking the magnitude of the complex field:

$$O_d^{(\ell)}(m,n) = |\mathbf{O}^{(\ell)}(m,n)|, \quad \ell = 1, 2, \ldots, L \qquad (5.21)$$

5.3.3 Corner Likelihood Response

The outputs of the filters may be used to generate a measure that can be associated with the likelihood of a particular location in an image being the corner of some structure. Such a measure may be formulated as

$$C^{(\ell)}(m,n) = \frac{\prod_{k=0}^{K/2-1} |f_k^{(\ell)}(m,n)|}{p + \left(\sum_{k=0}^{K/2-1} |f_k^{(\ell)}(m,n)|\right)^{K/2}} \qquad (5.22)$$

The map produced by Equation (5.22) uses the product of the magnitude of responses from individual filters sensitized to different directions, yielding a strong response only if all filters have a strong response. The design of the kernels is important here, because the energy is measured in a particular radial frequency band corresponding to the scale of the decomposition, so to prevent an arbitrary point-noise sources from yielding strong responses. Also, the denominator normalizes the response to local anisotropic energy. In realistic imagery, this map should be weighted by a measure of local signal energy, in order to avoid strong measures in low-signal regions. For example, one may choose to weight the corner response by anisotropic energy computed at the same, or another, scale.

To illustrate this idea, Figure 5.17(a) shows a scene containing objects on a flat surface. The measure, overlaid on Figure 5.17(b), is given by

$$C_{\text{Likelihood}}^{(2)}(m,n) = C^{(2)}(m,n) I^{(1)}(m,n) \qquad (5.23)$$

where $I^{(\ell)}$ is an isotropic measure of local energy in the ℓth sub-band of radial frequency. It is generated from filter outputs by

$$I^{(\ell)}(m,n) = \left(\sum_{k=0}^{K/2-1} |f_k^{(\ell)}(m,n)|^2\right)^{\frac{1}{2}} \qquad (5.24)$$

Bilinear interpolation is used to resize feature maps at different scales.

5.3.4 Phase Estimation

Phase estimates extracted from the steerable filters provide a means of feature extraction. Examples on the use of local phase (or symmetry) as an image feature can be found in [18, 25, 47, 50]. It is possible to apply the technique of steered phase estimation to construct a likelihood image for edges of particular structures. Figure 5.16 shows measures of steered phase extracted from different scales of

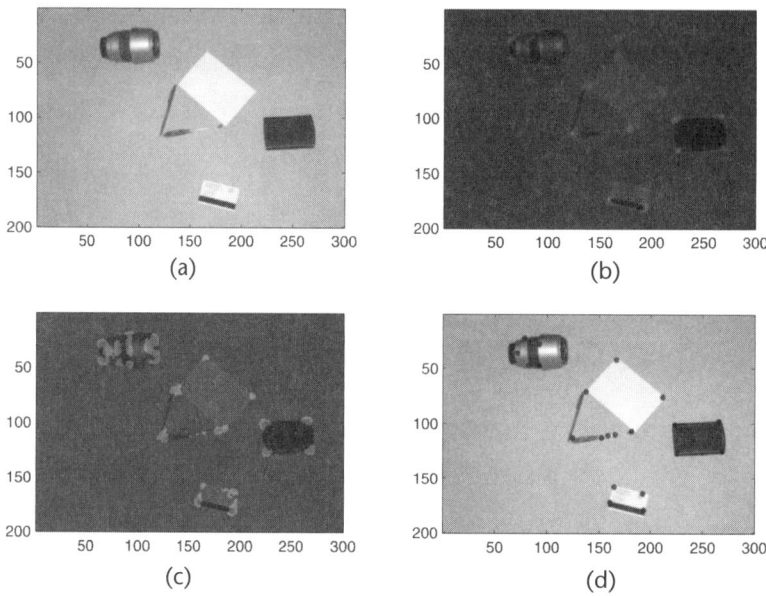

Figure 5.17 (a) Original image. (b) Measure $C^{(2)}_{\text{Likelihood}}$, superimposed on original image. (c) $C^{(2)}_{\text{Likelihood}}$ thresholded at 0.2, to yield a binary region mask. (d) After post processing applied to (c), which consists of connected component labelling, finding the centroid by binary moments, and sketching a circle around the centroids. http://www.bg.ic.ac.uk/staff/aab01/book

decomposition of a simple scene. The phase estimate $\psi_s^{(\ell)}(m,n)$ is obtained from the argument of the complex filter outputs $f_s^{(\ell)}(m,n,\phi_s)$,

$$\psi_s^{(\ell)}(m,n) = \text{angle}(f_s^{(\ell)}(m,n,\phi_s)) \qquad (5.25)$$

For a given problem, it is necessary to tune and combine the steered phase responses from different scales in order to achieve the desired result. A comprehensive example of generating edge responses incorporating non-maxima suppression by combining responses across scales is provided in [32].

5.4 Inference from V1-Like Representations

A more generic and formal framework for the detection of spatial structures in V1 representations may also be formulated, in which the information encoded by a population of simple or complex neurons is treated as being *observation data* from which inferences are then to be drawn. The formulation of shape detection is first expressed in terms of vector fields, but multiple scalar or vector fields may also be considered. In the analysis that follows, only vector field observations that arise from a single scale of a filter bank are considered, but the problem is formulated in such a way that it should be clear how to extend the notation and concepts across scales. The notation that is employed here is derived from that used in

5.4 Inference from V1-Like Representations

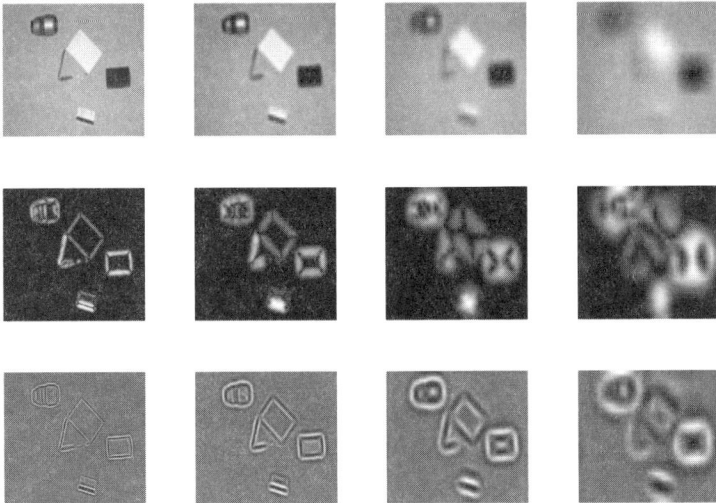

Figure 5.18 Top row: Lowpass channels at each of the four levels of the pyramid. Middle row: Intensity proportional to orientation dominance, $O^{(\ell)}(m,n)$, at each of the four scales shown in the top row. Bottom row: Intensity proportional to cosine of steered phase response at each of the four scales in the top row.

modern Bayesian statistics, and the reader is referred to the excellent book by Mackay [51].

5.4.1 Vector Image Fields

Primate vision employs various strategies for visual processing that are highly effective. The apparent separation of visual input into various streams, or channels, of representation is one obvious example. These interpretations, collectively referred to as "channel models" of vision, have been suggested for decades and are backed up by much physiological [52] and psychophysical [53] evidence. From a pattern recognition viewpoint, such ideas might not be surprising: the use of multiple channels is analogous to generating components of a feature space as might be constructed for pattern recognition purposes. In this section, the issue of constructing a posterior distribution for the position of a specific shape in an image from a general vector field is discussed as a means of obtaining shape-specific responses.

Gradient-like field estimates are thought to play a significant role in biological vision, and particularly for distinguishing form (shape). A probabilistic framework is suggested for locating planar shapes from gradient fields or, more generally, from vector fields as might be generated from a wavelet filter bank. By using a particular choice of conditional-Gaussian statistical model, we may show that a biological-like refinement to shape detection, based on using nonlinear combinations of the outputs of linear spatial filters, may be derived. This is extended to a phase-invariant version of shape detection and of keypoint detection toward the end of this chapter.

5.4.2 Formulation of Detection

This inference problem is now expressed as follows:

Q1: Given an estimate of vector fields of a digital image, what is the most likely location of a known planar shape, S, in the image?

This question may be answered probabilistically, for example, by estimating the posterior density function of shape location given observations:

$$f(S,\theta,M,\mathbf{t}_c|(\mathcal{B},\mathcal{X});\Theta_1) = \frac{f((\mathcal{B},\mathcal{X})|S,\theta,M,\mathbf{t}_c;\Theta_1)f(S,\theta,M,\mathbf{t}_c|\Theta_1)}{f((\mathcal{B},\mathcal{X})|\Theta_1)} \quad (5.26)$$

where θ represents a known rotation, M, a known scaling of the shape (zoom factor), and where $\mathbf{t}_c$ represents the possible locations of the shape in the visual field. Θ_1 represents a parameter vector of assumptions on viewing conditions, $\Theta_1 = \Theta_{0\setminus\{\theta,M\}}$, which is the top-level hypothesis[2] excluding rotation, θ and magnification, M. $(\mathcal{B},\mathcal{X})$ denotes the vector field $\mathcal{B}$ observed at known locations, $\mathcal{X}$. Although this chapter does not permit us to cover the estimation of all parameters, we do discuss, in Section 5.3, an example of the estimation of orientation fields at different scales. Rotation-invariant scale estimation techniques are also suggested in [54]. For the illustrations below, we consider cases in which M and θ are either known or unimportant.[3]

(5.26) is of the standard Bayesian form

$$\text{Posterior} = \frac{\text{Likelihood} \times \text{Prior}}{\text{Evidence}} \quad (5.27)$$

and so can be easily recognized as allowing the maximum a posteriori (MAP) solution to (5.26) of

$$\mathbf{t}_{\text{opt}} = \underset{\theta,M,\mathbf{t}_c}{\text{argmax}} \frac{f((\mathcal{B},\mathcal{X})|S,\theta,M,\mathbf{t}_c;\Theta_1)f(S,\theta,M,\mathbf{t}_c|\Theta_1)}{f((\mathcal{B},\mathcal{X})|\Theta_1)} \quad (5.28)$$

Alternatively, local maxima may also be located in $f(S,\theta,M,\mathbf{t}_c|(\mathcal{B},\mathcal{X});\Theta_1)$, for finding multiple instances of the shape.

We must now look for a plausible model for the likelihood function,

$$f((\mathcal{B},\mathcal{X})|S,\theta,M,\mathbf{t}_c;\Theta_1) \quad (5.29)$$

Since the observations represent samples of a vector field, it would appear appropriate to assume that the observed field differs from some *expected field* which is derived from knowledge of the operators used to estimate the field in individual images. Because there are many factors which will contribute to the fluctuation of the observed field, one might invoke the central limit theorem, and

---

2. The top-level, unquestioned hypotheses, when presented to the right of the "|" sign, are separated by a semicolon, ";", from other parameters that may be estimated from an earlier stage in the inference process. The top-level hypothesis Θ_0 includes all conditions under which the data is observed. $\Theta_{0\setminus\{A,B\}}$ denotes the full set of conditional hypothesis parameters excluding parameters A and B.
3. For example, for objects that are circular, with no angular variation, it is unimportant to know θ.

5.4 Inference from V1-Like Representations

opt for a Gaussian form of dispersion in the argument of the magnitude of the vector difference between the observed field and the expected field, conditional on position, $\boldsymbol{\mu}_b(\mathbf{x}^{(i)}|S,\mathbf{t}_c;\Theta_0)$. Accordingly, one may write, for each observation vector **b** at location $\mathbf{x}_i$, $i \in 1\ldots N$,

$$f((\mathbf{b}^{(i)},\mathbf{x}^{(i)})|S,\mathbf{t}_c;\Theta_0) = \frac{1}{K_2}\exp\left\{-||\mathbf{b}(\mathbf{x}^{(i)}) - \boldsymbol{\mu}_b(\mathbf{x}^{(i)}|S,\mathbf{t}_c;\Theta_0)||^2/2\sigma_b^2\right\} \quad (5.30)$$

where K_2 is a normalizing constant. All observation vector tuples are thus taken to be independent with respect to the specified conditions. By Equation (5.30), the observation vector tuples are assumed to be drawn from Gaussian distributions of possibly different means (conditioned on shape handle location, shape type, and all top-level hypotheses) but for the purpose of simplicity, one may assume that all tuples have identical variance.

The joint likelihood for observation of N pixels can then be written as

$$\begin{aligned}
f((\mathcal{B},\mathcal{X})|S,\mathbf{t}_c;\Theta_0) &= \frac{1}{K_2}\prod_{i=1}^{N}\exp\left\{-||\mathbf{b}(\mathbf{x}^{(i)}) - \boldsymbol{\mu}(\mathbf{x}^{(i)}|S,\mathbf{t}_c;\Theta_0)||^2/2\sigma_b^2\right\} \\
&= \frac{1}{K_2}\exp\left\{-\frac{1}{2\sigma_b^2}\sum_{i=1}^{N}||\mathbf{b}(\mathbf{x}^{(i)})||^2\right\} \\
&\quad \times \exp\left\{-\frac{1}{2\sigma_b^2}\sum_{i=1}^{N}||\boldsymbol{\mu}_b(\mathbf{x}^{(i)}|S,\mathbf{t}_c;\Theta_0)||^2\right\} \\
&\quad \times \exp\left\{\frac{1}{\sigma_b^2}\sum_{i=1}^{N}\mathbf{b}(\mathbf{x}^{(i)}) \cdot \boldsymbol{\mu}_b(\mathbf{x}^{(i)}|S,\mathbf{t}_c;\Theta_0)\right\}
\end{aligned} \quad (5.31)$$

The geometric operation of an inner product between vector field patches can be seen in the final exponential of Equation (5.31).

5.4.3 Sampling of $(\mathcal{B},\mathcal{X})$

In a digital image, the sampling of $\mathbf{x}^{(i)}$ is generally on the finite image grid. Under these conditions, the component of Equation (5.31)

$$\sum_{i=1}^{N}||\boldsymbol{\mu}_b(\mathbf{x}^{(i)}|S,\mathbf{t}_c;\Theta_0)||^2 \quad (5.32)$$

does not vary with the candidate shape position, $\mathbf{t}_c$.

Furthermore, recognizing that, for one complete frame of observation data

$$\frac{1}{2\sigma_b^2}\sum_{i=1}^{N}||\mathbf{b}(\mathbf{x}^{(i)})||^2 = \text{Constant} \quad (5.33)$$

and under the assumption of uniform object position prior and uniform evidence statistics, we have, on taking the natural logarithm of Equation (5.31)

$$\mathbf{t}_{\text{opt}} = \underset{\mathbf{t}_c}{\arg\max}\left[\sum_{i=1}^{N}\mathbf{b}(\mathbf{x}^{(i)}) \cdot \boldsymbol{\mu}(\mathbf{x}^{(i)}|S,\mathbf{t}_c;\Theta_0)\right] \quad (5.34)$$

First, an important observation, which is the geometric and the computational significance of Equation (5.34): for each candidate position vector, t_c, the function to be maximized is computed by a dot product between the boundary gradient field of the observed image space and the expected gradient field for the given shape around that candidate point. This may be viewed as a direct extension to correlation-based template matching, but applied to vector fields. There are some key extensions that will be introduced through proposing a more complex statistical model that then leads to a divisive normalization term, similar to a form of nonlinearity used to describe neuronal populations.

We note, to conclude this discussion, that the constraint of sampling on a regular grid is relaxed by the treatment given in Chapter 6. Moreover, the technique of normalized convolution, which is discussed in Chapter 6, provides an efficient extension of the vector field matching algorithms described in this chapter to irregularly sampled data.

5.4.4 The Notion of "Expected" Vector Fields

For a given shape, the expected vector field conditioned on shape position and rotation is in turn also dependent on the operators applied to the detection problem and on the shape itself. We shall use the first partial derivatives of a two-dimensional Gaussian, taken along the x and y Cartesian axis directions, to produce a simple vector field for illustration and analysis purposes:

$$g_x = -\frac{x}{s^2} g_0(x,y) \tag{5.35}$$

and

$$g_y = -\frac{y}{s^2} g_0(x,y) \tag{5.36}$$

where

$$g_0(x,y) = \frac{1}{2\pi s^2} \exp\left\{-\frac{x^2 + y^2}{2s^2}\right\} \tag{5.37}$$

and s controls the scale of this derivative operator. Consider a binary image, $b(\mathbf{x}^{(i)}|S, t_c = 0, A = 1; \Theta_2)$, of a unit intensity shape positioned at $t_0 = 0$, on a 0 intensity background. The hypothesis for this example, Θ_2, represents the assumptions of the top-level observation hypothesis, Θ_0, but excludes contrast parameter, A. Applying these gradient operators to this binary shape will yield a gradient field which is often estimated by

$$\boldsymbol{\mu}(\mathbf{x}^{(i)}|S, t_c = 0, A = 1; \Theta_2) = \Big(g_x \otimes b(\mathbf{x}^{(i)}|S, t_c = 0, A = 1; \Theta_2)\Big)\mathbf{u}_x \\ + \Big(g_y \otimes b(\mathbf{x}^{(i)}|S, t_c = 0, A = 1; \Theta_2)\Big)\mathbf{u}_y \tag{5.38}$$

where $\otimes$ denotes spatial convolution, $\mathbf{u}_x$ and $\mathbf{u}_y$ are the unit vectors in the (x,y) coordinate system, and the N locations $\mathbf{x}^{(i)}$ correspond to the sampling points

5.4.5 An Analytic Example: Uniform Intensity Circle

The precise form for the expected gradient field conditioned on shape position and rotation is dependent on the operators applied to the detection problem, and on the shape itself. Let us assume that the field is obtained from derivative of Gaussian gradient operators introduced above. The analogy in primate V1 would consist of sets of two antisymmetric receptive fields, retinotopically mapped, with orthogonal directions of peak sensitivity, and with relatively weak orientation selectivity.

For a unit intensity circle of radius R_a, centered at the origin of Cartesian space, we can express the gradient field as a magnitude and angle field in a polar coordinate system:

$$||\boldsymbol{\mu}(\mathbf{x}^{(i)}|S_0, \mathbf{t}_c = 0, A = 1; \Theta_2)|| = ||\boldsymbol{\mu}(\mathbf{r}^{(i)}|S_0, \mathbf{t}_c = 0, A = 1; \Theta_2)||$$
$$= \exp\left\{-(||\mathbf{r}^{(i)}|| - R_a)^2/2s^2\right\} \quad (5.39)$$

$$\angle\boldsymbol{\mu}(\mathbf{x}^{(i)}|S_0, \mathbf{t}_c = 0, A = 1; \Theta_2) = \angle\boldsymbol{\mu}(\mathbf{r}^{(i)}|S_0, \mathbf{t}_c = 0, A = 1; \Theta_2)$$
$$= \angle\mathbf{r}^{(i)} + \pi \quad (5.40)$$

Before continuing, we note that Equation (5.34) may be rewritten as

$$\mathbf{t}_{opt} = \underset{\mathbf{t}_c}{\operatorname{argmax}}\left[\sum_{i=1}^{N} ||\mathbf{b}(\mathbf{x}^{(i)})|| \cdot ||\boldsymbol{\mu}(\mathbf{x}^{(i)}|\mathbf{t}_c)||\cos(\theta_i)\right] \quad (5.41)$$

where $\theta^{(i)}$ is the angle between the observed gradient direction at the ith pixel and the *conditional expected* gradient direction, for example, specified by (5.40).

5.4.6 Vector Model Plausibility and Extension

One can treat a derivative of Gaussian as a simple form of wavelet, despite the support of these functions being essentially infinite in space. It is widely recognized that the univariate statistics of wavelet coefficient space are well approximated by a Laplacian density function. Recently, Sendur and Selesnick developed a bivariate Laplacian model for a wavelet coefficient space which they applied to denoising [55].

---

4. One may argue that discretization does not usually follow a derivative estimation in practice. However, the order in which this has been presented is in line with that used by the scale-space community. An argument to justify the approximation of continuous derivatives by discretized derivatives in practice is that the degree of smoothing may always be selected to reduce the aliasing error to an arbitrarily small amount.

Adopting the view of recent work on independent component analysis [56] and the statistics of natural scenes [41], the Laplacian distribution is merely a reflection of the sparseness of the "detail" coefficient space. In fact, a Laplacian-like PDF can arise from marginalizing a *conditional* bivariate Gaussian model over the location variable, t_c, even for a "standard" image + Gaussian noise model.

There are, perhaps, two reasons that such bivariate, conditional models have not been widely considered before in the literature: the first is the analytical difficulty of working with them, and the second is that experimental verification of the model is difficult. The approach which follows demonstrates the suitability of such models from an empirical point of view, indicates the relationship of such models to the biological context, and further evaluates them in a practical setting using the complex wavelet design of Section 5.2.

5.4.7 Vector Fields: A Variable Contrast Model

Between any two gradient images of the same shape, there will be a certain degree of contrast uncertainty. This contrast uncertainty will scale all values, including the x and y components of the gradient estimate around objects of interest, and might be considered a source of noise that violates independence between components or pixels of a single image. For example, returning to the case of the gradient template of a circle, one might expect that the expected gradient field depends on the contrast settings used during learning the shape. Thus, for a circle of *contrast parameter*, A, which may be negative, we have the more general form of (5.39) and (5.40):

$$\begin{aligned}
||\boldsymbol{\mu}_b(\mathbf{x}^{(i)}|S_0, \mathbf{t}_c = 0, A; \Theta_2)|| &= ||\boldsymbol{\mu}_b(\mathbf{r}|S_0, \mathbf{t}_c = 0, A; \Theta_2)|| \\
&= |A|\exp\left\{-(||\mathbf{r}|| - R_a)^2/2s^2\right\}
\end{aligned} \quad (5.42)$$

$$\begin{aligned}
\angle\boldsymbol{\mu}_b(\mathbf{x}^{(i)}|S_0, \mathbf{t}_c = 0, A; \Theta_2) &= \angle\boldsymbol{\mu}_b(\mathbf{r}^{(i)}|S_0, \mathbf{t}_c = 0, A; \Theta_2) \\
&= \angle\mathbf{r}^{(i)} + \frac{\pi}{2}[\text{sign}(A) + 1]
\end{aligned} \quad (5.43)$$

Although there will be uncertainty in this contrast value, some prior information about this uncertainty, and particularly about its role in providing "coupling" between the components of the gradient field, may be postulated. The following model for the auxiliary conditional gradient field is proposed through the auxilary "deviation" field,

$$\beta(\mathbf{x}^{(i)}|S, \mathbf{t}_c, A; \Theta_2) = A\mathbf{b}(\mathbf{x}^{(i)}) - \boldsymbol{\mu}_b(\mathbf{x}^{(i)}|S, \mathbf{t}_c, A = 1; \Theta_2) \quad (5.44)$$

where, without loss of generality, the contrast of the average template used to define the shape during learning is 1, alongside a prior for the contrast (amplitude

5.4 Inference from V1-Like Representations

scaling) parameter, A, that captures how an object's overall contrast might be distributed from one scene or location to another:

$$f_A(A;\Theta_2) = \sqrt{\frac{\alpha_A}{\pi}} e^{-\alpha_A(A-\mu_A)^2} \qquad (5.45)$$

and, the conditional PDF for the auxiliary conditional random vector field deviation,

$$f_\beta(\boldsymbol{\beta}(\mathbf{x}^{(i)})|S,\mathbf{t}_c,A;\Theta_2) = \frac{\alpha_\beta}{\pi} e^{-\alpha_\beta \|\boldsymbol{\beta}(\mathbf{x}^{(i)}|S,\mathbf{t}_c,A;\Theta_2)\|^2} \qquad (5.46)$$

Thus, the gradients of pixels are coupled not only through the conditional template, but also through a contrast parameter, A, which may appear in practice as a *hidden* variable.

5.4.8 Plausibility by Demonstration

In Figure 5.19(b), a sample is drawn from the gradient field model of small, bright circles of radius around three pixels against a darker background. Parameter μ_A was set to 1, α_β was set to 200, and $\alpha_A = 1$, that is, corresponding standard deviations of 0.05 and $\sqrt{1/2}$, respectively. The generation of samples from this field is performed as follows:

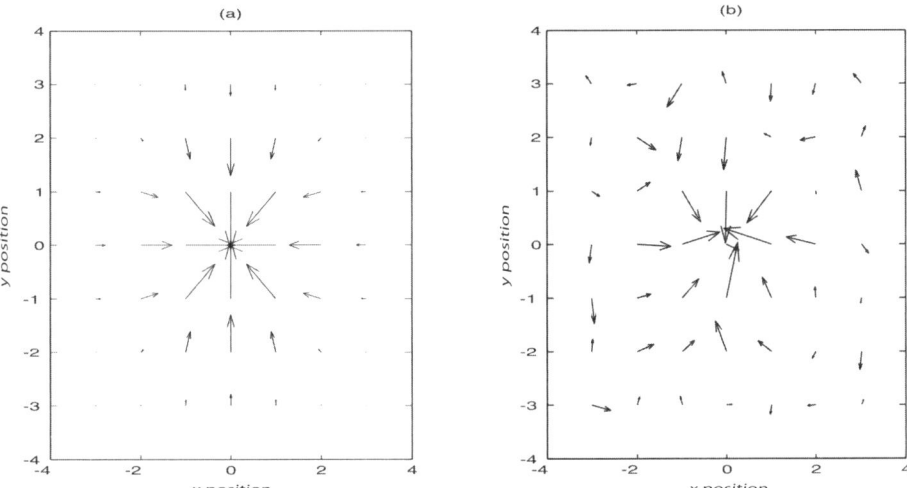

Figure 5.19 (a) Left. Illustration of a mean vector field conditional on a bright circle being present at location (0, 0) on a dark background. (b) Right. A sample of a vector field drawn from $f_\beta(\boldsymbol{\beta}(\mathbf{x}^{(i)})|S_0,\mathbf{t}_c = \mathbf{0}, A = 1;\Theta_2)$ using the conditional mean vector field to the left. S_0 is a unit intensity circle of radius three pixels. The nonzero correlation between spatially distinct pixels is often adequately captured by the conditional mean vector.

Algorithm 1 Algorithm for drawing realizations of a shape vector field drawn from the statistical model presented in the text, with known scale and magnification from a known *distribution* of contrasts (see text).

1. $A \sim f_A(A)$ using Equation (5.45).
2. $\boldsymbol{\mu}_b(.|.)$ is defined using Equations (5.42) and (5.43).
3. $\boldsymbol{\beta}(\mathbf{x}^{(i)}) \sim f_\beta(\boldsymbol{\beta}(\mathbf{x}^{(i)})|S, \mathbf{t}_c, A; \Theta_2), \forall i \in [1\ldots N]$
4. Using Equation (5.44), a realization of $\mathbf{b}(\mathbf{x}^{(i)}), \forall i \in [1\ldots N]$ may be created.
5. Go to Step 1 until $N_{\text{realizations}}$ are completed

The conditional mean vector field, $\boldsymbol{\mu}_b(\mathbf{x}^{(i)}|S_0, \mathbf{t}_c = 0, A = 1; \Theta_2)$, is illustrated in Figure 5.19(a). The reader should note that the samples drawn in Figure 5.19(a) include a variation in contrast which introduces coupling between x and y gradient components and *independent* noise on the x and y components.

5.4.9 Plausibility from Real Image Data

To validate the proposed model for gradient field statistics in the presence of objects, a simple experiment was conducted. An image of garden peas scattered across a rug, Figure 5.20(a), was acquired using a standard consumer digital camera (FujiFilm FinePix, 4700) on the "Normal Quality" setting at a resolution of 1,280 × 960. Normal artificial room lighting conditions were used, and the object–camera distance was of the order of 1.5 meters. The "V" channel of the HSV colorspace representation was extracted from the JPEG file of the camera. Using this channel, a thresholding operation was applied to binarize the image, then connected component labelling was used to find individual peas. The centroids of the binary pea images were found, and 33 × 33 patches centered on several pea centroids were extracted from gradient field estimates using simple Prewitt operators. A mean gradient field was computed, and the bivariate probability density function of the deviation of actual gradient estimates from this mean was estimated by histogram binning. The result is shown in Figure 5.20(b). The total number of pixels used to estimate this histogram (on a 20 × 20 grid of bins) is approximately 20,000.

To show that this approach leads to plausible PDFs that reflect those of real image data, one may use sampling principles [57] to marginalize the conditional density function, $f_\beta(\boldsymbol{\beta}(\mathbf{x}^{(i)})|S_0, \mathbf{t}_c, A; \Theta_2)$, over A using a standard normally distributed random number generator,[5] with $\mu_A = 1$, and $\alpha_A = 1$. The result of this is shown in Figure 5.20(d). This should be compared with the estimates from the "peas-on-rug" image of Figure 5.20(b).

Finally, one should note that Figure 5.20(b,d) may not appear very Laplacian-like because these are *only* estimated/simulated, respectively, in the immediate region of an object, as defined relative to the largest bounding box accommodating a single pea. The bivariate PDF of the joint gradient statistics for the *whole* image is indeed much more Laplacian-like (sparse).

---

5. Matlab 6.5 running under Windows 2000.

5.4 Inference from V1-Like Representations

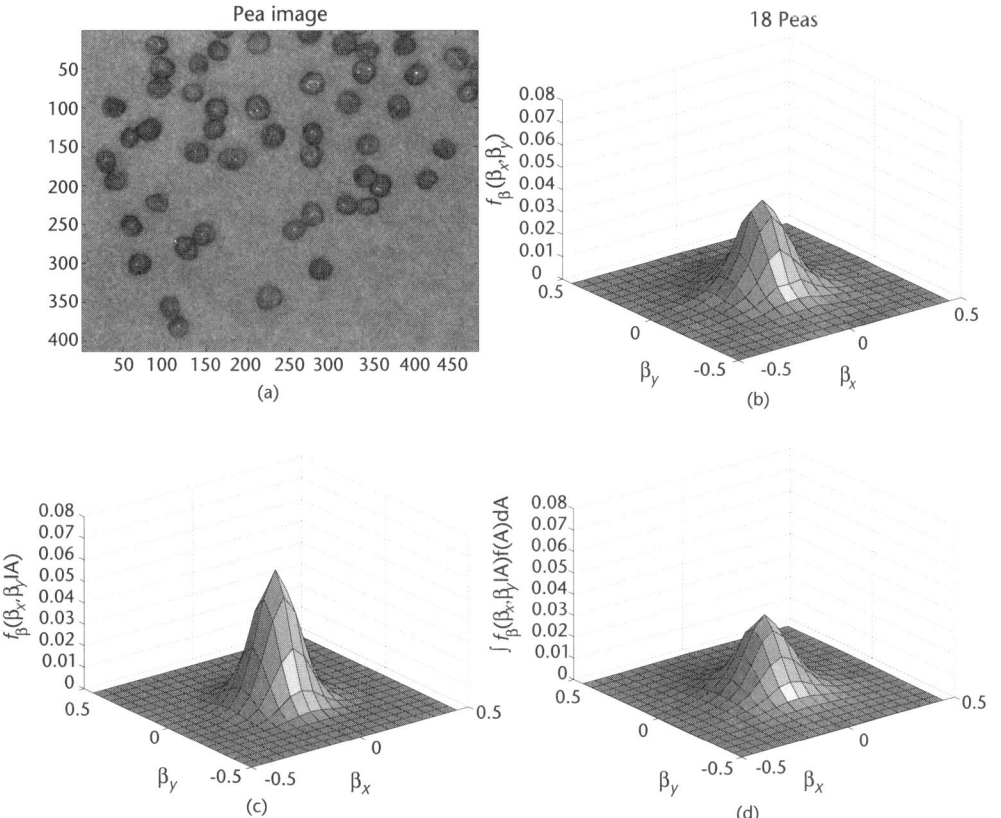

Figure 5.20 (a) Peas-on-rug image. (b) Experimentally estimated bivariate PDF of region containing single peas from the "peas-on-rug" image obtained by methods described in the text. (c) Sampling from $f_\beta(\beta_x, \beta_y | S_0, \mathbf{t}_c = \mathbf{0}, A; \Theta_2)$ for $A = 1$. (d) Sampling from $S_0, \mathbf{t}_c = \mathbf{0}, A; \Theta_2$ and the prior on A, $f(A)$, and marginalizing over A. Note that the model including marginalization yields a PDF (d) that is closer to that of the true gradient deviation data (b). Vertical axes on (b), (c), and (d) correspond to probability density.

5.4.10 Divisive Normalization

The analytic marginalization of (5.46) over A using the prior of (5.45) is also tractable, and leads to

$$f_\beta(\boldsymbol{\beta}(\mathbf{x}^{(i)})|S, \mathbf{t}_c; \Theta_2) = \int_{-\infty}^{\infty} f_\beta(\boldsymbol{\beta}(\mathbf{x}^{(i)})|S, \mathbf{t}_c, A; \Theta_2) f_A(A) dA \qquad (5.47)$$

where, using [58],

$$\int_{-\infty}^{\infty} \exp\{-p^2 x^2 \pm qx\} dx = \frac{\sqrt{\pi}}{p} \exp\left\{\frac{q^2}{4p^2}\right\}, \qquad p > 0 \qquad (5.48)$$

and (5.47) results in the following result for N *conditionally* independent pixel observations:

$$f_b(\{\mathbf{b}(\mathbf{x}^{(i)})\}_{i=1...N}|S,\mathbf{t}_c;\Theta_2) = \prod_{i=1}^{N} f_b(\mathbf{b}(\mathbf{x}^{(i)})|S_0,\mathbf{t}_c,\boldsymbol{\mu}_t)$$
$$= Z_N e^{\alpha_\beta(2\alpha_A\Phi_1+2\Phi_2-\Phi_3-\Phi_4-\alpha_A\Phi_5)} \quad (5.49)$$

where $Z_N = \prod_{i=1}^{N} \frac{\alpha_\beta}{\pi} \sqrt{\frac{\alpha_A}{Z_0^{(i)}}}$, with

$$Z_0^{(i)} = \alpha_A + \alpha_\beta \left[b_x^2(\mathbf{x}^{(i)}) + b_y^2(\mathbf{x}^{(i)}) \right] \quad (5.50)$$

and

$$\Phi_1 = \sum_{i=1}^{N} \frac{b_x(\mathbf{x}^{(i)})\mu_x(\mathbf{x}^{(i)}|\mathbf{t}_c) + b_y(\mathbf{x}^{(i)})\mu_y(\mathbf{x}^{(i)}|\mathbf{t}_c)}{Z_0^{(i)}} \quad (5.51)$$

$$\Phi_2 = \sum_{i=1}^{N} \frac{b_x(\mathbf{x}^{(i)})b_y(\mathbf{x}^{(i)})\mu_x(\mathbf{x}^{(i)}|\mathbf{t}_c)\mu_y(\mathbf{x}^{(i)}|\mathbf{t}_c)}{Z_0^{(i)}} \quad (5.52)$$

$$\Phi_3 = \sum_{i=1}^{N} \frac{\left[\alpha_A + \alpha_\beta b_y^2(\mathbf{x}^{(i)})\right]\mu_x^2(\mathbf{x}^{(i)}|\mathbf{t}_c)}{Z_0^{(i)}} \quad (5.53)$$

$$\Phi_4 = \sum_{i=1}^{N} \frac{\left[\alpha_A + \alpha_\beta b_x^2(\mathbf{x}^{(i)})\right]\mu_y^2(\mathbf{x}^{(i)}|\mathbf{t}_c)}{Z_0^{(i)}} \quad (5.54)$$

and

$$\Phi_5 = \sum_{i=1}^{N} \frac{b_x^2(\mathbf{x}^{(i)}) + b_y^2(\mathbf{x}^{(i)})}{Z_0^{(i)}} \quad (5.55)$$

The functions $\mu_x(\mathbf{x}^{(i)}|\mathbf{t}_c)$ and $\mu_y(\mathbf{x}^{(i)}|\mathbf{t}_c)$ are the x and y components of the mean gradient field, $\boldsymbol{\mu}(\mathbf{x}^{(i)}|\mathbf{t}_c)$, $b_x(\mathbf{x}^{(i)})$, and $b_y(\mathbf{x}^{(i)})$ are the x and y components of the observed vector gradient field at the ith pixel, located at $\mathbf{x}^{(i)}$.

Each of the terms Φ_1, Φ_2, Φ_3, and Φ_4 involves spatial convolution (or cross correlation) between the appropriate mask and simple functions of the gradient field components. Term Φ_1, for example, approaches Kerbyson and Atherton's [59] filtering technique if α_A is significantly greater than α_β. The terms that dictate the variation of this log-likelihood function over space define an *accumulator space* that is conceptually similar to that of the compact Hough transform corresponding to the spatial location parameters.

Finally, note that the term most similar to the vector correlation, introduced in Section ??, corresponds to Φ_1, and, as seen in (5.51), contains a *divisive normalization* term at *each* pixel location. The effectiveness of this modification to realistic image matching problems is evaluated in the following section.

5.5 Evaluating Shape Detection Algorithms

The performance of vector matching and contrast marginalized gradient-based detection technique has been tested on a range of image types and in various applications of computer vision and cannot be fully described here. Generally speaking, however, one needs to establish ground truth data in a detection scenario, and to assess the performance of any proposed algorithms in different conditions of contrast and noise. Below, we compare the performance of standard approaches, such as the compact Hough transform and spatial cross correlation, with the algorithms described above using simple models of antisymmetric receptive fields, based on a derivative of Gaussian estimators. We also demonstrate the performance of a phase-invariant shape detector in Section 5.55, using the GPU implementation of the complex wavelet V1 model described earlier.

5.5.1 Circle-and-Square Discrimination Test

To compare the performance of the techniques discussed above under different contrast conditions and noise levels, an image containing circles and squares was synthesized. Alternating circles and squares with variations in both size and contrast were also placed along the horizontal axis. The background intensity of the image was varied linearly along the horizontal axis and also Gaussian random noise was added.[6] The shapes in the image were generated using generalized functions, which can be tuned to simulate the blurring that might arise in a real imaging system.

To compare the discriminative performance of the techniques, a receiver operating characteristic (ROC) curve [60] may be plotted. The ROC is a plot of true positive rate (TPR) against the false positive rate (FPR). The true positives represent the correct detection of *squares* (shape S_2) with a side-length of 9 pixels, whereas the false positives represent the detection of all the other objects: circles with various diameters and squares with different side-lengths. ROC results are generated by applying a swept threshold in 100 steps across the range of accumulator space. The experiment was then repeated at a different noise level, with no change of parameter setting, apart from adapting to the range of amplitudes in accumulator space. One hundred threshold levels were applied to the accumulator space where TPR and FPR were calculated for each level. Values of σ were linearly selected in steps of 0.01 in this range. The maximum intensity in the noise-free image was just over 2, and the minimum intensity just less than -1.2. This equates to a maximum σ of around 0.32 in a unit-intensity range image. To illustrate the nature of the noise in some tests, Figure 5.21(a) shows the noise-free image, while Figure 5.21(b) shows the image at the maximum noise level used, in which discrimination of circles from squares is almost impossible.

The ROC plot shown in Figure 5.22 compares vector filtering (without contrast marginalization) with correlation filtering. As can be seen, the vector-based filtering approach is significantly more selective to desired shape characteristics than cross correlation,[7] and while this is to be expected, such an evaluation of selectivity

---

6. Uniform noise was also added in some tests.
7. The cross correlation uses a mean-subtracted template.

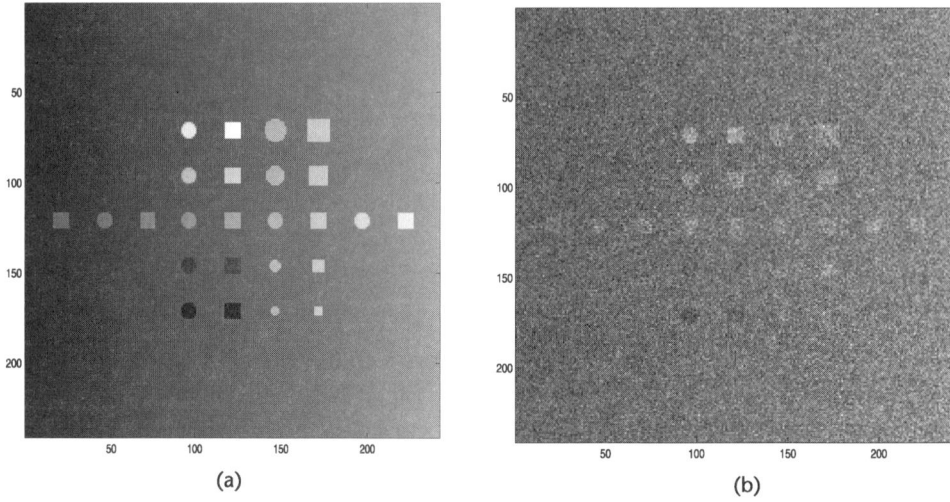

Figure 5.21 (a) Noise-free synthetic image. (b) Image at maximum noise level used for ROC estimation of Figures 5.22 and 5.23. See text for details.

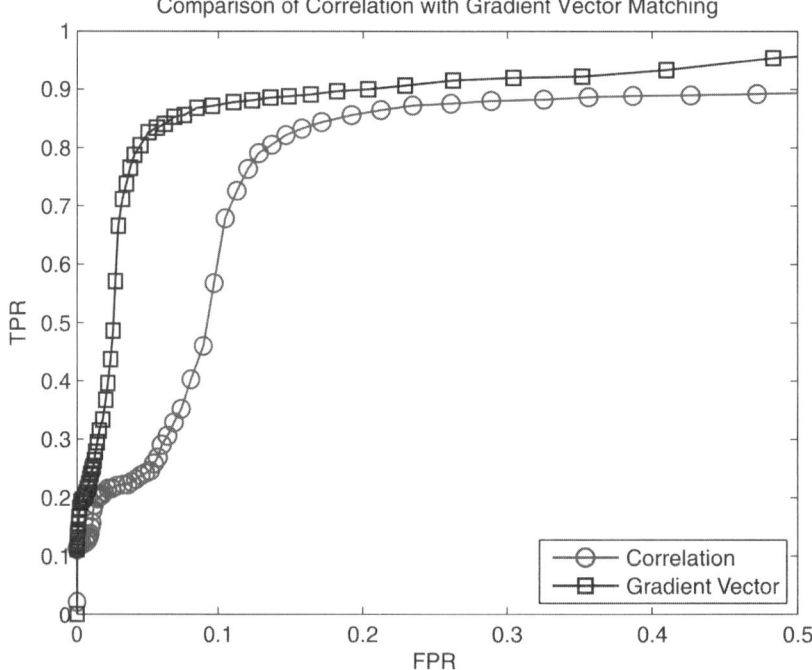

Figure 5.22 ROC comparing the performance of zero-mean cross correlation–based template matching with vector-based template matching of (5.34).

5.5 Evaluating Shape Detection Algorithms

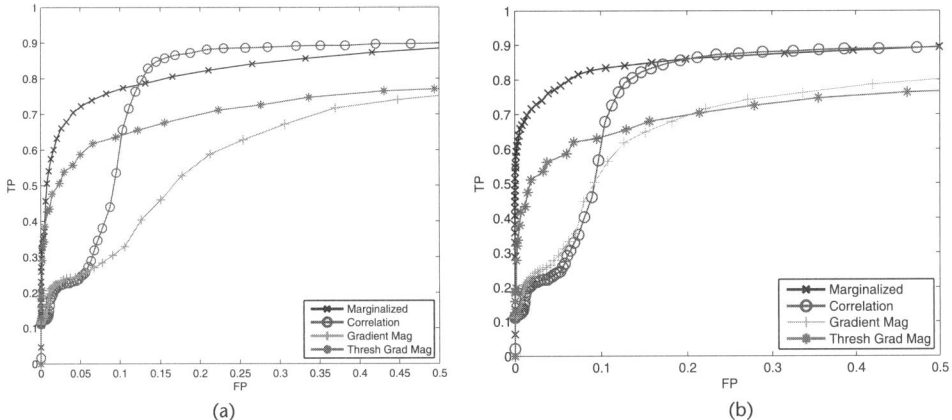

Figure 5.23 (a) contrast marginalized detection: ROC with $\alpha_A = 2$. (b) Contrast marginalized detection: ROC with $\alpha_A = 20$. Note that the effect of α_A is to increase the certainty in contrast. While correlation out-performs the marginalized gradient-based approach at low settings of the contrast certainty parameter, α_A, [see (a)] at thresholds associated with FP rates above ≈ 0.12, increasing this parameter [see (b)] consistently yields better performance at all thresholds. Noise realizations are different in (a) and (b).

demonstrates convincingly the superiority of gradient feature-based approaches to shape characterization over scalar field correlation.

The ROC plots in Figure 5.23(a,b) compare the contrast marginalized vector approach of Section 5.4.6, (5.49), with three other variants of shape detection techniques in images with random noise as described above; these are: magnitude-based compact Hough transform, a thresholded (binary) compact Hough transform, and a mean-subtracted cross correlation.

Figure 5.24(a) illustrates the test image accumulator space with *uniform* random noise between 0 and 1 added to a test image as described above. Profiles

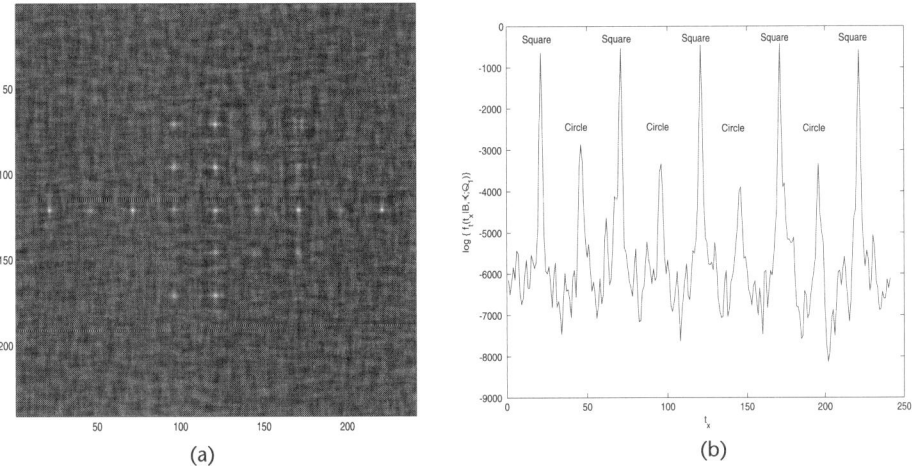

Figure 5.24 Left: Illustration of accumulator space: $\log(f(S_2, \mathbf{t}_c | \mathcal{B}, \mathcal{X}; \Theta_0))$ for nine pixel squares detection. Right: Profile through top shape response space at row=121.

through the center of the accumulator space illustrate that the system is robust to violations in noise distribution, yet maintains its shape selectivity.

The conclusion that can be drawn is that detection in the vector space is much more shape-selective than in scalar intensity space. This is in line with the general observation that gradient spaces are *sparse*. Sparsity is a property often associated with biological population encodings and is often cited as a justification for the receptive field patterns found in visual neuroscience [41]. Furthermore, the use of divisive normalization, a form of nonlinear wavelet space normalization that is similar to population encoding in neuronal systems, leads to much enhanced performance. This is highly significant in terms of the attainable performance of detection algorithms: when one has a large number of frames and pixels to examine, even small changes in detector performance can have a huge impact on the time spent examining incorrect detections, and the instance of missed detections can equally become quite large if one attempts to keep the false detections to a minimum.

5.6 Grouping Phase-Invariant Feature Maps

Due to the availability of phase-invariant responses from the complex steerable pyramid defined in Section 5.3, it is possible to create maps of orientation which can characterize objects in a phase-invariant fashion. The images below show an encoding trained on a spatially registered average of 400 faces obtained from the AR database [61]. These have been registered by manually identifying the tips of the noses of all subjects. The overlaid orientation field identifies the pattern of dominant orientation and orientation strength, $O^{(\ell=2)}(m,n)$, and this is illustrated in Figure 5.25(a).

For a new scene, the orientation dominance pattern may be compared with that of the template. This is done using a matching algorithm similar, but not identical, to that suggested by the divisive normalization derivation earlier in this chapter. Note that no explicit decisions are made about intermediate feature maps in order to achieve this. Grouping of features happens without any explicit decisions being made on edges of the face or their directions. First, note that the orientation field of the sketch of a face in Figure 5.25(a) is similar to that of the template built on face images: this is a property of the complex cell-like nature of the wavelet decomposition, which lends its phase-invariant character to the vector field representation of orientation. Second, note that the field is stable even if the face contrast is dramatically reduced and is blurred as in the case of Figure 5.25(a). Finally, the location of the detected face is shown in Figure 5.25(d). The details of the matching and combination of the complex wavelet channels incorporate other key innovations that are described in [62].

5.6.1 Keypoint Detection Using DTCWT

Focus-of-attention mechanisms in the human visual system are of great interest. Lim [63] performed studies on subjects viewing images of abstract art and analyzed, using linear discriminant analysis, the most discriminating features of

5.6 Grouping Phase-Invariant Feature Maps

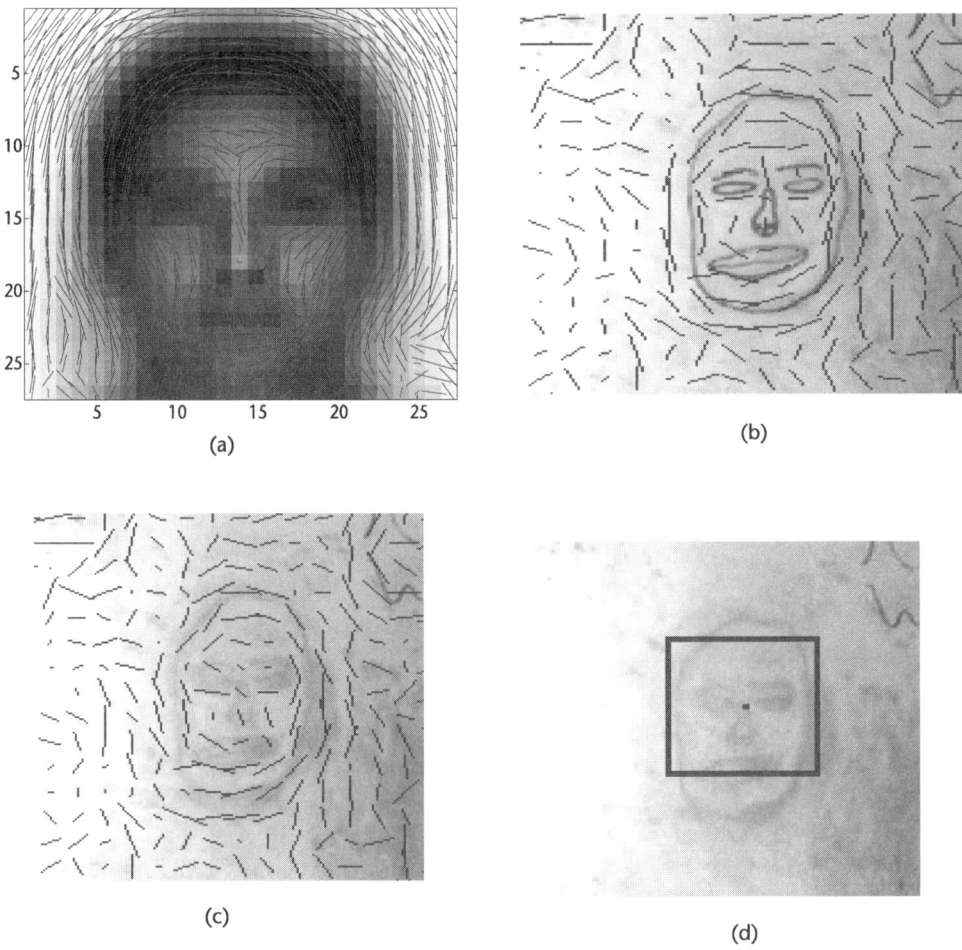

Figure 5.25 (a) Average face template and overlaid phase-invariant orientation map. (b) Orientation map derived from hand-drawn face on a whiteboard. Note that the map follows the contours in a similar fashion to the field of the template. (c) Even after smudging and reducing contrast, the orientation map stays quite stable. (d) Firing of detector corresponding to the most probable location of the face.

twenty-dimensional feature space constructed out of complex, steerable wavelet decompositions. His analysis showed that interiors of objects and corners of objects yielded the most significant predictive ability for fixation locations of subjects. Accordingly, Bharath and Kingsbury [64] proposed the use of complex wavelet decompositions for defining two classes of phase-invariant image keypoints: such keypoints define fixation locations around which analysis of visual structure might be locations around which analysis of visual structure might be locations around which analysis of visual structure might be performed, for example, as an input to a patch correspondence routine.

The principles described in the previous sections are not constrained to work with either derivative of Gaussian field estimators, or with the complex steerable

Figure 5.26 FMP and CM keypoint locations and scales; FMP keypoints are represented with × markers; CM keypoints are denoted by filled circles. Note the high proportion that falls onto facial features. http://www.bg.ic.ac.uk/staff/aab01/book

wavelet decomposition described in previous sections. As a demonstration of this, we consider the application of the principles described above to keypoint detection, widely used in computer vision, but implemented on the dual-tree complex wavelet transform of Kingsbury [28].

Two different keypoint types are created using the DTCWT: the first, referred to as filter magnitude product (FMP) keypoints, are similar to the corner likelihood measures described in Section 5.3.3, but were proposed independently by Fauqueur [29], while the second type, termed circular measure (CM) keypoints, are similar to the shape detection algorithm, but use a shape template designed to yield a strong response to circular groupings of tangential boundary responses in a multirate scheme, with a peak sensitivity fairly arbitrarily set at a radius of around $5 \times 2^{\ell}$ pixels, where 2^{ℓ} represents the level of the DTCWT.

Figure 5.26(a) and 5.26(b) demonstrates that the keypoints are located in sensible positions or direct the attention of a computer vision algorithm toward the interiors of objects, and also on the junctions or corners of objects. Note that the scales (indicated by the dashed circles) of CM keypoints group single or multiple objects together, including faces, at relatively coarse scales, while the FMP keypoints generally fall onto junctions or corners, or fall onto facial features.

5.7 Summary and Conclusions

This chapter has described practical schemes for mimicking early cortical visual processing, based around linear and nonlinear functions of wavelet transforms. Details of a complex steerable wavelet that is tunable were provided, implementations of which already exist in software, on FPGA devices (see Chapter 13), and on the nVidia GeForce Graphics platform. It is clear that although there are close analogies to spatial processing operators in image processing and computer vision, there remains a gap between the biological complexity and what can be applied practically. For example, temporal responses in existing image processing front ends are usually $\delta_1(k)$ functions (with k being frame/time index), whereas there are well-defined velocity tuning characteristics for most V1 neurons. This, we believe, represents a rich landscape for innovation in linear, spatiotemporal filter design.

The divisive normalization terms identified in Section 5.4 through a statistical model for vector fields mimics population coding mechanisms observed in biology and have been very successfully applied using derivative of Gaussian models of simple cells, the steerable complex wavelets described in this chapter, and the DTCWT platform of Kingsbury [28]. Evaluations applied to both test and natural images have verified that the performance of these detection schemes is comparable to other known low-to-midlevel techniques currently in widespread use, but has the additional advantage of being *generic*, that is, not application specific.

References

[1] Mallat, S., *A Wavelet Tour of Signal Processing*, 2nd ed. Academic Press, 1999.

[2] Burt, P. J. and Adelson, E. H., "The Laplacian pyramid as a compact image code," *IEEE Transactions on Communications*, Vol. COM-31,4, pp. 532–540, 1983. [Online]. Available: citeseer.ist.psu.edu/burt83laplacian.html

[3] Marr, D., *Vision*. Freeman, 1984.

[4] Enroth-Cugell, C. and Robson, J. G., "The contrast sensitivity of retinal ganglion cells of the cat," *Journal of Physiology (London)*, Vol. 187, 1966, pp. 517–552.

[5] Rodieck, R. W., "Quantitative analysis of cat retinal ganglion cell response to visual stimuli," *Vision Research*, Vol. 5, 1965, pp. 583–601.

[6] Marr, D. and Hildreth, E., "Theory of edge detection," *Proceedings of the Royal Society London B*, Vol. 207, 1980, pp. 187–217.

[7] Witkin, A. P., "Scale-space filtering," in *International Joint Conference on Artificial Intelligence*, 1983, pp. 1019–1021.

[8] Koenderink, J., "The structure of images," *Biological Cybernetics*, Vol. 50, no. 5, 1984, pp. 363–370.

[9] Lindeberg, T., "Scale-space theory in computer vision," *Monograph*, 1994. [Online]. Available: citeseer.ifi.unizh.ch/article/lindeberg94scalespace.html

[10] Lowe, D., "Distinctive image features from scale-invariant keypoints," *International Journal of Computer Vision*, Vol. 60, no. 2, 2004, pp. 91–110.

[11] Vaidyanathan, P. P., *Multirate Systems and Filter Banks*. Englewood Cliffs, NJ: PrenticeHall, 1993.

[12] Akansu A. N. and Haddad, P. R., *Multiresolution Signal Decomposition: Transforms, Subbands, Wavelets*, 2nd ed. Academic Press, 2001.

[13] Mallat, S. and Hwang, W. L., "Singularity detection and processing with wavelets," *IEEE Transactions on Information Theory*, Vol. 38, no. 2, 1992, pp. 617–643.

[14] Gabor, D., "Theory of communication," *Journal of the Institute of Electrical Engineers*, Vol. 93, 1946, pp. 429–457.

[15] Marcelja, S., "Mathematical description of the responses of simple cortical cells," *Journal of the Optical Society of America*, Vol. 70, no. 11, 1980, pp. 1297-1300.

[16] Daugman, J., "Uncertainty relation for resolution in space, spatial-frequency, and orientation optimized by two-dimensional visual cortical filters," *Journal of the Optical Society of America A-Optics Image Science and Vision*, Vol. 2, no. 7, 1985, pp. 1160-1169.

[17] ——, "High confidence visual recognition of persons by a test of statistical independence," *IEEE Transactions on Pattern Analysis and Machine Intelligence*, Vol. 15, Nov 1993, pp. 1148-1161.

[18] Granlund, G. and Knutsson, H., *Signal Processing for Computer Vision*. Kluwer Academic Publishers, 1994.

[19] Granlund, G. H., "In search of a general picture processing operator," *Computer Graphics and Image Processing (CGIP)*, Vol. 2, 1978, pp. 155-173. [Online]. Available: citeseer.ifi.unizh.ch/article/lindeberg94scalespace.html

[20] Bigun, J., Granlund, G. and Wiklund, J., "Multidimensional orientation estimation with applications to texture analysis and optical flow," *IEEE Transactions on Pattern Analysis and Machine Intelligence*, Vol. 13, Aug 1991, pp. 775-790.

[21] Daubechies, I., "Orthonormal basis of compactly supported wavelets," *Communications on Pure Applied Math*, Vol. 41, 1988, pp. 909-996.

[22] ——, "The wavelet transform, time-frequency localization and signal analysis," *IEEE Transactions on Information Theory*, Vol. 36, Sept 1990, no. 5, pp. 961-1005.

[23] Mallat, S., "Multifrequency channel decompositions of images and wavelet models," *IEEE Transactions on Acoustics, Speech, and Signal Processing*, Vol. 37, pp. 2091-2110, Dec 1989.

[24] Simoncelli, E., et al., "Shiftable multi-scale transforms," *IEEE Transactions on Information Theory*, Vol. 38, March 1992, no. 2, pp. 587-607.

[25] Freeman, W. T. and Adelson, E. H., "The design and use of steerable filters," *IEEE Transactions on Pattern Analysis and Machine Intelligence*, Vol. 13, 1991, no. 9, pp. 891-906.

[26] Perona, P., "Deformable kernels for early vision," *IEEE Transactions on Pattern Analysis and Machine Intelligence*, Vol. 17, May 1995, no. 5, pp. 488-499.

[27] Bovik, A., Clark, M., and Geisler, W., "Multichannel texture analysis using localized spatial filters," *IEEE Transactions on Pattern Analysis and Machine Intelligence*, Vol. 12, Jan 1990, pp. 55-73.

[28] Kingsbury, N. G., "Complex wavelets for shift invariant analysis and filtering of signals," *Journal of Applied and Computational Harmonic Analysis*, no. 3, May 2001, pp. 234-253.

[29] Fauqueur, J., Kingsbury, N. and Anderson, R., "Multiscale keypoint detection using the dual tree complex wavelet transform," in *IEEE International Conference on Image Processing*, 2006.

[30] Kingsbury, N., "Rotation invariant local feature matching with complex wavelets," in *EUSIPCO*, 2006.

[31] Selesnick, I. W., "The double-density dual-tree dwt," *IEEE Transactions on Signal Processing*, Vol. 52, no. 5, May 2004, pp. 1304-1314.

[32] Bharath, A. A. and Ng, J., "A steerable complex wavelet construction and its application to image denoising," *IEEE Transactions on Image Processing*, Vol. 14, no. 7, July 2005, pp. 948-959.

[33] Bharath, A. A., "A tiling of phase-space through self-convolution," *IEEE Transactions on Signal Processing*, Vol. 48, no. 12, December 2000, pp. 3581-3585.

[34] Portilla, J., et al., "Image denoising using Gaussian scale mixtures in the wavelet domain," *IEEE Transactions on Image Processing*, Vol. 12, no. 11, 2003, pp. 1338-1351.

[35] Ng, J., Bharath, A., and Chow, P., "Extrapolative spatial models for detecting perceptual boundaries in colour images," *International Journal of Computer Vision*, Vol. 73, no. 2, 2007, pp. 179-194.

[36] Seo, J. and Yoo, C., "Image watermarking based on invariant regions of scale-space representation," *IEEE Transactions on Signal Processing*, Vol. 54, no. 4, 2006, pp. 1537-1549.

[37] Martinez-Perez, M., et al., "Retinal blood vessel segmentation by means of scale-space analysis and region growing," in *Lecture Notes in Computer Science*, Vol. 1679, 1999, pp. 90-97.

[38] Martinez-Perez, M., et al., "Segmentation of blood vessels from red-free and fluorescein retinal images," *Medical Image Analysis*, Vol. 11, no. 1, Feb 2007, pp. 47-61.

[39] Martinez, L. and Alonso, J., "Complex receptive fields in primary visual cortex," *Neuroscientist*, Vol. 9, no. 5, Oct 2003, pp. 317-331.

[40] Dayan, P. and Abbott, L. F., *Theoretical Neuroscience*. MIT Press, 2001.

[41] Field, D. J., "Wavelets, vision and the statistics of natural scenery," *Philosophical Transactions of the Royal Society - A*, Vol. 357, no. 1760, 1999, pp. 2527-2542.

[42] Bharath, A. A., "Steerable filters from Erlang functions," in *Proceedings of the 9th British Machine Vision Conference*, 1998, pp. 144-153.

[43] Duits, R., et al., "α scale spaces on a bounded domain," in *Scale Space '03*, ser. LNCS, L. D. Griffin and M. Lillholm, Eds., Vol. 2695. Heidelberg: Springer, 2003, pp. 502-518.

[44] Lindeberg, T., "Principles for automatic scale selection," in *Handbook on Computer Vision and Applications*. Academic Press, 1999, Vol. 2, pp. 239-274.

[45] Hel-Or, Y. and Teo, P., "Canonical decomposition of steerable functions," *Journal of Mathematical Imaging and Vision*, Vol. 9, no. 1, 1998, pp. 83-95.

[46] Itti, L., Koch, C. and Niebur, E., "A model of saliency-based visual attention for rapid scene analysis," *IEEE Transactions on Pattern Analysis and Machine Intelligence*, Vol. 20, no. 11, Nov 1998, pp. 1254-1259.

[47] Morrone, M. C. and Burr, D. C., "Feature detection in human vision: A phase-dependent energy model," *Proceedings of the Royal Society of London, B*, Vol. 235, 1988, pp. 221-245.

[48] Ringach, D., Shapley, R. and Hawken, M., "Orientation selectivity in macaque v1: Diversity and laminar dependence," *Journal of Neuroscience*, Vol. 22, no. 13, 2002, pp. 5639-5651.

[49] Martínez-Perez, M. E., et al., "Segmentation of retinal blood vessels based on the second directional derivative and region growing," in *Proceedings of International Conference on Image Processing*, Vol. 2, 1999, pp. 173-176.

[50] Kovesi, P., "Phase congruency: A low-level image invariant," *Psychological Research Psychologische Forschung*, Vol. 64, no. 2, 2000, pp. 136-148.

[51] Mackay, D. J., *Information Theory, Inference and Learning Algorithms*. Cambridge: Cambridge University Press, 2003.

[52] Ng, J., Bharath, A. A. and Zhaoping, L., "A survey of architecture and function of the primary visual cortex (V1)," *EURASIP Journal on Advances in Signal Processing*, Vol. 2007, 2007, pp. Article ID 97961, 17 pages, doi:10.1155/2007/97961.

[53] Wandell, B., *Foundations of Vision*. Sinauer Associates Inc., 1995.

[54] Ng, J. and Bharath, A., "Steering in scale space to optimally detect image structures," in *ECCV '04*, ser. Lecture Notes in Computer Science, Vol. 3021. Springer, 2004, pp. 482-494.

[55] Sendur, L. and Selesnick, I. W., "Bivariate shrinkage functions for wavelet-based denoising exploiting interscale dependency," *IEEE Transactions on Signal Processing*, Vol. 50, no. 11, pp. 2744-2756, Nov 2002.

[56] Olshausen, B. and Field, D., "Sparse coding with an overcomplete basis set: A strategy employed by V1?" *Vision Research*, Vol. 37, 1997, pp. 3311–3325.

[57] Evans, M. and Swartz, T., *Approximating Integrals via Monte Carlo and Deterministic Methods*. Oxford University Press, 2000.

[58] Gradshteyn, I., Ryzhik, I. and Jeffrey, A., *Tables of Integrals, Series and Products: 5th Edition*. Academic Press, 1994.

[59] Kerbyson, J. and Atherton, T., "Circle detection using Hough Transform filters," in *Image Analysis and its Applications, 1995, IEE Conference Publication No. 410*, 1995, pp. 370–374.

[60] Peterson, W., Birdsall, T. and Fox, W., "The theory of signal detectability," *IEEE Transactions on Information Theory*, Vol. 4, Sept 1954, no. 4, pp. 171–212.

[61] Martinez, A. and Benavente, R., "The AR face database," Purdue University, Tech. Rep. 24, June 1998.

[62] Imperial Innovations Limited, "Device for computer vision," Patent Application, August 2006.

[63] Lim, M., "Visual features in human and machine vision," Ph.D. dissertation, Imperial College London, 2007.

[64] Bharath, A. A. and Kingsbury, N., "Phase invariant keypoint detection," in *International Conference on Digital Signal Processing (DSP 2007)*, Cardiff, Wales, 2007.

CHAPTER 6
Beyond the Representation of Images by Rectangular Grids

Maria Petrou

6.1 Introduction

The sampling of a scene by a set of sensors placed on a rectangular grid is usually unquestionable and taken for granted. The human visual system, however, does not rely on such a grid. The fundamental question we are trying to answer here is: can we do image processing when the data are not available on a regular grid? The answer to this question is yes. In fact, it has been shown by several researchers that the use of irregular grids may even lead to more robust representations. An example is the face authentication system presented in [21], which was based on representing the faces by values along tracing lines as opposed to values at sampling points, and in blind tests it performed two orders of magnitude better than other competing algorithms. Scientists in many disciplines had to deal with irregularly sampled data for many years, and they developed several techniques for processing such data. This chapter is an overview of some of these approaches.

Some techniques of image processing are readily applicable to irregularly sampled data, for example, the use of co-occurrence matrices for texture analysis and mathematical morphology for texture or shape analysis. However, linear image processing is largely based on the use of a regular grid, being convolution or transform based.

There are two approaches we may adopt to do image processing when the data is not regularly sampled.

- Before we apply any standard image processing algorithm we may interpolate the data so that they appear on a regular grid. The process of such an interpolation is known as *griding*. This approach is a two-step process. The extra step, namely that of interpolation, may introduce errors.
- To avoid the interpolation errors, one may try to perform the image processing operations directly on the available irregularly sampled data.

In this chapter we start by discussing issues related to linear image processing and then we touch upon some aspects of nonlinear image processing.

6.2 Linear Image Processing

Linear image processing is based on linear image transformations and convolutions. One way to deal with irregularly sampled data is to interpolate first onto a regular grid and apply subsequently the desired transformation. Alternatively, one may deal directly with the irregular samples bypassing the interpolation step. We shall discuss both approaches here.

6.2.1 Interpolation of Irregularly Sampled Data

The three most popular methods for this are:

- Kriging;
- Iterative error correction;
- Normalized convolution.

6.2.1.1 Kriging

In a nutshell, according to the method of Kriging, we give to the missing points values that are weighted linear combinations of the values of the points we have, choosing the weights so that the covariances of the data are preserved. Kriging is known to be the best linear unbiased estimator (BLUE).

Let us see how it works in detail. First, let us denote by $P_1, P_2, \ldots, P_n$, the irregularly placed data points; by V_i, the value of random variable V at point P_i ($V_i \equiv V(P_i)$); by m, the mean of values V_i; and by σ^2, their variance.

The problem we are called to solve is: estimate the value of variable V at point P_0.

According to Kriging estimation, the sought value $\hat{V}(P_0)$ is

$$\hat{V}(P_0) = \sum_{i=1}^{n} w_i(P_0) V(P_i) \tag{6.1}$$

where $w_i(P_0)$ are the weights with which the available values of the n points P_i will be combined. The residual error of this estimate is

$$R(P_0) \equiv \hat{V}(P_0) - V(P_0) \tag{6.2}$$

where $V(P_0)$ is the true value at position P_0. We must choose weights w_i so that the variance of error $R(P_0)$ is minimized.

It can easily be shown that the variance of the error function is given by

$$\tilde{\sigma}_R^2 = \sum_{i=1}^{n} \sum_{j=1}^{n} w_i w_j \tilde{C}_{ij} - 2 \sum_{i=1}^{n} w_i \tilde{C}_{i0} + \sigma^2 \tag{6.3}$$

where $\tilde{C}_{ij}$ is the covariance between random variables V_i and V_j, and $\tilde{C}_{i0}$ is the covariance between random variables V_i and V_0.

From the available data we can work out the covariance matrix C. We may then choose the weights so that the variance of the error is minimal, subject to the condition:

$$\sum_{i=1}^{n} w_i = 1 \tag{6.4}$$

We may use for this optimization problem the Lagrange parameter method, according to which we combine the function that has to be minimized with the constraint expressed as a function that has to be 0, using a parameter that is called the Lagrange multiplier.

So, according to this method, we must minimize

$$\hat{\sigma}_R^2 \equiv \sum_{i=1}^{n}\sum_{j=1}^{n} w_i w_j \tilde{C}_{ij} - 2\sum_{i=1}^{n} w_i \tilde{C}_{i0} + \sigma^2 + 2\mu \underbrace{\left(\sum_{i=1}^{n} w_i - 1\right)}_{\text{must be 0}} \quad (6.5)$$

where μ is the Lagrange multiplier.

Now we have $(n+1)$ unknowns, namely the n weights plus μ. We differentiate Equation (6.5) with respect to the $(n+1)$ unknowns and set these first partial derivatives to zero. We obtain

$$CW = D \quad (6.6)$$

where C is the covariance matrix of the available data,

$$C = \begin{pmatrix} \tilde{C}_{11} & \cdots & \tilde{C}_{1n} & 1 \\ \cdots & \cdots & \cdots & \cdots \\ \tilde{C}_{n1} & \cdots & \tilde{C}_{nn} & 1 \\ 1 & \cdots & 1 & 0 \end{pmatrix}_{(n+1)\times(n+1)} \quad (6.7)$$

W is the vector of the sought unknowns,

$$W = \begin{pmatrix} w_1 \\ \cdots \\ w_n \\ \mu \end{pmatrix}_{(n+1)\times 1} \quad (6.8)$$

and D is made up from the covariances of the unknown value we seek to estimate and the known values:

$$D = \begin{pmatrix} \tilde{C}_{10} \\ \cdots \\ \tilde{C}_{n0} \\ 1 \end{pmatrix}_{(n+1)\times 1} \quad (6.9)$$

The elements of D are chosen to agree with the elements of C for the same shift, estimated from the data. Once C and D are known, Equation (6.6) can be easily solved for the unknowns:

$$W = C^{-1}D \quad (6.10)$$

Estimation of the Covariance Matrix of the Data

The covariance matrix of the data is estimated from the so-called variogram (or "semivariogram" as some people call it[1]) of the data. The (semi)variogram is de-

---

1. We shall use here the term "variogram" to mean $\gamma(h)$ as defined by Equation (6.11).

fined as half the expected squared difference between two data points separated by a distance vector $\bar{h}$ with magnitude h and any orientation:

$$\gamma(h) \equiv \frac{1}{2} E\{[V(P_i) - V(P_i + \bar{h})]^2\} \tag{6.11}$$

Here E is the expectation operator. If $N(h)$ is the total number of distinct pairs of data points V_i and V_j, which are at a distance $d_{ij} = h$ from each other, we may write:

$$\gamma(h) = \frac{1}{2|N(h)|} \sum_{(i,j)|d_{ij}=h} (V_i - V_j)^2 \tag{6.12}$$

It can be easily shown that the variogram and the corresponding covariance function $\tilde{C}(h)$ are related by:

$$\gamma(h) = \sigma^2 - \tilde{C}(h) \tag{6.13}$$

To reduce the effect of noise, we use a parametric model to fit the variogram. The use of a model also has the advantage that it allows us to estimate the values of the correlation between the unknown value and the rest (i.e., the elements of vector D).

Variogram Models
It is understood that no model will fit the variogram for the full range of its argument. It is expected that as h becomes large, the pairs of samples will no longer be correlated and so the value of γ will tend to a constant characterizing the randomness of the uncorrelated values. In general, we accept that any model will fit over a finite range of h and so three characteristic parameters of the semivariogram are defined to express this.

- **Nugget** For a very small value of the distance, it is expected that the value of the semivariogram will reach zero but if the value of the variogram does not approach zero due to sampling error or some other factors, then this nonzero value is known as the nugget effect.
- **Range** As the distance between the data points increases, the value of the semivariogram also increases. After a certain point, the increase in the distance does not have any effect on the value of the variogram, that is, after this particular distance the value of the variogram becomes constant. This particular distance is known as the range.
- **Sill** The maximum value attained by the semivariogram at the range is known as the sill.

The variogram then may be fitted by a variety of models, the most commonly used of which are the following.
Fractal model
This is a very commonly used model, given by

$$\tilde{\gamma}(h) = \gamma_0 h^{2H} \Rightarrow \log \tilde{\gamma}(h) = \log \gamma_0 + 2H \log h \tag{6.14}$$

6.2 Linear Image Processing

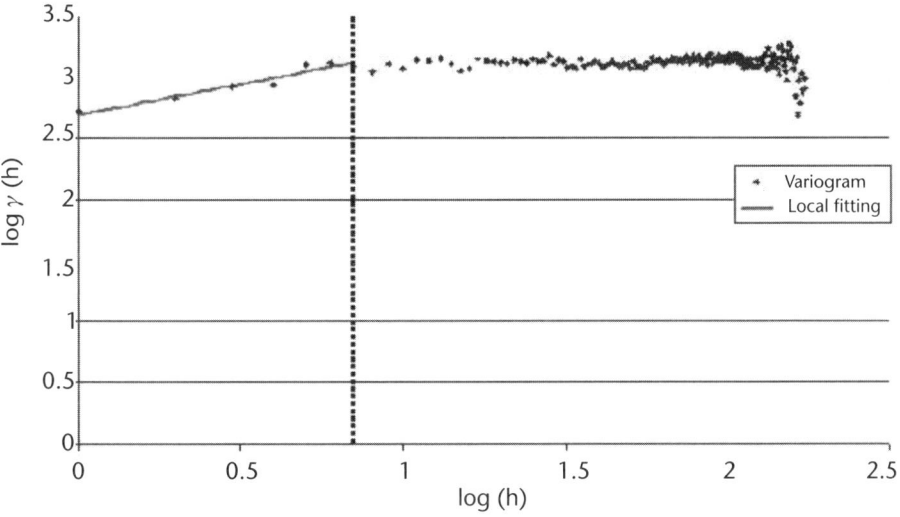

Figure 6.1 The variogram of an image and the fitted fractal model.

where γ_0 and H are model parameters to be estimated from the data, alongside the range h_R, the sill, and the nugget.

An example of variogram fitting with the fractal model is shown in Figure 6.1. The vertical dotted line in this figure is the range h_R, that is, the distance up to which the model fits.

Exponential model

$$\tilde{\gamma}(h) = C_0 + C_1\left(1 - \exp\left(\frac{-|h|}{h_R}\right)\right) \tag{6.15}$$

Spherical model

$$\tilde{\gamma}(h) = C_0 + C_1\left(1.5\frac{h}{h_R} - 0.5\left(\frac{h}{h_R}\right)^3\right) \tag{6.16}$$

Gaussian model

$$\tilde{\gamma}(h) = C_0 + C_1 - C_1\exp\left(\frac{-|h|^2}{h_R^2}\right) \tag{6.17}$$

Linear model

$$\tilde{\gamma}(h) = C_0 + C_1\frac{h}{h_R} \tag{6.18}$$

In the above models, the range h_R and model parameters C_0 and C_1 are estimated from the data.

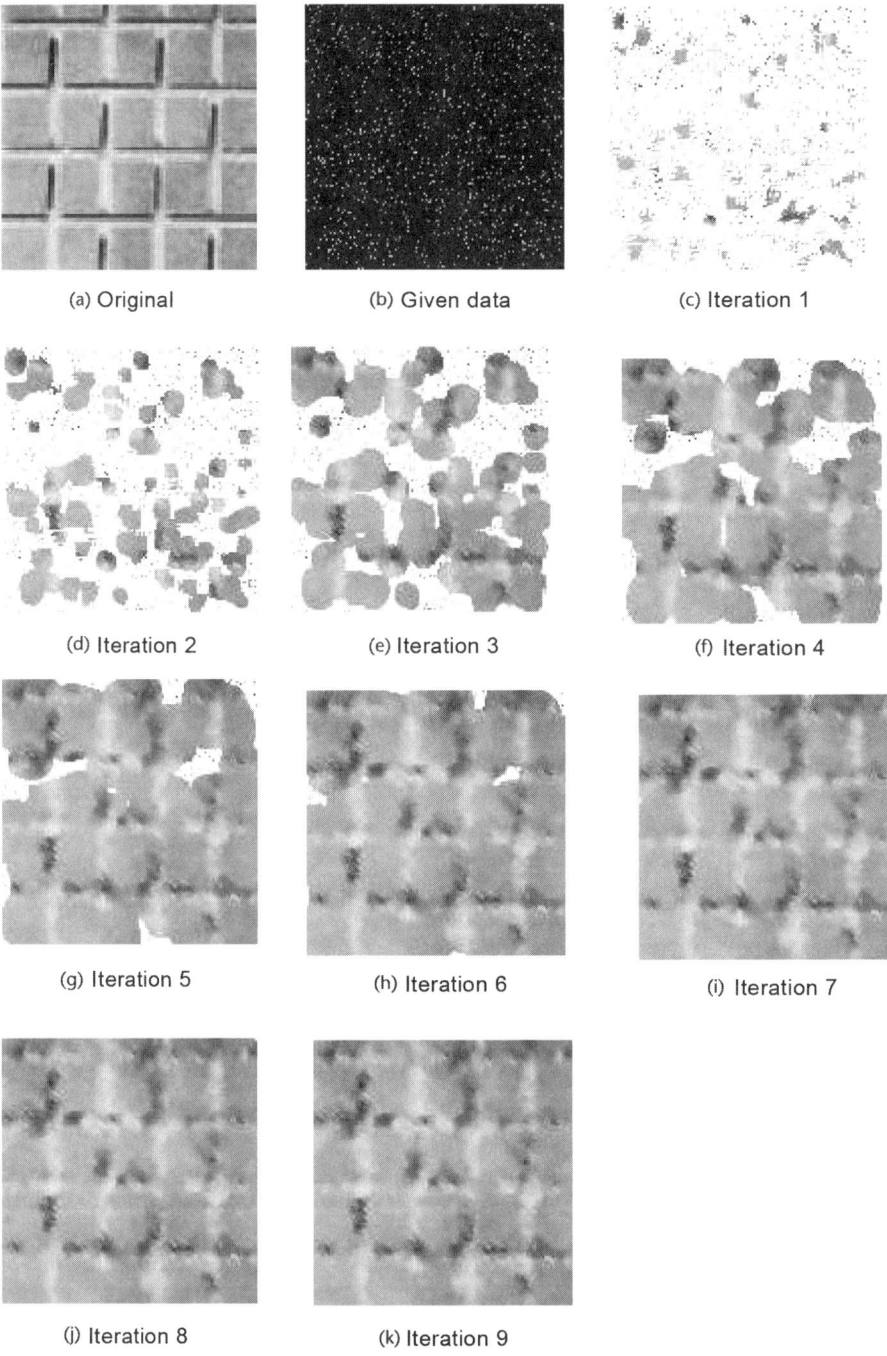

Figure 6.2 The original pattern may be recovered from only 6% of the original data points, using Kriging and recovering the pattern in the form of growing islands around the available data.

Application to Image Interpolation
Chandra et al. [13] used the Kriging approach to interpolate irregularly sampled images. Figure 6.2 shows some example results. To demonstrate the method, the original image was irregularly undersampled, keeping only 6% of the original number of pixels. The results shown were obtained with the fractal model. The maximum value of h over which the model was valid, that is, the range h_R, was used to define a local window around each position with missing value. Only data values inside the local window were used to estimate the missing value. As the data were quite sparse, many pixels with missing values did not have enough neighbors inside the local window to allow the estimate of their value. The algorithm was applied several times, each time estimating the values of pixels with at least five neighbors with values, and allowing the pattern to grow in the form of islands around the available values.

6.2.1.2 Iterative Error Correction

Under this umbrella there is a whole set of methods based on the theory of frames. According to this theory, under certain conditions, it is possible to determine frame functions from which a frame operator may be defined appropriate for reconstructing a signal from its samples, in an iterative way.

The problem may be stated as follows: given a sequence of sampled values of some unknown signal, design functions called *frames* such that the internal product, indicated by $< \cdot >$, of the unknown signal with the frame functions is equal to the values of the signal at the sampling points.

The so-called frame functions are like basis functions of the function space that cover the whole space and are used to express any signal as a linear combination of them. However, the differences between frames and basis functions are the following.

1. The coefficients of the expansion in terms of basis functions are unique; not so the expansion in terms of frames.
2. The number of frames is larger than the number of basis functions (frames do not constitute an orthogonal basis).

The frames-based approach will become clearer with an example. Let us denote by F the function we wish to reconstruct, by f the values of the function we have for certain of its arguments, and by A the frame operator we shall use. We start by applying operator A on f to obtain a first estimate of F, say, F_0:

$$F_0 = Af \qquad (6.19)$$

If F_n indicates the estimate of F at the nth iteration and f_n the estimated values at the positions where we know the values, we set

$$F_{n+1} = F_n + A(f - f_n) \qquad (6.20)$$

where operator A is understood to be applied to the points for which data is available (i.e., to the f part of F) yielding values for the full range of arguments of F. For an appropriate A it can be shown that

$$\lim_{n \to \infty} F_n = F \qquad (6.21)$$

with the estimation error after n iterations being

$$||F - F_n|| \leq \gamma^{n+1} ||F|| \qquad (6.22)$$

where $\gamma < 1$ is a constant and $||\ ||$ is an appropriate norm.

There are several different methods according to the type of the initial estimate F_0 and operator A used. Some of them are the following.

- The Wiley-Marvasti method [11]: the initial guess is a trivial interpolation, where the values of all unknown points are set equal to zero.
- The adaptive weights method [4]: the trivial interpolation is weighted by some factors that reflect the distance of the irregularly placed samples from the neighboring sampling points.
- The Voronoi method [22]: it uses the nearest neighbor interpolation.
- The piecewise-linear method [2]: it performs the interpolation with a linear continuous function defined between successive sampling points.
- The projection onto convex sets (POCS) method [12]: operator A here is not linear; the iterations are obtained by the application of successive projections. A projection is equivalent to lowpass filtering in the Fourier domain. A projection, therefore, limits the bandwidth of the signal.

The fastest algorithm from the above appears to be the adaptive weights algorithm. More details may be found in [20].

The way these approaches work will become clearer with an application example.

Let F be the image we wish to recover, defined over a regular grid, and f a set of N irregularly placed pixels for which we have values, identified by $i = 1, \ldots, N$. We use the following algorithm.

Step 1 Use the Voronoi method to interpolate the samples on a regular grid. This is equivalent to saying that every pixel of the regular grid is given the value of its nearest pixel with known value. [This defines operator A in Equation (6.19).] This way you obtain a first estimate of F, call it F_0. [This corresponds to Equation (6.19).]

Step 2 Calculate the Fourier transform of the interpolation you obtained this way.

Step 3 Discard the high frequencies. (You know that by performing nearest neighbor interpolation you mess up the high frequencies, so you might as well omit them as containing wrong information.)

Step 4 Calculate the inverse Fourier transform, to produce an estimate of the image, F_1.

Step 5 Having omitted the high frequencies, you damaged also the values at the pixel positions for which you knew the correct values at the beginning. Calculate the error committed at those positions $f(i) - f_1(i)$, for $i = 1, \ldots, N$. [Steps 2 to 5 correspond to computing the difference that appears on the right-hand side of Equation (6.20).]

Step 6 Interpolate the error using the Voronoi method, that is, apply nearest neighbor interpolation to the error field. [This corresponds to computing the second term on the right-hand side of Equation (6.20).]

Step 7 Use the estimated error at each pixel of the regular grid to produce an improved estimate F_2 of the image. [This corresponds to Equation (6.20).]

Go to step 2 and repeat until you achieve convergence.

Convergence is guaranteed if the maximal gap between two irregularly placed samples is smaller than the Nyquist limit of the imaged continuous scene. Typically, 20 iterations suffice for a good reconstruction.

6.2.1.3 Normalized Convolution

This is the only method of interpolation of irregularly sampled images that was developed by computer vision scientists directly [7].

Let $f(x,y)$ be an image and $g(x,y)$ a smoothing filter. Let (x_s, y_s), for $s = 1, \ldots, S$, be random positions for which the image values are known. We define the sampling mask $c(x,y)$ to have value 1 at pixels with known values and value 0 at pixels with unknown values:

$$c(x,y) = \begin{cases} 1 & \text{if } (x,y) = (x_s, y_s) \text{ for some } s \in [1,S] \\ 0 & \text{otherwise} \end{cases} \tag{6.23}$$

The algorithm then is as follows.

Step 1 Convolve $f(x,y)c(x,y)$ with $g(x,y)$:

$$C(x,y) \equiv (f(x,y)c(x,y)) * g(x,y) \tag{6.24}$$

This step effectively tells us to put 0s at the regular grid positions with missing values and convolve it with a smoothing filter.

Step 2 Convolve $c(x,y)$ with $g(x,y)$:

$$NC(x,y) \equiv c(x,y) * g(x,y) \tag{6.25}$$

This step effectively tells us to replace the known values with 1s in the regular grid and convolve with the same smoothing filter.

Step 3 Divide the two results point by point:

$$\tilde{f}(x,y) = \frac{C(x,y)}{NC(x,y)} \tag{6.26}$$

This is the estimated full image.

One may use as a smoothing filter $g(x,y)$ a lowpass filter, for example, the integral of the Canny edge enhancing filter [14]. The proposers of this method, Knutsson and Westin [7], used

$$g(x,y) = \begin{cases} r^{-\alpha} \cos^\beta \left(\dfrac{\pi r}{2 r_{max}} \right) & \text{if } r < r_{max} \\ 0 & \text{otherwise} \end{cases} \tag{6.27}$$

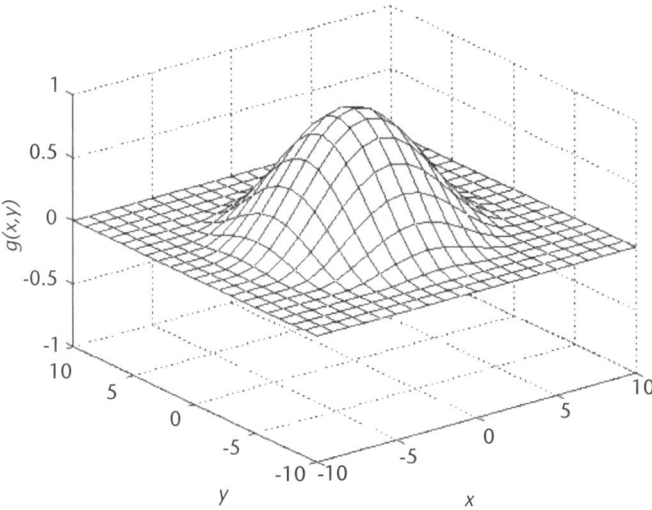

Figure 6.3 The lowpass filter used by Knutsson and Westin [7] for normalized convolution.

Figure 6.4 Reconstructing Lena from only 10% of the original number of pixels. (a) original image, (b) data available, (c) result of convolving (b) with $g(x,y)$, (d) normalized convolution.

where $r \equiv \sqrt{x^2 + y^2}$, and α and β are some positive integers. Filter $g(x,y)$ for $\alpha = 0$, $\beta = 2$, and $r_{max} = 8$ is shown in Figure 6.3.

Figure 6.4 shows an original image and its reconstruction from 10% of the original number of pixels using normalized convolution and the filter shown in Figure 6.3.

Figure 6.5 shows an example of an image reconstruction, starting with only 5% of its original pixels, and using iterative error correction and normalized convolution.

In general, the error iterative correction method produces better results than normalized convolution. However, normalized convolution is effectively a single pass algorithm, and it turns out that it can cope with lower rates of subsampling (bigger gaps in the data). In addition, normalized convolution opens up the way of performing linear image processing on irregularly sampled scenes, since linear image processing largely relies on convolutions. For example, gradient detection directly from irregularly sampled data is now straightforward. Figure 6.6 shows the result of computing the gradient magnitude map from the full Lena image and from its subsampled version, keeping only 10% of its pixels, and using normalized convolution. For both cases the Canny filter was used.

Note that the human brain does not seem to be doing edge detection, that is, nonmaxima suppression and thinning to the gradient magnitude, but rather

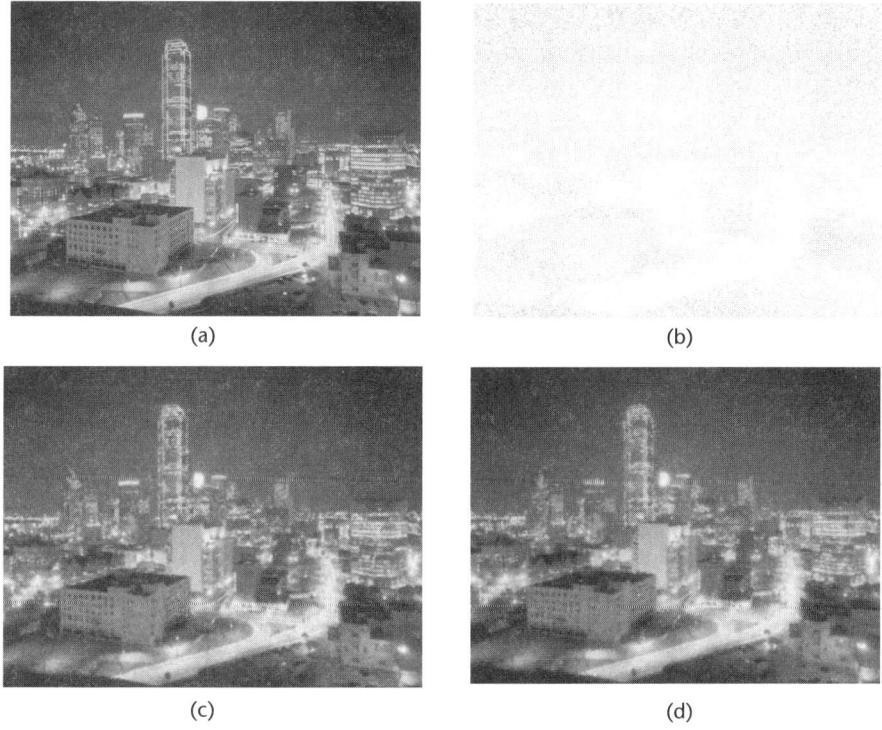

Figure 6.5 The original image shown in (a) was reconstructed from only 5% of the original data points shown in (b), using the iterative error correction method with Voronoi interpolation in (c) and normalized convolution in (d).

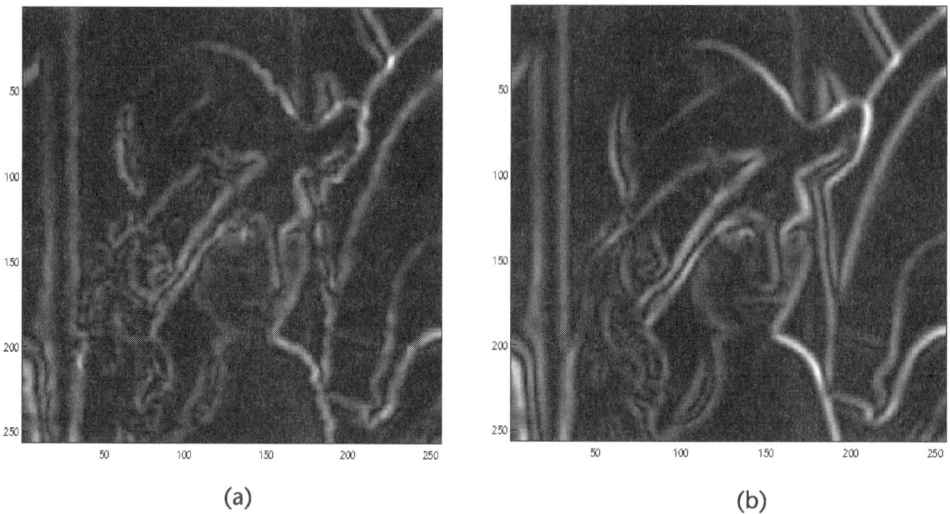

Figure 6.6 Gradient magnitude from 10% of the pixels (a) and from the full image (b) using the Canny filter.

gradient magnitude and orientation estimation followed by salienation. We shall discuss salienation in Section 6.3.1.

The ability to estimate gradients directly from the irregularly placed samples raises the question as to whether we can compute other features from irregularly placed samples that will allow the solution of other problems of image processing, like texture segmentation and shape recognition. Some important methods of image processing rely on the calculation of the Fourier transform of the data. Such methods are, for example, texture segmentation methods based on the use of Gabor functions and wavelets [15]. In the next section we discuss how Fourier components may be estimated from irregularly placed data directly, without griding.

6.2.2 DFT from Irregularly Sampled Data

We start by remembering how the continuous Fourier transform of a function $p(x)$ is defined:

$$P(\omega) = \int_{-\infty}^{+\infty} p(x) e^{-j\omega x} dx \qquad (6.28)$$

Here ω is the frequency. Its inverse is given by:

$$p(x) = \frac{1}{2\pi} \int_{-\infty}^{+\infty} P(\omega) e^{j\omega x} d\omega \qquad (6.29)$$

Assume now that function $p(x)$ is sampled at N regular intervals equal to T_S, to produce samples p_n, so that: $p_n \equiv p(x_n)$ where $x_n = nT_S$, for $n = 0, \ldots, N-1$. The total signal duration is $T \equiv NT_S$. The Fourier transform of p_n is defined only at certain regularly spaced frequencies, that is, function $P(\omega)$ is only computed for certain values ω_m, which are multiples of the dominant frequency $\frac{2\pi}{T}$: $\omega_m \equiv m\left(\frac{2\pi}{T}\right) = \frac{2\pi m}{NT_S}$, for $m = 0, \ldots, N - 1$. Then

$$P(\omega_m) \equiv \sum_{n=0}^{N-1} p(x_n) e^{-j\omega_m x_n} \qquad (6.30)$$

or:

$$P(\omega_m) = \sum_{n=0}^{N-1} p(x_n) e^{-j\left(m\frac{2\pi}{NT_S}\right)(nT_S)} = \sum_{n=0}^{N-1} p(x_n) e^{-jm\frac{2\pi}{N}n} \qquad (6.31)$$

So, the discrete Fourier transform (DFT) for regularly sampled data is given by:

$$P(m) = \sum_{n=0}^{N-1} p(n) e^{-j\frac{2\pi}{N}mn} \qquad (6.32)$$

Its inverse is:

$$p(n) = \frac{1}{N} \sum_{m=0}^{N-1} P(m) e^{j\frac{2\pi}{N}mn} \qquad (6.33)$$

Note that here we simply used indices m and n instead of ω_m and x_n, respectively.

Now consider the form (6.30) would have had if we did not know that samples x_n were placed regularly inside the range $[0, T]$:

$$P(m) = \sum_{n=0}^{N-1} p(n) e^{-j\frac{2\pi}{T}mx_n} \qquad (6.34)$$

So, the DFT for nonregularly placed samples is only different from the DFT for regularly placed samples in the fact that the actual spatial coordinate x_n of the nth sample appears in the exponent of the kernel rather than index n.

Once we know how to estimate frequency components from irregularly placed samples, effectively all texture segmentation methods are usable: we already mentioned that co-occurrence matrix methods as well as morphology-based methods (e.g., granulometry) are applicable to irregularly sampled textures; we next showed how Fourier-based methods like Gabor function-based and wavelet-based ones may also be made applicable. This means that a large part of image processing can be done directly from irregularly sampled scenes. However, a lot of image processing and pattern recognition still remain rooted on the use of a regular grid for data representation. This fundamental issue will be discussed in the next section.

6.3 Nonlinear Image Processing

The simplest form of nonlinear image processing is thresholding. One of the first algorithms a novice comes across, when studying image processing, is edge detection. After the calculation of the gradient magnitude at each pixel, the student learns that we perform nonmaxima suppression and thresholding, so that we produce a thin-edge map of the scene, keeping only gradient magnitudes that are local maxima in the direction of the gradient orientation and stronger than a certain threshold. It appears, however, that the human brain does not perform such operations. It rather enhances and keeps edges that are *salient* than simply have strong contrast. This process is thought to take place in the V1 part of the visual cortex [8, 9]. For completeness, we include here the basic model proposed and show what its effect is in comparison with the conventional edge detection algorithms as implemented by image processing packages.

However, nonlinear image processing goes beyond just edge detection: it reaches the very foundations of visual perception and recognition. One may use 1,000 words to describe a familiar face to a third person, and yet, the third party will not be able to recognize that face in a crowd, unless some very distinct mark, like a big black mole on the nose, characterizes that face. Even so, the person who is familiar with the face could have recognized it even if a narrow slot in the door opened and the face peeped through it for a moment, exposing just a small part of it. The fact that the same person can use 1,000 words without being able to uniquely identify the face and yet be able to recognize it from some momentary exposure to part of it, shows that pattern recognition relies on features that are not perceptually identifiable. And yet, a lot of pattern recognition algorithms rely on features that have a linguistic descriptor. We advocate here that pattern recognition has to break free of conventional thinking, starting from the point of data representation and moving in the direction of extracting image features that have the desirable properties for the task, but which do not necessarily make perceptual sense. Thus, we advocate here the *subconscious image processing*, and present a possible tool for that, that allows one to do exactly this: to move away from the conventional rectangular grid of data representation and to extract features from the image that are not constrained by human language. This tool will be presented in section 6.3.2.

6.3.1 V1-Inspired Edge Detection

A model of V1 as a salienator was proposed in [8]. This model is schematically shown in Figure 6.7. Each column in V1 has processors that receive information from the same area in the scene and respond selectively to different orientations (typically 16).

Each orientation sensor, designed to respond to edges at orientation θ at position i, consists of two neurons: the first communicates with the outside world and is associated with an excitation value (membrane potential) $x_{i\theta}$; the second does not receive input from the outside world, but it rather collects all inhibitory signals this sensor receives from other sensors and communicates them, as an inhibitory membrane potential $y_{i\theta}$, to the first neuron. The membrane potentials of the two neurons of sensor $i\theta$ vary with time, according to the following set of coupled differential equations:

$$\frac{dx_{i\theta}}{dt} = -\alpha_x x_{i\theta} - \sum_{\Delta\theta} \psi(\Delta\theta) g_y(y_{i,\theta+\Delta\theta}) \\ + J_0 g_x(x_{i\theta}) + \sum_{j \neq i, \theta'} J_{i\theta, j\theta'} g_x(x_{j\theta'}) + I_{i\theta} + I_0 \qquad (6.35)$$

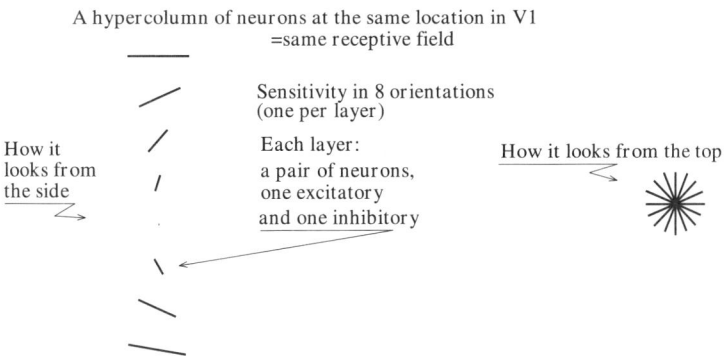

Figure 6.7 A column in V1 consists of neurons that receive input from the same locality, and they are tuned to respond to specific orientations.

$$\frac{dy_{i\theta}}{dt} = -\alpha_y y_{i\theta} + g_x(x_{i\theta}) + \sum_{j \neq i, \theta'} W_{i\theta,j\theta'} g_x(x_{j\theta'}) + I_c \tag{6.36}$$

where α_x and α_y are time constants of decay of the excitation and inhibition values of the sensor, respectively.

$J_{i\theta,j\theta'}$ and $W_{i\theta,j\theta'}$ are the synaptic strengths or *horizontal* connections with other sensors of the excitation and inhibition values of the sensor, respectively.

$g_x(x_{i\theta})$ is a sigmoid activation function, expressing the fact that in nature often things happen after the input variable reaches a certain threshold, beyond which it follows a linear model, until a second threshold is reached, beyond which the response saturates and no gain is received even if the input is increased.

$g_y(y_{i\theta})$ is a linearly increasing function, gently to begin with and more sharply later.

$\psi(\Delta\theta)$ is an even function modelling inhibition within the same column of orientation sensors. It is even because both orientations on either side of the orientation of maximum sensitivity of the sensor are equally suppressed by the inhibitory signals a sensor sends to the other sensors in the same column.

J_0, I_0, and I_c are some constants that allow us to control the model.

$I_{i\theta}$ is the external input to the sensor depending on the gradient map of the image,

$$I_{i\theta} = \hat{I}_{i\beta} \phi(\theta - \beta) \tag{6.37}$$

where $\hat{I}_{i\beta}$ is the gradient magnitude at pixel i and β is the orientation of the gradient vector. Note that the inclusion of the gradient magnitude, as a factor to the alignment of the orientation of a candidate edgel with the orientation selected by a neuron, is important, because it allows weak edge responses, largely due to noise, with unreliable estimation of their orientation, to have reduced effect in the relaxation process. Function $\phi(\theta - \beta)$ is an orientation tuning function between the orientation to which the sensor is sensitive and the orientation of the gradient vector of the pixel:

$$\phi(\theta - \beta) = e^{-\frac{|\theta-\beta|8}{\pi}} \tag{6.38}$$

The exact formulae of the various functions involved and the values of their parameters may be found in [9].

In order to be able to use such a postprocessing scheme for edge detection, we must translate the above formulae into the digital domain. Let us consider that indices ij indicate a particular pixel. With each pixel we associate two K-element long vectors x_{ijk} and y_{ijk}, with k taking values $1, 2, \ldots, K$. In an iterative scheme, where the iteration step is denoted by index n, the elements of x_{ijk} for each pixel (i,j) are updated as follows:

$$\begin{aligned} x_{ijk}^{n+1} &= -\alpha_x x_{ijk}^n - \sum_{l=1, l \neq k}^{K} \psi(l-k) g_y(y_{ijl}^n) \\ &\quad + J_0 g_x(x_{ijk}^n) + \sum_{(m,n) \neq (i,j)} \sum_{l=1}^{K} J_{ijkmnl} g_x(x_{mnl}^n) + I_{ijk} + I_0 \\ y_{ijk}^{n+1} &= -\alpha_y y_{ijk}^n + g_x(x_{ijk}^n) + \sum_{(m,n) \neq (i,j)} \sum_{l=1}^{K} W_{ijkmnl} g_x(x_{mnl}^n) + I_c \end{aligned} \tag{6.39}$$

Each index k (or l) is associated with one of the K orientations to which the continuous valued orientation of an edgel is quantized. This means that any two successive orientations indicated by angles θ_k differ by $180°/K$. Functions J_{ijkmnl} and W_{ijkmnl} express the excitatory and inhibitory signals exchanged between receptive fields (pixels) at locations (i,j) and (m,n) and tuned at orientations k and l, respectively. Figure 6.8 shows how such functions may look like, for

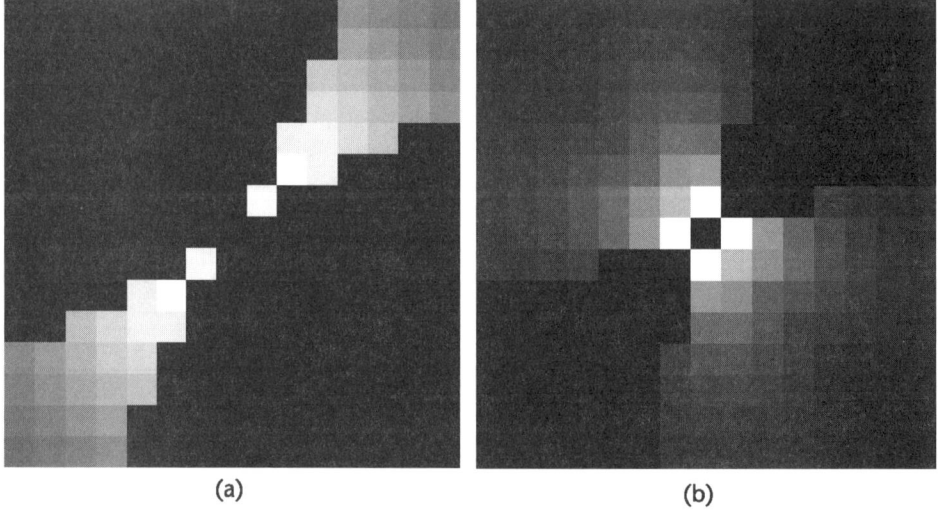

(a)   (b)

Figure 6.8 The excitatory (a) and the inhibitory (b) field of influence of a neuron tuned to respond maximally to an edgel oriented at $\pi/3$ degrees with respect to the horizontal axis.

6.3 Nonlinear Image Processing

orientation $\pi/3$ when $K = 12$ different orientations are used and the messages exchanged between the pixels (receptive fields) are within a local neighborhood of size 15×15.

So, the algorithm for edge detection now is as follows.

- **Step 1** Assign to each pixel (i,j) K possible sensors for K different orientations, each associated with two values, x_{ijk} and y_{ijk}.
- **Step 2** Give some initial values to all x_{ijk} and y_{ijk}, say, 0.5 and 0, respectively.
- **Step 3** Use Equation (6.39) to update the values of x_{ijk} and y_{ijk} iteratively.
- **Step 4** At each iteration step and for each pixel (i,j), select the maximum response $\hat{x}_{ij} \equiv \max_k x_{ijk}$ over all values of k. Threshold values $\hat{x}_{ij}$ to produce the current saliency map.
- **Step 5** Stop when the produced saliency map is satisfactory.

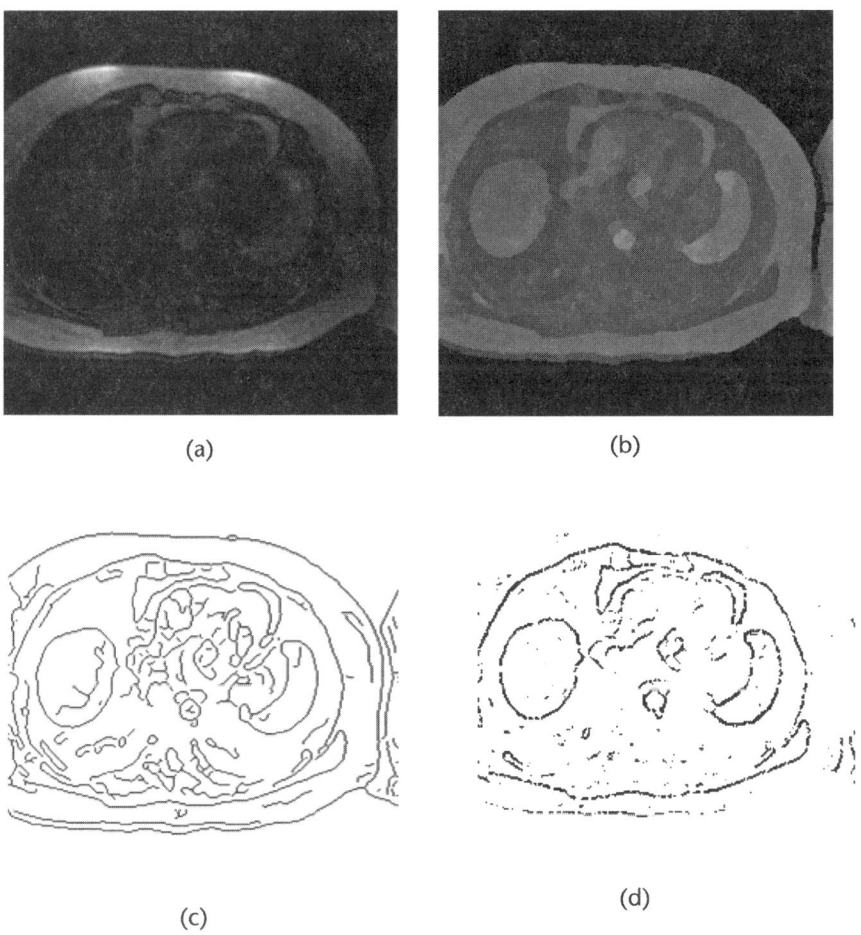

Figure 6.9 In (c) the output of Canny edge detector applied to image (b), which is an enhanced version of magnetic resonance image (a) and in (d) the output of the V1 model-based edge detection algorithm.

Figure 6.9 shows the original magnetic resonance image (MRI) of the liver of a patient and its enhanced version. A conventional edge detector was applied to it and one can see that lots of unwanted noisy edges were detected. This edge detector is based on local contrast strength. On the other hand, the output of the algorithm that is based on (6.39) and outputs for each pixel a saliency value x_{ij}, which is subsequently thresholded, shows the edges that are smooth, regular, and most likely meaningful.

We notice that the above algorithm is vague in terms of the stopping criterion. This is intentionally so. It has been shown that complex systems with mutually interacting units that do not interact in a symmetric way (i.e., A influences B in a different way from the way B influences A) do not converge to a solution [10]. Systems with symmetric interactions converge to a minimum energy state, but when that minimum is not the global minimum, they tend to get stuck to the wrong solution. We say that such systems "hallucinate" when they go wrong [10]. On the other hand, complex systems, the units of which exchange inhibitory as well as excitatory signals, tend to oscillate between good and bad solutions, plausible and not plausible interpretations of reality, low and high energy states. Most natural complex systems fall in this category: economic systems, where states of low or high unemployment or inflation are succeeding each other; closed population systems, where the population goes through cycles of growth and suppression; and so on. Figure 6.10 shows the output of the algorithm used to produce the result in Figure 6.9(d), on the same input image with the same parameters, as the iteration steps are allowed to proceed. We can see that the output oscillates between bad and good states. From this series of possible outputs, somehow the human brain can pick up the interpretation of the scene that is most appropriate. How we can emulate this in an algorithm is an open question. One possibility is to monitor the result by counting, for example, the edgels that are present, or the smoothness of the created shapes, and stop it automatically when such a criterion is met.

6.3.2 Beyond the Conventional Data Representations and Object Descriptors

Once we start questioning the use of the rectangular grid for data representation, we have to consider representations that may convey the same information as the rectangular grid ones but at the same time allow irregular sampling. There are indeed such representations, and we shall examine one of them here.

6.3.2.1 The Trace Transform

The conventional way of sampling a continuous scene is to use sampling points (regularly or irregularly placed) and measure some scene value at those points (for an optical image this value is the brightness of the scene at that point). We know, however, that the same information concerning the scene may be captured in alternative ways. For example, in computer tomography we measure the integral of the absorbency function of the material from which an object is made, along lines tracing the object, and produce the so-called sinogram, which is nothing else than the Radon transform of the measured function. The Radon transform may then be inverted to yield the values of the function at all points in space. So, whether we

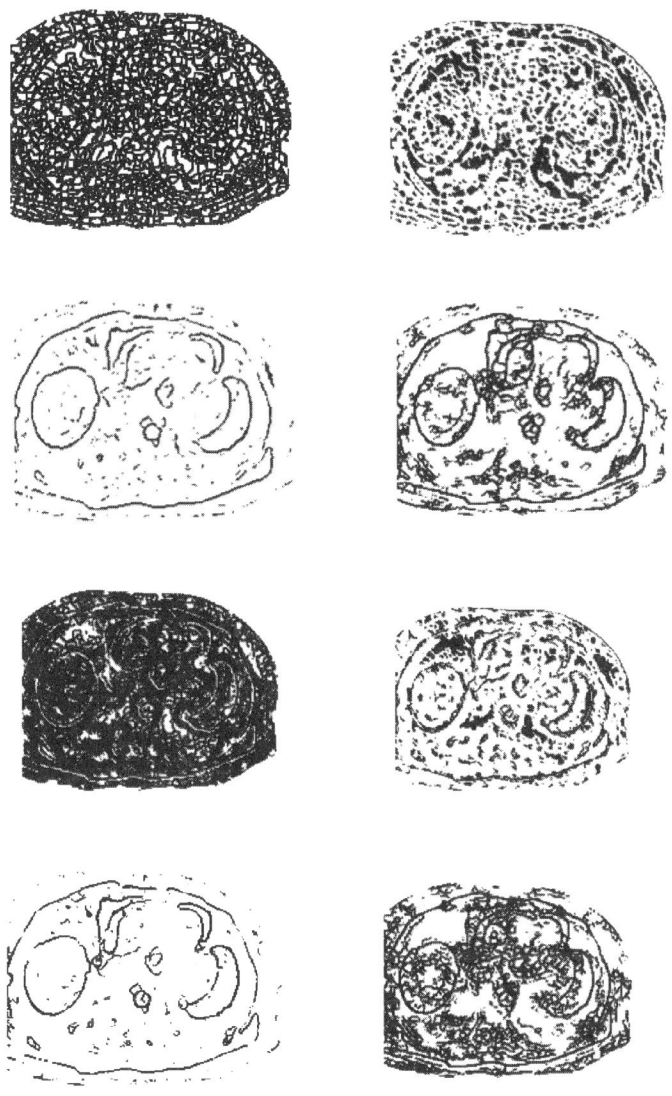

Figure 6.10 In a complex system with excitatory as well as inhibitory interactions, the state of the system passes through good and bad states as the iterations progress, without ever converging.

sampled the function at distinct points or we computed integrals of the function along tracing lines, we would have captured the same information concerning the object. The trace transform is a generalization of the Radon transform, where along the tracing lines one measures some functional of the scene function, not necessarily the integral. Each tracing line may be parameterized by two parameters, its distance p from the center of the scene and the orientation ϕ of the normal to the line from the center of the scene, with respect to some reference direction (see Figure 6.11). We may, therefore, represent whatever we measure along each

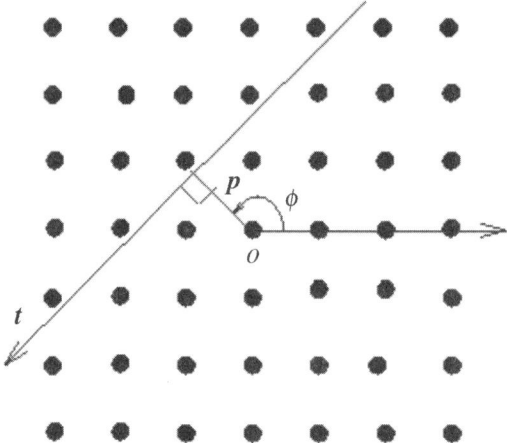

Figure 6.11 The trace transform samples the scene by computing a functional along tracing lines as opposed to using sampling points, depicted here by the black dots.

tracing line in a two-dimensional set of axes, along one of which we measure angle ϕ and along the other length p. This representation is the trace transform of the scene. Figure 6.12 shows the conventional representation of an object and its representation by the trace transform when the functional we compute along each tracing line is the integral of the brightness of the scene (Radon transform). These two representations contain exactly the same information about the scene. The second, however, allows one to perform tasks that are rather more difficult by the conventional representation, and it certainly allows us to do image processing without having to rely on the use of a regular rectangular grid.

In the next subsection we show how one may produce features from such a representation that characterize the depicted scene.

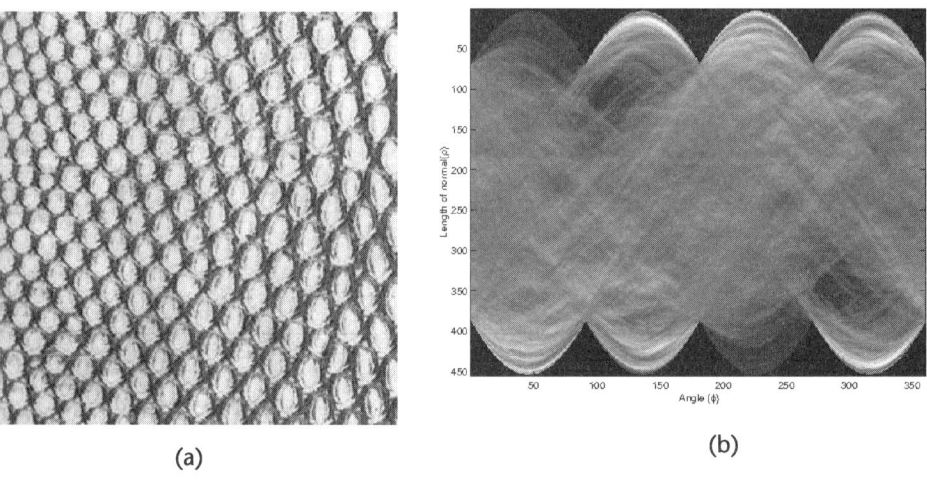

(a) (b)

Figure 6.12 Two representations of the same scene: the conventional sampling at regular points (a) and the representation by measurements along tracing lines (b).

6.3.2.2 Features from the Trace Transform

One may compute scene features using the trace transform as follows.

- Compute from the scene a functional T along tracing lines (ϕ, p). This produces the trace transform $\text{Result}_T(\phi, p)$.
- Compute on $\text{Result}_T(\phi, p)$ a functional P for all values of p, that is, compute a functional on the columns of the trace transform. This is equivalent to performing measurements from the scene along batches of parallel lines (all having the same ϕ). The result is the so-called circus function $\text{Result}_P(\phi)$.
- Compute on $\text{Result}_P(\phi)$ a functional Φ for all values of ϕ. The result is the so-called triple feature: a number that depends on the combination $TP\Phi$.

Tables 6.1 to 6.3 list functionals that may be used to produce triple features. Details on how to implement the trace transform in practice may be found in [19]. Chapter 13 of this volume presents a hardware implementation of the trace transform that allows it to be computed in real time.

It has been shown [6, 16] that, with the appropriate choice of functionals T, P, and Φ, the triple feature may be invariant to rotation, translation, scaling, or affine distortions in general. If one uses ten different functionals for each step, one may produce 1,000 features all characterizing the scene in an affine invariant way, none of which having a linguistic term to identify it, none of which corresponding to some perceptual concept we can express.

Table 6.1 Functionals That May be Used to Produce the Trace Transform

	T Functional
1	$\sum_{i=0}^{N} x_i$
2	$\sum_{i=0}^{N} i x_i$
3	2nd_Central_Moment/Sum_of_all_values
4	$\sqrt{\sum_{i=0}^{N} x_i^2}$
5	$\text{Max}_{i=0}^{N} x_i$
6	$\sum_{i=0}^{N-1} |x_{i+1} - x_i|$
7	$\sum_{i=0}^{N-1} |x_{i+1} - x_i|^2$
8	$\sum_{i=3}^{N-3} |x_{i-3} + x_{i-2} + x_{i-1} - x_{i+1} - x_{i+2} - x_{i+3}|$
9	$\sum_{i=0}^{N-2} |x_{i-2} + x_{i-1} - x_{i+1} - x_{i+2}|$
10	$\sum_{i=4}^{N-4} |x_{i-4} + x_{i-3} + \ldots + x_{i-1} - x_{i+1} - \ldots - x_{i+3} - x_{i+4}|$
11	$\sum_{i=5}^{N-5} |x_{i-5} + x_{i-4} + \ldots + x_{i-1} - x_{i+1} - \ldots - x_{i+4} - x_{i+5}|$
12	$\sum_{i=6}^{N-6} |x_{i-6} + x_{i-5} + \ldots + x_{i-1} - x_{i+1} - \ldots - x_{i+5} - x_{i+6}|$
13	$\sum_{i=7}^{N-7} |x_{i-7} + x_{i-6} + \ldots + x_{i-1} - x_{i+1} - \ldots - x_{i+6} - x_{i+7}|$
14	$\sum_{i=4}^{N-4} \sum_{k=0}^{4} |x_{i-k} - x_{i+k}|$
15	$\sum_{i=5}^{N-5} \sum_{k=0}^{5} |x_{i-k} - x_{i+k}|$
16	$\sum_{i=6}^{N-6} \sum_{k=0}^{6} |x_{i-k} - x_{i+k}|$
17	$\sum_{i=7}^{N-7} \sum_{k=0}^{7} |x_{i-k} - x_{i+k}|$
18	$\sum_{i=10}^{N-10} \sum_{k=0}^{10} |x_{i-k} - x_{i+k}|$
19	$\sum_{i=15}^{N-15} \sum_{k=0}^{15} |x_{i-k} - x_{i+k}|$
20	$\sum_{i=20}^{N-20} \sum_{k=0}^{20} |x_{i-k} - x_{i+k}|$

Table 6.2 Functionals That May be Used to Process the Columns of the Trace Transform and Produce the Circus Function

	P Functional
1	$\text{Max}_{i=0}^{N} x_i$
2	$\text{Min}_{i=0}^{N} x_i$
3	$\sqrt{\sum_{i=0}^{N} x_i^2}$
4	$\frac{\sum_{i=0}^{N} i x_i}{\sum_{i=0}^{N} x_i}$
5	$\sum_{i=0}^{N} i x_i$
6	$\frac{1}{N} \sum_{i=0}^{N} (x_i - \hat{x})^2$
7	c so that: $\sum_{i=0}^{c} x_i = \sum_{i=c}^{N} x_i$
8	$\sum_{i=0}^{N-1} |x_{i+1} - x_i|$
9	c so that: $\sum_{i=0}^{c} |x_{i+1} - x_i| = \sum_{i=c}^{N-1} |x_{i+1} - x_i|$
10	$\sum_{i=0}^{N-4} |x_i - 4x_{i+1} + 6x_{i+2} - 4x_{i+3} + x_{i+4}|$

Here $\hat{x}$ is the mean of the x_i values.

However, one may not necessarily be interested in invariant descriptors. One may wish to produce features to recognize certain image content, or monitor a situation. The idea then is to use many combinations of functionals and work out many triple features. For example, by using the functionals listed in Tables 6.1 to 6.3 one may produce $20 \times 10 \times 18 = 3{,}600$ features. From those features, one may use training data and select the features that exhibit the desired behavior to solve

Table 6.3 Functionals That May be Used to Process the Circus Function and Produce a Triple Feature

	Φ Functional
1	$\sum_{i=0}^{N-1} |x_{i+1} - x_i|^2$
2	$\sum_{i=0}^{N-1} |x_{i+1} - x_i|$
3	$\sqrt{\sum_{i=0}^{N} x_i^2}$
4	$\sum_{i=0}^{N} x_i$
5	$\text{Max}_{i=0}^{N} x_i$
6	$\text{Max}_{i=0}^{N} x_i - \text{Min}_{i=0}^{N} x_i$
7	i so that $x_i = \text{Min}_{i=0}^{N} x_i$
8	i so that $x_i = \text{Max}_{i=0}^{N} x_i$
9	i so that $x_i = \text{Min}_{i=0}^{N} x_i$ without first harmonic
10	i so that $x_i = \text{Max}_{i=0}^{N} x_i$ without first harmonic
11	Amplitude of the first harmonic
12	Phase of the first harmonic
13	Amplitude of the second harmonic
14	Phase of the second harmonic
15	Amplitude of the third harmonic
16	Phase of the third harmonic
17	Amplitude of the fourth harmonic
18	Phase of the fourth harmonic

a particular task. This method has been demonstrated on texture recognition [17], monitoring the level of use of a parking lot [6], face recognition [21], and so on. This is the approach we shall use in the next section to reverse engineer some properties of the human visual system.

6.4 Reverse Engineering Some Aspect of the Human Visual System

In this section we discuss how we may use the methodology described in the previous section to answer the question: which features do humans use to group textures in perceptual classes?

Many researchers have attempted to answer this question. However, the experiments performed often pre-empt the answer. Some mistakes often made are the following.

1. The subjects are asked to express what makes them consider two patterns as being similar. This already makes the fundamental mistake of restricting the characteristic features used by the brain to those that are consciously perceived and for which linguistic terms exist. For example, people use terms like "directionality," "blobbiness," "randomness," and so on. We have argued earlier that human visual perception must rely on subconsciously identified features, much more powerful than consciously identified ones.
2. The subjects are asked to answer questionnaires where boxes have to be ticked with already predecided answers. This not only restricts the characteristics to those that may be easily verbalized, but it also imposes on the subject the prejudice of the researcher.
3. The researcher computes several features from the images and then selects those that identify best the textures humans also identified. The problem is that the computed features either tend to be all of the same nature, for example, amplitude in certain different frequency bands, or features with linguistic definitions, like "fractal dimension," "lacunarity," and so on.

Here we avoid these pitfalls. We too start by computing features, however, we compute a very large number of features of a diverse nature by using the trace transform. We then select those that rank a set of textures in the same way as humans do. We test the usefulness of these features by using them to rank a set of

Table 6.4 Combinations of Functionals From Tables 6.1 to 6.3 That Produce the Same Ranking of Textures as Humans

T Functional	P Functional	Φ Functional
$\sum_{i=0}^{N-1} \|x_{i+1} - x_i\|$	$\text{Max}_{i=0}^{N} x_i$	$\sum_{i=0}^{N-1} \|x_{i+1} - x_i\|^2$
$\sum_{i=0}^{N-1} \|x_{i+1} - x_i\|$	$\text{Min}_{i=0}^{N} x_i$	$\sum_{i=0}^{N-1} \|x_{i+1} - x_i\|^2$
$\sum_{i=0}^{N-1} \|x_{i+1} - x_i\|$	$\text{Min}_{i=0}^{N} x_i$	Amplitude of the fourth harmonic
$\sum_{i=0}^{N-1} \|x_{i+1} - x_i\|$	$\sum_{i=0}^{N} i x_i$	$\sum_{i=0}^{N-1} \|x_{i+1} - x_i\|^2$
$\sum_{i=4}^{N-4} \sum_{k=0}^{4} \|x_{i-k} - x_{i+k}\|$	$\sum_{i=0}^{N-4} \|x_i - 4x_{i+1} + 6x_{i+2} - 4x_{i+3} + x_{i+4}\|$	Amplitude of the second harmonic

textures totally different from those used to select the features, and confirm that they indeed could reproduce the human ranking, even though the features were selected from totally different patterns. We believe that in this way, we identify genuinely generic features used by the human subconscious to decide the similarity of patterns. More details on these experiments may be found in [18]. Table 6.4 lists the combinations of functionals that produced triple features which allowed texture ranking according to the human visual system.

6.5 Conclusions

One of the fundamental characteristics of the human visual system is its ability to see straight lines even with peripheral vision that relies on sparse and irregularly placed sensors. This raises the question: can we perform image processing and pattern recognition from irregularly sampled data? Irregularly sampled data arise very often in practice and what we usually do is to resample them on a regular grid. This, however, introduces errors and it evades the real question. Various methods have been proposed for resampling (also known as "griding") and we reviewed some of them here. One of them, namely that of normalized convolution, offers a breakthrough in handling irregularly sampled data: it treats the missing data as points with 0 certainty and allows griding, but at the same time offers a way to perform convolution with irregularly placed samples. This leads to the possibility of performing a large part of image processing algorithms directly from such data, for example, gradient detection.

The second major point of this chapter concerns the gradient map postprocessing which leads to edge detection. Conventional edge detectors rely on the magnitude of the gradient vectors to perform thinning (nonmaxima suppression) and thresholding. The method we used here, based on the model of V1 for salienation, detects edges on the basis of collinearity and smoothness of their shape rather than contrast alone.

Finally, a more generic breakthrough in handling irregularly sampled data comes from the realization that there is no need to represent the data with values at sampling points. The trace transform allows the representation of a scene in terms of functionals computed along tracing lines. Although this is believed to be far from what the human vision does, it presents an alternative mechanism of vision, where, for example, the saccades performed by the eye are not "dead time" but mechanisms of alternative collection of information from the scene.

We may conclude by saying that the algorithms presented here have been mostly inspired by the study of the human visual system, but they do not necessarily try to imitate it. Some of them may be used as tools with which to poke the human brain in order to understand better the mechanisms it uses. We presented a method as unbiased as possible for the identification of features that the human brain uses to identify similar patterns. It turned out that these features are various forms of differentiators. We called this approach of producing features beyond the constraints imposed by human language and the conscious perception of pattern characteristics "*subconscious image processing*."

References

[1] Chandra, S., Petrou, M. and Piroddi, R., "Texture interpolation using ordinary Kriging." Pattern Recognition and Image Analysis, Second Iberian Conference, IbPRIA2005, Estoril, Portugal, June 7-9, J S Marques, N Perez de la Blanca and P Pina (eds), Springer LNCS 3523, Vol. II, 2005, pp. 183-190.

[2] Comincioli, V., 1995. "Analisi Numerica: Metodi, Modelli, Applicazioni." New York, McGraw Hill.

[3] Davis, J. C., 1973. *Statistics and Data Analysis in Geology*, John Wiley.

[4] Feichtinger, H. G. and Groechenig, K., "Theory and practice of irregular sampling," in *Wavelets, Mathematics and Applications*. Boca Raton, FL: CRC Press, 1994, pp. 305-363.

[5] Haas, T. C., "Kriging and automated variogram modelling within a moving window," *Atmospheric Environment*, Part A, Vol. 24, 1990, pp. 1759-1769.

[6] Kadyrov, A. and Petrou, M., "The trace transform and its applications." *IEEE Transactions on Pattern Analysis and Machine Intelligence, PAMI*, Vol. 23, 2001, pp. 811-828.

[7] Knutsson, H. and Westin, C.-F., "Normalised and differential convolution: methods for Interpolation and Filtering of incomplete and uncertain data," *IEEE Conference on Computer Vision and Pattern Recognition*, 1993, pp. 515-523.

[8] Li, Z., "A neural model of contour integration in the primary visual cortex," *Neural Computation*, Vol. 10, 1998, pp. 903-940.

[9] Li, Z., "Visual segmentation by contextual influences via intra-cortical interactions in the primary visual cortex," *Networks: Computation in Neural Systems*, Vol. 10, 1999, pp. 187-212.

[10] Li, Z., "Computational design and nonlinear dynamics of a recurrent network model of the primary visual cortex," *Neural Computation*, Vol. 13, No. 8, 2001, pp. 1749-1780.

[11] Marvasti, F., 2001. *Non-Uniform Sampling: Theory and Practice*. Dordrecht: Kluwer.

[12] Patti, A. J., Sezan, M. I. and Murat-Tekalp, A., "Super-resolution video reconstruction with arbitrary sampling lattices and non-zero aperture time." *IEEE Transactions on Image Processing*, Vol. 6, No. 8, 1997, pp. 1064-1076.

[13] Petrou, M., Piroddi R. and Chandra, S., 2004. "Irregularly sampled scenes." EUROPTO Symposium on Remote Sensing, Image Processing for Remote Sensing, Sept 2004, Grant Cannaria, SPIE 5573.

[14] M Petrou, 1994. "The differentiating filter approach to edge detection." *Advances in Electronics and Electron Physics*, Vol. 88, pp. 297-345.

[15] Petrou, M. and Garcia Sevilla, P., 2006. *Image Processing: Dealing with Texture*. John Wiley.

[16] Petrou, M. and Kadyrov, A., "Affine invariant features from the trace transform." *IEEE Transactions on Pattern Analysis and Machine Intelligence, PAMI*, Vol. 26, 2004, pp. 30-44.

[17] Petrou, M., Piroddi R. and Telebpour, A., "Texture recognition from sparsely and irregularly sampled data." *Computer Vision and Image Understanding*, Vol. 102, 2006, pp. 95-104.

[18] Petrou, M., Telebpour, A. and Kadyrov, A., 2007. "Reverse engineering the way humans run textures," *Pattern Analysis and Applications*, Vol. 10, pp. 101-114.

[19] Petrou, M. and Wang, F., 2008. "A tutorial on the practical implementation of the trace transform." Chapter 11 in the *Handbook of Texture Analysis*, M. Mirmehdi, X. Xie and J. Suri (eds), World Scientific.

[20] Piroddi. R. and Petrou, M, "Analysis of irregularly sampled data: a review." *Advances in Imaging and Electron Physics*, Vol. 132, 2004, pp. 109-165.

[21] Srisuk, S., Petrou, M., Kurutach, W. and Kadyrov, A., "A face authentication system using the trace transform." *Pattern Analysis and Applications*, Vol. 8, 2005, pp. 50-61.

[22] Strohmer, T., 1993. "Efficient methods for digital signal and image reconstruction from nonuniform samples." PhD thesis, Institut fuer Mathematik der Universitaet Wien.

CHAPTER 7
Reverse Engineering of Human Vision: Hyperacuity and Super-Resolution

Mirna Lerotic and Guang-Zhong Yang

7.1 Introduction

Visual acuity is a measure of the ability of an observer to resolve small spatial detail in a scene. The visual acuity of an observer is defined as a discrimination threshold in a simple visual task, such as line separation or resolution of dots. In certain tasks the resolution of the human eye reaches beyond the size of the photoreceptors on the retina. The ability of the human eyes to see beyond the resolution of photoreceptors on the retina is called *hyperacuity* [1].

It is well known that human eyes do not have a uniform visual response, in fact, the best visual acuity is only within a visual angle of one to two degrees. This is called foveal vision. For areas toward which we do not direct our eyes when observing a scene, we have to rely on a cruder representation of the objects offered by peripheral vision, of which the visual acuity drops off dramatically from the center of focus. The limited extent of the fovea demands the eyes being highly mobile and able to sweep across a large visual angle. As a result, the motion of the eyes is extremely varied in both amplitude and frequency. During the periods of visual fixation, small eye movements continuously alter the projection of the image on the retina. These fixational eye movements include small saccades, slow drifts, and physiological nystagmus. Existing research has shown that micromotions, including both microsaccades of the eyes [2] and subpixel (sub-sampling) movements of visual scene [3], can enhance visual resolution, although the mechanisms of the human visual system are unknown.

The role of fixational eye movements has been debated from the early history of eye movement research [2]. Steinman [4] advocated that microsaccades serve no useful purposes, whereas Ditchburn [5] claimed that the microsaccades not only aid human vision, but also are critical to normal vision. Recent experiments connect the rate of microsaccades with firing of neurons in macaque monkeys [6], which lead to a conclusion that microsaccades have a specific purpose in vision. To date, however, the role of fixational eye movements remains an open research question.

In this chapter, we describe a possible connection between super-resolution and hyperacuity. Similar to hyperacuity in humans, in computer vision, super-resolution is a method to achieve resolution better than that given by the pixel resolution of the camera. This is achieved through camera movement by merging a set of subpixel shifted low-resolution images into one high-resolution image. In the same way, human eyes sample a scene through fixational eye movements, and this additional subsampling can lead to hyperacuity, that is, resolution beyond the normal vision.

7.2 Hyperacuity and Super-Resolution

For conventional spatial-acuity tasks such as resolving two dots or grating resolution [7], visual acuity of the human eyes is in the range of 30 to 60 seconds of arc, which is the approximate diameter of the photoreceptors in fovea. On the other hand, there are specific tasks in which the threshold is as low as 2 seconds of arc. Observers are capable of detecting an 8 to 10 seconds of arc instantaneous displacement of a short line or bar, a 2 to 4 seconds of arc misalignment in a Vernier-type task [8], and a 2 to 4 seconds of arc difference in the separation of lines in a stereo-acuity task [9].

The ability of the human eye to resolve Vernier-type tasks beyond the intrinsic resolution of the photoreceptors in the eye is called hyperacuity [1,10–13]. Vernier task refers to the ability to detect a small lateral displacement of two offset lines. Figure 7.1 illustrates the cone distribution of the fovea and a Vernier-type stimulus. The human eye is able to resolve the positional difference of the bars even if the spacing between them is smaller than the diameter of a cone in the fovea.

Recently, Ruci et al. [14] demonstrated the benefits of fixational eye movements in visual discrimination. It was found that the percentages of correct discrimination were significantly lower under visual stabilization compared with normal vision, which allows fixational movement. For more details on fixational eye movements see Chapter 1 and [2,15–17].

Similar to human vision, super-resolution [3,18] is a method in computer vision for achieving resolution better than the pixel resolution of the camera. From a set of low-resolution images, a high resolution image is calculated where the additional

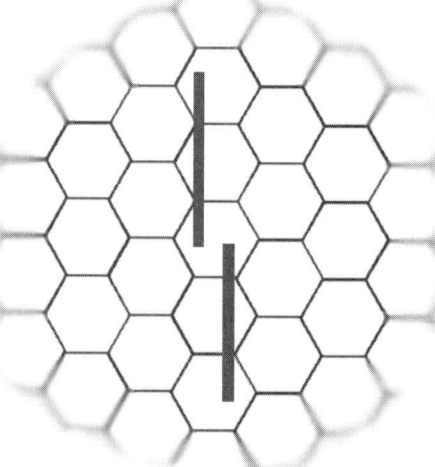

Figure 7.1 Vernier-type stimulus shown on top of the cone pattern in the fovea.

information needed to achieve high resolution is gained through overlapping and shifted views of the observed scene. The low-resolution images are shifted with respect to one another with subpixel precision. Since our eyes are never stationary, in human vision the subpixel shifts of the images are provided by the incessant eye movements.

7.3 Super-Resolution Image Reconstruction Methods

In computer vision, resolution beyond the sensor resolution is obtained through super-resolution [3]. The reduction of pixel size is possible if there are subpixel shifts between the low-resolution images used in calculations, thus providing additional "views" of the observed scene as shown in Figure 7.2(a). The simplest super-resolution algorithm consists of the registration of the low-resolution images to the subpixel precision, interpolation to the high-resolution grid, followed by image restoration [19]. More popular super-resolution approaches include projection onto convex sets (POCS) [20,21], frequency domain [22], constrained least squares (CLS) [23,24], and maximum a posteriori (MAP) [25] methods.

If one considers a high-resolution (HR) image $\mathbf{x}$ to be the true representation of the scene, then a low-resolution (LR) image $\mathbf{y}_i$ may be expressed as

$$\mathbf{y}_i = \mathbf{DM}_i\mathbf{Hx} + \mathbf{v}_i \qquad (7.1)$$

for $i = 1 \ldots K$, where K is the number of low-resolution images, $\mathbf{D}$ is a downsampling operator, and $\mathbf{M}_i$ is a warp operator which accounts for translation, rotation, and other transformations. In (7.1), $\mathbf{H}$ incorporates the point-spread function (PSF) of the sensor, motion blur, and blur caused by the optical system, and $\mathbf{v}_i$ describes the additive, Gaussian noise.

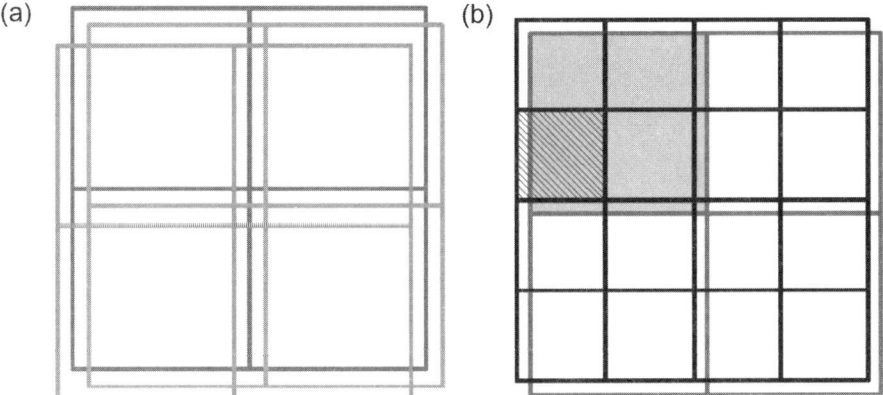

Figure 7.2 (a) Low-resolution images used in super-resolution image reconstruction have subpixel shifts between them. (b) The subpixel shifts of the low-resolution images add additional information for reconstruction, as each low-resolution pixel will contribute to more high-resolution pixels. The point-spread function of such an integrating sensor is calculated as the intersection of the areas of low-resolution and high-resolution pixels over the area of the low-resolution pixel.

This observation model can be rewritten in more convenient form by incorporating down-sampling, warp, and blur operators in a single matrix $\mathbf{W}_i$:

$$\mathbf{y}_i = \mathbf{W}_i \mathbf{x} + \mathbf{v}_i \tag{7.2}$$

The recovery of the HR image $\mathbf{x}$ from a set of subpixel shifted LR images $\mathbf{y}_i$ is the goal of all super-resolution algorithms.

7.3.1 Constrained Least Squares Approach

Super-resolution image $\mathbf{x}$ is a solution to (7.1). This is an ill-posed problem that can only be solved by regularization to stabilize the inversion. Using the constrained least squares method (CLS), the high-resolution image $\mathbf{x}$ is found by minimizing the following expression [23, 24]:

$$\min_{\mathbf{x}} \left[\sum_{i=1}^{K} \|\mathbf{y}_i - \mathbf{W}_i \mathbf{x}\|^2 + \alpha \|\mathbf{C}\mathbf{x}\|^2 \right] \tag{7.3}$$

where $\mathbf{C}$ is usually a highpass filter used to impose smoothness to the solution. The highpass filter regularization minimizes the amount of highpass energy in the restored image since most natural images can be assumed to be locally smooth

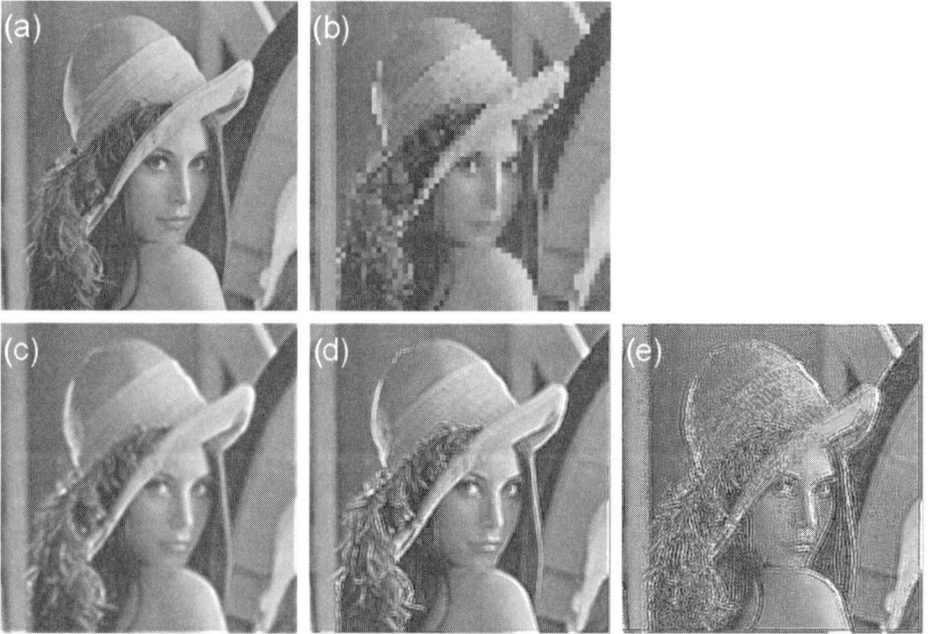

Figure 7.3 (a) Original image; (b) low-resolution image, down-sampled four times from the original image (shown on the original image scale for comparison); (c), (d), (e) super-resolution image reconstructions using the CLS method with different values of parameter α. In (c) α is equal to 0 and high-frequency content is lacking. In (d) α is equal to 0.1, and high-frequency content is recovered but the image suffers from minor artifacts around the edges. If parameter α is too large the artifacts will dominate the image as shown in (e) where the image was calculated with α equal to 0.2.

with limited high-frequency detail. In this work, the highpass filter used for image reconstructions using the CLS method is given by

$$\mathbf{C} = \begin{pmatrix} 0 & -\frac{1}{4} & 0 \\ -\frac{1}{4} & 1 & -\frac{1}{4} \\ 0 & -\frac{1}{4} & 0 \end{pmatrix} \qquad (7.4)$$

In (7.3), parameter α is the regularization parameter that balances the first term, called data fidelity, and the second term for the smoothness of the solution. For the problem formulated in (7.3), the solution is usually found iteratively

$$\hat{\mathbf{x}}^{n+1} = \hat{\mathbf{x}}^n + \beta \left[\sum_{i=1}^{K} \mathbf{W}_i^T (\mathbf{y}_i - \mathbf{W}_i \hat{\mathbf{x}}^n) - \alpha \mathbf{C}^T \mathbf{C} \hat{\mathbf{x}}^n \right] \qquad (7.5)$$

where β denotes the convergence parameter. The transposed operator $\mathbf{W}_i^T$ represents upsampling, warping, and the blur operator.

Image reconstructions by using the CLS method are shown in Figure 7.3 for different values of parameter α. If α is too small [e.g., Figure 7.3(c)], reconstruction lacks high-frequency detail, while if too large α introduces unwanted artifacts to the image [e.g., Figure 7.3(e)]. The best reconstruction was found using $\alpha = 0.1$. In

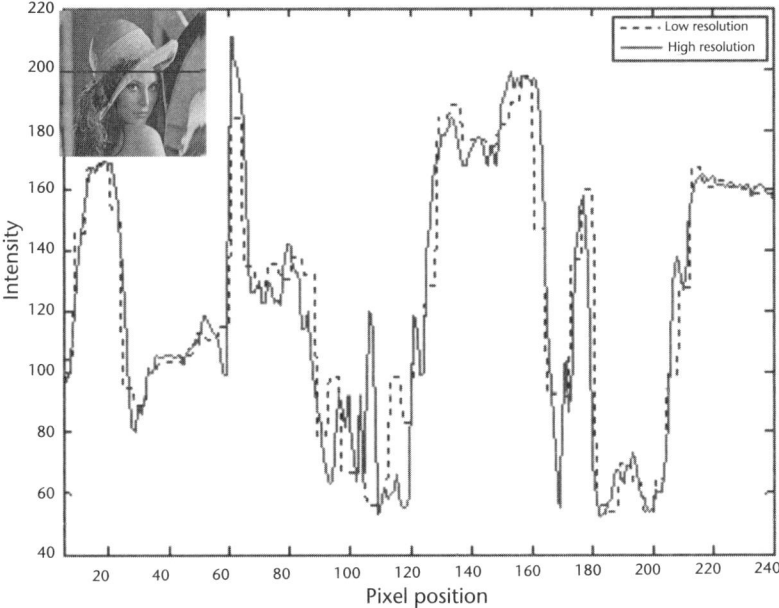

Figure 7.4 Line profiles of the original, high-resolution image and the down-sampled low-resolution image (scaled for comparison). Due to down-sampling high-frequency edge information is lost in the low-resolution profile. These line profiles were extracted from the images from the row illustrated on the insert.

practice, the best value of α can be found by balancing the desired high-frequency detail and unwanted artifacts.

The lack of detail in the down-sampled low-resolution images compared with the original high-resolution image is illustrated by line profiles shown in Figure 7.4.

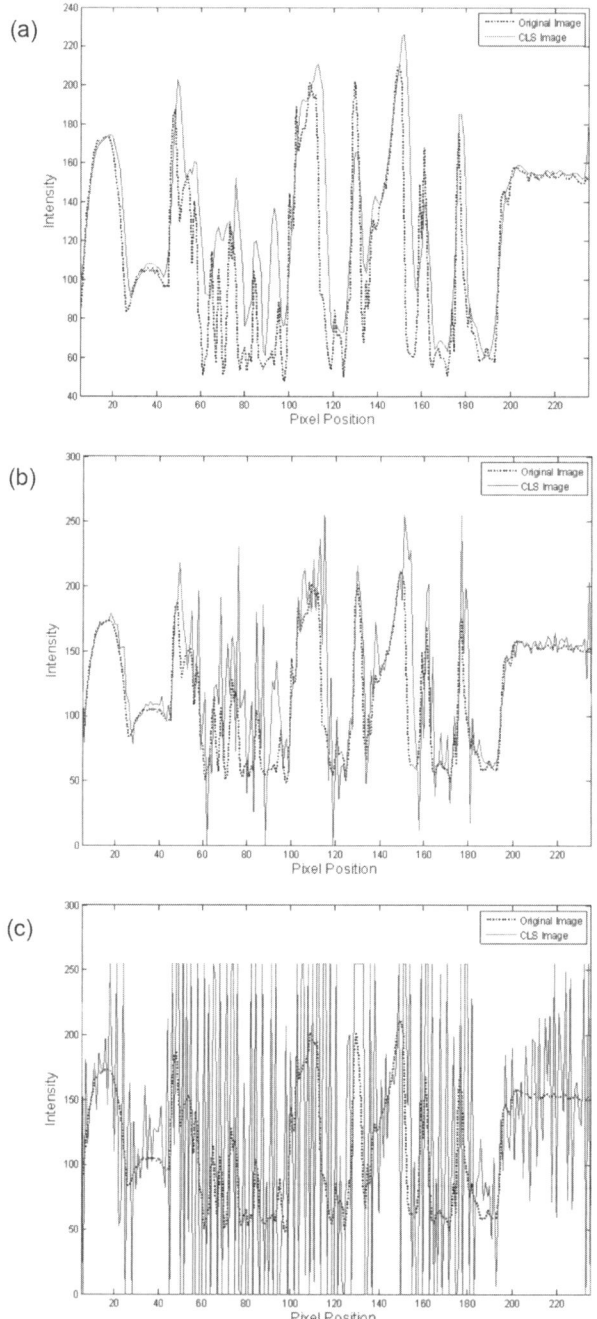

Figure 7.5 Line profiles of CLS reconstructions compared with the original image for α equal to 0, 0.1, and 0.2 are shown on (a), (b), and (c), respectively.

7.3 Super-Resolution Image Reconstruction Methods

The goal of super-resolution algorithms is to retrieve the edge information from the low-resolution images so that the reconstructed image represents the original scene. In Figure 7.5, reconstructions for different α values are compared with the ground truth, that is, the original image. The best reconstruction is the one with retrieved edge detail such as Figure 7.5(b) and without unwanted artifacts present in the reconstruction shown in Figure 7.5(c).

7.3.2 Projection onto Convex Sets

Another super-resolution reconstruction algorithm used to calculate an HR image is based on the method of projection onto convex sets (POCS) [20,21]. The POCS method is an iterative method which incorporates prior knowledge by restricting the solution to the intersection of closed convex sets. In each iteration, a solution $\mathbf{x}^{n+1}$ is derived such that

$$\mathbf{x}^{n+1} = P_m P_{m-1} \ldots P_2 P_1 \mathbf{x}^n \tag{7.6}$$

where P_i are projection operators that project the arbitrary solution $\mathbf{x}^n$ onto m closed convex sets. Each convex set represents a constraint on the solution such as amplitude, energy, or reference-image constraint [20]. In other words, the projection operators impose prior knowledge about the HR image on the solution.

Tekalp et al. [21] introduced the following closed convex set constraint for each pixel of each low-resolution frame $\mathbf{y}_i$

$$C_i = \left\{ \mathbf{x}_{k,l} \colon \left| \mathbf{r}^{(\mathbf{x})}_{i;m,n} \right| \leq \delta_0 \right\} \tag{7.7}$$

where δ_0 defines a bound that represents the statistical confidence in observed LR frames and is usually estimated from the noise present in the LR images. In (7.7), $\mathbf{r}^{(\mathbf{x})}_i$ in a sense represents the difference between the HR image $\mathbf{x}$ convolved with a point-spread function (PSF) and the ith LR image $\mathbf{y}_i$

$$\mathbf{r}^{(\mathbf{x})}_{i;m,n} = \mathbf{y}_{i;m,n} - \sum_{k,l} \mathbf{x}_{k,l} \mathbf{h}_{i;m,n;k,l} \tag{7.8}$$

In (7.8), $\mathbf{h}_i$ is a PSF of the sensor for the ith frame, given by the ratio of the area where LR pixel (m,n) and HR pixel (k,l) overlap divided by the total area of the LR pixel, as illustrated in Figure 7.2(b). The projection of an HR image $\mathbf{x}$ onto convex set C_i defined by (7.7) can be derived as [21]

$$x^{n+1}_{k,l} = \begin{cases} \dfrac{r^{(x)}_{i;m,n} - \delta_0}{\sum_{o,p} \mathbf{h}_{i;m,n;o,p}} \mathbf{h}_{i;m,n;k,l} & \text{if } \mathbf{r}^{(x)}_{i;m,n} > +\delta_0 \\ 0 & \text{if } -\delta_0 \leq \mathbf{r}^{(x)}_{i;m,n} \leq \delta_0 \\ \dfrac{r^{(x)}_{i;m,n} + \delta_0}{\sum_{o,p} \mathbf{h}_{i;m,n;o,p}} \mathbf{h}_{i;m,n;k,l} & \text{if } \mathbf{r}^{(x)}_{i;m,n} < -\delta_0 \end{cases} \tag{7.9}$$

In addition to the described constraints imposed by the data, there are other constraints that can be imposed from the prior knowledge about the solution space. One such constraint is the amplitude constraint set C_A

$$C_A = \{\mathbf{x}_{k,l} : \alpha \leq \mathbf{x}_{k,l} \leq \beta, \beta > \alpha\} \qquad (7.10)$$

with the projection of an arbitrary function $\mathbf{x}_{k,l}$ onto constraint set C_A given by

$$\mathbf{x}_{k,l} = \begin{cases} \alpha & \text{if } \mathbf{x}_{k,l} < \alpha \\ \mathbf{x}_{k,l} & \text{if } \alpha \leq \mathbf{x}_{k,l} \leq \beta \\ \beta & \text{if } \mathbf{x}_{k,l} > \beta \end{cases} \qquad (7.11)$$

In this way, the amplitude constraint can keep the solution within the known amplitude limits.

The energy constraint C_E limits the maximum permissible energy E in the reconstructed image

$$C_E = \left\{\mathbf{x}_{k,l} : \|\mathbf{x}\|^2 \leq E\right\} \qquad (7.12)$$

The projection onto the energy constraint set C_E is equal to

$$\mathbf{x}_{k,l} = \begin{cases} \mathbf{x}_{k,l} & \text{if } \|\mathbf{x}_{k,l}\|^2 \leq E \\ \sqrt{\dfrac{E}{\|\mathbf{x}_{k,l}\|^2}} \mathbf{x}_{k,l} & \text{if } \|\mathbf{x}_{k,l}\|^2 > E \end{cases} \qquad (7.13)$$

There are other constraints, such as the reference image and bounded support constraints. Figure 7.6 illustrates the reconstruction of the POCS HR images by the use of the projection operator presented in Equation (7.9) together with amplitude and energy constraints. Line profiles for different values of δ_0 are shown in Figure 7.7. The POCS technique is a superior CLS reconstruction method because super-resolution images calculated using POCS contain a high degree of recovered edge detail not present in the CLS reconstruction.

Figure 7.6 Figure 6 Super-resolution image reconstructions calculated using the POCS algorithm with parameter δ_0 equal to 0, 5, and 20 shown on (a), (b), and (c), respectively. If δ_0 is set too low, the reconstructed image will be noisy and artifacts will be visible around the edges [see image (a)]. On the other hand, if δ_0 is set too high, high-frequency detail will not be recovered [see image (c)].

7.3 Super-Resolution Image Reconstruction Methods

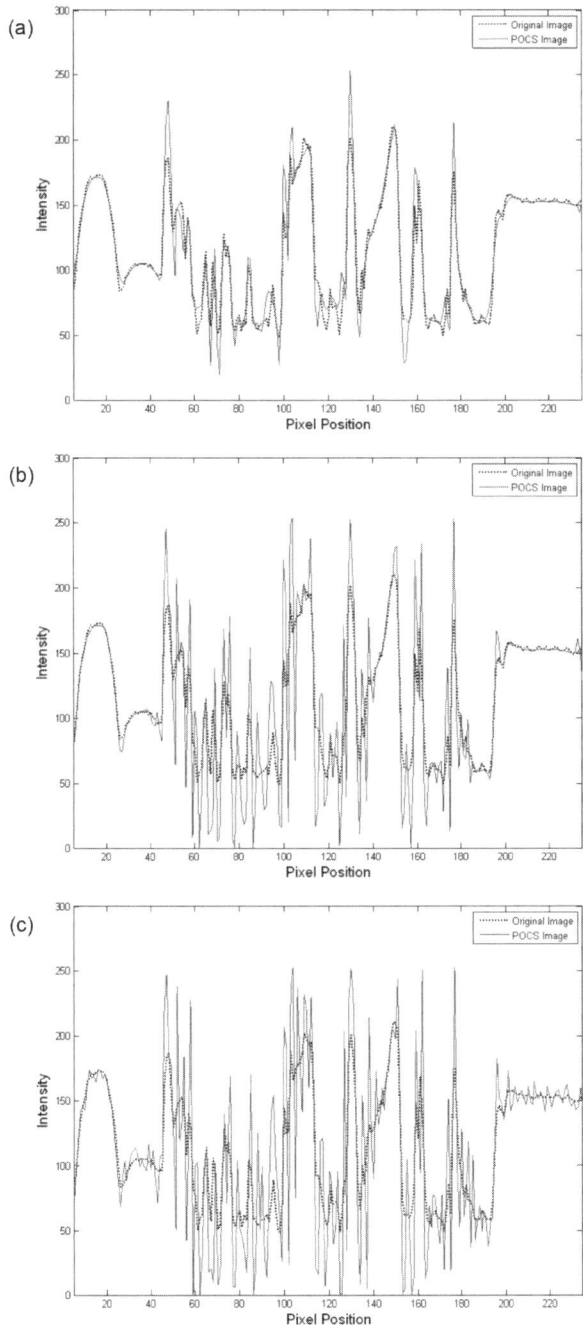

Figure 7.7 Line profiles of POCS reconstructions compared with the original image, for δ_0 equal to 20, 5, and 0 are shown on (a), (b), and (c), respectively.

7.3.3 Maximum A Posteriori Formulation

For super-resolution reconstruction, a regularization approach based on Bayesian estimation may also be used. As an example, maximum a posteriori (MAP) is used to estimate an unobserved quantity on the basis of observed data [25]. The MAP estimate of the high-resolution image $\hat{\mathbf{x}}$ is given by

$$\hat{\mathbf{x}} = \underset{\mathbf{x}}{\operatorname{argmax}} \Pr(\mathbf{x}|\mathbf{y}_i) \tag{7.14}$$

where $\Pr(\mathbf{x}|\mathbf{y}_i)$ is the likelihood function describing "the likelihood" of the high-resolution image $\mathbf{x}$ given the low-resolution observed images $\mathbf{y}_i$. Using Bayes rule this becomes

$$\hat{\mathbf{x}} = \underset{\mathbf{x}}{\operatorname{argmax}} \frac{\Pr(\mathbf{y}_i|\mathbf{x})\Pr(\mathbf{x})}{\Pr(\mathbf{y}_i)} \tag{7.15}$$

or, more conveniently, we minimize the negative log of the numerator since the denominator is not a function of $\mathbf{x}$,

$$\hat{\mathbf{x}} = \underset{\mathbf{x}}{\operatorname{argmax}} \{-\log \Pr(\mathbf{y}_i|\mathbf{x}) - \log \Pr(\mathbf{x})\} \tag{7.16}$$

The conditional density in the first term, under the assumption that pixels in a low resolution image deviate independently from the predictions of the imaging model according to a multivariate Gaussian, is equal to

$$\Pr(\mathbf{y}_i|\mathbf{x}) = \frac{1}{(2\pi)^{\frac{N}{2}} \sigma^N} \exp\left\{-\frac{1}{2\sigma^2} \sum_{i=1}^{K} \|\mathbf{y}_i - \mathbf{W}_i \mathbf{x}\|^2\right\} \tag{7.17}$$

and represents the data fidelity. For the prior term in Equation (7.15) often the choice is a Gaussian or Markov random field (MRF) prior.

Farsiu et al. [26] developed a bilateral-TV prior based on the combination of a bilateral filter and total variance (TV) regularization. In their approach for color super-resolution they use bilateral-TV on the luminance components, while for the chrominance they use simpler regularization based on the L_2 form.

7.3.4 Markov Random Field Prior

The prior term in the likelihood function contains the information that we know about the image, or an image model. One of the prior functions used to model the image is an MRF with the Gibbs probability density function

$$\Pr(\mathbf{x}) = \frac{1}{(2\pi)^{\frac{N_1 N_2}{2}} |\mathbf{C}|^{\frac{1}{2}}} \exp\left(-\frac{1}{2}\mathbf{x}^T \mathbf{C}^{-1} \mathbf{x}\right) \tag{7.18}$$

where $\mathbf{C}$ is the covariance matrix, and $N_1 \times N_2$ is the image size. If covariance matrix $\mathbf{C}$ is decomposed into a sum of products, Equation (7.18) can be written as

7.3 Super-Resolution Image Reconstruction Methods

$$\Pr(\mathbf{x}) = \frac{1}{(2\pi)^{\frac{N_1 N_2}{2}} |C|^{\frac{1}{2}}} \exp\left(-\frac{1}{2\lambda} \sum_{c \in C} \mathbf{x}^T \mathbf{d}_c \mathbf{d}_c^T \mathbf{x}\right) \quad (7.19)$$

where $\mathbf{d}_c$ is a coefficient vector for clique c. The coefficients are chosen to impose the assumed smoothness of the high-resolution images, so often first- or second-order derivative discrete approximations are used as a measure of smoothness. Finite difference approximation to second-order derivatives is given by

$$\begin{aligned}
\mathbf{d}_{1;k,l}^T \mathbf{x} &= \mathbf{x}_{k,l+1} - 2\mathbf{x}_{k,l} + \mathbf{x}_{k,l-1} \\
\mathbf{d}_{2;k,l}^T \mathbf{x} &= \frac{1}{2}\left(\mathbf{x}_{k-1,l+1} - 2\mathbf{x}_{k,l} + \mathbf{x}_{k+1,l-1}\right) \\
\mathbf{d}_{3;k,l}^T \mathbf{x} &= \mathbf{x}_{k-1,l} - 2\mathbf{x}_{k,l} + \mathbf{x}_{k+1,l} \\
\mathbf{d}_{4;k,l}^T \mathbf{x} &= \frac{1}{2}\left(\mathbf{x}_{k-1,l-1} - 2\mathbf{x}_{k,l} + \mathbf{x}_{k+1,l+1}\right)
\end{aligned} \quad (7.20)$$

where the first and third equations have vertical and horizontal direction, respectively, while the second and third are derivatives along the diagonals.

As an example, image reconstruction by using the above MAP-MRF method is shown in Figure 7.8. It can be seen that the quality of the reconstruction is good, the high-frequency detail present in the original image shown in Figure 7.3(a) is retrieved, and the reconstructed image is free of artifacts. MAP-MRF images have a higher degree of reconstructed edge detail compared with POCS and CLS, and the artifacts present on the POCS reconstructions are not present in them. The line profile of MAP reconstruction shown in Figure 7.9 does not differ much from the original image profile.

Convergence of the algorithm is illustrated in Figure 7.8(b) using image entropy. Entropy is a quantity that describes "randomness" of an image and represents the amount of information present in the image. It is defined as [27]

$$E = -\sum_i P_i \log P_i \quad (7.21)$$

where P_i is a probability that state i occurs and can be calculated using the image histogram. Entropy of image reconstructions calculated with MAP-MRF for a

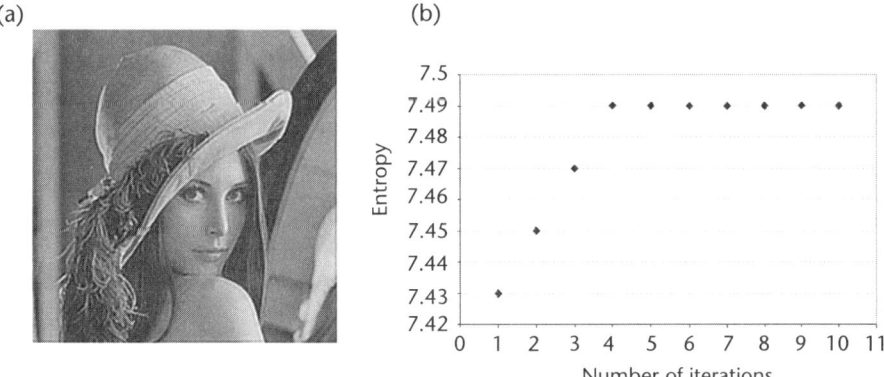

Figure 7.8 (a) Super-resolution image reconstruction using the MAP-MRF method. (b) Convergence of the MAP-MRF method is illustrated using entropy vs the number of iterations.

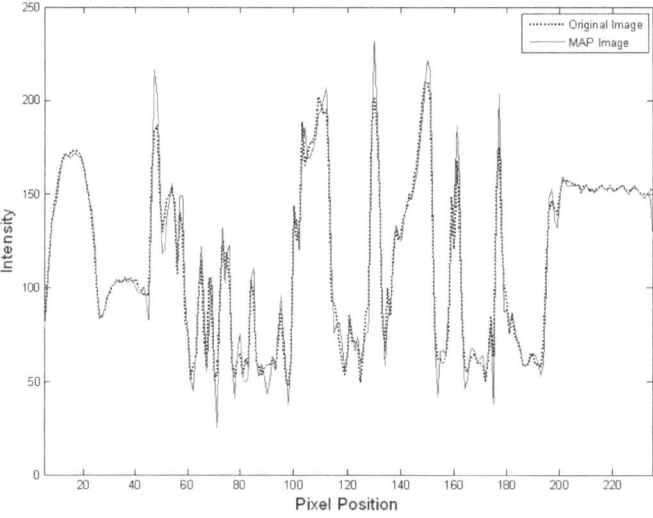

Figure 7.9 Line profiles of MAP reconstructions compared with the original image.

Figure 7.10 (a) Original image; (b), (c), (d) super-resolution image reconstructions using the MAP, POCS, and CLS method, respectively.

different iteration number is plotted in Figure 7.8(b). Maximal entropy value is reached in only 4 iterations, which can be of great importance when one considers real-time implementations.

7.3.5 Comparison of the Super-Resolution Methods

To provide a more in-depth comparison of the performance of the described super-resolution techniques, reconstructed images using MAP, POCS, and CLS are shown in Figure 7.10(b, c, and d) respectively. The original high-resolution image is shown in Figure 7.10(a). It is evident that the image reconstructed using the MAP method is the best among the results presented—the image sharpness is enhanced with minimal artifacts. Although the POCS reconstruction also appears to be very sharp, there are noticeable artifacts around the edges. The poorest reconstruction was obtained by using the CLS method due to the lack of high-resolution details and the presence of artifacts.

Convergence of these methods is also compared using entropy of the reconstructed images. Figure 7.11 shows the entropy values reconstructed using MAP, POCS, and CLS for different numbers of iterations. CLS and MAP converged to the maximal entropy value in just four iterations, whereas POCS converged in nine iterations. For the reconstruction in the subsequent sections, the MAP super-resolution method will be used as it provides the sharpest, artifact-free result.

7.3.6 Image Registration

In super-resolution imaging, fractional pixel image registration accuracy is crucial to the subsequent reconstruction result. Many different image registration techniques over the years have been proposed, and they may be divided into phase and

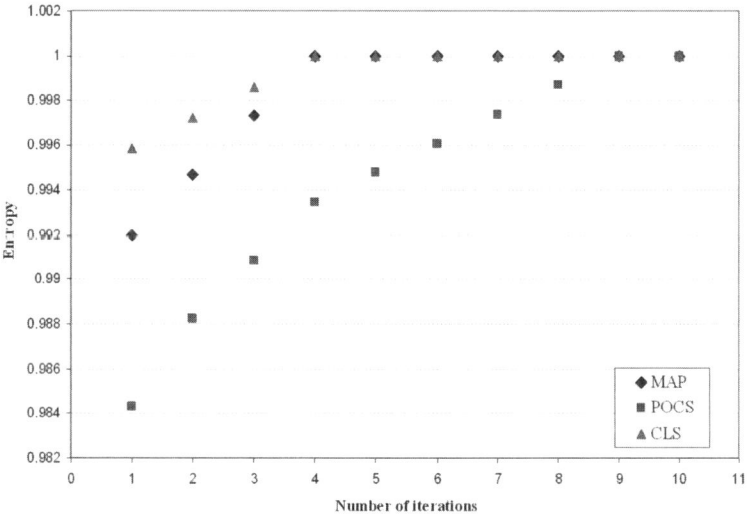

Figure 7.11 Entropy values of the images reconstructed using MAP, POCS, and CLS for different numbers of iterations.

Figure 7.12 Images illustrating the effects of poor registration. The super-resolution reconstruction in (a) was calculated using nonregistered images, that is, the shifts between the images were wrong. For comparison the image in (b) is calculated using the registered low-resolution images.

spatial domain methods. Phase correlation methods [28] are based on the fact that in the frequency domain two shifted images differ only by a phase shift. Correlation of the Fourier transforms of the images is used to calculate shifts between the images. Spatial domain methods are based on the direct manipulation of the image data. One of the popular spatial domain methods is block matching [29], where blocks of pixels are compared between the two images, minimizing a disparity measure. Keren *et al.* [30] developed an iterative motion estimation algorithm based on the Taylor series expansion. A pyramidal scheme is used to increase the precision for large motion parameters. This method is proved to be robust and accurate, and therefore is used for image registration in subsequent sections of this chapter. Figure 7.12 illustrates the effects of poor registration on the super-resolution reconstruction.

7.4 Applications of Super-Resolution

7.4.1 Application in Minimally Invasive Surgery

One of the recent applications of super-resolution imaging is in surgical robotics. Robotic-assisted minimally invasive surgery (MIS) is one of the most promising advances in surgical technology in recent years [31]. MIS is a new kind of surgery performed through natural body openings or small artificial incisions. Figure 7.13 illustrates a typical MIS setup in which surgical tools such as laparoscopic instruments and miniature cameras are inserted into the patient's inflated abdomen through small incisions. The benefits of minimally invasive procedures over traditional surgery include significant reduction in operative blood loss, scarring, and postoperative pain, which yields faster recovery and rehabilitation time, leading to a shorter hospital stay. While being beneficial to the patient, MIS introduces additional strain to the operating surgeon.

7.4 Applications of Super-Resolution

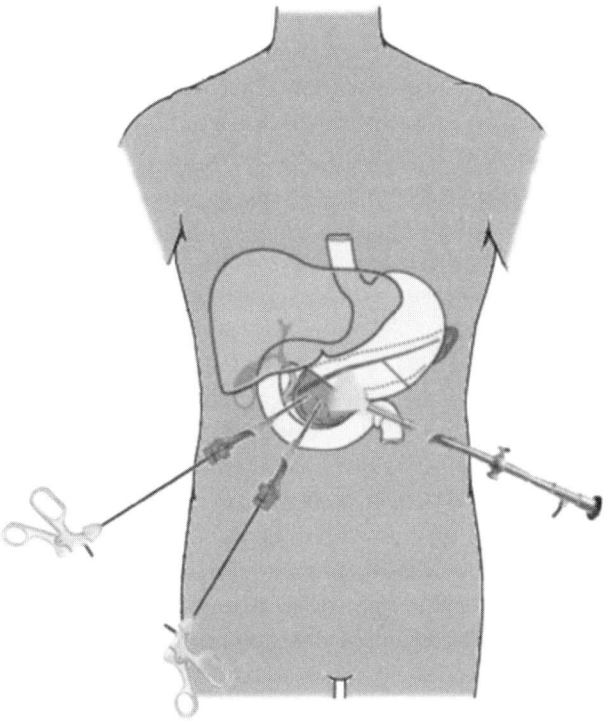

Figure 7.13 Diagram of a typical minimally invasive surgery setup.

While MIS speeds up recovery times, the procedure itself can be time consuming due to a difficult operating environment. The deployment of MIS is associated with the complexity of instrument controls, restricted vision and mobility, difficult hand–eye coordination, and the lack of tactile perception, which require a high degree of operator dexterity [32]. During minimally invasive surgery, change of the field of view (FOV) necessitates adjustment of the camera position and often refocusing. This is largely due to the restriction on the number of trocar ports used during MIS and the small FOV required for achieving a large magnification factor for performing microscale tasks such as coronary anastomosis. The consequence of restricted vision significantly affects the visual-spatial orientation of the surgeon and the awareness of the peripheral sites. During MIS, constant change of the orientation and focal length of the laparoscope camera is cumbersome and can impose extra visual and cognitive load to the operating surgeon in realigning the visual pathways and anatomical landmarks. In certain cases, this can also contribute to surgical errors as view rotation is known to affect surgical performances. The limited FOV has also made intra-operative surgical guidance difficult due to the paucity of unique surface features for aligning pre- or intraoperative three-dimensional image data.

By using super-resolution, large magnification of the region of interest can be provided to the surgeon without the need to move the camera, such that the perceived resolution is larger than the intrinsic resolution of the laparoscope camera. In this case, a super-resolution image reconstruction method can be used to add artificial zoom to the MIS surgery. Instead of moving the camera a surgeon could select a desired region and get a high-resolution image of that region. Another approach could be to use eye tracking to locate the region the surgeon is focusing on and increase the resolution of that selection, leading to foveal vision in MIS. A detailed description of eye tracking can be found in Chapter 8. An example image taken during the MIS surgery is shown in Figure 7.14.

The region highlighted by the white square in Figure 7.14 is the region where the surgeon performing the operation needs a closer view. This region can be manually selected by the surgeon or from the gaze location of the surgeon by using eye tracking. Figures 7.15(a, c) show example images of two sets of 30 low-resolution images extracted from the highlighted region. Subpixel shifts are provided by the endoscopic camera movements, which, in this example, span a few pixel lengths over the region covered by the sets.

Super-resolution reconstructions, shown in Figure 7.15(b, d), are calculated by using a color MAP algorithm. It is evident that the high-resolution images appear sharper and the high-frequency content is higher than in the low-resolution images. This application can aid MIS surgery in providing artificial zoom of the chosen area, removing the need to move the camera.

The described super-resolution method uses images of a static scene, and the transformation incorporated into the image model mainly involves global affine transformations. For nonstatic scenes in which parts of the scene move relative to each other, the quality of the super-resolution reconstruction is decreased. This can be problematic in the MIS application since the tissue warps in a relatively fast timescale, so it becomes important to have a high enough frame rate to capture a set of images before the scene is changed. To overcome this problem, a high-speed camera can be used. The frame rate of the camera needs to be sufficiently high to

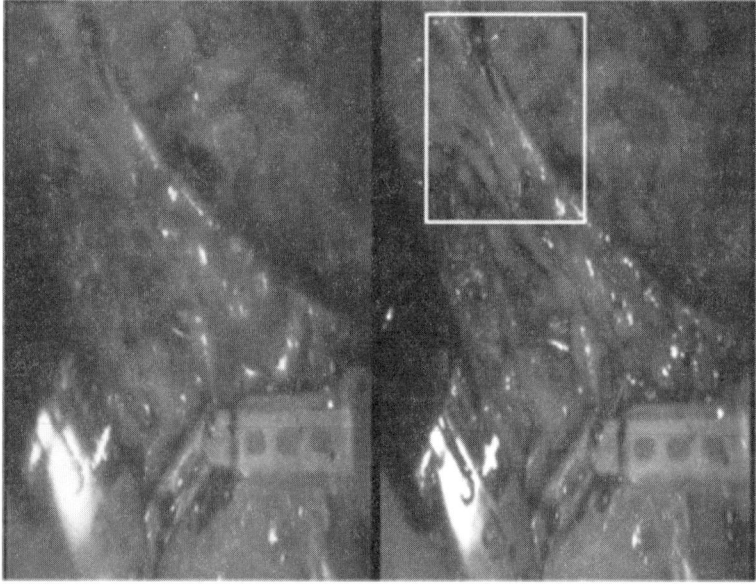

Figure 7.14 Image taken during TECAB surgery on a stereoscope. The white square represents the region of interest for which a closer view is shown in Figure 7.15. http://www.bg.ic.ac.uk/staff/aab01

7.4 Applications of Super-Resolution

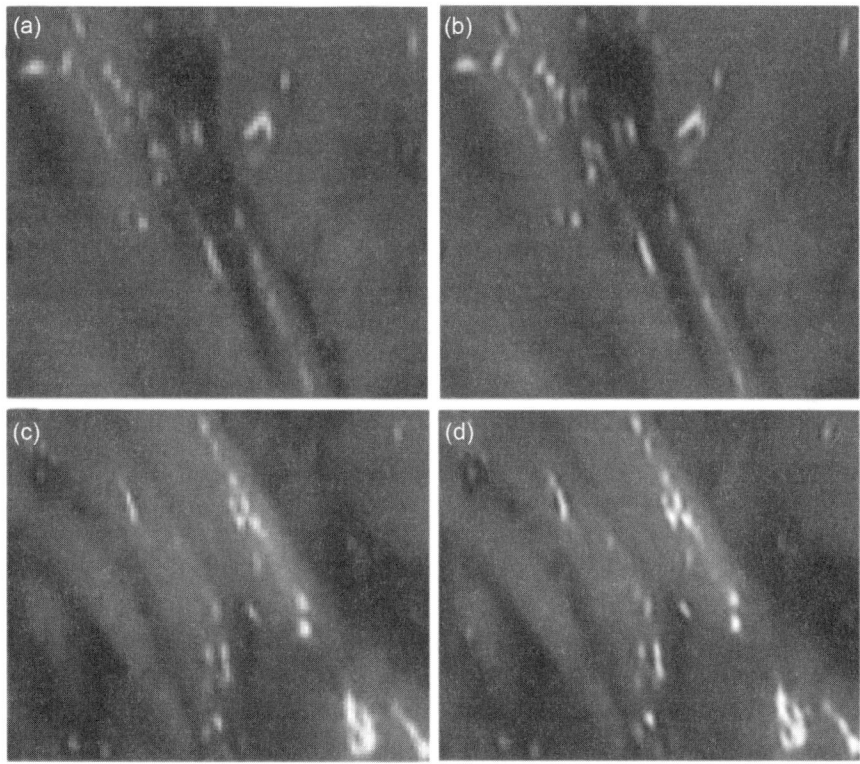

Figure 7.15 Two sets of low-resolution images, examples of which are shown in (a) and (c), were taken from the "white square" region shown in Figure 7.14. Higher-resolution super-resolution image reconstructions of sets (a) and (c) are shown on (b) and (d), respectively. http://www.bg.ic.ac.uk/staff/aab01

be able to capture a desired number of low-resolution frames before deformation occurs.

7.4.2 Other Applications

Thus far, super-resolution has been used in a range of applications including:

- Forensic science [33, 34];
- Video surveillance [35];
- Medical imaging [36, 37];
- Astronomy [38, 39];
- Satellite imaging [40, 41].

In video surveillance and forensic applications, super-resolution is used to improve the quality of the captured images such as those of licence plates of cars or of human faces. In medical imaging, such as computed tomography (CT) and magnetic resonance imaging (MRI), the resolution of the images (mainly through-plan resolution) can be pushed beyond the limited resolution of the instruments using super-resolution since the acquisition of multiple images is possible. In astronomy and satellite imaging, observed objects are often too far or too faint to be imaged with the desired resolution, and therefore super-resolution is used to create high-resolution images of them.

7.5 Conclusions and Further Challenges

The super-resolution image reconstruction method is used to increase the intrinsic resolution of a camera by calculating a high-resolution image from low-resolution images. Three different super-resolution algorithms, constrained least square, projection onto convex sets, and maximum a posteriori, were described and compared, of which the MAP-MRF method was shown to be the best among the three.

The super-resolution method presents a novel approach to add an artificial zoom in MIS, such that the perceived resolution of the foveal FOV is greater than the intrinsic resolution of the laparoscope camera. The basic motivation of the technique is to investigate the use of fixational movements to robotic-assisted MIS such that the perceived resolution of the foveal FOV is greater than the intrinsic resolution of the laparoscope camera. This will allow the use of a relatively large FOV for microsurgical tasks where super-resolution may be applied to fixation points in response to real-time eye tracking.

It should be noted that when faced with a dynamic scene, the edges of the super-resolution reconstructed image will appear blurry and the probability of artifacts increases. Typical dewarping algorithms cannot be used because they tend to "wash away" the information needed for reconstruction. This problem could be overcome by using a high frame-rate camera or by combining the existing super-resolution algorithms with dewarping.

In addition to dynamic super-resolution, other future research opportunities [42] include improving robustness and calculation speed of the method and investigations of better subpixel registration techniques. Moreover, as super-resolution becomes more and more mainstream, there is a need to make the technique automatic and relieve the user from the need to tune the model parameters with each application. Some work has been done in incorporating prior information into super-resolution in cases where the subject in the images is known, such as faces [43]. In this case, the need for multiple images is replaced by a model of the subject. So far this method has successfully been applied to human faces but it promises to lead to many different applications.

References

[1] Westheimer, G., "Visual acuity and hyperacuity," *Investigative Ophthalmology & Visual Science*, Vol. 14, No. 8, 1975, pp. 570–572.

[2] Martinez-Conde, S., S. L. Macknik and D. H. Hubel, "The role of fixational eye movements in visual perception," *Nature Reviews Neuroscience*, Vol. 5, No. 3, 2004, pp. 229–240.

[3] Park, S. C., M. K. Park and M. G. Kang, "Super-resolution image reconstruction: a technical overview," *Signal Processing Magazine, IEEE*, Vol. 20, No. 3, 2003, pp. 21-36.

[4] Steinman, R. M., et al., "Miniature eye movement," *Science*, Vol. 181, No. 102, 1973, pp. 810-819.

[5] Ditchburn, R. W., "The function of small saccades," *Vision Research*, Vol. 20, No. 3, 1980, pp. 271-272.

[6] Martinez-Conde, S., S. L. Macknik and D. H. Hubel, "Microsaccadic eye movements and firing of single cells in the striate cortex of macaque monkeys," *Nature Neuroscience*, Vol. 3, No. 3, 2000, pp. 251-258.

[7] Williams, D. R., "Visibility of interference fringes near the resolution limit," *Journal of the Optical Society of America. A, Optics, Image Science, and Vision*, Vol. 2, No. 7, 1985, pp. 1087-1093.

[8] Levi, D. M., P. V. McGraw and S. A. Klein, "Vernier and contrast discrimination in central and peripheral vision," *Vision Research*, Vol. 40, No. 8, 2000, pp. 973-988.

[9] Geisler, W. S., "Physical limits of acuity and hyperacuity," *Journal of the Optical Society of America. A, Optics, Image Science, and Vision*, Vol. 1, No. 7, 1984, pp. 775-782.

[10] Fahle, M. and T. Poggio, "Visual hyperacuity: spatiotemporal interpolation in human vision," *Proceedings of the Royal Society of London. Series B, Biological Sciences*, Vol. 213, No. 1193, 1981, pp. 451-477.

[11] Fahle, M. W., "Parallel, semiparallel, and serial processing of visual hyperacuity," *Human Vision and Electronic Imaging: Models, Methods, and Applications*, Santa Clara, CA: 1990, pp. 147-159.

[12] Westheimer, G., "The resolving power of the eye," *Vision Research*, Vol. 45, No. 7, 2005, pp. 945-947.

[13] Hadani, I., et al., "Hyperacuity in the detection of absolute and differential displacements of random dot patterns," *Vision Research*, Vol. 20, No. 11, 1980, pp. 947-951.

[14] Rucci, M. and G. l. Desbordes, "Contributions of fixational eye movements to the discrimination of briefly presented stimuli," *Journal of Vision*, Vol. 3, No. 11, 2003, pp. 852-864.

[15] Moller, F., et al., "Binocular quantification and characterization of microsaccades," *Graefe's Archive for Clinical and Experimental Ophthalmology*, Vol. 240, No. 9, 2002, pp. 765-770.

[16] Carpenter, R. H. S., *Movements of the Eyes*, London: Pion Limited, 1977.

[17] Ditchburn, R. W., *Eye Movements and Visual Perception*, Oxford: Clarendon Press, 1973.

[18] Chaudhuri, S., *Super-Resolution Imaging*, Kluwer Academic Publishers, Norwell, MA, 2002.

[19] Ur, H. and D. Gross, "Improved resolution from subpixel shifted pictures," *CVGIP: Graphical Models and Image Processing*, Vol. 54, No. 2, 1992, pp. 181-186.

[20] Stark, H. and P. Oskoui, "High-resolution image recovery from image-plane arrays, using convex projections," *Journal of Optical Society of America A*, Vol. 6, No. 11, 1989, pp. 1715-1726.

[21] Tekalp, A. M., M. K. Ozkan and M. I. Sezan, "High-resolution image reconstruction from lower-resolution image sequences and space-varying image restoration," *IEEE International Conference on Acoustics, Speech, and Signal Processing, ICASSP-92*, 1992, pp. 169-172.

[22] Tsai, R. Y. and T. S. Huang, "Multipleframe image restoration and registration," In *Advances in Computer Vision and Image Processing*, Greenwich, CT: JAI Press Inc., 1984, pp. 317-399.

[23] Banham, M. R. and A. K. Katsaggelos, "Digital image restoration," *Signal Processing Magazine, IEEE*, Vol. 14, No. 2, 1997, pp. 24-41.

[24] Hong, M.-C., M. G. Kang and A. K. Katsaggelos, "An iterative weighted regularized algorithm for improving the resolution of video sequences," *Proc. 1997 IEEE International Conference on Image Processing*, 1997, pp. 474–477.

[25] Schultz, R. R. and R. L. Stevenson, "Extraction of high-resolution frames from video sequences," *Image Processing, IEEE Transactions on*, Vol. 5, No. 6, 1996, pp. 996–1011.

[26] Farsiu, S., M. Elad and P. Milanfar, "Multiframe demosaicing and super-resolution of color images," *Image Processing, IEEE Transactions on*, Vol. 15, No. 1, 2006, pp. 141–159.

[27] Gonzalez, R. C. and R. E. Woods, *Digital Image Processing*, New Jersey: Prentice-Hall, Inc., 2002.

[28] Kuglin, C. and D. Hines, "The phase correlation image alignment method," *Proceedings of the International Conference in Cybernetics and Society*, 1975, pp. 163–165.

[29] Jain, J. and A. Jain, "Displacement measurement and its application in interframe image coding," *Communications, IEEE Transactions on (legacy, pre-1988)*, Vol. 29, No. 12, 1981, pp. 1799–1808.

[30] Keren, D., S. Peleg and R. Brada, "Image sequence enhancement using sub-pixel displacements," *Proceedings CVPR '88, Computer Society Conference on Computer Vision and Pattern Recognition*, Ann Arbor, MI, 1988, pp. 742–746.

[31] Ballantyne, G. H., "Robotic surgery, telerobotic surgery, telepresence, and telementoring," *Surgical Endoscopy*, Vol. 16, No. 10, 2002, pp. 1389–1402.

[32] Tendick, F., et al., "Sensing and manipulation problems in endoscopic surgery: experiment, analysis, and observation," *Presence*, Vol. 2, No. 1, 1993, pp. 66–81.

[33] Schultz, R. R., "Super-resolution enhancement of native digital video versus digitized NTSC sequences," 2002, pp. 193–197.

[34] Guichard, F. and L. Rudin, "Velocity estimation from images sequence and application to super-resolution," 1999, pp. 527–531.

[35] Cristani, M., et al., "Distilling information with super-resolution for video surveillance," *Proceedings of the ACM 2nd International Workshop on Video Surveillance & Sensor Networks*, New York, NY, 2004, pp. 2–11.

[36] Greenspan, H., et al., "Super-resolution in MRI," *Proceedings of the 2002 IEEE International Symposium on Biomedical Imaging*, 2002, pp. 943–946.

[37] Kennedy, J. A., et al., "Super-resolution in PET imaging," *Medical Imaging, IEEE Transactions on*, Vol. 25, No. 2, 2006, pp. 137–147.

[38] Willett, R. M., et al., "Wavelet-based superresolution in astronomy," *Proceedings of the Astronomical Data Analysis Software and Systems (ADASS) XIII*, Strasbourg, France, 2004, pp. 107.

[39] Cheeseman, P., et al., "Super-resolved surface reconstruction form multiple images," *Technical Report FIA-94-12, NASA Ames Research Center, Artificial Intelligence Branch*, 1994.

[40] Tao, H., et al., "Superresolution remote sensing image processing algorithm based on wavelet transform and interpolation," *Image Processing and Pattern Recognition in Remote Sensing*, Hangzhou, China, 2003, pp. 259–263.

[41] Aguena, M. L. S. and N. D. A. Mascarenhas, "Multispectral image data fusion using POCS and super-resolution," *Computer Vision and Image Understanding*, Vol. 102, No. 2, 2006, pp. 178–187.

[42] Farsiu, S., et al., "Advances and challenges in super-resolution," *International Journal of Imaging Systems and Technology*, Vol. 14, No. 2, 2004, pp. 47–57.

[43] Baker, S. and T. Kanade, "Hallucinating faces," *Fourth International Conference on Automatic Face and Gesture Recognition*, Grenoble, France, 2000, pp. 83–88.

CHAPTER 8
Eye Tracking and Depth from Vergence

George P. Mylonas and Guang-Zhong Yang

Original materials derived from the following papers:
Yang, G.-Z., Dempere-Marco, L., Hu, X.-P., Rowe, A. "Visual search: Psychophysical models and practical applications." *Image and Vision Computing*, 2002; 20:291–305

Mylonas, G. P., Darzi, A., Yang, G.-Z. "Gaze Contingent control for minimally invasive robotic surgery." *Computer Aided Surgery*, September 2006; 11(5):256–266

8.1 Introduction

Visual search is the act of searching for a target within a scene. If the scene is between 2 degree and 30 degree, the eyes will move across it to find the target. If the scene is larger still, the head moves as well. The number of visual search tasks performed in a single day is so large that it has become a reactive rather than a deliberative process for most normal tasks. Human eyes do not have a uniform visual response. In fact, the best visual acuity is only within a visual angle of 1 to 2 degree. This is called foveal vision. The fovea, also known as foveal centralis, forms part of the retina and is located in the center of the *macula* region. The center of the fovea is the "foveal pit," which contains a higher concentration of cone cells compared to the rest of the retina and no rods. During a fixation, the eye lens focuses the fixated area onto the fovea. For areas that we do not direct our eyes toward when observing a scene, we have to rely on a cruder representation of the objects offered by nonfoveal vision, of which the visual acuity drops off dramatically from the center of focus. When we try to understand a scene, we do not scan it randomly. Instead, we fixate our eyes on particular areas and move between them. The intrinsic dynamics of eye movement are complex, but the following movements are common:

- *Saccadic:* This is the voluntary rapid eye movement to direct the fovea to a specific point of interest.
- *Miniature:* These are a group of involuntary eye movements that cause the eye to have a dither effect. They include drift and microsaccades.
- *Pursuit:* It is a smooth involuntary eye movement that acts to keep a moving object foveated.
- *Compensatory:* These movements are similar to the pursuit movement but they maintain a fixation while the head is moving.
- *Vergence:* This is how the eyes converge to a single point, relating to focusing on a near or far object.

- *Optokinetic:* This is an involuntary sawtooth movement that the eye performs when observing repeated moving patterns.

Saccadic eye movement is the most important to consider when studying visual search. A saccade is a rapid voluntary movement from one point to another. It has a fast acceleration at approximately $40,000°/s^2$ and a peak velocity of $600°/s$. Saccades are observed in visual searches in the range of 1 to 40 degree. The objective of a saccade is to foveate a particular area of interest in a search scene. The fovea has an operation range of 0.6 to 1 degree. Miniature eye movements, on the other hand, are small and involuntary movements of less than 1 degree. There are three different types of miniature eye movements: eye-drift, microsaccade, and eye-tremor. The effect of these small eye movements imposes a lower limit on the accuracy of eye-tracking techniques, and an accuracy of more than 0.5 degree is generally thought not to be required.

8.2 Eye-Tracking Techniques

Most experiments involving eye tracking have mainly been conducted in clinical psychology/psychiatry, with schizophrenia being one of the most frequently considered topics. The objective of eye-tracking systems is to determine when and where fixations occur. The time order of the list of fixations represents the actual visual search that takes place. There are various methods of tracking the eye position relative to the head. Assuming the head position is known, then this can be used to follow the gaze of a subject. Figure 8.1 gives an example of two different types

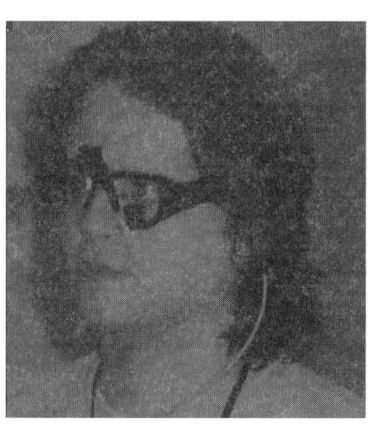
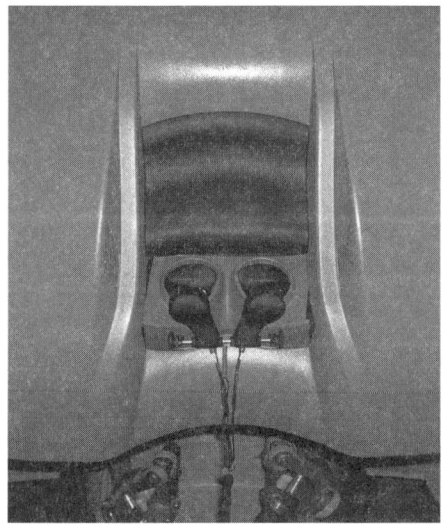

Figure 8.1 Example of a head-mounted eye tracker (left) and a binocular eye tracker mounted on a robotic surgical console (right).

of eye-tracking systems; the one shown on the left is a wearable eye-tracking device whereas the other uses remote optics, and therefore offers a much more natural working environment. There are currently three major categories of techniques for measuring eye movements: electro-oculography (EOG), scleral contact lens/search coils, and techniques based on reflected light such as video-oculography (VOG).

EOG is based on the measurement of skin electric potential difference around the eye, which varies as the eye rotates. Electrodes can be placed above, below, and to the sides of the eyes, with the potential difference across the eye indicating the eye position. This method can achieve an accuracy of $\pm 1.5°$. Since it effectively measures the position of the eye relative to the head, head tracking must be used if the gaze position is to be determined.

Scleral contact lens and search coils rely on either a mechanical or optical reference object mounted on contact lenses worn directly on the eyes. The basic principle of the technique dates back to over a century ago, although at that time the reference object was placed directly onto the cornea. The current implementation of the system commonly involves the measurement of the response of wire coils mounted on the contact lens moving through an electromagnetic field. Despite the invasive nature of the method, it is one of the most accurate measurement techniques for determining the relative position of the eye and the head.

The third category, which is based on reflected light, is the most commonly used eye-tracking technique. It involves the measurement of distinguishable features of the eyes during visual search. These features include limbus, pupil, and corneal reflection. Anatomically, the front view of the eye can be divided into three parts: the sclera, the iris, and the pupil. The interface of the sclera and the iris is called the limbus. The limbus moves proportionally to the rotation of the eye, and its position can be tracked optically with the use of photodiodes. This technique, however, is only effective for horizontal tracking. For vertical tracking, the eyelid blocks the limbus and thus prevents its true position from being obtained. The technique can nevertheless be extended to tracking the pupil position as the pupil/iris interface is better defined, and the eyelid does not obscure the pupil. Both of these techniques require the head to be fixed in position or the use of head-mounted optical sensors.

In visual search, it is important to determine the point in space that is being looked at (the point-of-regard) rather than the position of the eye relative to the head. To accomplish this, it is necessary to have a combined analysis of the above features. Most nonintrusive video-based approaches, such as video-oculography, rely on the measurement of both pupil and corneal reflection, commonly referred to as "pupil minus corneal reflection (P-CR)" technique. The geometrical structure of the eye is formed in such a way that the corneal reflection of a fixed light source, known as Purkinje reflection, is relatively stable while the eyeball, and hence the pupil, rotates in its orbit. In fact, since the radius of curvature of the cornea is about half that of the eye, the corneal reflection during saccade moves in the direction of eye movement but only about half as far as the pupil moves. Since the relative position between the center of the pupil and the corneal reflection remains constant with minor head movements, it is therefore possible to use the corneal reflection as the reference point to determine the point-of-regard without physically fixing the position of the head. To avoid the problems of eyes being out of focus or even falling out of the viewing angle of the camera due to head movement, systems based on the pupil/corneal reflection method usually require automatic eye tracking combined with autofocusing features or small camera aperture. The range with which some modern systems can operate is normally between ± 40 degree as further eye movements will cause the corneal reflection to fall outside the spherical part of the cornea, thus requiring more complicated geometrical correction steps. Furthermore, if a larger amount of head movement is to be accommodated, this has to be done by using a head-tracking device or specially developed image-based algorithms.

Eye illumination in video-oculography is typically achieved by using infrared light sources. Most visible wavelengths are absorbed in the pigment epithelium but incident radiation in the infrared creates several Purkinje reflections on the boundaries of the lens and cornea. Given a fixed light source, there are, in fact, four reflections associated with it. Among these four reflections, two of them are caused by the reflection of the front and rear surfaces of the cornea, and the other two from the reflection of the lens. The first Purkinje image is often referred to as the "glint." By measuring the position of the first and fourth Purkinje reflections, it is possible to separate translational and rotational eye movements. Although this technique offers more accurate measurement results, the fourth Purkinje image is normally very weak due to a change of the refraction index at the rear of the lens. Therefore, the surrounding lighting must be strictly controlled in such experiments. Figure 8.2 schematically illustrates how the four Purkinje reflections are formed, as well as the relative positions of the pupil and the first Purkinje reflection when the eye fixates to the center and the four corners of a computer screen.

Another effect of infrared illumination of the eye is that the relatively dark iris becomes bright. Additionally, the retina may or may not act as a retroreflector, depending on the relative configuration of the camera and infrared light source. If the optical axis of the camera imaging the eye is coaxial with the infrared illuminator,

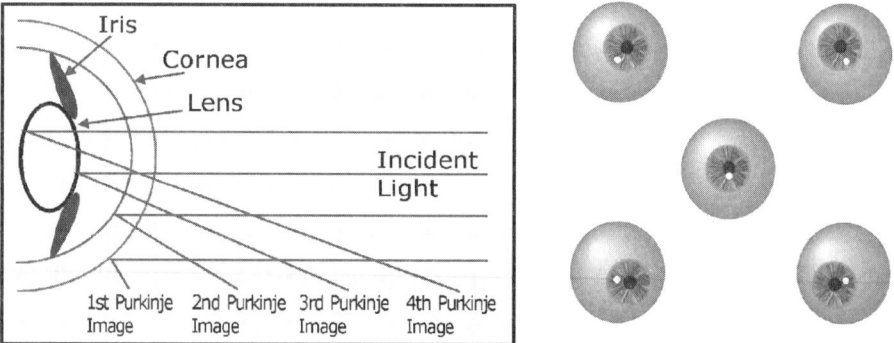

Figure 8.2 The basic principle of how the four Purkinje images are formed (left). The relative positions of the pupil and the first Purkinje reflection during visual search as seen from the camera when the observer fixates to the center and the four corners of a screen.

the retroreflected light back-illuminates the pupil. The result is a "bright-pupil" image, similar to the "red-eye" effect in flash photography. In contrast, if the axis of the camera is not coincident with the illuminator, the reflected light does not enter the camera, resulting in a "dark-pupil." In either case, the pupil produces high contrast with the iris, and, by setting a threshold value at an appropriate level, the pupil region can be segmented. With image processing, the center of both the pupil and the glint can be identified and localized [1]. The two centers define a vector, which can be mapped to a unique eye gaze direction through an appropriately designed calibration task.

8.3 Applications of Eye Tracking

8.3.1 Psychology/Psychiatry and Cognitive Sciences

So far, psychology, psychiatry, and cognitive sciences are the most active research areas that involve eye tracking. The main objectives of these experiments are to obtain a detailed understanding about the cognitive process of visual search [2, 3] and examine the developmental differences of oculomotor functioning at different stages of our lives. Particular interest is being paid to cognitive processes in infants since even those a few months old can use eye gaze to note what is of interest to them, either by directional looking or pupil enlargement. In parallel to this, experiments have also been undertaken to analyze visual perception in animals, mainly monkeys, so as to improve our understanding of visual recognition under different stimuli [4, 5], as well as gaining further insight into the neural codes that relate to visual perception [6] and neural mechanisms causing strabismus.

It has been found that eye-tracking techniques are particularly useful in diagnosing reading disabilities. Research has shown that there exist characteristic patterns of eye movements which are symptomatic of particular reading disabilities [7–9]. Among them, people who suffer from dyslexia show an abnormal pattern of the saccadic eye movements when they are reading [10]. Particularly, when the number and duration of fixations during reading are compared between dyslexics and normal readers, it indicates that the number of forward saccades is higher and their size smaller for dyslexics. Furthermore, their fixation durations are longer than those from normal readers [11]. Eye-tracking techniques can also be used to investigate psycholinguistic issues in cognitive processes (e.g., contextual representation in interpretation, models of idiom processing, spoken word recognition, and ambiguity resolution) [12–17].

Eye-movement abnormalities have been actively used to investigate psychiatric disorders. This line of research was triggered by Philip Holzman in the mid-1970s after noticing the pursuit eye-movement disturbances not only in schizophrenic patients, but also in their unaffected relatives. In recent years, much work has been done to investigate the neurobiological basis of affective disorders, obsessive-compulsive disorder, dementias, attention deficit disorder, and autism [18]. For patients suffering from mental diseases, eye-tracking experiments have been widely used in assessing schizophrenia [19–24], Alzheimer's [25, 26], and bipolarity [27].

8.3.2 Behavior Analysis

Since the early days of visual tracking research, it has been recognized that the knowledge of visual search patterns of experienced observers can potentially be useful for improving the learning process of trainees by adopting a learning-by-imitation procedure. Such an approach has been found to be particularly effective in medical imaging where the interpretation of image details is largely subjective. Eye tracking provides an effective means of gaining information about how experts assimilate different visual clues to reach a confident diagnosis [28–34]. It also offers an objective way of assessing the competence of trainees in fulfilling the recognition tasks. Furthermore, gaze-contingent processing also enables the exploitation of implicit knowledge for teaching, joint problem solving, or in a global framework, the communication of practical knowledge and skills. Due to the widespread use of on-line tutorials, eye-tracking techniques can also play an important role in assessing how subjects interact with the tutorial in distance learning systems.

Capturing visual attention has been found to be particularly useful for reducing errors in critical tasks such as those undertaken by control operators and clinicians [35, 36]. Stresses related to fatigue and noise are major factors involved in making poor judgments. Eye tracking is becoming an integral part of driving simulators as it provides vital information about the reasoning, prediction, and level of alert of the driver [37]. For investigating the safety implications of using mobile phones while driving, as there exists a shift of attentional resources [38], measures derived from eye tracking, such as the number of fixations and duration of fixations, have been used together with response time measures to assess the effect of a cellular phone conversation on the performance of driving tasks such as searching for traffic signals.

Another important area that involves visual tracking is usability studies. The term usability refers to the extent to which the users can operate a product to achieve specific goals [39]. For example, eye-tracking techniques enable the characterization of the user's behavior in using a particular design of computer interface. By analyzing parameters such as the time spent looking at each region, the dwell time on certain icons, and the latency in finding a specific function, it is possible to derive the general friendliness and intuitiveness of the interface [40]. The value of this approach for optimizing the design of human–computer interfaces has received extensive interest, and the Special Interest Group on Computer-Human Interaction SIGCHI '99 conference contained a workshop dedicated to this issue.

Based on the same principles, it is also possible to use eye-tracking systems to determine what attracts people's attention. For advertising, it is particularly useful for understanding where people devote their visual interest. The material involved can either be static images or video sequences [41–47]. A study on consumers' eye-movement patterns on yellow pages' advertising was undertaken by Lohse [48], which revealed several interesting behaviors of the general population. Among them it has been found that consumers noticed over 93% of the quarter page display advertisements whereas almost 75% of the plain listings were completely ignored. Some other interesting results arise from on-line advertising and the realization that Internet users do not tend to fixate on web advertisements but rather seem to avoid them. Furthermore, eye-tracking techniques can be useful for assessing advertisement strategies. In particular, they have been considered to assess how people react to "explicit" advertisements (i.e., advertisements that depict the target product together with a related headline in a semantically straightforward way) and "implicit" advertisements (i.e., the images and text depicted are not directly related to the product) [49].

8.3.3 Medicine

It has been over 20 years since the initial use of eye tracking for analyzing problems associated with medical image understanding. An early study in scan-path analysis of radiographs by Kundel in 1978 [31] provided information on the nature of errors in detecting pulmonary nodules in chest X-ray images. It was found that 30% of false-negative errors were due to incomplete scanning, 25% were due to failure to recognize visual features, and 45% due to wrong decisions. Under the current climate of clinical governance in medicine, the potential implication of using visual tracking for quality assurance and training is extensive, as it provides an objective way of addressing some of the problems mentioned earlier. For example, the issue of incomplete scanning can be avoided by using eye tracking to highlight areas that are missed during visual examination. For training, the scan paths of consultant radiologists can be replayed to trainees to understand the intrinsic relationships between different feature points that attract fixations.

In addition to the global-focal model, Nodine and Kundel [50] also proposed a visual dwell time algorithm, which has shown to be of value for avoiding the failure of recognizing abnormal visual features. This approach used the scan path coordinates and the associated dwell time to identify hot spots in the image. These hot spots signify areas toward which attention has been drawn during the visual search and are subsequently prompted to the interpreter at the end of examination to reconfirm their visual significance. This dwell time analysis was subsequently improved by Nodine [51], who provided a more in-depth analysis of eye position recordings based on fixation clustering. In Figure 8.3, high-resolution computed tomography (HRCT) images were examined by experienced radiologists. Their fixation scan path and the dwell time of each fixation are overlaid for subsequent examination. Nodine also investigated the effect of training on the visual search performance [52]. The analysis of time-to-hit data (i.e., time needed for the observers to fixate on the target region) indicated that the importance of the search was not to sample the image exhaustively, nor to find a given target, but rather to find something peculiar or quaint about the image. It is through experience that trained observers become more aware of the presence of such peculiarities in the images. Such awareness brings visual attention to this quaint region for interpretation and, subsequently, relies on a knowledge base to recognize atypical variants for diagnostic purposes.

The issues arising from these studies indicate the significance of gaining more experience in interpreting visual input once the eyes fixate onto a target candidate. Covering a specific region of interest does not by itself guarantee detection. It is possible to detect features of interest on images with peripheral vision. However, the likelihood of detection decreases with the eccentricity of the feature of interest with regard to the center of the foveal attention. Overall, it can be concluded that

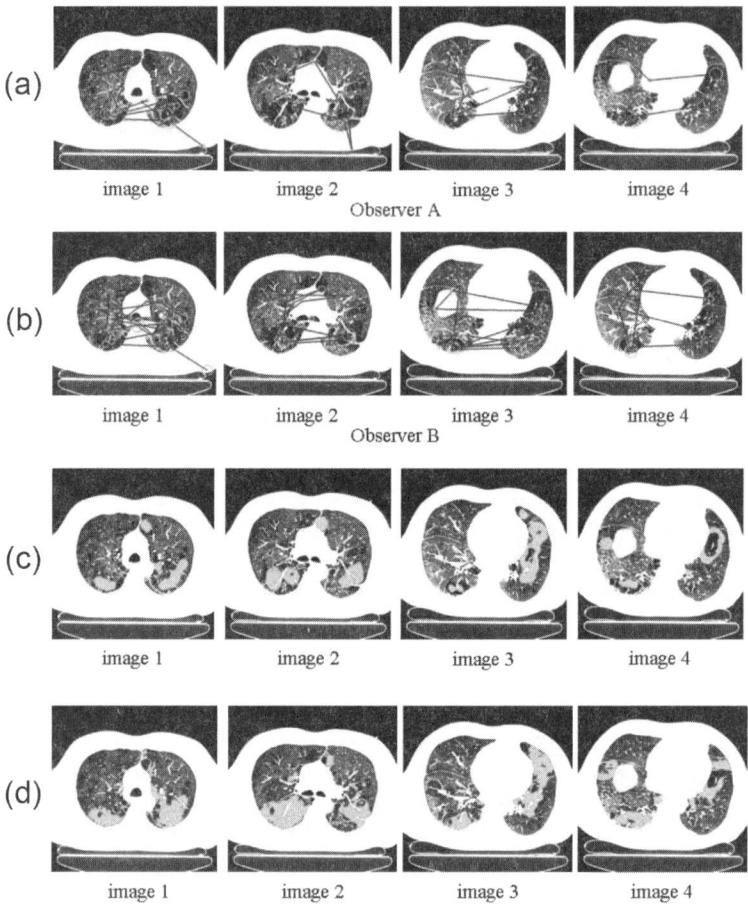

Figure 8.3 (a, b) Four HRCT lung images of a patient with centrilobular emphysema and the corresponding visual search scan paths recorded from two experienced radiologists. The size of the circles represents the dwell time of each fixation and the lines between fixations indicate the saccadic eye movements. (c, d) The derived saliency map from the two observers. It is evident that despite the apparent difference of the scan-path patterns, there is a strong correlation between the derived saliency maps suggesting a common perceptual behavior captured by eye tracking.

skilled visual search comes under more cognitive control with training. Furthermore, visual attention becomes more perceptually attuned to the detection and decision-making requirements of the task [53–55].

Beyond radiology, eye tracking has also been used to investigate pharmacological effects on patients, assist surgical operations, and map human brain activity with functional MRI (fMRI). Pokorny [56] presented an early study on the influence of drugs on eye-tracking tasks. Later, Fant [57] found that both acute and residual subjective, physiologic, and performance effects of consuming any drug can be assessed by eye-tracking tests. In laser refractive surgery, patient eye and head movements are major sources of error due to the change in relative position of the surgical beam and the target [58]. Active eye-tracking systems have been incorporated into the laser delivery systems, which are used to cancel out the effect of patient eye movement [58–61] and therefore help to minimize post-operative glare. In fMRI, eye-movement monitoring has been integrated into the system to examine the relationship between visual perception and regional brain activities. The use of eye-tracking techniques is also useful in the study of diseases such as vertigo symptoms [62], multiple sclerosis [63], and insulin-dependent diabetes mellitus [64] through visuomotor responses of the patients.

8.3.4 Human–Computer Interaction

It is expected that the major impact of accurate, compact, and easy-to-use eye-tracking systems is likely to be in the design of human–computer interaction. Military research sponsored some of the early work in eye-tracking systems. In conventional applications, one of the most common uses of eye-tracking techniques is for assisting disabled people. To this end, the gaze may be used to select a given action from a computer screen, enabling them to interact and control their environment. In these systems, careful considerations must be made so as to avoid undesired activations [65–67]. A usual way to type text using the gaze as input is to display a keyboard on the computer screen and use dwell time as a selection technique. It is unlikely that text input with eye gaze is to become a substitute to keyboard typing for general use except for certain special areas, such as mobility enhancement for disabled people. By incorporating eye tracking, Hyrskyhari [68] describe a translation aid, iDict, to improve the reading of foreign language documents with the assistance of gaze path information. The system contains built-in information that includes general knowledge of the reading process, lexical and syntactical language analysis of the text that helps in the identification of potentially unknown words or syntactically complex structures, and user's profile. The operation of iDict is based on tracking the reading process of an on-screen document and detecting where the user seems to have troubles. The system infers what help is needed and delivers it to the user. This application is targeted to a wide audience, whereas most current applications are designed for people with disabilities.

In the last few years, eye-tracking techniques have been attracting increasing interest from the computer graphics community due to the rich sets of visual stimuli, ranging from two-dimensional imagery to three-dimensional immersive virtual scenarios that they generate. Visual attention and perception can influence the way scenes are constructed. One of the possibilities opened to the community is the use of eye tracking for gaze-contingent applications for which the resolution and quality of an image or animation are adjusted according to the viewer's saccades and fixations. Another line of research not much studied yet is the analysis of the perception of synthesized images and animations with the objectives of optimizing the realism of such perception and improving the efficiency of the graphics algorithms [69].

Gaze-contingent displays rely on the observation that most of the resources that are used to produce large and high-resolution displays are actually wasted, particularly when dealing with single-user displays. This is due to the physiology of the human visual system that can only resolve detailed information within a very small area at the fovea. Therefore, by knowing where the gaze of the observer is directed at a given instant, it would be convenient to display high-resolution information in that region. This requires a dynamic updating of the image in real time based on eye-gaze direction data. The production of a display completely indistinguishable from a full-resolution display, however, is a very challenging task since the update rate of the display needs to be extremely high and the degradation in the peripheral fields has to be kept minimal. The effect of varying the spatiotemporal parameters that affect the performance of eye-contingent multiresolution displays has been discussed by Loschky [70] and Parkhurst [71].

In telecommunication and telework tasks, it is important to know what the state of attention of the participants is. In this context, visual attention indicated by gaze direction plays an important role [72–76]. An example of such an application is the GAZE Groupware system. This is an eye-controlled audio–visual conferencing system that, based upon eye-tracking techniques, allows users to see who is talking to whom about what in a three-dimensional meeting room on the Internet [77]. Another interesting application of eye tracking is the so-called interest- and emotion-sensitive media [78,79], which enables audience-directed script navigation by considering the users' interests and emotional states. The potential of this technology has motivated an intensive research interest [80]. Both IBM and MIT have proposed prototype systems that allow the viewer to interact with a multithreaded documentary using a multimodal interface [81, 82].

In the field of affective computing, Partala [83] investigated pupillary responses to emotionally provocative sound stimuli. The results suggest that the pupil size discriminates among different kinds of emotional stimuli. Furthermore, eye-tracking techniques may be used to improve a human interface. Chino [84] simulated a "meta-communication" by requiring from the user intending to input voice commands to a given system to previously gaze at the agent in order to request to talk. This allows the system to accept only intended voice input. It is interesting that eye tracking is also finding its way to consumer products. Canon has devices (35-mm camera and video camera) which use this technology to focus in the direction of the user's gaze rather than in the direction in which the camera is pointed. Until recently, the greatest number of focusing points available on a 35-mm camera was five. The Canon EOS-3 35-mm camera enables photographers to enjoy 45 focusing points. All focusing points are linked directly to Canon's eye-controlled autofocus system [85].

8.4 Gaze-Contingent Control for Robotic Surgery

One of the most interesting applications of eye tracking is in instrument control in minimally invasive surgery (MIS). Endoscopy, including bronchoscopy and laparoscopy, is the most common procedure in MIS, which is carried out through natural body openings or small artificial incisions. If handled properly, endoscopes are completely harmless to patients. MIS and diagnostic endoscopy can achieve their clinical goals with minimal inconvenience to patients. Compared to conventional techniques, patient trauma and hospitalization can be greatly reduced and diagnostic accuracy and therapeutic success increased. However, the complexity of the instrument controls, restricted vision and mobility, difficult hand–eye coordination, and the lack of tactile feedback require a high degree of manual dexterity of the operator.

Medical robotics and computer-assisted surgery are new and promising fields of study, which aim to augment the capabilities of surgeons by taking the best from robots and humans. With robotically assisted MIS, dexterity is enhanced by microprocessor-controlled mechanical wrists, which permit motion scaling and reduce gross hand movements. This technology also allows the performance of microscale tasks that would otherwise be impossible. Current robotic systems allow the surgeon to operate while seated at a console viewing a magnified stereo image of the surgical field. Hand and wrist maneuvers are then seamlessly translated into precise, real-time movements of the surgical instruments inside the patient. The continuing evolution of technology, including force feedback and virtual immobilization through real-time motion adaptation, will permit more complex procedures, such as beating heart surgery, to be carried out using a static frame of reference. The use of robotically assisted MIS also provides an ideal environment for integrating patient-specific preoperative data for performing image-guided surgery and active constraint control, all of which can be conducted without the need of the surgeon to remove his or her eyes from the operating field of view.

One of the major applications of robotically assisted MIS is the performance of totally endoscopic coronary artery bypass grafts (TECAB) on a beating heart. The main challenge of beating heart surgery is the destabilization introduced by cardiac and respiratory motion, which significantly affects precise tissue–instrument interaction and the execution of complex grafts. Despite the use of mechanical stabilization, the motion of the epicardial surface hinders delicate tasks such as small vessel anastomosis [86], which is compounded by the high viewing magnification factor used during the procedure. A number of three-dimensional soft-tissue structural recovery techniques have been proposed for improved surgical guidance and motion compensation. These include the use of multiview geometry based on novel computer vision techniques with or without the use of fiducial markers [87–90]. Despite the success achieved with these techniques, particularly for off-line processing, their potential clinical value with real-time *in situ* depth recovery is hindered by the morphological complexity and highly deformable nature of the soft tissue, coupled with the specularity and inter-reflections under common MIS conditions.

In the remaining sections, we investigate the use of eye gaze for simplifying, as well as enhancing, robotic control in surgery. More specifically, we demonstrate that eye gaze derived from binocular eye tracking can be effectively used to recover three-dimensional motion and deformation of the soft tissue during MIS procedures. Compared to the use of other input channels, eye gaze is the only input modality that implicitly carries information on the focus of the user's attention at a specific point in time. This allows seamless *in vivo* registration of the motion and deformation fields within the anatomical area of interest that is directly under fixation. In this case, it is only necessary to accurately track deformation fields within a relatively small area that is directly under foveal vision. Simple rigid body motion of the camera can therefore be used to provide a perceptually stable operating field of view. Given the complexity of robotic control in surgical environments, this approach also facilitates the effective hand–eye coordination for improved surgical performance. Detailed phantom assessment of the accuracy and temporal response of the system, as demonstrated on a laboratory-based robotic arm, is presented. Preliminary *in vivo* results for TECAB procedures are also provided, demonstrating the capability of the system in extracting coupled cardiac deformation due to cardiac and respiratory motion.

8.4.1 Ocular Vergence for Depth Recovery

One of the strongest depth cues available to humans is the horizontal disparity that exists between the two retinal images. There is a close relationship between the horizontal disparity and depth perception, which varies with the viewing distance. More specifically, as the fixation point moves away from the observer, the horizontal disparity between the two retinal images is diminished and vice versa. In order to extract quantitative information regarding the depth of the fixation point, ocular vergence needs to be measured, thus providing a veridical interpretation of stereoscopic depth [91]. By using any of the eye-tracking methods described earlier, the combined tracking of both eyes provides the ocular vergence measure, which in turn determines the fixation point.

In order to perform the *gaze-contingent* experiments, a stereo viewing environment that is similar to the daVinci surgical robot (Intuitive Inc., CA) was created. The system consists of a stereoscopic console and an industrial robot geared with a customized stereo-camera rig. The stereo console allows the user to examine three-dimensional video captured by the two cameras on the robot. The optical path that permits stereo viewing is illustrated in Figure 8.4, where two TFT monitors are used to display the live video feeds. The purpose of the mirrors in use is to scale down the images to a size that matches the interpupilary distance, thus facilitating the fusion of the two views into a single three-dimensional image. By using two eye-tracking cameras built into the stereoscope, it is possible to quantify the ocular vergence and determine the depth of the fixation point of the user while observing the stereo images (Figure 8.4, right). Since the two gaze vectors are expected to be epipolar, we can determine the fixation point as the intersection of the two. To establish the mapping between pupil-glint vectors and points in three-dimensional space and also to correct for subject-specific variations of the eye geometry, calibration is required prior to each eye-tracking session.

For assessing the accuracy of the proposed three-dimensional depth recovery framework through ocular vergence, a phantom heart model was created by using thixotropic silicone mold rubber and prevulcanized natural rubber latex with rubber mask greasepaint to achieve a specular appearance and high visual fidelity, as can be seen in Figure 8.5(a). The phantom is deformable by means of a system of 4 oil-filled pistons with controllable injection levels. In this way, the amount of deformation can be accurately controlled and reproduced. It is worth noting that the materials used for making the phantom model are CT and MR compatible. Figure 8.5(b) depicts the reconstructed phantom heart from a series of CT slices

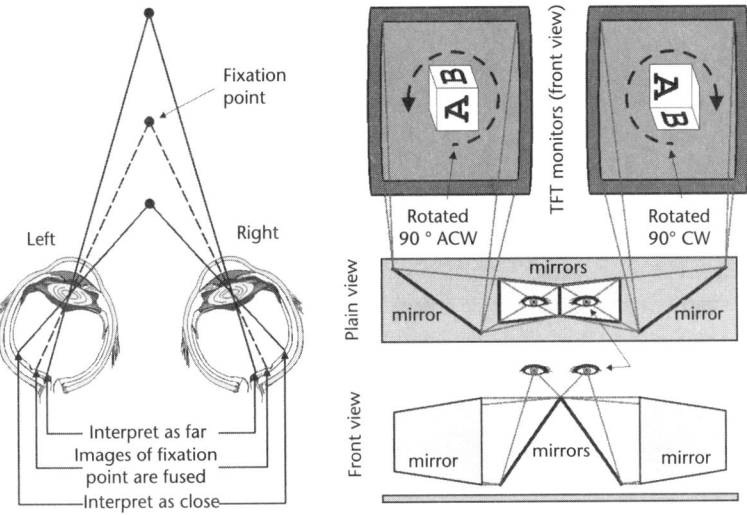

Figure 8.4 (a)The relationship of the horizontal disparity between the two retinal images and depth perception which varies with the viewing distance. Ocular vergence is quantified allowing for the three-dimensional fixation to be determined. (b) Schematic illustration of the stereoscopic viewer with binocular eye tracking. While a subject fuses the parallax images displayed on the monitors, both eyes are tracked.

at different deformation levels. Figure 8.5(c) shows the entire phantom experiment setup of the robotic arm. To determine the exact position of the epicardial surface, an Aurora (Northern Digital Inc., Ontario) 5-DOF electromagnetic catheter-tip tracker was used. The device has an accuracy of 0.9 to 1.3 mm RMS, depending on the distance of the tool from the magnetic field generator and the amount of ferromagnetic interference in the vicinity.

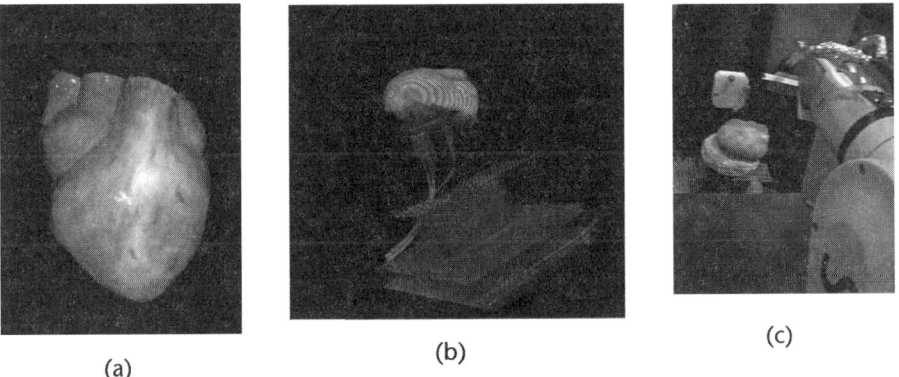

Figure 8.5 (a) The silicon phantom heart. (b) The reconstructed phantom from a series of CT slices. The supporting base is also visible here along with the oil-filled syringes used to deform the heart in a reproducible manner. In (c) the robot with the mounted optical-tracker retroreflectors and the stereo-camera rig. http://www.bg.ic.ac.uk/staff/aab01

8.4.2 Binocular Eye-Tracking Calibration

In order to establish the relationship between pupil-glint vectors and fixations in three-dimensional space, initial subject-specific eye-tracking calibration is required. There are different methods of binocular calibration and the preference depends on the degree of flexibility required.

Free calibration based on radial basis spline: This calibration approach does not require any prior knowledge of the extrinsic or intrinsic parameters of the stereo laparoscope. Calibration is performed by the use of a wireframe three-dimensional calibration grid with known dimensions. The grid is positioned in front of the stereo laparoscopic cameras, and the user is required to fixate on a number of targets (nodes) on the grid while pupil-glint vectors are recorded.

Calibration for the corneal reflection and pupil center vectors is calculated using radial basis spline [92]. Let $\mathbf{X} = (X^1, X^2, X^3)$ be a 1-to-1 three-dimensional vector value function of a four-dimensional eye-coordinate vector $\mathbf{p} = (p^1, p^2, p^3, p^4)$. Vector $\mathbf{X}$ corresponds to the three-dimensional coordinates of a calibration target in the world frame of reference. Vector $\mathbf{p}$ encapsulates the left and right eye two-dimensional vectors defined by the centers of the pupil and the glint. Assuming further that vector function $\mathbf{X}$ can be decoupled into three independent scalar functions of vector $\mathbf{p}$, each scalar component of function $\mathbf{X}$ can thus be continuously interpolated with radial basis spline, that is:

$$X^d(\mathbf{p}) = \sum_{i=1}^{N} b_i \Phi_i(\mathbf{p}) + \mathbf{a} \cdot \begin{bmatrix} \mathbf{p} & 1 \end{bmatrix}^T \tag{8.1}$$

where b_i is the radial basis coefficient of the corresponding basis function Φ_i and vector $\mathbf{a}$ is a global affine coefficient. The spline parameters $\mathbf{a}$ and b are determined by solving the following system of linear equations:

$$\begin{bmatrix} \Phi_1(\mathbf{p}_1) & \cdots & \Phi_N(\mathbf{p}_1) & & \\ \vdots & & \vdots & \mathbf{P} & \\ \Phi_1(\mathbf{p}_N) & \cdots & \Phi_N(\mathbf{p}_N) & & \\ & \mathbf{P}^T & & 0 & \end{bmatrix} \begin{bmatrix} b_1 \\ \vdots \\ b_N \\ a^1 \\ \vdots \\ a^5 \end{bmatrix} = \begin{bmatrix} X_1^d \\ \vdots \\ X_N^d \\ 0 \\ \vdots \\ 0 \end{bmatrix} \tag{8.2}$$

$$\text{and} \quad \mathbf{P} = \begin{bmatrix} 1 & p_1^1 & p_1^2 & p_1^3 & p_1^4 \\ \vdots & \vdots & \vdots & \vdots & \vdots \\ 1 & p_N^1 & p_N^2 & p_N^3 & p_N^4 \end{bmatrix}^T$$

The radial basis function Φ_i was defined as that of the Euclidean distance in four-dimensional from a given point $\mathbf{p}$ to the ith control point $\mathbf{p}^i$, that is,

$$\Phi_i(\mathbf{p}) = r_i^2 \ln r_i^2 \quad \text{where} \quad r_i = \|\mathbf{p} - \mathbf{p}_i\| \tag{8.3}$$

The necessary condition for the coefficient matrix in Equation (8.2) not being ill conditioned is that any two given points in the four-dimensional space are not coplanar. This criterion can be ensured by sampling the *a priori* spline control point value, (X^1, X^2, X^3), as a unique point in the calibrated three-dimensional volume. This was performed in this study by presenting the observer with 27 nodes in sets of 9, displayed in one of the three calibration planes. Fewer nodes can be used for shorter calibration times with a trade-off on the accuracy of the resulting calibration.

Free calibration is an attractive option when the intrinsic and extrinsic parameters of the stereo endoscope are not known. The downside is that positioning the grid in such a way that all the nodes are visible may require some fiddling. If a robotically controlled endoscope with position feedback is used, then it is possible to use one or more nodes on a single planar grid and then simulate the other nodes or planes using controlled translations of the endoscope.

Calibration from known intrinsic and extrinsic stereo endoscope parameters: As opposed to the free calibration method described above, this approach takes into consideration the intrinsic and extrinsic stereo endoscope parameters. Prior to the eye-tracking calibration, the two endoscope cameras need to be calibrated, which can be achieved by using one of the many existing techniques [93]. With this method, each eye is calibrated separately. The aim is to find the gaze direction of each eye and then localize the two-dimensional fixation point as the intersection of each eye's gaze vector with the respective stereoscope screen plane. Localization of the three-dimensional fixation point is then achieved by stereo triangulation of the pair of two-dimensional fixation points on each of the two views of the stereoscope screens [94, 95].

Radial basis splines are used for each eye individually this time. By letting $\mathbf{X} = (X^1, X^2)$ be a 1-to-1 two-dimensional vector value function of a two-dimensional pupil-glint vector $\mathbf{p} = (p^1, p^2)$, each scalar component of function $\mathbf{X}$ can then be continuously interpolated with the radial basis spline according to Equation (8.1). The *a priori* spline control point values (X^1, X^2) correspond to the coordinates of the calibration markers on the viewing screen's frame of reference. These are collected by presenting each individual eye with a two-dimensional grid of markers lying along conjugate epipolar lines on the two screen planes. During calibration, care should be taken so that the subject is always able to fuse the two markers as a single one, avoiding diplopia, which would invalidate the calibration. In general, this is achieved by displaying the targets at the same coordinates on both screens. Controlled shift is introduced if necessary such that binocular viewing is made comfortable. With this approach, as few as four markers near the corners of the screens are enough for calibration. However, better accuracy is achieved when more—preferably 9—are used.

The main advantage of this method is that it does not have to be performed again if some of the parameters of the stereo laparoscope are changed, as this is often the case during an operation. Triangulation of the fixation point in this case is automatically readjusted according to the updated intrinsic and extrinsic parameters of the stereo laparoscope. This can be particularly useful in conjunction with active camera self-calibration methods [96]. Also, by obtaining the calibration parameters of the stereo system, image rectification is possible so distortions from the camera lenses are corrected.

Head shift cancellation: During robotic surgery operations, the surgeon will often lift his or her head from the stereoscopic console view-port. Also, even short periods of viewing will result in the surgeon's head shifting relative to the position where eye-tracking calibration was performed. For maintaining accuracy of eye tracking throughout the operation, it is necessary to compensate and allow for free head motion. P-CR-based eye tracking can compensate for relative camera and head movements under the assumption that when the camera is translated with respect to the eye, the pupil and the corneal reflection are translated by the same amount. This is true to some extent but principally requires the head movement to be small and the distance between the eye and the infrared light source and camera to be relatively large. Given the nature of the stereoscopic viewers used in systems like the daVinci, the binocular eye tracker has to be maintained close to the user's eyes.

In [97], a method is presented that allows distinguishing eye movements with respect to the head from camera movements with respect to the head. The method is based on the observation that during eye movements, the eye rotates within the socket such that the center of the pupil with respect to the eye camera frame moves a greater distance than the corneal reflection does. On the other hand, during a camera movement (or interchangeably a head movement), the difference in displacement of the center of the pupil and the corneal reflection is much smaller. This technique derives two gain parameters, the *camera-gain* and the *eye-gain*. These values represent the fraction of a one-unit pupil movement that the corneal reflection moves during a camera/head movement and during an eye movement, respectively. The two gains can be obtained through calibration. As far as the *eye-gain* is concerned, this is simultaneously derived through binocular calibration as was described in the previous section. Calculation of the *camera-gain* requires an extra calibration step where data is collected during camera movement while no eye movements occur. It should be noted that the *camera-gain* does not seem to be subject specific and works well among different subjects. This means that calibration has to be performed only once, unless something changes with the hardware configuration.

8.4.3 Depth Recovery and Motion Stabilization

In order to demonstrate the practical value of the proposed concept, two experiments were conducted; one involved the use of binocular eye tracking for gaze-contingent depth recovery from soft tissue, and the other used the same concept for adaptively changing the position of the camera to cancel out cyclic motion of the tissue to stabilize the foveal field of view.

For the depth recovery experiments, both real scenes captured by the stereo camera and computer-generated surfaces were used. This is necessary since in the master control console of a robotic system both live video and synthetically generated images are often present. It is therefore important to establish that similar depth reconstruction behavior can be achieved. Five subjects were asked to observe the two images by following a suggested fixation path. The fixation points were acquired from the eye tracker during the task and the acquired depth coordinates were recorded.

For depth recovery on a synthetic tissue, the subjects were asked to follow a predefined path by fixating onto image features of their preference. No visual markers were introduced, and they were relatively free to select and fixate on image features of their preference. Figure 8.6 presents the corresponding depths recovered from these subjects where the ground-truth data is provided as a comparison. It is evident that a relatively close correlation has been achieved, demonstrating the feasibility of veridical reconstruction of the real depth.

To assess the binocular system frequency response, experiments were carried out involving six subjects. The first set of experiments investigates the oscillatory response of the binocular visual system over a frequency range. The subjects involved were asked to keep fixating onto a feature of their preference on the surface of the phantom heart model. In parallel with binocular eye tracking, the robot is set to oscillations of gradually increasing frequencies along the z-axis (depth). While the three-dimensional fixation point of a subject is tracked, the position of the robotic cameras is also recorded by using the Polaris optical tracker. After data collection, ARMA modelling is used to derive the coefficients of the parametric system that describe the transfer function of the system. Table 8.1 summarizes the response of the visual system in oscillation along the z-axis, which indicates that it is accurate up to frequencies of about 1.8 Hz. Beyond this limit, there is considerable attenuation and noticeable phase shift.

To further examine the ability of the proposed binocular eye-tracking framework in recovering tissue deformation, the deformable phantom model was used along with an Aurora catheter-tip tracker positioned on the epicardial surface. The subjects were asked to keep fixating on a surface feature close to the Aurora sensor. While the phantom was subjected to different levels of deformation, both the fixation point and the position of the electromagnetic sensor were tracked. In

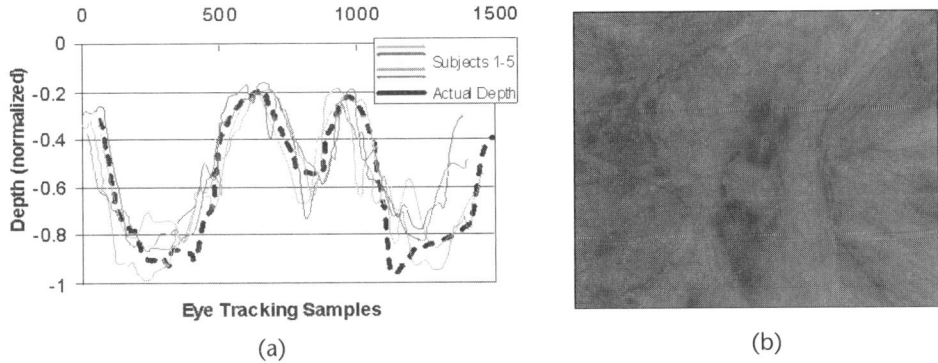

Figure 8.6 (a) A comparison of the recovered depths by the five subjects studied against the actual depth of the virtual surface depicted. (b) The synthetic tissue with known surface geometry used for the experiment. http://www.bg.ic.ac.uk/staff/aab01

Table 8.1 Error Analysis Comparing the Oculomotor Response of the 6 Subjects Over a Range of Frequencies

Frequency (rad/s)	Amplitude		Phase (rad)	
	Mean	std	Mean	std
0.1	0.950	0.118	−0.031	0.016
1.0	0.918	0.132	−0.307	0.120
3.0	0.822	0.204	−0.871	0.098
6.0	0.650	0.156	−1.494	0.150
10.0	0.577	0.125	−2.394	0.278
12.0	0.520	0.126	−2.955	0.316

Figure 8.7, the gaze-contingent recovered deformation is compared with the actual levels reported by the sensor, demonstrating the practical accuracy of the method.

To evaluate the *in vivo* value of the proposed framework, data from a robotic-assisted TECAB procedure with the daVinci robot was used. The video footage of the operation was played back in the stereoscopic viewer while a subject was eye-tracked. The purpose of the experiment was to demonstrate how cardiac and respiratory motion could be recovered and decoupled by using eye tracking. The image in Figure 8.8 shows a snapshot of the two views from a 40-sec long footage. A large portion of the view was occupied by the daVinci robotic Endowrist grasping the cardiac stabilizer just before it was positioned in place. During the entire footage, the laparoscope was stationary, and the deformed tissue area under foveation appears at the bottom left portion of the image. What appears on the video sequence is the deforming epicardial surface with the respiratory motion, principally manifested along the horizontal axis superimposed by the cardiac motion. The graphs provided in Figure 8.8 show the collected eye-tracking data on the x-, y-, and z-axes (depth). Independent component analysis with the extended Infomax algorithm was then used to decouple respiratory from cardiac motion [98].

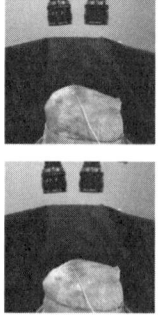
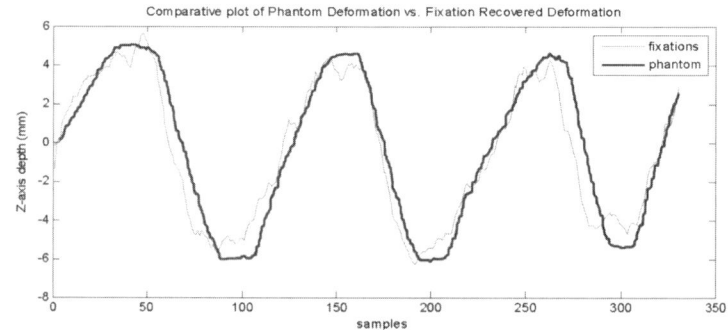

Figure 8.7 The recovered deformation from ocular vergence for the phantom model, where the actual deformation of the phantom heart surface as measured by an Aurora catheter-tip electromagnetic tracker is compared with the gaze-contingent reconstructed deformation.

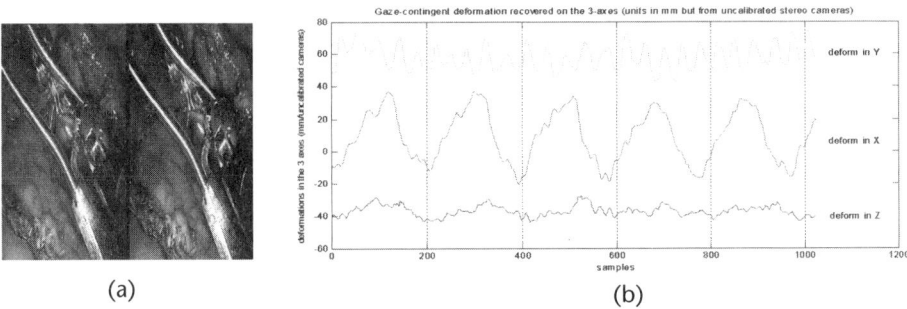

Figure 8.8 (a) the recovered epicardial surface deformation for a TECAB procedure. (b) a snapshot of the binocular views used in the experiment. On the right the eye-tracking–acquired tissue deformations on the x-, y-, and z-axes.

8.5 Discussion and Conclusions

Visual search is a common task that people perform throughout their daily lives. The number of applications that it can be applied to is potentially large. The examples described earlier by no means attempt to be a complete survey of the host of applications being carried out in this field. Nonetheless, it is hoped that they provide a flavor of what is the current state of the art of this captivating area. The use of eye tracking for objective assessment of image interpretation skills in engineering and medicine, designing improved human–computer interfacing for interacting with virtual and actual environments, incorporating visual search modelling systems into the behavior of foraging robots, and defining new frameworks for computer-assisted image understanding is only the tip of an iceberg that is gradually gathering its momentum moving toward more widespread practical applications.

Also, deploying robots around and within the human body, particularly for robotic surgery, presents a number of unique and challenging problems. These arise from the complex and often unpredictable environments that characterize the human anatomy. The ethical and legal barriers imposed on interventional surgical robots give rise to the need for a tightly integrated docking between the operator and the robot, where interaction in response to sensing is firmly under the command of the operating surgeon. In this chapter, we have demonstrated the potential value of gaze-contingent control in robotic-assisted MIS. Both spatial and temporal accuracies of the method in terms of deformation tracking and motion stabilization have been assessed with detailed phantom validation. It is worth noting that for the depth-recovery experiment, there are some discrepancies noticeable in the comparative depth-extraction plots, both among the participating subjects and the provided ground truth. The reason for these discrepancies is due to the fact that fixation onto a predefined scan path is practically difficult and deviation due to fixation jitter is difficult to avoid. The results derived from the subjects studied, however, demonstrate the relative consistency of the three-dimensional depth recovered.

For the motion-stabilization experiments outlined in this chapter, it is important to note that delays in the video relay, the robot communication, and motor controls can potentially introduce instability to the feedback loop. The use of adaptive optics for motion stabilization can potentially minimize this effect. Another option would be to investigate the use of prediction schemes, which would compensate for delay-induced instabilities [99, 100]. This is particularly useful for compensating cyclic tissue deformation, which is characteristic for cardiac motion. It should be noted that there are a number of other issues that need to be considered for the future integration of the proposed gaze-contingent framework. These include the dynamics of vergence [101, 102] and subject/scene-specific behavior of the eye [103, 104]. Other issues related to monocular preference [105], visual fatigue [106], and spatial errors that can arise when projecting three-dimensional space on a two-dimensional window [107] will also have to be considered. Nevertheless, the studies presented in this chapter represent the first attempt in using eye tracking and depth from vergence for robotic-assisted MIS. The basic concept outlined is also applicable to other applications that involve binocular viewing consoles.

References

[1] Morimoto CH, Mimica MRM. "Eye gaze tracking techniques for interactive applications," *Computer Vision and Image Understanding*, 98 (2005) 4-24.

[2] Wolfe JM, Gancarz G. "Guided search 3.0," in: V. Lakshminaragaman (Ed.), *Basic and Clinical Applications of Vision Science*, Kluwer Academic Publishers, Dordrecht, 1996, pp. 189-192.

[3] Kutas M, Federmeier KD. "Minding the body," *Psychophysiology* 35 (2) (1998) 135-150.

[4] Schweigart G, Mergner T, Barnes G. "Eye movements during combined pursuit, optokinetic and vestibular stimulation in macaque monkey," *Experimental Brain Research* 127 (1) (1999) 54-66.

[5] Lisberger SG. "Postsaccadic enhancement of initiation of smooth pursuit eye movements in monkeys," *Journal of Neurophysiology* 79 (4)(1998) 1918-1930.

[6] Martinez-Conde S, Macknik SL, Hubel DH. "Microsaccadic eye movements and firing of single cells in the striate cortex of macaque monkeys," *Nature Neuroscience* 3 (3) (2000) 251-258.

[7] Rayner K, McConkie GW. "What guides a reader's eye movements?" *Vision Research* 16 (1976) 829-837.

[8] Just MA, Carpenter PA. "A theory of reading: from eye fixations to comprehension," *Psychological Review* 87 (4) (1980) 329-354.

[9] Just MA, Carpenter PA. "Eye fixations and cognitive processes," *Cognitive Psychology* 8 (1976) 441-480.

[10] Fischer B, Hartnegg K, Mokler A. "Dynamic visual perception of dyslexic children," *Perception* 29 (5) (2000) 523-530.

[11] De Luca M, et al. "Eye movement patterns in linguistic and non-linguistic tasks in developmental surface dyslexia," *Neuropsychologia* 37 (12) (1999) 1407-1420.

[12] Titone DA, Connine CM. "On the compositional and noncompositional nature of idiomatic expressions," *Journal of Pragmatics* 31(12) (1999) 1655-1674.

[13] Tanenhaus MK, Spivey-Knowlton MJ. "Eye-tracking," *Language and Cognitive Processes* 11(6) (1996) 583–588.

[14] Pickering MJ, Traxler MJ, Crocker MW. "Ambiguity resolution in sentence processing: evidence against frequency-based accounts," *Journal of Memory and Language* 43 (3) (2000) 447–475.

[15] Dahan D, et al. "Linguistic gender and spoken-word recognition in French," *Journal of Memory and Language* 42 (4) (2000) 465–480.

[16] Garrod S, Tenas M. "The contribution of lexical and situational knowledge to resolving discourse roles: bonding and resolution," *Journal of Memory and Language* 42 (4) (2000) 526–544.

[17] Matsunaga S. "Universality in reading processes: evidence from an eye-tracking study," *Psychologia* 42 (4) (1999) 290–306.

[18] Sweeney J. "Eye movements and psychiatric disorders," *Conference Program and Abstract from the 11th European Conference on Eye Movements*, Turku, Finland, 2001, p. K2 (Abstract).

[19] Konrad HR, et al. "Eye tracking abnormalities in patients with cerebrovascular-disease," *Laryngoscope* 93 (9) (1983) 1171–1176.

[20] Levy DL, et al. "Eye tracking and schizophrenia: a selective review," *Schizophrenia Bulletin* 20 (1) (1994) 47–62.

[21] Levy DL, et al. "Quantitative characterization of eye tracking dysfunction in schizophrenia," *Schizophrenia Research* 42 (3) (2000) 171–185.

[22] Hooker C, Park S. "Trajectory estimation in schizophrenia," Schizophrenia Research 45 (1–2) (2000) 83–92.

[23] Nicolson R, et al. "Premorbid speech and language impairments in childhood-onset schizophrenia: association with risk factors," *American Journal of Psychiatry* 157 (5) (2000) 794–800.

[24] Holzman PS. "Eye movements and the search for the essence of schizophrenia," *Brain Research Reviews* 31 (2–3) (2000) 350–356.

[25] Muller G, et al. "Impaired eye tracking performance in patients with presenile onset dementia," *International Journal of Psychophysiology* 11(2) (1991) 167–177.

[26] Muller G, Weisbrod S, Klingberg F. "Test battery for the objectification and differential-diagnosis in the early-stage of presumed presenile-dementia of Alzheimer-type," *Zeitschrift fur Gerontologie* 26 (2) (1993) 70–80.

[27] Holzman PS, Obrian C, Waternaux C. "Effects of lithium treatment on eye-movements," *Biological Psychiatry* 29 (10) (1991) 1001–1015.

[28] Nodine CF, Kundel HL. "The cognitive side of visual search in radiology," in: JK O'Regan, A Levy-Schoen (Eds.), *Eye Movements: From Physiology to Cognition*, Elsevier, Amsterdam, 1986.

[29] Cowley HC, Gale AG, Wilson ARM. "Mammographic training sets for improving breast cancer detection," in: HL Kundel (Ed.), *Proceedings-SPIE The International Society for Optical Engineering*, vol. 2712, 1996, pp. 102–113.

[30] Cowley HC, Gale AG. "Non-radiologists and CAD systems in breast cancer screening," *Computational Imaging and Vision* (13) (1998) 371–374.

[31] Kundel HL, Nodine CF, Carmody DP. "Visual scanning, pattern recognition and decision-making in pulmonary nodule detection," *Investigative Radiology* 13 (1978) 175–181.

[32] Niimi R, et al. "Eye-tracking device comparisons of three methods of magnetic resonance image series displays," *Journal of Digital Imaging* 10 (4) (1997) 147–151.

[33] White KP, Hutson TL, Hutchinson TE. "Modeling human eye behavior during mammographic scanning: preliminary results," *IEEE Transactions on Systems Man and Cybernetics, Part A Systems and Humans* 27 (4) (1997) 494–505.

[34] Krupinski EA. "Visual scanning patterns of radiologists searching mammograms," *Academic Radiology* 3 (1996) 137–144.

[35] Saito S. "Does fatigue exist in a quantitative measurement of eye-movements?" *Ergonomics* 35 (5–6) (1992) 607–615.

[36] Larson GE, Perry ZA. "Visual capture and human error," *Applied Cognitive Psychology* 13 (3) (1999) 227–236.

[37] Tock D, Craw I. "Tracking and measuring drivers' eyes," *Image and Vision Computing* 14 (8) (1996) 541–547.

[38] Scialfa CT, McPhee L, Ho G. "The effects of a simulated cellular phone conversation on search for traffic signs in an elderly sample," *Proceedings of Eye Tracking Research and Applications Symposium, ACM*, 2000, pp. 45–51.

[39] International Organization for Standardization (adapted from ISO 9241-11), http://www.iso.org.

[40] Bend CR, Ottens D, Horst R. "Use of the eyetracking system in the usability laboratory," *35th Annual Meeting of the Human Factors Society, Proceedings of the Human Factors Society*, New Orleans, LA, 1991, pp. 461–465.

[41] Goldberg JH, Probart CK, Zak RE. "Visual search of food nutrition labels," *Human Factors* 41(3) (1999) 425–437.

[42] Pieters R, Rosbergen E, Hartog M. "Visual attention to advertising: the impact of motivation and repetition," *Advances in Consumer Research* 23 (1996) 242–248.

[43] Pieters R, Rosbergen E, Wedel M. "Visual attention to repeated print advertising: a test of scanpath theory," *Journal of Marketing Research* 36 (4) (1999) 424–438.

[44] Pieters R, Warlop L, Hartog M. "The effect of time pressure and task motivation on visual attention to brands," *Advances in Consumer Research* 24 (1997) 281–287.

[45] Pieters R, Warlop L. "Visual attention during brand choice: the impact of time pressure and task motivation," *International Journal of Research in Marketing* 16 (1) (1999) 1–16.

[46] Fox RJ, et al. "Adolescents' attention to beer and cigarette print ads and associated product warnings," *Journal of Advertising* 27 (3) (1998) 57–68.

[47] Krugman DM, et al. "Do adolescents attend to warnings in cigarette advertising: an eye-tracking approach," *Journal of Advertising Research* 34 (6) (1994) 39–52.

[48] Lohse GL. "Consumer eye movement patterns on yellow pages advertising," *Journal of Advertising* 26 (1) (1997) 61–73.

[49] Radach R, Vorstius C, Radach K. "Eye movements in the processing of advertisements: effects of pragmatic complexity," *Conference Program and Abstract from the 11th European Conference on Eye Movements*, Turku, Finland, 2001, p. S49 (Abstract).

[50] Nodine CF, Kundel HL. "A visual dwell algorithm can aid search and recognition of missed lung nodules in chest radiographs," in: D. Brogdan (Ed.), *First International Conference on Visual Search*, Taylor and Francis, London, 1990, pp. 399–406.

[51] Nodine CF, et al. "Recording and analyzing eye-position data using a microcomputer workstation," *Behavior Research Methods, Instruments and Computers* 24 (3) (1992) 475–485.

[52] Nodine CF, et al. "Nature of expertise in searching mammograms for breast masses," *Academic Radiology* 3 (1996) 1000–1006.

[53] Dempere-Marco L, et al. "The use of visual search for knowledge gathering in image decision support," *IEEE Transactions on Medical Imaging*, 21 (7) (2002) 741–754.

[54] Dempere-Marco L, et al. "Analysis of visual search patterns with EMD metric in normalized anatomical space," *IEEE Transactions on Medical Imaging* 25 (8) (2006) 1011–1021.

[55] Xiao-Peng Hu, Dempere-Marco L, Guang-Zhong Yang. "Hot spot detection based on feature space representation of visual search," *IEEE Transactions on Medical Imaging* 22 (9) (2003) 1152-1162.

[56] Pokorny R, et al. "The influence of thugs on eye tracking tasks," *Electroencephalography and Clinical Neurophysiology* 64 (1) (1986) 12.

[57] Fant RV, et al. "Acme and residual effects of marijuana in humans," *Pharmacology, Biochemistry and Behavior* 60 (4) (1998) 777-784.

[58] Taylor NM, et al. "Determining the accuracy of an eye tracking system for laser refractive surgery," *Journal of Refractive Surgery* 16 (5) (2000) S643-S646.

[59] Pallikaris IG, et al. "Photorefractive keratectomy with a small spot laser and tracker," *Journal of Refractive Surgery* 15 (2) (1999) 137-144.

[60] Krueger RR. "In perspective: eye tracking and autonomous laser radar," *Journal of Refractive Surgery* 15 (2) (1999) 145-149.

[61] Tsai YY, Lin JM. "Ablation centration after active eye-tracker–assisted photorefractive keratectomy and laser in situ keratomileusis," *Journal of Cataract and Refractive Surgery* 26 (1) (2000) 28-34.

[62] Ura M, Pfaltz CR, Allum JHJ. "The effect of age on the visuoocular and vestibuloocular reflexes of elderly patients with vertigo," *Acta Oto-Laryngologica* 481 (Suppl.) (1991) 399-402.

[63] Howard ABD, Good DC, Konrad HR. "Eye tracking tests and evoked-responses in early multiple-sclerosis," *Electroencephalography and Clinical Neurophysiology* 58 (2) (1984) 40.

[64] Virtaniemi J, et al. "Voluntary eye-movement tests in patients with insulin-dependent diabetes-mellitus," *Acta Oto-Laryngologica* 113 (2) (1993) 123-127.

[65] Hutchinson TE, et al. "Human computer interaction using eye-gaze input," *IEEE Transactions on Systems, Man, and Cybernetics* 19 (6) (1989) 1527-1534.

[66] Frey LA, White Jr. KP, Hutchinson TE. "Eye-gaze word processing," *IEEE Transactions on Systems, Man, and Cybernetics* 20 (4) (1990) 944-950.

[67] Colombo C, DelBimbo A. "Interacting through eyes," *Robotics and Autonomous Systems* 19 (3-4) (1997) 359-368.

[68] Hyrskyhari A, et al. "Design issues of iDict: a gaze-assisted translation aid," *Proceedings of the Eye Tracking Research and Applications Symposium*, Palm Beach Gardens, FL, November 6-8 2000, pp. 9-14.

[69] O'Sullivan C, et al. "Eye-tracking for interactive computer graphics," *Conference Program and Abstract from the 11th European Conference on Eye Movements*, Turku, Finland, 2001, p. S45 (Abstract).

[70] Loschky LC, McConkie GW. "User performance with gaze contingent multiresolutional displays," *Proceedings of the Eye Tracking Research and Applications Symposium*, Palm Beach Gardens, FL, November 6-8 2000, pp. 97-103.

[71] Parkhurst D, Culurciello E, Niebur E. "Evaluating variable resolution displays with visual search: task performance and eye movements," *Proceedings of the Eye Tracking Research and Applications Symposium*, Palm Beach Gardens, FL, November 6-8 2000, pp. 105-109.

[72] Fuchs H. "Beyond the desktop metaphor: toward more effective display, interaction, and telecollaboration in the office of the future via a multitude of sensors and displays," *Advanced Multimedia Content Processing* 1554 (1999) 30-43.

[73] Velichkovsky B. "Communicating attention: gaze position transfer in cooperative problem solving," *Pragmatics and Cognition* 3 (2) (1995) 199-222.

[74] Velichkovsky B, Hansen JP. "New technological windows into mind: there is more in eyes and brains for human computer interaction," *Proceedings of ACM CH '96 Conference*, 1996, pp. 496–503.

[75] Vertegaal R, et al. "Why conversational agents should catch the eye," in: *CHI'00 Extended Abstracts on Human Factors in Computing Systems*, ACM, New York, pp. 257–258.

[76] Vertegaal R, Van der Veer GC, Vons H. "Effects of gaze on multiparty mediated communication," *Proceedings of Graphics Interface, Canadian Human Computer Communications Society*, Montreal, Canada, 2000, pp. 95–102.

[77] Vertegaal R. "The GAZE groupware system: mediating joint attention in multiparty communication and collaboration," *Proceedings of ACM CHI'99, Conference on Human Factors in Computing Systems*, Pittsburgh, PA, 1999.

[78] Glenstrup AJ, Engell-Nielsen T. "Eye controlled media: present and future state," University of Copenhagen, Bachelor's Degree Thesis, DIKU (Institute of Computer Science) Denmark, 1995.

[79] Hansen JP, Andersen AW, Roed P. "Eye-gaze control of multimedia systems," *Proceedings of the Sixth International Conference on Human Computer Interaction*, Tokyo, Japan, vol. 1, Elsevier, Amsterdam, 1995, pp. 37–42.

[80] Starker I, Bolt RA. "A gaze-responsive self-disclosing display," *Proceedings of SIGCHI 90*, ACM Press, New York, 1990, pp. 3–9.

[81] BlueEyes: creating computers that know how you feel, http://www.almaden.ibm.com/cs/BlueEyes/index.html.

[82] Bers J, et al. "CyberBELT: multi-modal interaction with multi-threaded documentary," *CHI'95 Mosaic Of Creativity*, short paper, May 7–11, 1995

[83] Partala T, Jokiniemi M, Surakka V. "Pupillary responses to emotionally provocative stimuli," *Proceedings of the Eye Tracking Research and Applications Symposium*, Palm Beach Gardens, FL, November 6–8 2000, pp. 123–129.

[84] Chino T, Fukui K, Suzuki K. "'GazeToTalk': a nonverbal interface with meta-communication facility," *Proceedings of the Eye Tracking Research and Applications Symposium*, Palm Beach Gardens, FL, November 6–8 2000, p. 111.

[85] Canon, Canon's eye-controlled autofocus system, http://www.usa.canon.com.

[86] Wimmer-Greinecker G, et al. "Current status of robotically assisted coronary revascularization," *Am J Surg* 188 (4A Suppl) (2004) 76S–82S.

[87] Cuvillon L, et al. "Toward robotized beating heart TECABG: assessment of the heart dynamics using high-speed vision," in: Duncan JS, Gerig G, (Eds.), *Proceedings of the 8th Medical Image Computing and Computer-Assisted Intervention, (MICCAI 2005)*, Palm Springs, CA, October 26–29 2005, Vol. 2, pp. 551–558.

[88] Stoyanov D, Darzi A, Yang G-Z. "Dense depth recovery for robotic assisted laparoscopic surgery," in: Barillot C, Haynor DR, Hellier P, (Eds.), *Proceedings of the 7th Medical Image Computing and Computer-Assisted Intervention (MICCAI 2004)*, Rennes-St-Malo, France, September 26–29, 2004, Vol. 2, pp. 41–48.

[89] Stoyanov D, Darzi A, Yang G-Z. "A practical approach towards accurate dense 3D depth recovery for robotic laparoscopic surgery," *Computer Aided Surgery* 10 (4) (2005) 199–208.

[90] Stoyanov D, et al. "Soft-tissue motion tracking in robotic MIS procedures," in: Duncan JS, Gerig G, (Eds.), *Proceedings of the 8th Medical Image Computing and Computer-Assisted Intervention, (MICCAI 2005)*, Palm Springs, CA, October 26–29, 2005, Vol. 2, pp. 139–146.

[91] Mon-Williams M, Tresilian JR, Roberts A. "Vergence provides veridical depth perception from horizontal retinal image disparities." *Experimental Brain Research* 133 (2000) 407–413.

[92]   Bookstein, FL. "Principal warps: thin plate splines and the decomposition of deformations, *IEEE Transactions on Pattern Analysis Machine Intelligence* 11 (6) (1989) 567-585.

[93]   Tsai R. "A versatile camera calibration technique for high-accuracy 3D machine vision metrology using off-the-shelf TV cameras and lenses," *Robotics and Automation* 3 (4) (1987) 323-344.

[94]   Duchowski A, et al. "3-D eye movement analysis," *Behav Res Methods Instrum Comput* 34 (4) (2002) 573-591.

[95]   Horn B. *Robot Vision*, MIT Press, Cambridge, MA 1986.

[96]   Stoyanov D, Darzi A, Yang G-Z. "Laparoscope self-calibration for robotic assisted laparoscopic surgery," in: Duncan JS, Gerig G, (Eds.), *Proceedings of the 8th Medical Image Computing and Computer-Assisted Intervention, (MICCAI 2005)*, Palm Springs, CA, October 26-29, 2005, Vol. 2, pp. 114-121.

[97]   Kolakowski S, Pelz J. "Compensating for eye tracker camera movement," *ETRA 2006 Proceedings*, San Diego, CA, March 27-29 2006.

[98]   Yang GZ, Hu X. "Multi-sensor fusion," in: G.-Z. Yang (Ed.), *Body Sensor Networks*, Springer, 2006 pp. 239-281.

[99]   Ortmaier T, et al. "Motion estimation in beating heart surgery," *IEEE Transactions on Biomedical Engineering*, 52 (10) (2005) 1729-1740.

[100] Ginhoux R, et al. "Active filtering of physiological motion in robotized surgery using predictive control," *IEEE Transactions on Robotics* 21 (1) (2005) 67-79.

[101] Howard IP, Allison RS, Zacher JE. "The dynamics of vertical vergence," *Experimental Brain Research* 116 (1) (1997) 153-159.

[102] Kawata H, Ohtsuka K. "Dynamic asymmetries in convergence eye movements under natural viewing conditions," *Japanese Journal of Ophthalmology* 45 (5) (2001) 437-444.

[103] Stork S, Neggers SF, Müsseler J. "Intentionally-evoked modulations of smooth pursuit eye movements," *Human Movement Science* 21 (3) (2002) 335-348.

[104] Rottach KG, et al. "Comparison of horizontal, vertical and diagonal smooth pursuit eye movements in normal human subjects," *Vision Research* 36 (14) (1996) 2189-2195.

[105] van Leeuwen AF, Collewijn H, Erkelens CJ. "Dynamics of horizontal vergence movements: interaction with horizontal and vertical saccades and relation with monocular preferences," *Vision Research* 38 (24) (1998) 3943-3954.

[106] Takeda T, et al. "Characteristics of accommodation toward apparent depth," *Vision Research* 39 (12) (1999) 2087-2097.

[107] Wann JP, Rushton S, Mon-Williams M. "Natural problems for stereoscopic depth perception in virtual environments," *Vision Research* 35 (19) (1995) 2731-2736.

[108] Yang GZ, et al. "Visual search: psychophysical models and practical applications," *Image and Vision Computing* 20 (2002) 291-305.

[109] Mylonas GP, Darzi A, Yang GZ. "Gaze-contingent control for minimally invasive robotic surgery," *Computer Aided Surgery* 11 (5) (2006) 256-266.

CHAPTER 9
Motion Detection and Tracking by Mimicking Neurological Dorsal/Ventral Pathways

M. Sugrue and E. R. Davies

9.1 Introduction

The near-perfect ability to detect, track, and understand moving objects is one of the human visual system's most fundamental and useful skills. Automatic tracking and surveillance are of great economic and social interest, as is evident from the 4 million or so CCTV cameras adorning our city streets. However, there is quite some way to go before the capabilities of such automatic systems can match those of the human visual system (HVS). It turns out that one of the keys to the design of effective tracking mechanisms is the way motion is detected and processed and how complementary information from both the motion and appearance of the object are integrated.

In the HVS, motion is detected by special neurons with a spatiotemporal profile. These neurons form part of a distinct "motion channel," while a "form channel" processes information on the instantaneous appearance of the object. Information from the two pathways is then integrated into a mental model of the target being tracked.

In computer vision, there are currently two broad paradigms for visual object tracking, each with its own problems. In the background modelling paradigm, each individual pixel is statistically modelled over time. Objects are detected as clumps or "blobs" of sufficiently changed pixels. The fundamental problem is that one-dimensional statistics cannot distinguish between noise, which may be scattered, and true motion, which is spatially coherent.

The second tracking paradigm is to forgo motion detection entirely and use location prediction combined with an object appearance model. A weakness of this approach is that the appearance model must be acquired somehow before tracking begins. During tracking, if the object changes appearance (nonrigid objects such as pedestrians deform as they move), the appearance model may fail frequently. Thus, at the very time when motion information would be most useful and appearance information least reliable, this paradigm relies totally on the latter.

In this chapter we present a new paradigm that is inspired by the tracking architecture of the HVS. As noted above, this employs a dual-channel form–motion approach where each channel is used to compensate for the other's weaknesses. The motion channel is computed using spatio-temporal wavelet decomposition, providing true motion detection, which is both more robust and computationally cheaper than background modelling. We also avoid the use of complex location prediction, instead relying on a logical sorting mechanism to maintain object tracking through occlusion and interactions.

In addition, we propose a new comprehensive theoretical framework for visual object tracking. All tracking systems can be understood as a combination of three modules: a motion detection (MD) module, an appearance model (AM) module, and a location prediction (LP) module. These three modules exist in differing proportions in all systems: on this basis, we propose a rationale for efficient system design.

A fourth module deals with object behavior analysis (BA) and is actually the ultimate goal of surveillance. There is currently little consensus on how to approach this final task. We present an approach which relies on directly accessing motion channel information. At every stage we present example images and quantitative results as evidence of the success of our methodology.

9.2 Motion Processing in the Human Visual System

In 1942 anatomist Gordon Lynn Walls observed "If asked what aspect of vision means the most to them, a watchmaker may answer 'acuity,' a night flier 'sensitivity,' and an artist 'colour.' But to animals which invented the vertebrate eye and hold the patents on most of the features of the human model, the visual registration of motion was of the greatest importance" [1]. Next to the simple detection of light and dark, the ability to see motion may be the oldest and most basic of visual capabilities [2].

Most of our understanding of how the HVS works is derived from experiments on animals. Dittrich and Lea (2001) provide a very accessible introduction to motion processing in birds, noting that many birds have the ability to define and distinguish patterns and objects using only motion information. The oft-mentioned ability of hawks to spot small prey at great distances only applies if the prey is in motion. Conversely, that many creatures have a "freezing" instinct in case of danger suggests that motion is a key detection and recognition cue of great evolutionary importance [3].

Studies of the brains of monkeys [4] have revealed that motion processing is widely distributed from the retina to the visual cortex [5]. However, many of the most interesting and important structures seem to be centered on the middle temporal (MT) and V5 regions [1]. Friston and Büchel (2000) discuss the processing connections between early V2 and the suspected site of motion processing in the V5/MT area [6]. Britten and Heuer (1999) explore what happens when one motion-sensitive cell in the MT region is faced with overlapping motion data for two objects [7]. Priebe and Lisberger (2004) explore how visual tracking is related to the contrast of the object. Unsurprisingly, higher-contrast objects are easier to track [8]. Bair et al. (2001) study the connection between MT response and stimulus duration [9]. Young et al. (2001) propose that motion-sensitive cells have spatiotemporal "difference of offset Gaussian" (DoOG) profile where speed and direction selectivity is controlled by the "tilt" of the discontinuity [2].

Although much motion processing has been traced to the MT area [10], it is known that there are neurons in all regions of the HVS that are sensitive to combinations of particular features and motion. Overall, motion detection is a direct experience, uniquely specified by the visual system. This can be seen most clearly in the "blindsight" condition, where a patient with neurological damage may be blind to stationary forms and objects but may still be aware of motion, while being unable to identify the source. Other interesting points include MT neurons sensitive to changes in object direction. Wertheim (1994) reviews egomotion in the HVS and the need for a reference signal (i.e., assumed fixed points in image space) as opposed to extraretinal signals (a nonimage knowledge of motion and eye position) [11]. Li (2002) discusses the need for on-retina visual bandwidth reduction, noting that the retina strips perhaps 80% of redundant information before transmission to the optic nerve [12]. Nadenau et al. (2000) [13] and Momiji (2003) [14,15] provide detailed reviews of how models of the HVS and retina have been used to optimize image compression.

Grossberg et al. (2001) note that visual motion perception requires the solution of two competing problems of "motion integration" and "motion segmentation": they state, "The former joins nearby motion signals into a single object, while the latter keeps them separate as belonging to different objects" and suggest a neural model to explain this ability [16].

Neri et al. (1998) attempt to explain the HVS's remarkable ability to detect "biological motion," that is, to distinguish cyclical motion of living things from that of rigid inanimate objects [17]. Thornton et al. (2002) report on a study for biological motion using light point displays, and discover that recognition is highly attention dependent—suggesting high-level processing [18].

Martin and Aggarwal (1978) propose a dual-channel approach whereby the "peripheral" region detects motion and directs an "attentive" tracker which incorporates object detail [19]. Giese and Poggio (2003) model biological motion recognition using a dual "form and motion" channel processing architecture. They base this on neurological data suggesting that the dorsal region of the HVS, which is specialized for motion detection, and the ventral region, which is specialized for object recognition, work in parallel to achieve complex visual tasks. They conclude with a number of open questions: *How is the information from the two pathways combined? Does form or motion recognition dominate for certain stimulus classes?* [20].

9.3 Motion Detection

Motion detection is a component of many tracking systems. There are broadly three categories of motion detection:

1. *Pixel or feature optical flow approach.* Every instance of a chosen feature type is detected in each frame and matched to the next. Moving objects are detected as connected regions of moving features. As frames will normally contain many features of very similar appearance, this is possible only by using constraint models to reduce ambiguity. These techniques are not well suited for surveillance applications because objects are small with respect to the scene and there are many motion discontinuities [21].

2. *Pixelwise statistical background modelling and subtraction.* A model of the stationary background is prepared. Regions of the current frame that are significantly different (in color or intensity) from this background are tracked as objects. When calculating the background model, each pixel is computed separately, using the statistical assumption that the most common value is that of the stationary background. The choice of the particular statistical method used is where the various published methods differ. This choice depends on the application, the noise properties of the video, and the available computational resources.
3. *Motion "distillation."* Spatiotemporal edge-based detection, which is the subject of this work: see particularly Section 9.3.3 below.

The earliest and simplest background modelling statistic is frame differencing, which was first reported in 1970 for satellite image analysis and was soon also applied to video [19, 22]. Jain and Nagel (1979) reported that, quite understandably, this had only limited applicability in surveillance "because images can be extracted only when they have been displaced completely from their position in a specified reference frame, which is usually the first frame of the sequence" [23]. The approach was largely forgotten until the early 1990s when computer power had improved to permit the use of more interesting statistics. Long and Yang's (1990) influential paper set out a number of methods involving the computation of a running average of pixel values in order to achieve "stationary background generation" [24]. They tested the method on both indoor and outdoor scenes. However, the videos were quite short (up to ~ 70 frames), and it is likely that they didn't experience many problems using the mean as their statistic. Many researchers continue to use temporal mean filters because of their simplicity [25–27]. However, simple statistics are not robust in complex lighting situations and thus require ad hoc postdetection noise removal steps [28].

A better statistic, although more costly, is the temporal median filter (TMF). In 1995 Gloyer et al. [29] and McFarlane and Schofield [21] independently reported TMF-based background models. Mean filtering has been shown to be more effective in the case of white Gaussian noise, while median filtering is better in the case of salt-and-pepper noise and burst noise [30]—common in video. Furthermore, changes due to a moving object appear like outlier noise when viewed on a pixelwise basis, allowing better background generation in the presence of moving objects—an important strength for practical surveillance systems. Wren's *Pfinder* (1997) has been described [31, 32] as a form of mode filter, calculating a "running Gaussian average" for each pixel [33]. This has a memory advantage over the TMF, in that only two parameters (μ and σ) of the Gaussian PDF are stored for each pixel. Other contemporary approaches include pixelwise linear prediction methods using the Kalman [34, 35] or Wiener [36] filters. These approaches suffer from expensive memory and processing requirements [37].

The observation that video noise is rarely monomodal led to Stauffer and Grimson's (1999, 2000) Gaussian mixture model (GMM) approach [38, 39]. In their scheme, each pixel is modelled as three Gaussian distributions of values over time. An iterative "expectation-maximization" (EM) algorithm is used to update

9.3 Motion Detection

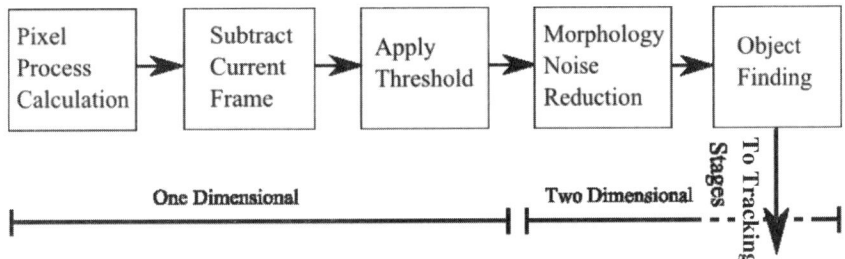

Figure 9.1 Logical scheme of motion detection by background modelling methods. The first three steps are one-dimensional pixelwise processes. "Pixel Process" refers to the chosen statistical model, such as median filtering or the Gaussian mixture model. The final two steps are two-dimensional. After motion detection is completed, the system moves on to tracking and behavior analysis stages.

each pixel model in every frame. Incoming values that are outside the immediate range of all the Gaussians are defined as possible foreground.

Figure 9.1 provides a general background modelling scheme. An often overlooked, yet vital and expensive, stage of background modelling and subtraction follows the production of the binary foreground map, but precedes the tracking stage. Here it is necessary to search the foreground map for objects, using connectivity and eliminating very small objects as probable noise. It is very common for pixelwise detection methods to result in "holes" or incorrect splits in the object silhouette, where parts of the object are similar in color or intensity to the background. This may be tackled using a series of morphological dilation and erosion steps on the foreground mask.

The background modelling approach can be thought of as an indirect route to spatiotemporal motion detection. The first step is a one-dimensional pixelwise statistical process. This is followed by a two-dimensional object detection stage based on morphology. The one-dimensional step detects *change* rather than true motion. The reason for this can be seen in Figure 9.2: when only one pixel is considered it is impossible to distinguish the case of localized random noise from the case of an object moving across the background. Only reference to the two-dimensional data can allow this distinction.

9.3.1 Temporal Edge Detection

Can we detect motion directly, without statistics? Video is usually thought of as a sequence of frames, as with a traditional roll of film. A different approach is to model video as a generalized three-dimensional spatiotemporal structure, notionally "stacking the frames" into an $x-y-t$ column as in Figure 9.3. This reveals some interesting and useful aspects. Stationary image features will form straight lines parallel to the t-axis, while features in motion will form lines with a nonparallel component.

An ordinary edge detector can be used to enhance these "temporal" edges. In the past several researchers have reported a variation of simple image differencing where the difference of edgemaps is taken. This approach has improved

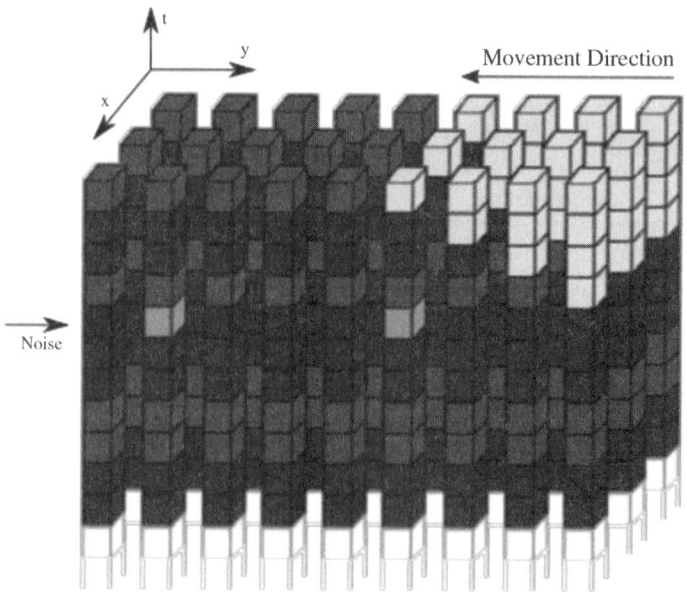

Figure 9.2 Video viewed as one-dimensional pixel statistics. On a pixelwise basis it is impossible to distinguish the isolated noise points (left) from a spatially coherent motion (right). Only subsequent two-dimensional processes can achieve this. An alternative approach is to apply three-dimensional, spatiotemporal motion detection from the outset.

illumination invariance but has disadvantages when nonglobal threshold levels are required [40].

A slightly different approach is to use a Sobel or other edge detector in the $x–t$ and $y–t$ planes. Using a filter oriented perpendicular to the t-axis highlights features in motion. Use of a 3×3 edge detector on a surveillance video in this fashion quickly extracts the moving objects in a robust way [41] (Figure 9.4). A threshold level is required to binarize the output for blob extraction and tracking. The

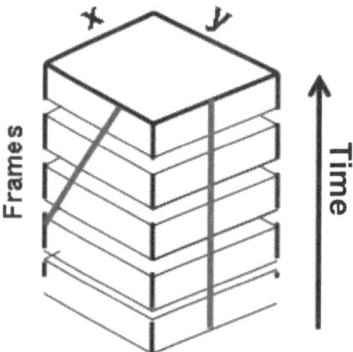

Figure 9.3 Result of "stacking the frames" into a $x–y–t$ column or "video cube". In this view, edges parallel to the t-axis are stationary while edges with a non-parallel component are in motion. A spatiotemporal edge detector tuned to detect these edges will thus also detect motion.

9.3 Motion Detection

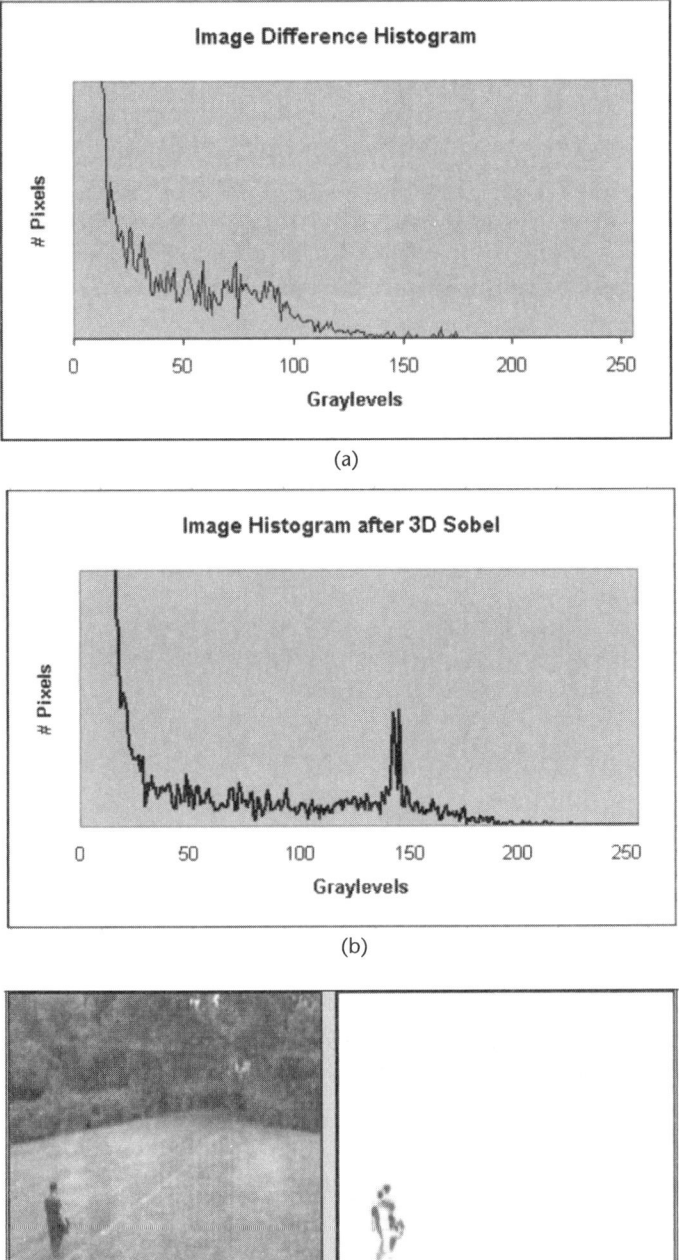

Figure 9.4 (a) image difference histogram. Noise and motion pixels grouped together. (b) sobel edge detection histogram. Motion pixels are shifted out and can easily be detected. (c) an example of motion detection based on Sobel temporal edge detection. © IEE 2005.

histograms in Figure 9.4 compare the outputs of image differencing and the Sobel method. With image differencing, as with all pixelwise statistical methods, the output histogram shows that pixel changes due to noise and motion are grouped together in a large peak near the zero position; the spatial properties of the Sobel give greater weight to pixel change due to motion, resulting in a peak at a much larger value. A threshold level may be chosen robustly (in the sense that the threshold is noncritical) in the large gap between the noise peak centered near 0 and the motion peak at larger values, while for image differencing, morphology and size filtering will be required to extract the target.

9.3.2 Wavelet Decomposition

Rather than use two two-dimensional edge detectors, we can construct a single three-dimensional spatio-temporal filter. Wavelets, introduced briefly in one-dimensional and two-dimensional form in Chapter 5 can also be extended into three-dimensional forms. The wavelet transform is a particularly useful mathematical technique for analyzing signals which can be described as aperiodic, noisy and discontinuous. The transform is composed of two parts, the "scaling" function lowpass filter and the "wavelet" function highpass filter. By repeated application of the scaling and wavelet filters, the transform can simultaneously examine both the position and the frequency of the signal.

In its continuous, general form the one-dimensional wavelet function is represented by:

$$\Psi_{a,b}(t) = \frac{1}{\sqrt{a}} \Psi\left(\frac{t-b}{a}\right) \tag{9.1}$$

where a represents the wavelet scale and b the location in the signal. This may be translated into a discrete wavelet transform by setting a discrete bounding *frame* and using a type of logarithmic scaling known as a *dyadic grid*. The result is a general discrete scaling function ϕ, and a wavelet function ψ, each in turn dependent on the separately chosen *mother wavelet* Ψ.

The choice of Ψ depends on the application. Commonly used functions include the Gaussian, the "Mexican Hat" function (a second derivative Gaussian), and the series of Daubechies wavelets. The JPEG 2000 standard, for example, uses the Daubechies wavelets in a two-dimensional fashion to separate images into a series of feature maps at different scales, which may then be efficiently encoded and compressed.

One way of viewing the behavior of the wavelet transform is as an edge detector [42], because a highpass filter detects discontinuities. To decompose a two-dimensional signal such as an image, a one-dimensional lowpass filter ϕ and a highpass filter ψ are combined using the tensor product to produce four filters, each sensitive to edge features at different orientations:

$$\begin{aligned}
\text{two-dimensional scaling function:} \quad & \phi(x,y) = \phi(x)\phi(y) \\
\text{two-dimensional horizontal wavelet:} \quad & \psi_h(x,y) = \phi(x)\psi(y) \\
\text{two-dimensional vertical wavelet:} \quad & \psi_v(x,y) = \psi(x)\phi(y) \\
\text{two-dimensional diagonal wavelet:} \quad & \psi_d(x,y) = \psi(x)\psi(y)
\end{aligned} \tag{9.2}$$

Figure 9.5 Scale pyramid output of wavelet image decomposition. The top row represents the output of the horizontal wavelet, the side column of the vertical wavelet, and the diagonal of the diagonal wavelet. The image in the top left column is the result of the final scaling function.

These filters are applied sequentially to the image. First, the three oriented wavelet filters are used to extract features at the highest detail scale. Next, the scaling filter is used to reduce the scale of the image, followed by another pass of the wavelet filters. This is repeated to the desired scale, producing a "feature pyramid" (Figure 9.5).

This system can be extended to three-dimensional spatiotemporal wavelet decomposition using tensor products. The result is eight spatiotemporal filters, ranging from the $s-t$ scaling function $\phi(x)\phi(y)\phi(t)$ to the $s-t$ diagonal wavelet $\psi(x)\psi(y)\psi(t)$.

9.3.3 The Spatiotemporal Haar Wavelet

To apply these wavelets to real data, a specific mother wavelet function must be chosen. The requirements of the mother wavelet are application specific while bearing in mind practical considerations such as computational cost. In this case, the goal is not to encode or record the image data, but merely to detect temporal discontinuities at particular $s-t$ orientations. The widely used Daubechies category of wavelet has the ability to detect polynomial signal patterns. The wavelet order is linked to the polynomial order to be detected. Daubechies-4 (D4) detects second-order polynomials. D6 is sensitive to third-order polynomials, and so on. The higher the wavelet order, the greater the computational cost of implementation. The simplest Daubechies wavelet is D2 and is also known as the Haar wavelet. Here we will demonstrate that this simplest wavelet, extended to three spatiotemporal dimensions, is a very powerful and efficient motion-detection tool.

The one-dimensional Haar mother wavelet is composed of just two coefficients. Tensor products of one-dimensional wavelets can produce a series of two-dimensional and three-dimensional scaling functions and wavelet functions. The three-dimensional row of Figure 9.6 shows an example of two of the eight spatiotemporal Haar wavelets, where black represents "−1" and white "+1." This wavelet function has its step-function oriented perpendicular to the t-axis.

Spatiotemporal wavelet decomposition of a signal produces a feature pyramid as with two dimensions. The input signal is convolved with both the scaling filter and the wavelet filter. The output of the scaling filter is reduced by a factor of two in each dimension. This output is again convolved with the scaling and wavelet filters [43].

The output of each wavelet convolution is a product of the speed and contrast of the edge feature. The equation of the wavelet filter computation is:

$$\sum_{t=t_0}^{t} \sum_{i} \sum_{j} x_{tij} - \sum_{t=t_1}^{t} \sum_{i} \sum_{j} x_{tij} = W \qquad (9.3)$$

where x_{tij} represents the video pixel data at the point (t,i,j) in spatio-temporal space. If part $t = t_0$ is equal to part $t = t_1$ then the wavelet output W will be zero. This condition may occur if there is no movement within the data analyzed, that is, if the $s-t$ orientation θ of any edge is parallel to the t-axis. We can define the filter output as:

$$W \propto \cos\theta \times \text{edge contrast} \qquad (9.4)$$

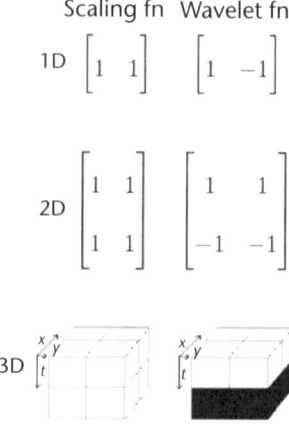

Figure 9.6 Coefficients for the Haar wavelet transform for different dimensions. For three-dimensions, white is used to represent a value of "1" and black "−1." Wavelet decomposition is carried out by sequentially convolving the signal data with first the scaling function and then the wavelet function for each desired scale level. © IEEE 2006.

Because of the fact that at a local level it is impossible to know whether a dark object is moving against a light background, or vice versa, direction information is ambiguous. For detection, where only binary motion data is required, W can be normalized to remove this ambiguity. In Section 9.5 we use the raw motion information for behavior classification.

Figure 9.7 details a comparison of segmentation results for several videos using three motion-detection methods—two traditional background modelling techniques, the TMF and the GMM, together with the $s-t$ Haar method. For comparison, we have chosen the TMF because it has comparable algorithmic complexity to our own method; it may be set up to operate over a small number of frames and it has a simple bootstrap. The GMM approach is chosen as it is an accepted, particularly widely used background modelling method. Here both the TMF and the new method are computed over eight frames (the new method is computed to the third decomposition level, for which the number of frames is 2^3), and both also have an 8-frame bootstrap. The GMM requires a larger number of frames in order for the Gaussians to stabilize on a particular distribution, and in our implementation we use a 20-frame bootstrap. The output from all three methods is presented without any subsequent morphological or noise-reduction steps, so as to be sure of comparing like with like. (Naturally, any method can be enhanced to improve performance, but here we focus on *intrinsic* performance for clarity.)

The videos are surveillance style and have been chosen for their differing degrees of noise, characterized by the median value of pixel variance over time, and they depict behaviors of pedestrian targets. The first case presented here (Figure 9.7, Video 1) is a simple motion segmentation task of an outdoor scene with diffuse lighting. The median pixel intensity variance is 1.65. The TMF results clearly show the difficulties of that method. The target is incorrectly segmented with leading and trailing edges separated. This problem is due to the slow speed of the target with respect to the temporal window size of 8 frames; the middle of the target has been absorbed into the background model. Both the GMM and new method give more accurate results.

Video 2 in Figure 9.7 shows a variety of target pedestrian behaviors. The pixel variance is 3.05. In frame 664, the target is walking directly across the frame. There are slight shadows that interfere with segmentation and more random image noise than Video 1. This noise shows clearly in both median and GMM background subtraction results because these methods behave as change detectors. Pixels are segmented if the contrast with the background model is above some threshold. (With the TMF this threshold is global: with the GMM it is pixelwise and adaptive.) Morphological closing (which would have to be anisotropic, and would to a fair extent be an ad hoc measure) could improve the background subtraction results, but this somewhat expensive step is unnecessary for the new method. However, it is commonly necessary for many implementations, such as [38].

The random, structureless nature of the video noise means that the edge detector of the new method reacts less strongly and is automatically removed by the scaling process. In this frame there are also two other small regions of motion. The TMF detects only one of these, while the GMM catches both, but in neither of these cases is a strong signal obtained. However, neither method is capable of

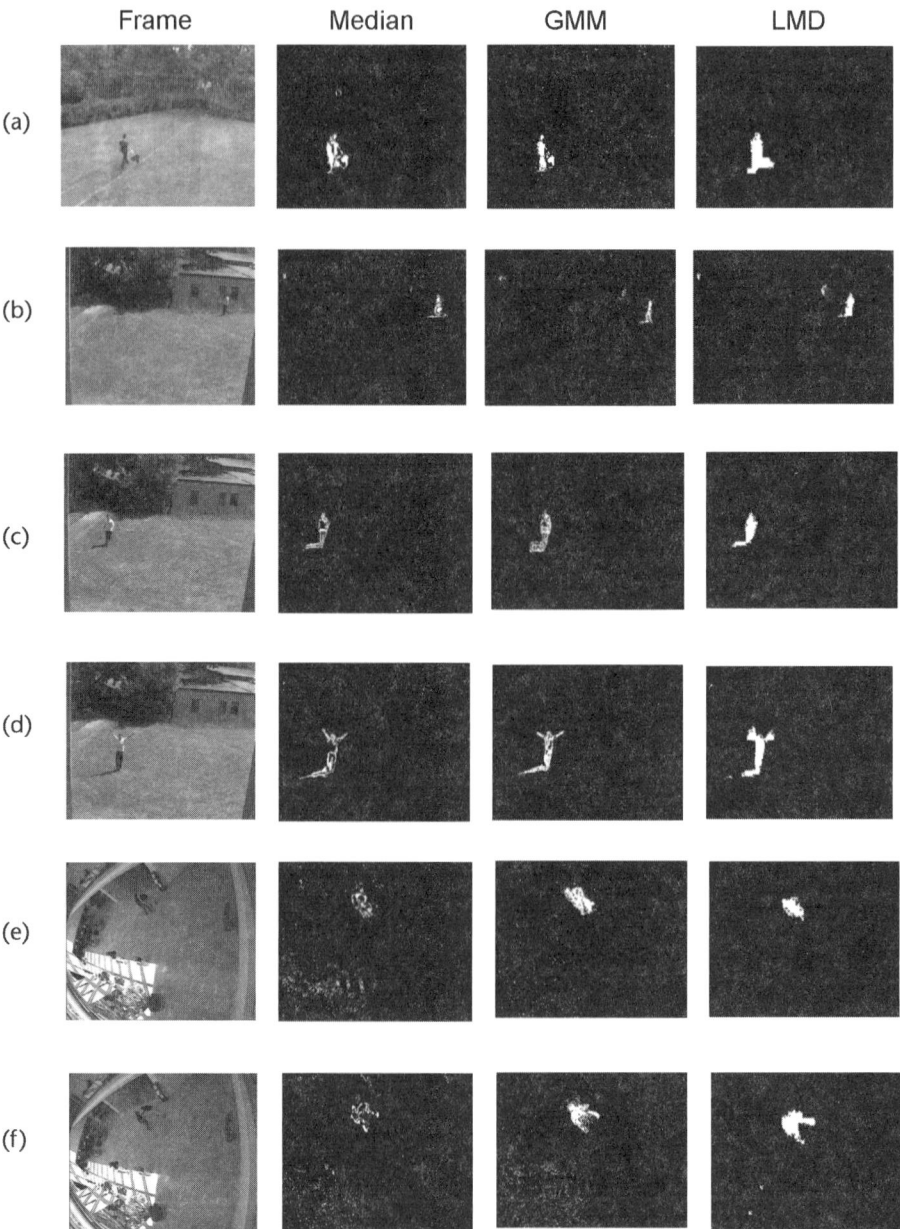

Figure 9.7 Comparisons of thresholded detection results for Haar wavelet and traditional background modelling. Three videos with different noise qualities are analyzed. The first (a) low noise is captured using a webcam on an overcast day. The second (b, c, d) is a long, higher-resolution camcorder video on a sunny day with changing lighting and shadows. The target pedestrian walks and runs in different directions. Frame 5,187 (d) shows the pedestrian walking slowly toward the camera and waving his hands. The final video (e, f) is taken from the CAVIAR project [44] and has considerably higher lighting noise than the other two. © IEEE 2006.

cleanly distinguishing these objects from noise, and it is likely that subsequent noise-reduction steps will remove them entirely. The new method clearly detects both small moving regions while robustly eliminating all noise.

Frame 5,136 is of a target pedestrian walking slowly toward the camera—a case rarely dealt with in the background modelling literature. Because of the slow motion relative to the image frame, the center of the target becomes absorbed into the background (to a large degree with the TMF, but less so with the GMM), and the background subtraction results show large gaps. In Frame 5,187 the pedestrian is waving his arms; this is discussed further in the future work section below. The new method results for Frame 5,187 show slight "smearing" of the arms because of their rapid motion, which is a characteristic by-product of this method. (It must be emphasized that motion analysis is necessarily carried out over time, and thus refers to a range of positions: the complete picture at any moment can therefore only be ascertained by combining form and motion information.) Again, the new method also detects a small moving object in the top left corner, which is completely missed by both the median and GMM methods.

The final frames are from the "fights & runs away 1" sequence in the CAVIAR database [44]. This video shows the highest degree of noise of these examples with pixel variance as high as 10 in the brightly lit bottom left quadrant of the video. Again, the new method has a cleaner response because this noise is random—and thus changing but not moving. In all cases the new method demonstrates a far greater robustness to noise than either the TMF or the GMM.

Table 9.1 presents quantitative object detection results for the s–t Haar method. Object detection rates for each video were compared with the manually established ground truth. In Table 9.1, TB stands for true blobs, which is the true number of moving objects in each frame. "FP" and "FN" stand for false positive and false negative. In test videos 1 and 2, all false positives were observed to be due to moving tree branches and all real moving objects were detected. In the CAVIAR video a small number of false positives were caused by shadows and a poster moving in air currents. The 24 false negatives were due to a single person moving slowly in a dark region of the video (though there is clearly some argument whether these false negatives should be counted as true positives). The numerical results for the TMF and GMM filters would be very poor without some postdetection noise-reduction step, and are thus not fairly comparable.

Table 9.1 Performance of Motion Channel with the Three Videos from Figure 9.7

	Frames	TB	FP	FN	Precision
Video 1	538	378	2	0	99.24%
Video 2	5,458	2,712	138	0	94.91%
CAVIAR	550	938	20	24	95.30%

"Frames" indicates the length of each test video in frames. "TB" indicates the total number of real moving objects in each frame (ground truth). "FP" is false positives, objects detected where there are none. False positives are mainly due to motion clutter. "FN" is false negatives, moving objects not detected. The precision results presented compare favorably with those for background modelling and are achieved at considerably less computational cost. © IEEE 2006.

9.3.4 Computational Cost

Here we offer a precise derivation of the computational costs of the system measured in operations per pixel. For comparison, the computational requirement of the TMF method of background modelling is ~ 256 operations per pixel (because of the need to analyze the intensity histogram). The GMM technique is more expensive, requiring a lengthy initialization step, followed by evaluation of an exponential function for each Gaussian and each pixel. For this system, the temporal window concept is replaced by the idea of a decomposition level D. The number of starting frames for this decomposition level is $2D$. On the first iteration, two frames are convolved with two filters—the scaling function and the wavelet function—to produce one scaled frame and one motion frame (of size $\frac{n}{2} \times \frac{n}{2}$, where n is the frame width). The next stage repeats this for two scaled frames from stage one, resulting in output frames of size $\frac{n}{4} \times \frac{n}{4}$. The third decomposition stage uses two second-level scaled frames and produces frames of size $\frac{n}{8} \times \frac{n}{8}$. The total number of pixels in the system is given by:

$$N = \sum_{i=0}^{D} \left(\frac{n}{2^i}\right)^2 \frac{2^D}{2^i} = 2^D n^2 \sum_{i=0}^{D} 2^{-3i} \quad (9.5)$$

For the 3D Haar wavelet transform, the number of operations required to decompose the signal to level D is given by (9.5) but with the expression summed to $D-1$. The number of operations per pixel is:

$$\frac{1}{2^D n^2} \sum_{i=0}^{D-1} \left(\frac{n}{2^i}\right)^2 \frac{2^D}{2^i} = \sum_{i=0}^{D-1} 2^{-3i} \quad (9.6)$$

This has a minimum value of 1 operation per pixel when $D = 1$ and the maximum is found using the limit as D goes to infinity:

$$\lim_{D \to \infty} \left(\sum_{i=0}^{D-1} 2^{-3i} \right) = \frac{8}{7} \simeq 1.14 \quad (9.7)$$

This measure of computational load, of less than about 1.14 operations per pixel per filter, is close to the minimum possible and is a major speed improvement when compared with that of the TMF or other methods.

Frame rate comparisons of techniques are problematic because of differences in implementation, input, and equipment. However, we can report that our implementation of the motion distillation scheme runs at 62 fps on a P4 machine, while the TMF and GMM run, respectively, about 10 and 80 times slower. Input frame size is 720×576.

9.4 Dual-Channel Tracking Paradigm

After detecting an object it must then be tracked. In the background modelling paradigm this involves correctly integrating all the separate frame-by-frame detections due to a single real object into a notional object stored in memory. There is then a problem of maintaining object identity through occlusions.

9.4 Dual-Channel Tracking Paradigm

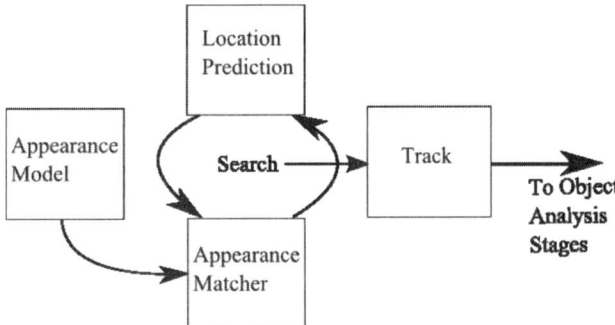

Figure 9.8 General scheme for foreground, "predict and match" tracking schemes. These methods are based on iteration between detection based on appearance and future location prediction. The appearance model must be provided. The object track is passed on to the behavior analysis modules.

There is another tracking paradigm which emphasizes object detection using an appearance model over, or to the exclusion of, motion detection. This paradigm, which may be thought of as "foreground tracking," involves the initial input of a starting location and object appearance. In order to avoid a costly global search, a location predictor is used to initiate search in subsequent frames at the target's probable position. Figure 9.8 gives a general schematic of this paradigm.

9.4.1 Appearance Model

Appearance models, or observation models, used in visual tracking are very similar to techniques for object recognition in static images. The most basic appearance models are those of optical flow and feature tracking techniques (single gray-level pixel values or corners plus relational constraints). Early Kalman filters generally used template matching [45] or color histograms [46]. Other reported possibilities include Harris "interest points" [47]. (Remarkably, some of the techniques in this area have not changed greatly in recent years.)

The appearance model used by Isard and Blake (1996) was based on curve segments [48], while their 1998 work [49] combined this with a color matcher. Other methods use corners or line segments, but these tend to be more clutter sensitive. Clutter was an important issue dealt with in Isard and Blake's work. In one example, they tracked a particular leaf moving among other similar leaves. Isard and Blake did not attempt to construct an appearance model capable of distinguishing its target from neighbors, but chose to overcome the problem using a multimodal tracking model.

Rui and Chen (2001) [50] and others [51, 52] used color histogram matching in their particle filters. This method has the advantage over shape matchers of being rotation invariant. Nummiaro et al. (2002) also claim that color histograms are robust to partial occlusion [51]. However, this is clearly conditional on the target's structure. Nummiaro's application was face tracking where the target was of approximately uniform color. Zhou et al. (2004) discuss techniques of adaptive appearance models for use with particle filtering [53].

The choice of model is highly data and target dependent. Because of deformations, nonrigid pedestrians are generally harder to track than rigid vehicles. Histogram matching gives more robust results for pedestrians, while template matching is more suitable for vehicles. Another important question is the source of the appearance model. Surveillance systems must be able to acquire the target appearance model on-line as new objects enter the scene. However, without a motion-detection stage or a preprogrammed appearance model, the target cannot initially be detected.

In our system, the use of a robust motion-detection system allows us to avoid a complex and costly appearance model. As discussed below, the appearance model need only be invoked to resolve ambiguous cases such as dynamic occlusions. The appearance model we have employed is based on color histogram matching. The color information for each sprite (see below) is recorded when it is initially detected and is updated in each frame.

9.4.2 Early Approaches to Prediction

Many tracking schemes reported in the literature include an inertia-based predictive mechanism. Location prediction serves to limit the search area. A global scene search is prohibitively costly. Commonly, prediction is based on the Kalman filter or the particle filter. Location prediction suffers from two fundamental flaws with regard to pedestrian tracking. First, it forces a total reliance on the appearance model (AM) algorithm, even while the object is in motion; also, pedestrians actually change shape and appearance in order to move and an AM algorithm sufficiently complex to predict this change may also be very costly. Second, AM algorithms rely on limits and assumptions of expected object motion. Pedestrians frequently behave in a highly complex way, changing direction, starting, stopping, meeting other pedestrians, and so on. Particle and Kalman filter–based methods commonly lose track of their target when it makes an unusual and sudden movement or when the object rotates or deforms, even though this very motion information could be used to compensate for AM weakness.

As object models become more complex, it becomes necessary to avoid global search and to predict the object motion and then limit the search to that area. The first successful predictor tried was the Kalman filter [54, 55]—a linear filter borrowed from the world of radar and control theory. However, it was quickly discovered that nonlinearity was extremely advantageous in visual tracking, and the extended Kalman filter (EKF), unscented Kalman filter (UKF), sequential Monte Carlo, Markov chain Monte Carlo, condensation/particle filter, and EKF- and UKF-directed particle filters have all been tried. It is often not made sufficiently clear that the filter stage serves only as a location predictor in order to direct the search. A separate appearance model, a priori or learned, must be tested at each candidate location suggested by the predictor. A Kalman filter is a type of on-line least-squares analysis system and thus gives a linear response [54–56]. Harris (1992) [57] and Blake et al. (1993) [58] were among the first to employ the Kalman to predict probable object locations or search candidates in visual tracking applications [59, 60], although it had been used for many years for object tracking in radar. The algorithms are initiated with a target model and location, search for this model in the next frame, and use this position change to predict the next location. Search is begun at the candidate location and is worked outwards until a good match is found. The error (difference between predicted location and actual location) is fed back to the tracker to improve the next prediction. The Kalman fails when the data is nonlinear or when the noise is non-Gaussian—which translates into poor performance in cluttered scenes [61]. Results can be improved with the EKF, which uses a Taylor series expansion to linearize nonlinear functions [62, 63], and the UKF, which uses the "unscented transform" [64, 65] to delinearize predictions [66, 67]. These improved versions of Kalman still react poorly in cluttered conditions due to their mono-modality and tend to get stuck on local minima.

In 1996, Isard and Blake proposed a tracker based on sequential Monte Carlo methods. Imperfections of the motion model would be overcome by introducing a random component into the search strategy. Each search candidate area (which would later become known as a particle) was assigned a weight based on confidence. Multiple particles of different weights can be maintained simultaneously to model a multimodal data field and thus "multiple hypotheses" of the target location, resulting in better performance in cluttered scenes. The particle location and weighting are altered, using results fed back from the appearance model. Particle filters generally give very good results but do fail in cases of sudden change in motion. They are, of course, only as good as the appearance model used in the search stage.

Isard and Blake (1998) extended their earlier work to give the "ICondensation" algorithm [49], including "importance sampling," to allow the combination of multiple separate appearance models (in this case, shape and color). Other work has involved directing each particle using an EKF [68] or UKF [50, 69] type of filter to improve location estimation. Particle filtering with "sampling importance resampling" has also been explored [52, 70]. Choo and Fleet (2001) reported a particle filter extended for tracking very complex target models, calling it a "hybrid Monte Carlo" particle filter [71]. Hue et al. (2000) give a detailed discussion of tracking multiple objects with particle filters [72].

9.4.3 Tracking by Blob Sorting

In this section we present our solution to the postdetection tracking problem. We use the terms "sprite" and "blob" to mean similar but distinct things. A blob is a connected region corresponding to the area of motion in the current frame. A blob may represent a whole moving object, part of an object if only part of an object is moving, or several overlapping objects, in the case of groups of objects moving together. A blob has no knowledge of real objects, either spatially or in time. However, a sprite represents an attempt to integrate a notional "object" from a number of blobs over many frames. This attempt to integrate blobs into sprites is the problem of postdetection tracking. "Object" refers to the real-world target being tracked.

This problem can be tackled by considering a series of questions. When a blob is detected, does it belong to (1) a currently active sprite, (2) an inactive but visible sprite, (3) an invisible or occluded sprite, or (4) a new sprite? Conversely, if a sprite cannot be matched to any current blob, is this because the object has (1) stopped moving, (2) been occluded, or (3) left the scene?

A further complication is due to the weakness of relying on motion detection. If an object stops moving then no blob will be detected. However, we may assume that a stationary object will not change its appearance radically from frame to frame, as such a sudden transformation would be detected by the motion-detection stage. Thus, in a "missing blob" situation, we may distinguish between the "occluded object" case and the "stationary object" case by testing the appearance model at the last known object location.

This problem statement suggests the dual-channel tracking architecture described in the neurological literature: see the paper by Giese and Poggio [20]. As with the HVS, one channel contains motion information only and the other contains instantaneous form or appearance information. Neither on its own is sufficient.

In our system we rely primarily on the motion channel, as computed using $s-t$ Haar wavelet decomposition, for initialization and detection. Detected blobs are sorted and matched using a two-stage rule-based matching scheme. Table 9.2 presents these rules as they are applied in the software. Blobs detected in the motion channel are sorted first using an area overlap test, followed by an appearance model test to resolve ambiguity. Single blob to multiple sprite matches are treated

Table 9.2 Sorting Rules of Blob Identity Maintenance

Motion Channel: Test Blobs	*Form Channel: Update Sprites*
if a blob was detected by MD stage	if the sprite was matched
if OT gives a unique match $\geq T_O$	if $\exists$ a unique match
match blob to sprite	**update sprite with blob data**
if OT gives a unique match $< T_O$	if $\exists$ one sprite to multiple blobs
perform AM test	**split sprite**
if $\exists$ a match $\geq T_{AM}$	if $\exists$ multiple sprites to one blob
match blob to sprite	**merge sprites**
else **register no match**	else $\exists$ no blob match
if OT gives multiple matches	perform AM test on current frame
perform AM test	if object present
if $\exists$ one response $\geq T_{AM}$	**register sleep sprite**
match blob to sprite	else **register occluded sprite**
else **register merged sprite**	
else $\exists$ no match	
perform AM test	
if $\exists$ a response	
register occluded sprite	
else **register new sprite**	

First, blobs extracted from the motion channel are sorted and matched to sprites. Then the list of sprites in memory is updated and checked against the form channel if needed. T_O is the overlap threshold and T_{AM} is the appearance model matching threshold.

Table 9.3 Qualitative Sorting Results of Blob Identity Maintenance

	Video 2	CAVIAR	Video 3
Unique match $\geq T_O$	96%	80%	78%
Unique match $< T_O$	1%	3%	2%
Multiple matches	1%	6%	12%
AM split	0%	1%	0.5%
Match from memory	1%	1%	1%
No match	1%	9%	6.5%

Results show occurrence of particular event classes in the sample videos. These results are compiled from a sample of 1,935 events in three videos.

as dynamic occlusions.[1] Unmatched blobs are considered new objects. Next, the sprite list is searched for unmatched sprites. These can be due to either occlusion or stopped ("sleeping") sprites. The form channel is then accessed to resolve the ambiguity. Table 9.3 provides quantitative results for the application of these rules for three videos. These results show that the quick overlap test is sufficient for the large majority of cases, with only a few percent of cases remaining ambiguous and requiring reference to the form channel.

Figure 9.9 shows a schematic diagram of this tracking system in action. Each ellipse represents a detected blob in the motion channel. All objects are moving left to right according to the frame numbers underneath. In example A, the object moves without stopping or occlusion. In this case, the system matches incoming blobs to the stored sprite in memory using overlap alone when there is a unique match above an area threshold. Table 9.3 demonstrates that this initial, quick test is sufficient for between 78% and 96% of cases (depending on video content). B shows the case where the tracked object passes behind a stationary object in frame 2; C is the case where the object stops moving in frame 2. The system has no scene knowledge and so cannot predict when objects may be occluded or stopped. In frame 2 the sprite will remain unmatched to a blob in both cases and the system must distinguish whether the sprite is occluded or sleeping. The sprite is sorted using the rules in the right-hand column of Table 9.2. The system uses the appearance model of each sprite, acquired during the tracking phase, and tests the current frame at the last known position of the sprite. In terms of dual-channel tracking, this step represents referencing the form channel.

Figure 9.10 presents some example pictorial tracking results for several cases of single tracked objects (frame 720), multiple tracked objects (frames 125 and 134), and dynamically occluded objects (125). Frame 2,230 shows a pedestrian who has entered and exited the scene five times, both walking and running. The system correctly recognized the pedestrian each time he re-entered the scene and connected the previously detected paths.

---

1. In this chapter, the term occlusion always means complete occlusion. The distinction between complete and partial occlusion is largely an issue for the appearance model and is outside the scope of the chapter.

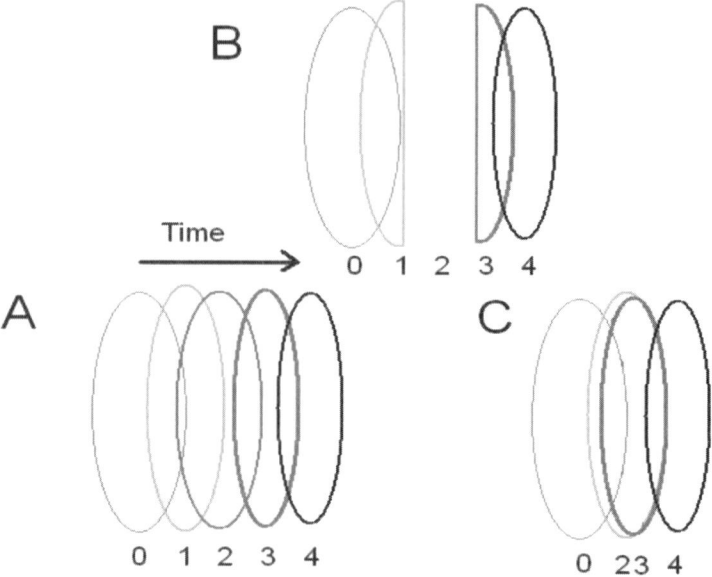

Figure 9.9 Schematic view of the motion channel output over time. Example A shows an object moving without stopping or being occluded. Examples B and C require reference to the form channel to decide whether the object has been occluded or has stopped moving.

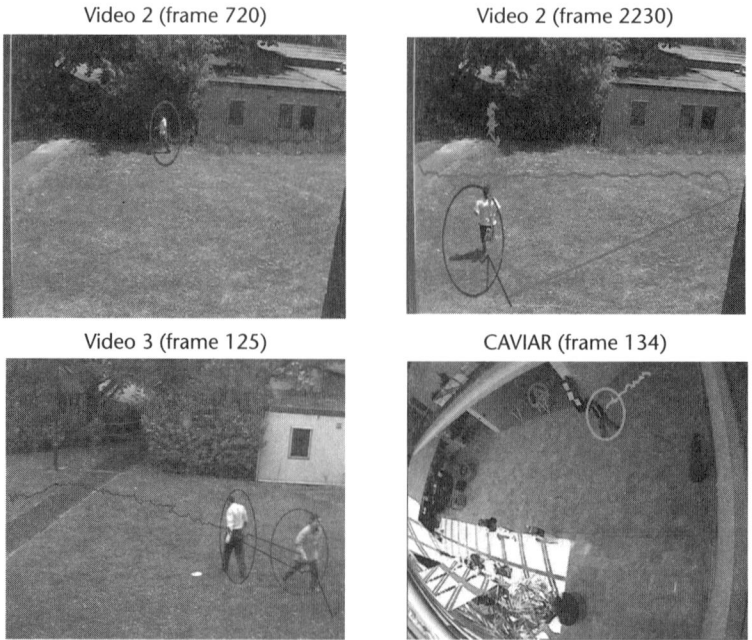

Figure 9.10 Variety of sprite tracking results. Video 2 shows a single pedestrian who randomly enters and exits the scene running, walking, and jumping. Video 3 contains three dynamic occlusions. The final video is from the CAVIAR database [44].

9.5 Behavior Recognition and Understanding

Of course, object detection and tracking are not the only end goals of a video surveillance system. Ideally, the system should be able to recognize what those objects are and what they are doing. There is currently no consensus as to how this should be achieved.

We may wish to distinguish cars from people, groups of people, different activities such as running, or to detect unusual behaviors. Alternatively, we may wish to log volumes of traffic (either vehicular or pedestrian) and derive statistics based on this data. We may divide surveillance systems into three broad categories. The first uses only the object track and path information while the second focuses on other available information. The third only involves identifying individuals using gait information and will not be discussed further here.

Malik et al. (1995) [73] developed a traffic monitoring system which detects crashes and congestion using vehicle behavior. Makris and Ellis [74, 75] developed a system that learns common pedestrian routes by merging paths together. Scene entry and exit points can be detected. Dee and Hogg (2004) [76] used an a priori scene model to break tracks into a tree of goals and subgoals. If a pedestrian deviates from a likely path, his or her behavior is flagged as suspicious. Owens et al. (2002) [77] use techniques of novelty detection [78, 79] to automatically pick out abnormal pedestrian routes. Niu et al. (2004) [80] use HMMs to detect complex interactions, such as stalking.

Heikkilä and Silvén (2004) [81] combine templates and speed information to distinguish pedestrians from cyclists. Zhou and Aggarwal (2001) [82] use the nearest-neighbor algorithm with multiple features to distinguish four categories of object (vehicle, bicycle, single person, and people group) and report a 5% misclassification rate. Selinger and Wixson (1998) [83] distinguish people from cars using the periodic deformations of people's blobs. Bobick and Davis (1998) [84] integrate the object's motion into a "motion history image" and then apply template matching (referred to as "temporal templates") to detect specific behaviors. Once objects have been classified, their interactions may be understood. Collins et al. (2000) [85] use hidden Markov models to classify events such as "human entered a vehicle" and "human rendezvous." Kojima et al. (2002) [86] seek to develop a system for automatically describing human behavior patterns after they have been successfully extracted. A more general approach is to produce an "activity graph" as a guide to potentially interesting events for a human CCTV operator. Graves and Gong (2005) [87] do this by matching iconic representations of human poses to the video.

Here we take a different approach. The HVS can recognize some objects and events solely from motion information. Recognizing a spider in the corner of one's eye from its movement, before it has been detected and tracked, is a common experience.

Section 9.3.3 described how the output of the motion detection stage is governed by Equation (9.4) and is a combination of edge speed and contrast. For the detection stage, the filter output was normalized; however, it contains richer information than the binary detection stage requires. Ideally, we would like to remove the contrast dependence, leaving only optical flow–style speed information. One way to achieve this would be to use the output of several filters with orthogonal direction sensitivity. However, this is both difficult and unnecessary. A different approach is to normalize the filter output across the detected object. This is done by taking the ratio of the sums of positive and negative filter outputs, as in (9.8):

$$R = \frac{\sum W_+}{\sum W_-} \qquad (9.8)$$

This ratio is invariant to object size. It is also constant for a moving rigid object regardless of the object shape, color, or speed. However, a nonrigid object such as a pedestrian will result in a ratio that changes cyclically in time.

Figure 9.11 gives example motion channel outputs for a car (left) and a pedestrian (right). The original frames (top) are presented with motion channel outputs (where red represents W_+ and blue W_-) for several frames (inserts). The graphs compare the total filter output (middle) and ratio (bottom) for the two objects. By analyzing the periodic nature of each signal, the system can correctly identify the object on the left as a car and the one on the right as a pedestrian. No scene model or calibration is necessary.

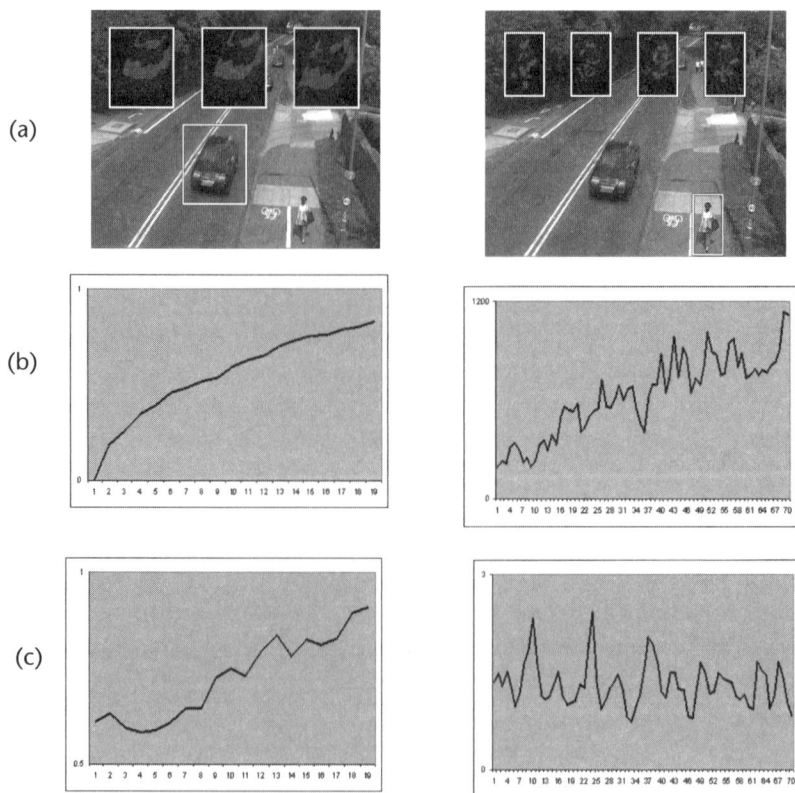

Figure 9.11 (a) Examples of motion channel outputs for a car (left) and a person (right) over time. (b) Motion channel output totals. (c) The ratio profiles for these objects.

9.6 A Theory of Tracking

Table 9.4 Classification of Objects and Behavior Using Motion Channel Information

	Sample Size	Misclassifications	Accuracy
Cars	14	1	93%
Pedestrians	33	1	97%
of which:			
Walking	25	2	92%
Running	4	1	75%
Strange behavior	4	1	75%

In the first two rows the misclassifications refer to pedestrians being distinguished from cars; in the last three rows, the misclassifications refer to the different actions of the pedestrians.

Further, by dividing the object into subsections and analyzing the resulting ratios, the system can distinguish between different pedestrian behaviors. Strange behavior, particularly involving aggressive hand movements that may be of interest to CCTV operators, can be detected by examining the motion output for the upper body. Running can be distinguished from walking by the speed of movement; however, this requires a few minor assumptions about the scene. Table 9.4 shows recognition results for this method. These results use data extracted from all previously mentioned videos. The system is shown to distinguish between cars and pedestrians with an accuracy of 95%, which compares favorably with Zhou and Aggarwal. This can be combined with the information gleaned from the tracking stage to provide a wide event descriptive power. This includes detection of object interactions, such as people meeting, loitering, people interacting with vehicles, and so on.

9.6 A Theory of Tracking

From the above presentation we can define some important principles for visual tracking, which together form a theoretical framework that can be applied equally to computer-based and biological systems.

> **Principle 1—Motion detection.** It is possible to detect a moving object without a priori information using as a basis only the assumption that an object moves with spatial coherence.

> **Corollary.** It is impossible to detect a nonmoving object without a priori information.

Foreground techniques such as particle filters require an appearance model of the object to be tracked. Techniques that include a motion-detection stage can track without this model, allowing arbitrary objects to be tracked. Conversely, objects only become detectable while in motion; when they stop moving, only comparison with a stored appearance model can detect them. We can also surmise that in a surveillance application all new objects will enter the scene in motion, and continue to move for some time before they stop. This provides a window of opportunity to detect the new object and acquire its appearance model for later use in case of stopping or occlusion.

It is also known that primitive animals can only detect prey while it is in motion. Motion detection probably evolved from the most primitive light and dark sensing cells when animal brains became powerful enough to correlate spatially coherent changes.

> **Principle 2—Motion vs change.** Motion can be distinguished from noise and change only by its spatial coherence.

The pixel process calculations of background modelling cannot distinguish noise or isolated change from spatially coherent motion. This can only be achieved by an additional two-dimensional object finding step. The spatiotemporal technique presented here can achieve true motion detection in one step.

> **Principle 3—Clutter.** Output clutter is produced by a failure of the appearance model.

Clutter can be understood as detection noise. A perfect appearance model (AM), which will uniquely detect the target object, will return no clutter. In real-world systems clutter is a problem. Particle filters compensate for clutter-prone AMs by maintaining a multimodal prediction stage, but at significant computational cost. Our system deals with this problem by relying on the motion channel while objects are in motion and when the objects are likely to be subject to changes in appearance.

> **Principle 4—Proportional component system design.** If any two of the three tracking components work perfectly the third is unnecessary.

Visual tracking systems can be viewed as having three components in various proportions—a motion detection stage, an appearance model stage, and a prediction stage. Principle 4 is that if any two of these components work perfectly, the third is unnecessary.

If the system has perfect motion detection, and perfect prediction while the object is occluded, there is no need for an appearance model to confirm the object's identity on reappearance. If the appearance model is perfect (i.e., produces no clutter or false positives) and the location prediction is perfect, then tracking will continue perfectly without any need for the crutch of a motion detection stage. This is the aim in particle filtering systems. Finally, if motion detection is perfect and the appearance model is perfect, then prediction is unnecessary as an occluded object which reappears elsewhere in the scene will be detected by its motion, perfectly matched by the appearance model to its original identity.

In the real world, perfect components are impossible. However, Principle 4 can be used as a guide to direct improvement efforts. On which stage would efforts to improve the system as a whole be best employed? For a background modelling approach, should improvements be directed toward the appearance model to reduce clutter, or toward the prediction module to reduce search area? Some methods, such as particle filtering, contain no motion detection stage. Would including one give more profitable returns than improving prediction or the appearance model?

In our work we have emphasized motion detection and the appearance model, while limiting prediction to search constraints and a series of logical sorting rules. Our results show the method to be quite robust for the videos tested. The strength of our spatiotemporal motion detection method allows us to avoid using a complex appearance model. Work on this module may improve performance in longer videos with large numbers of objects.

For many years the science of visual tracking has worked with largely ad hoc techniques, using "what works" without a general theoretical framework that may predict what works and why it works. It is hoped that the principles presented above may form the nucleus of a future complete theory of visual tracking.

9.7 Concluding Remarks

In this chapter we have discussed some of the problems with the current state of the art in motion detection and visual tracking. We have shown that the one-dimensional pixel processes used in background modelling do not produce true motion detection and cannot detect objects when they cease to move. The opposite approach—"foreground" methods such as particle filtering—cannot track objects unless an a priori appearance model is provided. The latter method is also forced to rely totally on the appearance model, even when the object is in motion and may be changing appearance (as pedestrians do) when the appearance model is at its weakest.

Our solution to these problems is to take, with inspiration from the HVS, a crutch from each side—the motion detection concept from background modelling on the one side and the appearance model-based foreground tracking on the other. We structured the system to switch from one side to the other, ensuring that the strongest approach is used at the appropriate time. Figure 9.12 illustrates this

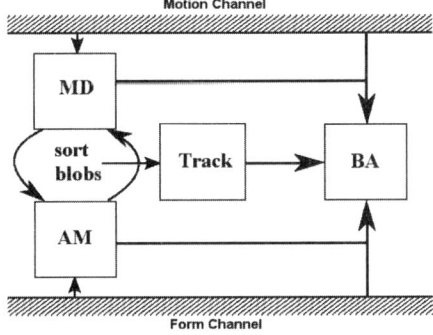

Figure 9.12 A schematic representation of the dual-channel tracking paradigm. Information is simultaneously accessed from both the motion channel and the form channel to produce detection, tracking, and behavior analysis.

architecture. We also present, for the first time, a theoretical framework which explains why such an approach is optimal.

The original motivation of our research was to see what could be learned from the HVS for visual tracking and implemented on today's computer hardware. As described in Section 9.2, while the broad architecture of the HVS is now becoming clear, the detail is generally not yet well understood. What stood out for us in this study was the dual-channel form–motion approach, and the method of spatiotemporal motion detection.

This project was originally intended as an exercise in reverse engineering. Thus an apology is offered, if any apology is needed, for not adhering to reverse engineering more strictly. In fact, because of the profound structural differences between the massively parallel "wetware" of the brain and the serial processes of the computer hardware, practicality requires that this implementation cannot be a direct "rewiring" from brain to machine. Naturally, it may be that, over time, the electrical representation of the human nervous system could be turned into an advantage when producing more advanced machine vision systems. Likewise, further detail of the wiring diagram in the HVS could help with this aim. However, here we have described the first steps of engineering motion analysis by reverse engineering some important aspects of the HVS structure, and fortunately these seem to have been very successful in producing viable robust solutions.

Finally, the behavior analysis method presented here explores a new concept based on direct access to motion channel information. As a demonstrator, we focus on detecting cyclical motion changes in order to distinguish between pedestrians and vehicles. However, the motion information is very rich and much more can be achieved. When the motion channel information is viewed on screen, a human can quickly recognize complex behaviors and sometimes even the mood of the target pedestrian. If the information is there, it should also be possible for a computer to detect these behaviors. We anticipate that the dual form–motion paradigm can profitably be applied to behavior analysis as well as tracking. This may require complex learning algorithms and behavior databases (which are outside the scope of this work) but we find the potential exciting.

Acknowledgments

The authors are grateful to Research Councils UK for grant GR/R87642/02, awarded under the Basic Technology scheme. Figure 9.4 is reproduced from [41] with permission from the IEE. Table 9.1, Figures 9.6 and 9.7, and some of the text are reproduced from [43] with permission from the IEEE. The authors acknowledge use of frames from the CAVIAR sequences [44] in Figures 9.7 and 9.10.

References

[1] R. Sekuler, S. N. J. Watamaniuk, and R. Blake. "Perception of visual motion." In H. Pasher, editor, *Stevens Handbook of Experimental Psychology*, pages 1–56, J. Wiley, 2001.

[2] R. A Young, R. M. Lesperance, and W. W. Meyer. "The Gaussian derivative model for spatial vision. I. Retinal mechanisms." *Spatial Vision*, 14:261-319, 2001.

[3] W. H. Dittrich and S. E. G. Lea. Avian visual cognition: motion discrimination and recognition. Technical report, Dept. Psychology, Tufts University, 2001.

[4] M. N. Shadlen and W. T. Newsome. "Motion perception: seeing and deciding." *Proceedings of the National Academy of Sciences, USA*, 93(93):628-633, 1996.

[5] M. J. Farah. *The Cognitive Neuroscience of Vision*. Blackwell, Oxford, 2000.

[6] K. J. Friston and C. Büchel. "Attentional modulation of effective connectivity from V2 to V5/MT in humans." *Proceedings of National Academy of Sciences, USA*, 97(13):7591-7596, 2000.

[7] K. H. Britten and H. W. Heuer. "Spatial summation in the receptive fields of MT neurons." *Journal of Neuroscience*, 19(12):5074-5084, 1999.

[8] N. J. Priebe and S. G. Lisberger. "Estimating target speed from the population response in visual area MT." *Journal of Neuroscience*, 24(8):1907-1916, 2004.

[9] W. Bair, E. Zohary, and W. T. Newsome. "Correlated firing in macaque visual area MT: time scales and relationship to behavior." *Journal of Neuroscience*, 21(5):1676-1697, 2001.

[10] E. P. Simoncelli and D. J. Heeger. "A model of neuronal responses in visual area MT." *Vision Research*, 38(5):743-761, 1998.

[11] A. H. Wertheim. "Motion perception during self-motion: the direct versus inferential controversy revisited." *Behavioural and Brain Sciences*, 17(2):293-355, 1994.

[12] Z. Li. "Optimal sensory encoding." In M. A. Arbib, editor, *The Handbook of Brain Theory and Neural Networks: Second Edition*, pages 815-819. MIT Press, Cambridge, MA, 2002.

[13] M. J. Nadenau, et al. "Human vision models for perceptually optimized image processing—a review." *Proceedings of IEEE*, page 15, Sept 2000.

[14] H. Momiji, et al. Modelling retinal functions: Fovea. Technical report, Imperial College London, 2003.

[15] H. Momiji. Review of mathematical models of retinal functions motivated by physiology. Technical report, Imperial College London, 2003.

[16] S. Grossberg, E. Mingolla, and L. Viswanathan. "Neural dynamics of motion integration and segmentation within and across apertures." *Vision Research*, 41:2521-2553, 2001.

[17] P. Neri, M. C. Morrone, and D. C. Burr. "Seeing biological motion." *Nature*, 395:894-896, Oct 1998.

[18] I. M. Thornton, R. A. Rensink, and M. Shiffrar. "Active versus passive processing of biological motion." *Perception*, 31:837-853, 2002.

[19] W. N. Martin and J. K. Aggarwal. "Survey: dynamic scene analysis." *Computer Graphics Image Processing*, 7:356-374, 1978.

[20] M. A. Giese and T. Poggio. "Neural mechanisms for the recognition of biological movements." *Nature Reviews Neuroscience*, 4:179-192, March 2003.

[21] N. J. B. McFarlane and C. P. Schofield. "Segmentation and tracking of piglets in images." *Machine Vision Applications*, 8:187-193, 1995.

[22] J. A. Leese, C. S. Novak, and V. R. Taylor. "The determination of cloud pattern motion from geosynchronous satellite image data." *Pattern Recognition*, 2:279-292, 1970.

[23] R. C. Jain and H. H. Nagel. "On the analysis of accumulative difference pictures from image sequences of real world scenes." *IEEE Transactions on Pattern Analysis Machine Intelligence*, 1(2):206-213, 1979.

[24] W. Long and Y. H. Yang. "Stationary background generation: an alternative to the difference of two images." *Pattern Recognition*, 23(12):1351-1359, 1990.

[25] A. Bevilacqua, L. Di Stefano, and A. Lanza. "An effective multi-stage background generation algorithm." In *Proceedings of IEEE Conference on Advanced Video and Signal Based Surveillance*, pages 388–393, 15–16 Sept 2005.

[26] D. Minnen, I. Essa, and T. Starner. "Expectation grammars: leveraging high-level expectations for activity recognition." In *Proceedings of IEEE Conference on Computer Vision Pattern Recognition (CVPR)*, volume 2, pages 626–632, June 2003.

[27] T. Horprasert, D. Harwood, and L. S. Davis. "A statistical approach for real-time robust background subtraction and shadow detection." In *Proceedings of ICCV FRAME-RATE Workshop*, pages 1–19, 1999.

[28] A. Bevilacqua. "Effective object segmentation in a traffic monitoring application." Proc. of IAPR Indian Conf. on Computer Vision Graphics and Image Processing, ICVGIP 2002, Vol. 1, pp. 125–130.

[29] B. Gloyer, et al. "Video-based freeway-monitoring system using recursive vehicle tracking." *Proceedings of SPIE Image and Video Processing III*, 2412:173–180, March 1995.

[30] D. S. Kalivas and A. A. Sawchuk. "Motion compensated enhancement of noisy image sequences." In *International Conference on Acoustics Speech Signal Processing*, volume 4, pages 2121–2124, 1990.

[31] M. Piccardi. "Background subtraction techniques: a review." In *International Conference on Systems, Man and Cybernetics*, volume 4, pages 3099–3104, Oct 2004.

[32] R. Cucchiara, M. Piccardi, and A. Prati. "Detecting moving objects, ghosts, and shadows in video streams." *IEEE Transactions Pattern Analysis Machine Intelligence*, 25(10):1337–1342, Oct 2003.

[33] C. Wren, et al. "Pfinder: Real-time tracking of the human body." *IEEE Transactions on Pattern Analysis Machine Intelligence*, 19(7):780–785, 1997.

[34] K. P. Karmann and A. von Brandt. "Moving object recognition using an adaptive background memory." In V. Cappellini, editor, *Time-Varying Image Processing and Moving Object Recognition 2*, pages 289–296. Elsevier, Florence, Italy, 1990.

[35] D. Koller, et al. "Towards robust automatic traffic scene analysis in real-time." In *Proceedings of IEEE International Conference on Pattern Recognition (ICPR)*, pages 126–131, Oct 1994.

[36] K. Toyama, et al. "Wallflower: principles and practice of background maintenance." In *Proceedings of International Conference on Computer Vision (ICCV)*, volume 1, pages 255–261, 1999.

[37] A. M. McIvor. Background subtraction techniques. Technical report, Reveal Ltd, 2000.

[38] C. Stauffer and W. E. L. Grimson. "Adaptive background mixture models for real-time tracking." In *Proceedings of IEEE Conference on Computer Vision Pattern Recognition (CVPR)*, volume 2, pages 246–252, 1999.

[39] C. Stauffer and W. E. L. Grimson. "Learning patterns of activity using real-time tracking." *IEEE Transactions on Pattern Analysis Machine Intelligence*, 22(8):747–757, 2000.

[40] P. L. Rosin. Thresholding for change detection. Technical report ISTR-97-01, Brunel University, 1997.

[41] M. Sugrue and E. R. Davies. Tracking in CCTV video using human visual system inspired algorithm. In *Proceedings of IEE International Conference on Visual Information Engineering (VIE 2005), University of Glasgow*, pages 417–423, 2005.

[42] P. S. Addison. *The Illustrated Wavelet Transform Handbook*. IoP, Bristol, UK, 2002.

[43] M. Sugrue and E. R. Davies. "Motion distillation for pedestrian surveillance." In *Proceedings of 6th IEEE International Workshop on Visual Surveillance (VS 2006), Graz, Austria*, pages 105–112, 13 May 2006.

[44] CAVIAR. "Caviar test case scenarios." In *EU CAVIAR project/IST 2001 37540, http://homepages.inf.ed.ac.uk/rbf/CAVIAR/ (website accessed 3 December 2006)*, 2001.

[45] M. R. Dobie and P. H. Lewis. "Object tracking in multimedia systems." In *International Conference on Image Processing and Its Applications*, pages 41–44, 1992.

[46] J. Yang and A. Waibel. Tracking human faces in real-time. Technical report CMU-CS-95-210, School of Computer Science, Carnegie Mellon University, Nov. 1995.

[47] P. Gabriel, et al. "Object tracking using color interest points." In *Proceedings of IEEE Conference on Advanced Video and Signal Based Surveillance*, Vol. 1, pages 159–164, 2005.

[48] M. Isard and A. Blake. "Contour tracking by stochastic propagation of conditional density." In *Proceedings of European Conference on Computer Vision (ECCV)*, pages 343–356, 1996.

[49] M. Isard and A. Blake. "Icondensation: unifying low-level and high-level tracking in a stochastic framework." *Lecture Notes in Computer Science*, 1406:893–908, 1998.

[50] Y. Rui and Y. Chen. "Better proposal distributions: object tracking using unscented particle filter." In *Proceedings of IEEE Conference on Computer Vision Pattern Recognition (CVPR)*, Vol. 2, pages 786–793, 2001.

[51] K. Nummiaro, E. B. Koller-Meier, and L. Van Gool. "A color-based particle filter." In *1st International Workshop on Generative-Model-Based Vision GMBV'02, in conjuction with ECCV'02*, pages 53–60, 2002.

[52] H. Nait-Charif and S. J. McKenna. "Tracking the activity of participants in a meeting." *Machine Vision Applications*, 17(2):1432–1769, 2006.

[53] S. Zhou, R. Chellappa, and B. Moghaddam. "Visual tracking and recognition using appearance-adaptive models in particle filters." *IEEE Transactions Image Processing*, 13:1491–1506, 2004.

[54] N. Funk. "A study of the Kalman filter applied to visual tracking." Technical Report CMPUT 652, University of Alberta, 2003.

[55] G. Welch and G. Bishop. An introduction to the Kalman filter. Technical report TR 95-041, University of North Carolina, 2003.

[56] A. S. Bashi. A practitioner's short guide to Kalman filtering. Technical Report SAGES, University of New Orleans, 1998.

[57] C. Harris. "Tracking with rigid models." In A. Blake and A. Yuille, editors, *Active Vision*, pages 59–73. The MIT Press, Cambridge, MA, 1992.

[58] A. Blake, R. Curwen, and A. Zisserman. "A framework for spatio-temporal control in the tracking of visual contours." *International Journal on Computer Vision*, 11(2):127–145, 1993.

[59] T. J. Broida, S. Chandrashekhar, and R. Chellappa. "Recursive 3-d motion estimation from a monocular image sequence." *IEEE Transactions Aerospace and Electronic Systems*, 26(4):639–656, 1990.

[60] A. Azarbayejani and A. P. Pentland. "Recursive estimation of motion, structure, and focal length." *IEEE Transactions on Pattern Analysis Machine Intelligence*, 17(6):562–575, 1995.

[61] M. Isard and A. Blake. "Condensation—conditional density propagation for visual tracking." *International Journals on Computer Vision*, 29(1):5–28, 1998.

[62] J. Lou, T. Tan, and W. Hu. "Visual vehicle tracking algorithm." *Electronics Letters*, 38(17):1024–1025, Aug 2002.

[63] D. Buzan, S. Sclaroff, and G. Kollios. "Extraction and clustering of motion trajectories in video." In *Proceedings of IEEE International Conference on Pattern Recognition (ICPR)*, volume 2, pages 521–534, 2004.

[64] L. Angrisani, M. Dapos Apuzzo, and R. Schiano Lo Moriello. "The unscented transform: a powerful tool for measurement uncertainty evaluation." In *Proceedings of IEEE International Workshop on Advanced Methods for Uncertainty Estimation in Measurement*, Vol. 1, 2005, pp. 27–32.

[65] E. A. Wan and R. van der Merwe. "The unscented Kalman filter." In S. Haykin, editor, *Kalman Filtering and Neural Networks*, pages 221–280. Wiley, New York, 2001.

[66] S. J. Julier and J. K. Uhlmann. A general method for approximating nonlinear transformations of probability distributions. Technical report, Dept. Engineering Science, University of Oxford, 1996.

[67] B. Stenger, P. R. S. Mendonca, and R. Cipolla. "Model-based hand tracker using an unscented Kalman filter." In *Proceedings of the British Machine Vision Association Conference (BMVC)*, volume 1, pages 63–72, 2001.

[68] J. F. G. de Freitas, et al. "Sequential Monte Carlo methods to train neural network models." *Neural Computation*, 12(4):955–993, 2000.

[69] R. van der Merwe, et al. "The unscented particle filter." *Advance Neural Information Processing Systems*, 13:584–590, 2001.

[70] M. S. Arulampalam, S. Maskell, N. Gordon, and T. Clapp. "A tutorial on particle filters for on-line nonlinear/non-Gaussian Bayesian tracking." *IEEE Transactions Signal Processing*, 50:174–188, 2002.

[71] K. Choo and D. J. Fleet. "People tracking using hybrid Monte Carlo filtering." In *Proceedings of International Conference on Computer Vision (ICCV)*, volume 2, pages 321–329, 2001.

[72] C. Hue, J. Le Cadre, and P. Perez. Tracking multiple objects with particle filtering. Technical report, Inst. Nat. Recherche Informatique Automatique (INISA), Oct 2000.

[73] J. Malik, et al. "Smart cars and smart roads." In *Proceedings of the British Machine Vision Association Conference (BMVC)*, Vol. 2, pages 367–382, 1995.

[74] D. Makris and T. Ellis. "Path detection in video surveillance." *Image Vision Computing*, 20(12):895–903, Oct 2002.

[75] D. Makris. Visual learning in surveillance systems. Transfer report, Dept. Electrical, Electronic and Information Engineering, City University, London, 2001.

[76] H. Dee and D. Hogg. "Detecting inexplicable behaviour." In *British Machine Vision Conference*, volume 2, pages 477–486, Sept 2004.

[77] J. Owens, A. Hunter, and E. Fletcher. "Novelty detection in video surveillance using hierarchical neural networks." In *ICANN*, pages 1249–1254, 2002.

[78] M. Markou and S. Singh. "Novelty detection: a review—part 1: statistical approaches." *Signal Processing*, 83:2481–2497, 2003.

[79] M. Markou and S. Singh. "Novelty detection: a review—part 2: neural network based approaches." *Signal Processing*, 83:2499–2521, 2003.

[80] W. Niu, et al. "Human activity detection and recognition for video surveillance." In *Multimedia and Expo*, volume 1, pages 719–722, June 2004.

[81] J. Heikkilä and O. Silvén. "A real-time system for monitoring of cyclists and pedestrians." *Image Vision Computing*, 22(7):563–570, July 2004.

[82] Q. Zhou and J. K. Aggarwal. "Tracking and classifying moving objects from video." In *Proceedings of Performance Evaluation of Tracking and Surveillance (CVPR PETS)*, pages 46–54, December 2001.

[83] A. Selinger and L. Wixson. "Classifying moving objects as rigid or non-rigid." In *Proceedings of DARPA Image Understanding Workshop*, Vol. 1, pp. 341–347, 1998.

[84] A. F. Bobick and J. W. Davis. Real-time recognition of activity using temporal templates. Technical Report 386, MIT Media Lab, Perceptual Computing Section, 1998.

[85] R. T. Collins, et al. A system for video surveillance and monitoring. Technical Report CMU-RI-TR-00-12, Robotics Institute, Carnegie Mellon University, 2000.

[86] A. Kojima, T. Tamura, and K. Fukunaga. "Natural language description of human activities from video images based on concept hierarchy of actions." *International Journals on Computer Vision*, 50(2):171–184, 2002.

[87] A. Graves and S. Gong. "Surveillance video indexing with iconic patterns of activity." In *IEE International Conference on Visual Information Engineering (VIE 2005)*, pages 409–416, 2005.

PART III
Hardware Technologies for Vision

CHAPTER 10

Organic and Inorganic Semiconductor Photoreceptors Mimicking the Human Rods and Cones

Patrick Degenaar, Konstantin Nikolic, Dylan Banks, and Lichun Chen

10.1 Introduction

Let us consider some of the basic principles of photodetection, using a simple example (Nikolic and Forshaw, 2005). The Sun is our main source of natural light on the Earth and radiates approximately as a black body with a temperature of about 5,800K. The spectrum of such radiation has a broad peak at about 500 to 600 nm—exactly at the wavelengths where the human eye is most sensitive. The Earth's atmosphere is most transparent to light between 300 and 700 nm, again exactly in the range of the human eye's sensitivity to electromagnetic radiation (Pye, 2003). If you are in the Sahara Desert, then approximately up to 1 kW/m² of sunlight, mostly in the visible part of the spectrum, falls on the sand. However, if you live in London, for example, then several factors intervene to reduce this illumination level, typically by a factor of about 100 (Figure 10.1).

Now imagine that you are using a digital camera to take a picture of the Tower of London. Light falling on any part of the object is reflected, and let us assume a reflectivity of about 50% (I_{refl}), into all directions (~2π steradians). Thus, only a small fraction of the reflected light actually enters the camera lens and falls on any one of the detector elements. From Figure 10.2, it is evident that the amount of light falling on one detector is given by:

$$P_{coll} = I_{refl} \left(\frac{\text{Solid angle of lens}}{2\pi} \right) D^2 = I_{refl} \left(\frac{A/u^2}{2\pi} \right) D^2$$

$$= I_{refl} \frac{A}{2\pi} \left(\frac{d}{f} \right)^2 \quad (10.1)$$

where the meaning of the symbols is shown in Figure 10.2. Let us now insert some representative values into the equation above. Suppose that the area of the lens is $A = 1$ cm², that the focal length of the lens is $f = 1$ cm, and that the detector is a square of a side $d = 10$ μm. Then the amount of light falling on a detector per second is $P_{coll} \sim 10^{-10}$ W, which represents approximately 3×10^8 photons per second. In an exposure time of 1/30th of a second, and with 100% quantum efficiency, approximately 10 million electrons would be generated in a detector. The rms value for the noise is just the square root of that number. Therefore,

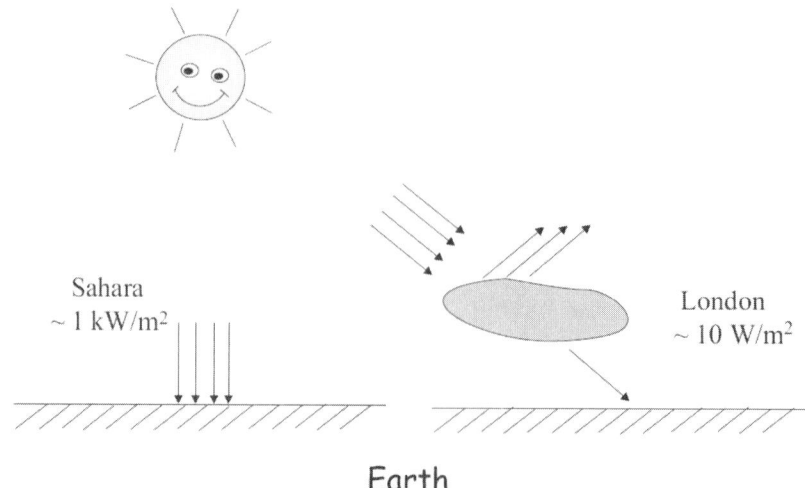

Figure 10.1 The total power of the sunlight per unit area: when light is falling perpendicularly to the surface of the Earth, it is approximately up to 1kW/m². However, for places like London, in spring, on a moderately cloudy day, the light intensity could be 100 time less.

these conditions would yield a signal-to-noise ratio (SNR) of about 3,000:1. If we now go indoors, where the room illumination is typically just 1% of the outdoor illumination, then only 100,000 electrons are generated per detector of size 10 μm × 10 μm. The SNR would be 300:1. One possible way of increasing the SNR, suggested by (10.1), would be to reduce the focal length. Microlenses are being integrated onto CMOS (complementary metal oxide semiconductor) chips, but at the moment it is extremely hard to make lenses with $f < 1$ μm, that is, less than the wavelength of light.

Now let us try to miniaturize the detector, to 1 μm × 1 μm or even down to 10 nm × 10 nm. The average number of electrons that is generated per exposure, in outdoor (indoor) conditions drops to 100,000 (1,000) and 10 (0.1) electrons,

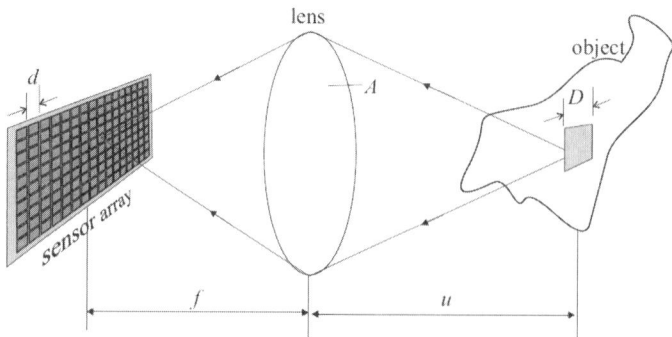

Figure 10.2 The image of a part of an object is formed by a lens on an individual detector in the sensor. Note: $D/u = d/f$.

respectively. The SNR drops to 30:1 and less than one, respectively, in indoor conditions. With an SNR of 30:1 it is hard to make good pictures with a digital camera. This analysis points to the size limit of the detectors to be of the order of 100 nm × 100 nm; for some practical applications, one of the crucial factors would be how small the focal length could be. Obviously, for indoor use, one would have to increase the exposure time or use an electronic flash. The 10 nm × 10 nm detector would, even when placed in the Sahara, receive on average only 1,000 photons per exposure. Of course it should be noted that the diffraction limit ($\sim 0.5\ \lambda/NA$) removes any advantage of having pixels smaller than a micron for most imaging systems.

Is the human retina, containing rods and cones, any more or less efficient than the digital camera detector? The focal length of the eye is 22 mm, and the pupil is up to 7 mm in diameter. By using the formula above, it is easy to show that a typical rod (of diameter 2 μm) will collect approximately 30,000 (300) photons outdoors (indoors) per exposure of 1/30th of a second. With a quantum efficiency of about 15% (i.e., on average only every 6th photon which goes through the retina is detected) that gives only 50 detected photons per exposure for the indoor conditions and that means an SNR of 7:1. Hence, one would expect the eye to give a very noisy signal in ordinary room illumination. However, that is not the case. Some of the amazing abilities of human photosensitivity lie with the averaging functions of the horizontal cells in the retina, but much lies with the dynamic scheme which can amplify the photonic signal by many orders of magnitude with a video rate frequency response. This amplification scheme allows us to detect down to a handful of photons.

A comparable solid state device to the human photoreceptor, such as the avalanche photodiode, simply cannot compete over the range of operation for which the human rods and cones provide response. We can learn many lessons from the light collection architecture of the human eye. Such lessons are not only of academic interest. Over the last decade there has been an explosion of developments in electronic imaging systems. CCD-based systems have now largely matured. There is now a strong drive to develop more intelligent imaging systems, whether for security purposes or simply to make more compact imaging systems for mobile telephony.

In this chapter we will examine the nature of the light detection method in the human eye, and look at equivalents in silicon imaging technology. We will then examine novel light-sensing technology based on organic semiconductor materials and the possibility of making 3-dimensional intelligent imaging structures thereof.

10.2 Phototransduction in the Human Eye

10.2.1 The Physiology of the Eye

In order to detect the whole image formed by the eye lens, the photoreceptors are distributed at the back of the eye, in the region called the retina (Figure 10.3). The human eye contains two types of photoreceptor: cones and rods (Kolb et al.). Both types of cells have a very elongated shape, with the final (photosensitive) segment

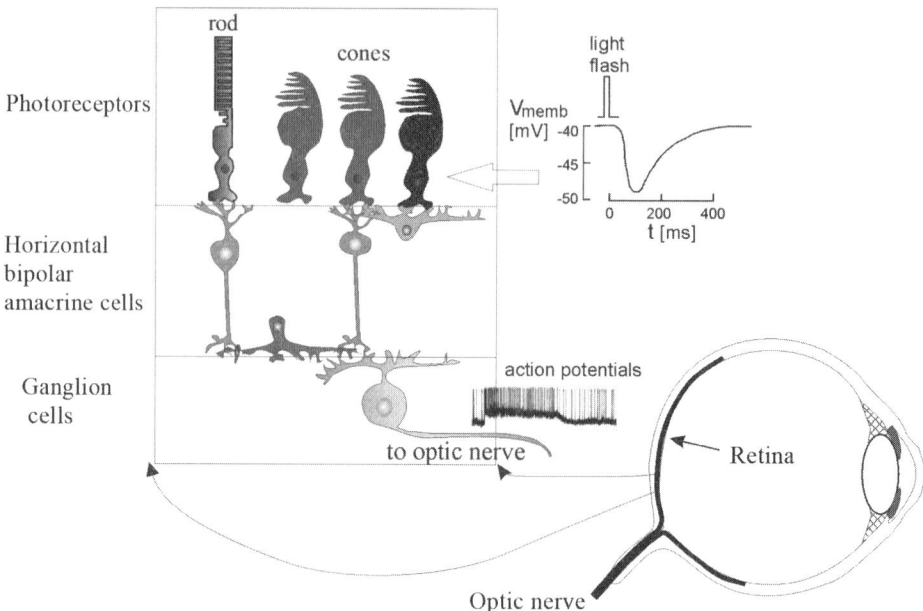

Figure 10.3 At the back of the eyeball of vertebrates there is a thin layer of neural tissue called the retina. A rough sketch of a small piece of the cross section of the human eye retina is shown on the left. A cone receptor potential in response to a light flash is shown and demonstrates the hyperpolarization of the cell membrane as a response to light. However, the ganglion cells are spiking neurons and code the information into action potentials which are sent down the optic nerve to the brain.

in the shape of a rod, of a cross section diameter of about 2 micrometers, or a cone, with the maximum diameter of approximately 6 micrometers (Kolb et al.). The total number of cones in the human eye is approximately 6 million. They are largely concentrated around a small area called the fovea (located at the eye lens focus). Additionally, there are about 120 million rods which are more evenly distributed outside of the fovea.

Cones are responsible for the color vision and need higher light levels for activation. They have three different types of photosensitive molecules (opsins), with sensitivity peak values in the short (blue), medium (green), and long (red) wavelengths. The rods are responsible for low light vision and also have a pigment whose sensitivity depends on the light wavelength. However, as they are not connected to the color processing neurons, they are achromatic. Rods also show a relatively slow response, that is, they need a long integration time.

It is believed that the relative abundance of cones is (very approximately): ~10% "blue, ~30% "green," and ~60% "red." However, since the spectral sensitivity of each cone is relatively broad, the overall sensitivity of the human eye for photopic vision, shown in Figure 10.4, has a peak around the green color (about 550 nm). The sensitivity to light photons is due to a chromophore molecule called retinal (Kolb et al.) embedded in the opsin molecules.

The rods are approximately 10,000 times more sensitive to light than the cones and they are responsible for low light–level vision. They are able to detect light

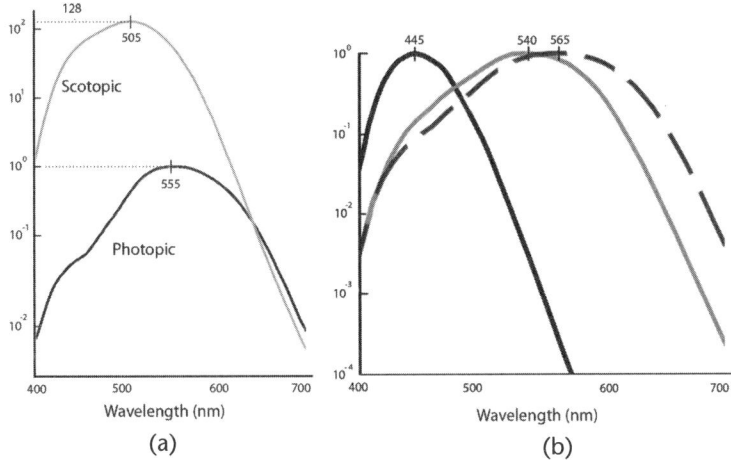

Figure 10.4 Relative spectral sensitivity of the human eye. (Left) Scotopic (low light levels, 10^{-6} to 10^{-2} cd/m^2, that is, rod cells) vs photopic vision (normal daylight vision 1 to 10^6 cd/m^2, cones). (Right) The logarithmic variation of the 3 photopic cone responses. All the variance is in relative terms, but it can clearly be seen that the scotopic is 2 orders of magnitude more sensitive than the photopic.

as faint as 10^{-6} cd/m^2, (see Kolb et al.). As the illumination increases above the level of normal indoor lighting (~100 cd/m^2), the rods' response saturates. Color vision operates within the 10 to 100 cd/m^2 range, due to cone stimulation. Color vision reaches its peak capability in the 100 to 1,000 cd/m^2 region. Beyond this, as the luminance continues to increase to very high levels, visual performance deteriorates. When the light intensity reaches ~10^7 cd/m^2 and beyond, retinal damage can occur. The sensitivity difference between rods and cones explains the entire 12 orders of magnitude range of visual sensitivity. The spectral sensitivities of rods and cones are shown in Figure 10.4. See also Chapter 2, for accurate models of cone wavelength characteristics.

The process of photoreception in the eye begins with the photon absorption by a retinal molecule. The molecule undergoes a shape change, which in turn levers a larger change in the host opsin protein (Filipek et al., 2003). This process triggers a signal that is passed through the processing layers of the retina and eventually farther to the visual cortex. Further details on the anatomy and computational models for describing dynamic photoreceptor response are given in Chapter 2.

10.2.2 Phototransduction Cascade

The actual transduction process by which a photon creates a change in the electrical potential across the photoreceptor cell membrane is complex and has been investigated for decades (Filipek et al., 2003). All major processes in the human phototransduction cascade are now believed to be identified, though investigations into the finer dynamics are still being carried out.

The phototransduction cascade starts with the activation of rhodopsin molecules (in rods, similar opsin molecules are in cones). Rhodopsin belongs to the

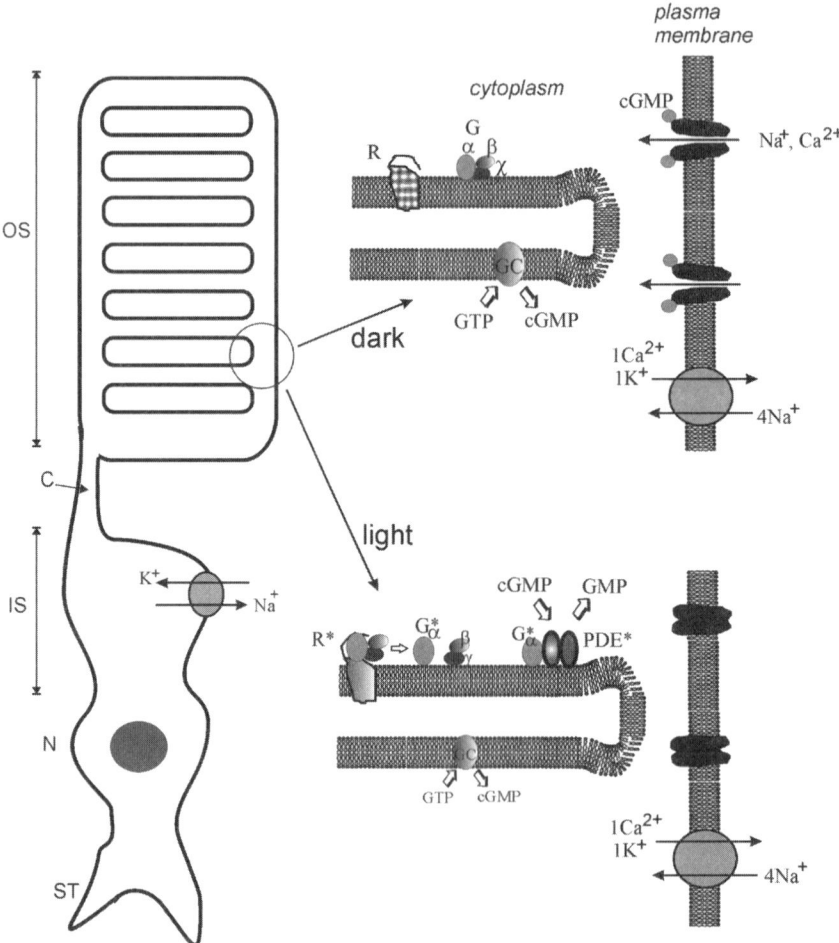

Figure 10.5 Schematic representation of vertebrate rod photoreceptor cell and phototransduction cascade. Major anatomic parts of the cell include: outer segment (OS), which is the photosensitive part and contains stacks of a number of membrane discs, narrow connecting passage – cilium (C), the inner segment (IS), which contains the cell's biosynthetic machinery, K^+/Na^+ ion exchanger, nucleus (N), and the synaptic terminal (ST). On the right is shown enlarged parts of the disc membrane together with the relevant membrane molecules and ligands which take part in the phototransduction cascade: rhodopsin (R), the G-protein (α, β, and γ subunits), phosphodiesterase (PDE). Activated forms are noted with stars (e.g., activated rhodopsin is R*). PDE* decomposes cGMP, whereas guanylate cyclase (GC) generates cGMP from GTP. Also shown is a part of the plasma membrane, together with the cGMP-gated ion channels and the ion exchange pump $Ca^{2+}/K^+/Na^+$.

group of G-protein coupled (GPC) receptors and rhodopsin actually represents probably the best-studied GPC receptor. We should note here that the following description of the phototransduction cascade represents an oversimplification of various processes of protein activation/deactivation and there is a vast amount of detail provided in literature (Lamb and Pough, 1992). Figure 10.5 illustrates the cascade steps, which are described below.

10.2.2.1 Light Activation of the Cascade

Absorption

The absorption of a light photon happens on a relatively small chromophore (light-absorbing molecule) called retinal. After absorbing light the retinal undergoes an isomeric transformation from a *cis* to the all-*trans* form. The all-*trans* form of retinal activates changes to the conformational shape of its rhodopsin host, which can then activate G-proteins. Photoisomerization of mammalian retinal is a very fast process and, unlike in insects, is not reversible. It needs to be physically ejected from its binding site and recycled elsewhere.

G-Protein Activation

An active rhodopsin molecule opens up a cavity where a G-protein can dock and get activated (through a GDP/GTP exchange). The G-protein consists of three subunits: alpha (G_α, also called transducin), beta (G_β), and gamma units (G_γ). After the active G-protein is released, the activated transducin (G_α-GTP) separates from the rest of the G-protein and binds to a PDE molecule. The rhodopsin molecule continues to activate G-proteins as long as it remains in the active state (the active lifetime is of the order of 0.1 to 1 second) and during that time hundreds of G-proteins can be activated.

Phosphodiesterase Activation and cGMP Conversion

The activated transducin then activates phosphodiesterase (PDE) via targeting of its γ subunit. Active PDE (PDE*), allows for cyclic guanosine monophosphate (cGMP) to be converted into GMP.

Ion Channel Closing

cGMP molecules bind to, and open, the ion channels, which are embedded in the rod/cone outer segment plasma membrane (Kaupp and Seifert, 2002). Activation of PDE causes a drop in cGMP concentration and closes the ion channels, blocking the entry of sodium ions. However the K$^+$/Na$^+$ exchanges in the IS continue to pump out the sodium ions, thus causing the *hyperpolarization* of the membrane, which spreads over the entire cell. It is thus the level of hyperpolarization that contains the information about the intensity of the incoming light.

The ion channels are nonselective cation channels, meaning that both Na$^+$ and Ca^{2+} ions can go through these channels when they are open. The current flowing through the cell due to ion channels is determined by intercellular calcium as well as cGMP, as given by the Hill equation (Nikonov et al., 2000):

$$J_{cG} = J_{cG,\max} C_{cGMP}^{n_c} / \left(C_{cGMP}^{n_c} + K_{cG}^{n_c} \right) \qquad (10.2)$$

where C_{cGMP} is the cGMP concentration, K_{cG} is the channel activation constant (which depends on Ca concentration), and n_c is the Hill coefficient of cGMP channel activation. Usually $n_c = 3$, that is, cooperative binding of three cGMP molecules, is needed in order to open a channel (Baylor, 1987).

When the change in the potential across the plasma membrane reaches the synaptic end of the photoreceptor cell, the standard mechanism of neural synaptic transmission is employed to pass this signal to the next layer of retinal cells.

10.2.2.2 Deactivation of the Cascade

Rhodopsin Deactivation

Photoexcited rhodopsin needs to be deactivated before it can go through a recovery process so as to be sensitive to another photon. Deactivation starts with a series of phosphorylations on the C-terminal side by rhodopsin kinase (RK) and then rhodopsin binds arrestin, which blocks the binding of G-proteins (Hamer et al., 2003), thus stopping the cascade. Within a few seconds after the deactivation of the excited rhodopsin, the bond between retinal and its host is broken and all-*trans* retinal diffuses away into the epithelium layer. There all-*trans* is transformed back into 11-*cis*-retinal. A new supply of retinal enters the eye from blood via the retinal pigment epithelium cells. As DNA does not encode retinal, it has to be fabricated from vitamin A.

The lifetime of the rhodopsin excited state is directly affected by the activity of the RK and Ca^{2+} concentration. Ca^{2+} inhibits the phosphorylation of excited rhodopsin, probably by binding to RK (Nikonov et al., 2000). For example, τ_{R^*} for salamander rods could change from $\tau_{R^*} = 0.35s$ in the dark, to $\tau_{R^*} = 0.085s$ under strong illumination (and very low Ca concentration).

Deactivation of PDE

Deactivation of active transducin–PDE* complex starts with GTP hydrolysis (GTP turns into GDP), which is possible since transducin is able to hydrolize the GTP attached to it. However, the GTPase activity of transducin is very slow so the rate is significantly accelerated by a multiprotein complex (Burns and Arshavsky, 2005), giving the time constant of PDE* deactivation of the order of 1s.

Replenishing cGMP

Guanylate cyclase (GC) in the membrane reconstructs new cGMP. Thus, with the end of the cascade, the concentration of cGMP increases, which in turn opens more ion channels depolarizing the cell. Intracellular calcium plays an inhibitory role on the catalytic activity of GC. Thus, after large flashes of light, there will be a lower concentration of calcium and, hence, a greater activity of GC.

10.2.3 Light Adaptation of Photoreceptors: Weber-Fechner's Law

Photoreceptor cells show light adaptation to the background ambient light by adjusting the molecular/ionic concentrations in the process of protein translocation and by adjusting enzyme activity rates of the phototransduction cascade. Much of this is mediated by intracellular levels of calcium. The cell calcium concentration is changing because there is Ca influx from the J_{cG} current, since a fraction of that current (15% to 20%) is due to Ca ions, and outflow of Ca happens through the Ca^{2+}-K^+/Na^+ ion pump. Also, some proteins bind and release Ca acting as Ca buffers. There are at least four processes that are directly regulated by calcium.

The Weber-Fechner's law is an empirical law which establishes a relationship between the physical intensity of a stimulus and the sensory experience and states that sensation increases as logarithm of stimulus intensity (Fain et al., 2001). In other words: change in perception is proportional to the relative change of stimulus (not the absolute value). Experimentally, it has been found that, indeed, the flash sensitivity of eye photoreceptor cells approximately follows this law over several orders of magnitude of the background illumination (Fain et al., 2001; Hamer et al., 2005). Thus, if the receptor sensitivity in dark is S_D then a cell's flash sensitivity at the background illumination I_B is given by (Fain et al., 2001):

$$S/S_D = I_0/(I_0 + I_B) \tag{10.3}$$

where I_0 is the background illumination needed to reduce sensitivity to one-half of the sensitivity in a dark adapted cell.

10.2.4 Some Engineering Aspects of Photoreceptor Cells

The number of rhodopsin molecules in one rod cell varies between the species, but a crude estimate could be $\sim 10^9$ (Hamer, 2000). The probability of a spontaneous (i.e., thermal) activation of a rhodopsin molecule is extremely low at $\sim 10^{-11} \text{s}^{-1}$ (Hamer, 2000; Lamb, 1994). Next in the cascade is the G-protein, and there are of the order of $\sim 10^8$ such molecules in a cell. However, the energy barrier for G-protein activation is much lower than for rhodopsin (spontaneous activation rate at body temperature is $k_{G,\text{spont}} \sim 2 \times 10^{-5} \text{s}^{-1}$) (Hamer, 2000; Lamb, 1994). The number of PDE molecules is roughly $\sim 0.5 \times 10^7$ (i.e., about 5 million) (Hamer, 2000), but each molecule is fully activated only when two transducin molecules are bound to it. Therefore we have effectively the following ratio between the membrane molecular species in the phototransduction cascade:

$$\text{R} : \text{G} : \text{PDE} = 100:10:1 \tag{10.4}$$

How does this affect the amplification process from an engineering point of view? First, the number of rhodopsin molecules is maximized in order to increase the probability of capturing photons. At the same time rhodopsin is very stable to thermal excitations and hence the thermal background noise is minimal. This is one of the reasons that rod cells are able to detect extremely low light levels: they are capable of detecting single photons. After rhodopsin activation, we need as many G molecules as possible in the vicinity so that this signal can be amplified by their activation. But, since G-proteins are less stable, there will always be a certain number of these molecules which are being thermally activated. This brings noise, that is, unwanted activation in the system. Hence, it is best to reduce the total number of G-proteins in order to reduce this noise. Therefore, some optimal number of G-proteins is required. The ratio between active and inactive G-proteins in the dark is:

$$G_d^*/G = \tau_{G*} \times k_{G,\text{spont}} = 3 \times 10^{-5} \tag{10.5}$$

The number of G-protein molecules activated in the dark, G_d^*, should be less than the number of activated molecules during the lifetime of one active rhodopsin molecule if the capture of a photon is to be detectable. Since the experiments show that this number is of the order of one thousand, then the total number of G-protein molecules in a rod cell should be of the order of $G = 1{,}000/(3 \times 10^{-5}) \sim 3 \times 10^7$ or less, which is in good agreement with experimentally found values (Hamer, 2000).

Eventually PDE has an even smaller activation energy barrier and hence it should be present at a lower concentration. However, the PDE concentration should be high enough to allow for good coupling between G* and PDE in order to keep up with the rate of G activation. Therefore the ratio between R, G, and PDE is an optimization problem in which the increase of concentration of G and PDE increases the gain, but also increases the noise. This is an optimization problem which has been shown to have an optimal signal-to-noise ratio between 60% to 80% of the maximum possible output level and this is roughly in agreement with the R:G:PDE ratio found in photoreceptor cells (Lamb, 1994).

In conclusion we can say that the phototransduction cascade has a large amplification: 1 photon $\rightarrow$ activates $\sim$1,000 G* $\rightarrow$ closes hundreds of channels $\rightarrow$ blocking the influx of $\sim 10^6$ Na$^+$ ions $\rightarrow$ causing the membrane hyperpolarization of about $\sim$1 mV. This is a huge gain, which comes at the expense of the energy provided by adenosine triphosphate (ATP) molecules, which participate in the ion transport of the ion exchange pumps in the cell. In addition, thermal energy is used for Brownian motion of molecules that need to collide in order to interact. Also, all enzymes require a certain body temperature to operate efficiently.

As we shall see in the subsequent photodiode sections of this chapter, photodiodes in general do not have gain. Thus, for a 1 to 1 photon to electron conversion and a detection limit of about 100 fA, the minimum detectable light intensity is about a million photons. Avalanche photodiodes, which have internal gain mechanisms, are capable of detecting single photons. However, they do so at high energy cost and do not have an ability to adapt their gain mechanism. Thus, unlike in rods and cones, their response tends to be saturated by noise amplification at higher light intensities.

10.3 Phototransduction in Silicon

The basis for semiconductor phototransduction is the conversion of photons into electrical charge. This charge is then sensed in the form of a change in resistance, voltage, or current, which can be detected by electronic circuits.

When light is absorbed by a material, an electron is excited from a lower-energy orbital to a higher-energy orbital. In metals the electronic effect is negligible due to the pre-existence of large amounts of mobile charge. In insulators, the generated charge has such low mobility that it usually recombines long before extraction could occur, and thus cannot be electronically detected. It is only in semiconductor materials that it is possible to efficiently detect photogenerated charges. The most commonly used semiconductor material is silicon, but gallium arsenide is commonly used for the telecoms wavelengths and organic semiconductors show a lot of promise.

The performance characteristics of photodetectors include:

- *Quantum efficiency:* The number of charges generated per incoming photon.
- *Noise:* The quantity of background noise produced by the detector.
- *Dynamic range:* The range between the upper and lower limits of detectability.
- *Linearity:* The linearity (or nonlinearity) in the response to light intensity.
- *Chromatic response:* The variance of the response with wavelength.
- *Bandwidth:* The speed of the device response over a given dynamic range.

The factors above will determine the specific photoresponse of an individual photodetector and is important in determining its functional use. Generally high quantum efficiency, low noise, high linearity, dynamic range, and bandwidth are desirable for the chromatic region where the detector is required to function. The response capabilities will vary with the quality of the semiconductor, the device structure, structural properties, and the optical interface. In the case of the latter, for example, antireflection coatings can increase the coupling efficiency to 95%.

The main classes of photodetectors are the following:

1. *Photoconductors* Photoconductors or photoresistors are usually formed from an undoped semiconductor layer. The layer will be highly resistive as undoped semiconductors have very little free charge. Incoming light will generate free charges due to trap filling and thus the conductivity will increase. Sensing is carried out by applying a voltage and measuring the current output. The response can be efficient, and subbandgap (infrared) detection is possible, but thermal noise can be a problem, and device speeds can be slow.
2. *Photodiodes* Photodiodes are generally formed by putting a p-doped and n-doped semiconductor in contact with each other. This arrangement will generate a built-in electric field at the p–n junction. When light is shone on the device it will generate charges that are swept to the electrodes under the influence of the internal field. The subsequent photocurrent is thus a measure of the light intensity. Response is generally linear and response time is fast.
3. *Avalanche photodiodes* These are heavily reverse-biased photodiodes. The strong bias results in photogenerated charges multiplying themselves via impact ionization. The resultant amplification can be high, but it amplifies the noise as well as the photocurrent. Avalanche photodiodes function well at low light intensities, but noise amplification exceedes signal amplification at higher intensities. Additionally, the high voltage required for their operation makes them more difficult to integrate with standard CMOS technology.
4. *Phototransistors* Phototransistors are transistors where a light path can access the active channel. The transistor channel is thus modulated by a logarithmic function of absorbed light intensity rather than gate voltage. Gain can be high, but the response is nonlinear.

5. *Photogates* Photogates are formed from metal oxide semiconductor layers. As light is absorbed in the semiconductor, charges are generated that accumulate like a capacitor. The charge can be read as a voltage and reset simultaneously. Photogates and related CCD structures have the advantage of integration time, but are difficult to implement into CMOS processing.

Photosensing devices are connected to electronic circuits, which amplify and process the electrical signals emanating from the detectors. For imaging-sensing technologies they are then implemented on two-dimensional arrays. The two most common technologies are the CCD (charge coupled device) and CMOS (complementary metal oxide semiconductor).

Digital image sensors have developed as successors of (color) films, with the advantage of having the captured image converted into electronic signals, which can be conveniently handled, processed, and stored by the use of microelectronic processors and memories. The development of image sensors has become one of the main driving forces in the semiconductor industry, and the market for these types of devices is expanding.

10.3.1 CCD Photodetector Arrays

The basic detector element in a CCD chip is a metal oxide semiconductor (MOS) structure, (see Figure 10.6). There is a charge-depleted area where new electron-hole pairs are generated by incoming photons. The electrons are then quickly pulled into a potential well near the oxide. The number of collected electrons in the well is then proportional to the level of the illumination of that element. The CCD sensor consists of a regular two-dimensional array of identical sensing elements.

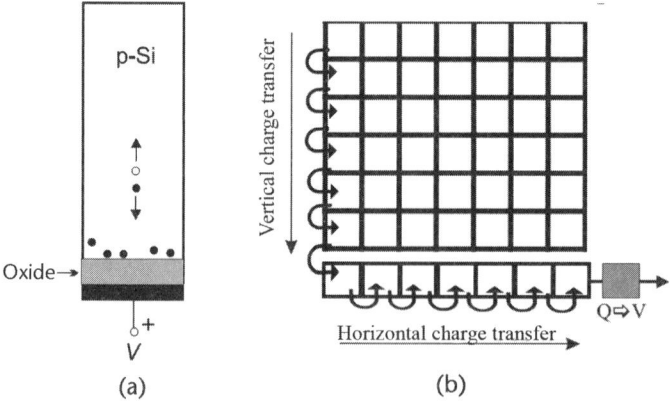

Figure 10.6 (a) Single detector in a CCD sensor is a metal oxide semiconductor structure, where the metal electrode is positively biased and attracts electrons to the oxide semiconductor interface (Waser, 2003). (b) A two-dimensional array of detectors in a CCD chip. Stored electrons are shifted from cell to cell along each column (in parallel)—slow vertical transport. Then charge is serially shifted in the horizontal direction along the last row of cells (fast). At the end the charge is converted into an electrical signal (voltage), amplified, and taken out of the CCD chip.

10.3 Phototransduction in Silicon

The sensor resolution increases as the size of individual detectors decreases, but the sensitivity of detectors increases if the detector area increases.

CCD and CMOS sensors have several common features, such as the principle of the photon detection (electron-hole creation), use of space charge region in silicon, and corresponding electric field for the separation of the generated charge.

The CCD chip technology is well developed and sensors generally display high quantum efficiency, very low dark current (10 to 20 pA/cm^2), and noise, and consequently a very high dynamic range of up to nine orders of magnitude (Magnan, 2003). Very large detectors are possible with this technology. At the time of writing 39 megapixel professional cameras are available, and minimum pixel sizes are down to 1.5 μm. Increasingly, CCD technology is being used in high-end image capture systems where high pixel numbers are required, but image processing is not.

The inherent drawback in CCD technology is the drive voltages of 10V to 15V required to extract the charge. Such high voltages are incompatible with standard CMOS transistors, resulting in all significant image processing having to be done off chip.

10.3.2 CMOS Photodetector Arrays

CMOS technology has been the basis of the microprocessor industry. Thus, the integration of light-sensing elements on silicon allows for image acquisition and processing to be performed on the same chip. As CMOS is a very mature process, fabrication facilities are readily available to developers and subsequent fabrication costs are relatively cheap.

The most common photodetector implemented on CMOS arrays is the photodiode. The photodiode consists of p-type and n-type silicon semiconductor layers sandwiched together with electrodes to extract charge. The basic concept of operation is shown in Figure 10.7. At the interface of the p and n semiconductors, negative electrons will diffuse into the p region and positive holes will move to the n region. This will leave a center with depleted charges and an electric field due to the accumulation of positive charges in the n region and negative charges in the p region. This structure is known as a *diode* and is, of course, an important building block for transistors. When light hits the depletion region, an electron

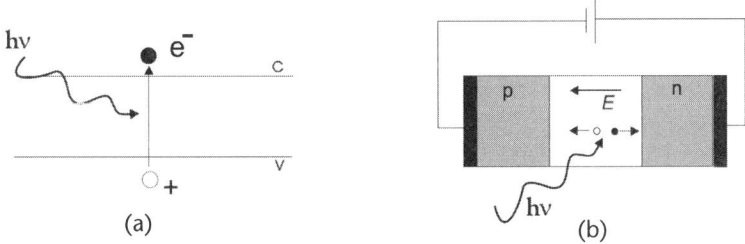

Figure 10.7 (a) A photon of sufficient energy can excite an electron in the valence band and cause its transition into the conduction band. (b) An electron-hole pair generated in the depletion region of a p–n junction is quickly separated by the electric field.

is stimulated into the conduction band leaving a hole in the valence band. The electron is then swept to the p region and the hole is swept to the n region where they can be extracted by the contacting electrodes. The resultant photocurrent is in the opposite direction of what is normally possible via a diode, allowing it to be detected.

Different photodiode designs will change the spectral response characteristics and quantum efficiency. Also, the overall size will vary the total photocurrent. A typical response profile from a 50 μm $\times$ 50 μm photodiode fabricated in 0.18 micron CMOS can be seen in Figure 10.8.

The response is linear for an incident power from 50 pW to greater than 100 nW. The minimum detectable light is set by the dark current, which is determined by the bias. While increasing the bias increases the frequency response and quantum efficiency, it also increases the dark current by a much greater factor due to increased leakage current across the diode. In the above configuration the photodiode is reverse biased by 1.5V, leading to a dark current of 8 fA/μm^2. The photodiode quantum efficiency is 76% for 530 nm wavelength. The effective output current characteristic from this photodiode is between 1 pA to 10 nA corresponding to 25 nWcm^{-2} to 2.6 mWcm^{-2}. These 5 decades of intensity variation correspond to the difference between Moonlight and a well-lit room.

A typical CMOS detector array and the circuit diagram of a CMOS photodiode pixel are shown in Figure 10.9. The CMOS architecture allows individual random addressing of each pixel. CMOS image sensors are usually chips that contain not only the light detectors (pixels), but may also have integrated amplifiers, A/D converters, digital logic, memory, and so on. Since CCD chips require some special and more expensive processes, the CMOS chips, which can utilize the standard MOSFET, a well-matured production technology, have became the preferred

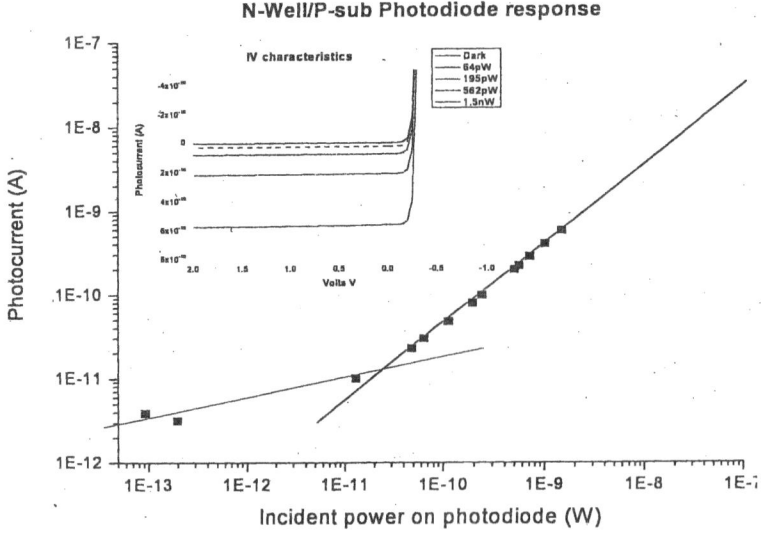

Figure 10.8 The response of a p-n junction photodiode consisting of an n-well on a p-substrate, fabricated in a 0.18 micron (UMC) technology. Below certain intensity, leakage effects will dominate the signal.

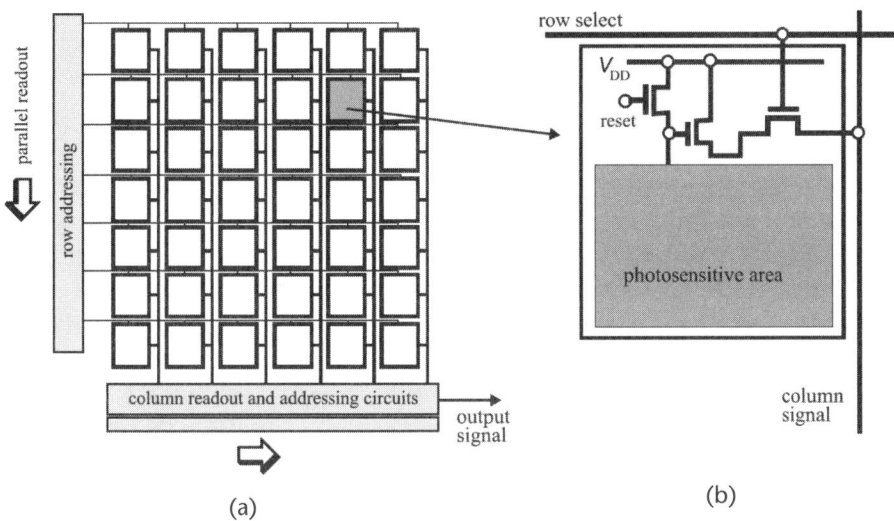

Figure 10.9 (a) The CMOS detector array architecture and (b) a symbolic scheme of a 3T (three transistor) CMOS photodiode pixel after (Magnan, 2003).

choice for mass market applications, such as digital cameras, mobile phones with camera, bar code readers, security cameras, and so on. CMOS chips are also becoming competitive in many high-performance applications. Vast investment into CMOS image sensors, driven by the mass market of digital photography and mobile phone cameras, gave impressive improvements for CMOS technology quantum efficiency, low dark currents (below 100 pA/cm^2), and good dynamic range of up to seven orders of magnitude (Magnan, 2003).

10.3.3 Color Filtering

There are two main techniques for obtaining color-selective photoreceptors in silicon devices: the first is to use color filters (Blanksby and Loinaz, 2000; Blanksby et al., 1997; Adams et al., 1998; Poenar et al., 1997) currently available from several foundries, where color filters and lenses are evaporated on top of the die photodiodes (Figure 10.10). The second is to use the natural absorption profile of silicon itself to develop an intrinsic chromatic filter (Simpson et al., 1999).

Color filtering is less effective than results from present imaging systems might indicate. The filters have to be thin so as not to distort the optics. Thus, further image processing is required to achieve proper color correction. Interestingly, none of the filters in the visual range absorbs well in the near infrared; a separate infrared filter is required. Some cameras provide a night vision mode by removing this infrared filter and adding up the RGB pixels into a single monochromatic intensity.

Color absorption within silicon is strongly wavelength dependent (Figure 10.11) after (Sze, 1981), which can be used to advantageous effect for color segmentation systems. Silicon has a higher absorption coefficient for blue wavelengths of light than it does for red (varying by four orders of magnitude). Much

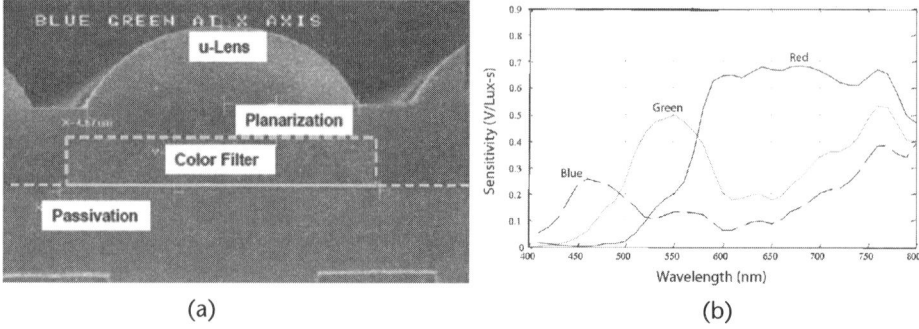

(a) (b)

Figure 10.10 Color filtering in CMOS. Color filters can be placed at the top layer of a CMOS device along with microlenses to improve the effective fill factor. In all cases an additional IR filter is required to remove Infrared wavelengths. This can usually be implemented in the lens. (Pictures courtesy of a major CMOS fabrication facility.)

of the blue radiation is absorbed within the first 300 nm of silicon while red wavelengths can penetrate to depths of over a micron. Given that the depletion region of the p–n junction is the most efficient region for photogenerated charge separation, the depth of the junction will determine the chromatic response of the photodiode. If the junction is close to the surface, it will have a blue shifted response, if it is deep, it will have a red shifted response.

This intrinsic filter methodology is most effectively implemented with an epitaxial process (Bartek et al., 1994), allowing lower doping concentrations at the p–n junction than a standard CMOS process and also allowing the junction depth

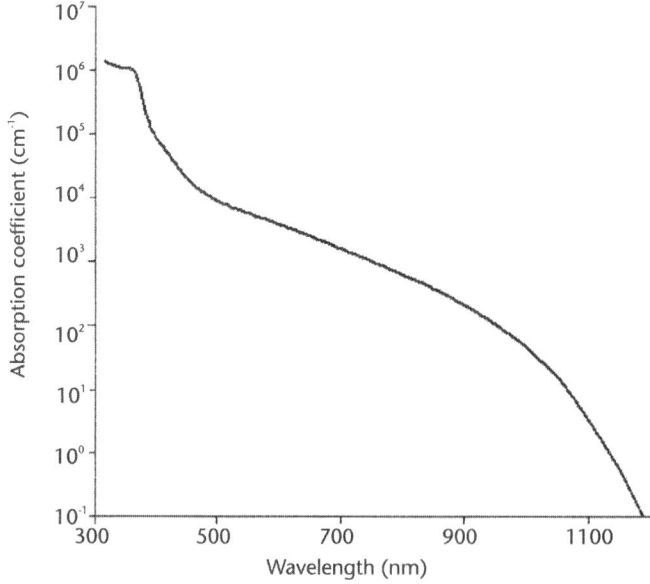

Figure 10.11 The absorption coefficient vs photon energy at different temperatures. (Data courtesy of Virginia semiconductor.)

10.3 Phototransduction in Silicon

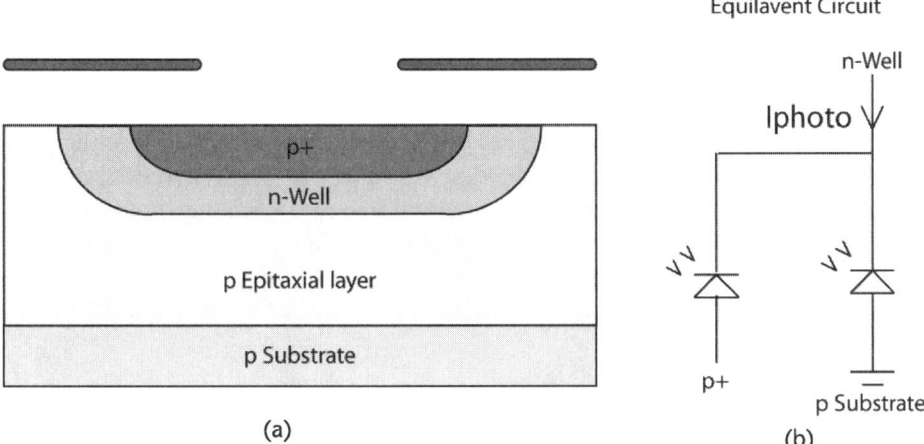

Figure 10.12 Image showing standard CMOS double junction (left). Formed from a p+ source implant of a standard pMOS transistor in an n-well. The top junction depth is approximately 200 nm while the n-well/p-substrate junction is approximately 1,500 nm deep. The equivalent circuit is shown on the right.

and active area to be accurately controlled. With this technique, silicon devices can produce effective color-selective photodiodes. Unfortunately, epitaxial growth is not currently available within standard CMOS processes, but there is some option to control junction depth within standard CMOS processes. CMOS technologies have a number of available photodiode structures, with differing junction depths. These may be deep, shallow junctions or double junctions (Figure 10.12), depending on the doping configuration. The depths of these junctions will be process specific as will be the absorption coefficients which depend on doping concentrations.

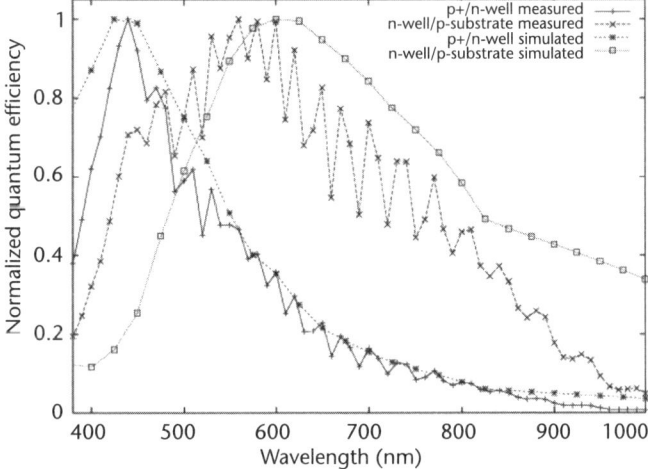

Figure 10.13 The plot shows normalized wavelength-specific photocurrent responses for a standard CMOS p–n–p double-junction photodiode. In reality the responses are quite different and require compensating amplification or scaling.

It is possible to use these standard photodiode junction depths to perform color segmentation (Lu et al., 1996; Findlater et al., 2003; Simpson et al., 1999) (Figure 10.13). The results shown here clearly indicate that the intrinsic properties of silicon may be utilized for color segmentation.

Some triple well processes are also available, but these are either epitaxial nonstandard processes, such as that used by Foveon where epitaxial growth is required, or biCMOS (Chouikha et al., 1998). These techniques have the advantage of a three color segmentation and in the case of vertically stacked pixels with epitaxial processes, no loss of spatial resolution.

Both of these techniques may reduce the quantum efficiency of the effective photoreceptor by up to 50% for color-specific wavelengths and 85% for white light through absorbance of photons within the filters and the silicon.

10.3.4 Scaling Considerations

How is the scaling down of the CMOS technology affecting the CMOS image sensor technology? An analysis by Wong (Wong, 1996) has first shown that the scaling down of the pixel area would follow the decrease of the minimum feature size down to the 0.25 μm technology, reaching 5 μm × 5 μm (see Figure 10.14). Any further decrease in the pixel size seems now to be unnecessary due to the diffraction limit of the camera lenses, and signal-to-noise issues. Nevertheless, some mobile phone cameras have pixel dimensions of 2 μm × 2 μm in order to achieve megapixel resolutions.

In general, image sensors will benefit from the decrease of the minimum feature size mainly through the increased fill-factor and increased circuitry density (and therefore more functionality) in a single pixel. The fill-factor is important, since

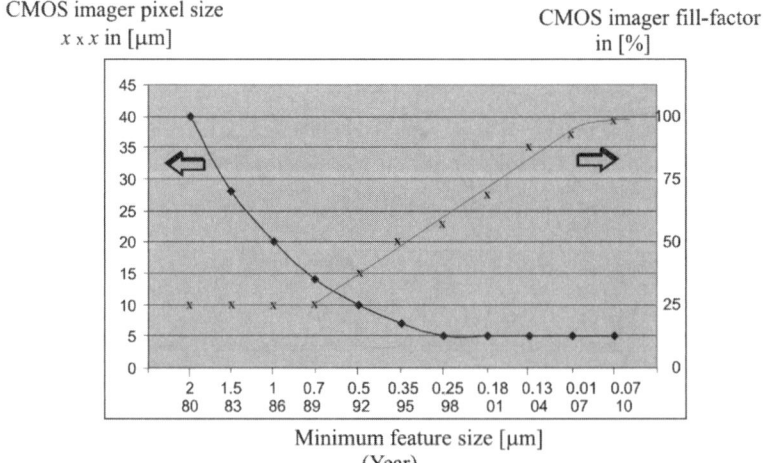

Figure 10.14 (The CMOS pixel size and fill-factor as a function of the minimum feature size (with the year of the first shipment for DRAM application).) This graph is compiled from (Magnan, 2003) and table 1 in (Wong, 1996). See also (Nikolic and Forshaw, 2005).

the effective quantum efficiency of detectors is equal to the product of internal quantum efficiency and fill-factor.

The major limitation from scaling down CMOS technology results from the increased doping levels, which are required to achieve such scaling. Increased doping levels will make the depletion region between the p and n regions more abrupt. As the depletion region is responsible for separating electron-hole pairs, a reduction in its width will reduce the number of pairs being separated and, thus, the photocurrent response to the incoming light. In addition, the increased doping levels and smaller dimensions lead to increased tunnelling currents in the photodiodes and the transistors. This will in turn increase the dark current and make low light operation more difficult. Various pixel architectures have been suggested as a way of combating these problems (Wong, 1996).

10.4 Phototransduction with Organic Semiconductor Devices

The downside of developing intelligent retina-like imaging chips with CMOS technology is that CMOS is an inherently two-dimensional technology. As a result, there is direct competition for space between light-sensing and image-processing structures. In contrast the human eye has a three-dimensional architecture, as described in Chapter 2.

CMOS is ideal for creating processing architectures, as it is a mature technology and relatively cheap and efficient. It is, however, difficult to produce three-dimensional structures with CMOS technology as it is difficult to grow good quality crystalline silicon on top of amorphous silicon dioxide dielectric layers.

There are a range of technologies available which can be used to create light-sensing structures. However, only a few are capable of being integrated as a post-processed layer on top of CMOS. One promising technology is based on organic semiconductor material. In this section we shall therefore review the basis for organic semiconductor technology before later describing its vertical integration onto CMOS imaging devices. Organic photosensor devices have human-like spectral sensitivity, good quantum efficiency, and no wavelength-dependent interference effects, as they can be placed on top of the CMOS dielectric stack. Furthermore, postprocessing of organic semiconductor material need not be much more complicated than postprocessing of organic absorption filters carried out in color CMOS imaging processes. To this end, we look at three organic photodiode structures that could be used to make color CMOS/organic hybrid imaging chips.

The principle of incorporating molecular semiconductors into light-sensitive electronic devices is not a new concept. The solid-state photovoltaic effect was first observed in a selenium cell in 1873, and selenium's photoconductive characteristics have been researched since. Until the silicon photocell was invented at Bell Labs in 1954, organic materials formed the principal focus of photovoltaic research. Present-day xerographic devices largely use organic molecular photoconductive materials, and the organic light emitting diode, LED, field is growing at an ever-increasing rate. The revived concept of using organic materials as photovoltaic devices and solar cells is relatively recent. However, significant progress has been made in the understanding of their photovoltaic properties.

10.4.1 Principles of Organic Semiconductors

Carbon literally forms the backbone of organic semiconductor molecules. Conjugated materials have alternating double and single bonds between carbon atoms forming a conjugated backbone. The single bonds, known as σ-bonds, form the structural backbone, but the second bonds, known as π-bonds, are much weaker and more fluid. When electrons are not confined between two carbon atoms, but are spread over a larger chain, they are described as delocalized. Such delocalization can lead to localized charge states which travel with ease along the polymer chain. Figure 10.15 shows an example of a conjugated polymer, polyacetylene.

The delocalised π-bonds result in the formation of π and π* (bonding and antibonding*) molecular orbitals. However, these orbital's must obey the Pauli exclusion principle and as such cannot occupy the same energy levels. They thus split (degenerate) into sublevels, giving rise to a semiconductor band structure. As more molecules join the chain, electron delocalization increases and the bands widen, decreasing the energy gaps between each bonding state. In silicon the lower-energy band is known as the valence band and the higher is known as the conduction band. In organic systems they are known as the highest occupied molecular orbital (HOMO), and the lowest unoccupied molecular orbital (LUMO), respectively.

As with silicon, organic semiconductor materials will absorb light with greater photon energy than the energy difference between the HOMO and LUMO bands. The band gap typically varies between 1.6 and 4 eV, which is higher than that for silicon but matches well with the visual spectrum, which is between 1.6 and 3 eV. Organic photodetectors have large variations in optical band gaps, allowing effective color segmentation of the visible spectrum. This creates smaller energy (absorption) band gaps, changing the absorption properties of the molecule—larger delocalization molecular chains have lower-energy band gaps and may absorb longer wavelength (lower-energy) photons.

The absorption of a photon results in a transition between the HOMO and the LUMO. This generates a bound electron-hole pair (exciton), just as in silicon. Within organic materials, however, electrons and holes are not simply a missing electron, but are physically manifested as polarons that look like electronic defects in the polymer chain. Also, unlike in silicon, which is a three-dimensional lattice, the charge states are confined along the one-dimensional chain. For them to split,

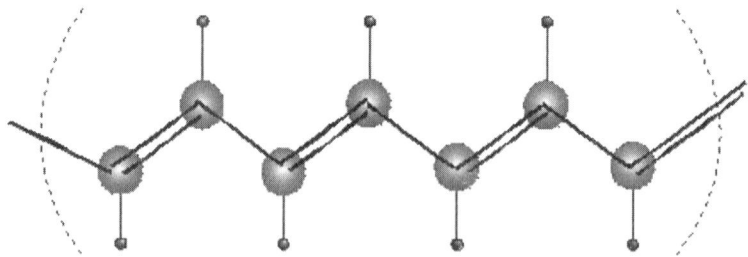

Figure 10.15 Polyacetylene shows the classic conjugated carbon bonding scheme which allows for electronic conduction.

additional energy is required for them to jump to adjacent chains. As a result, photon absorption results in bound electron-hole pairs known as excitons, with binding energies between 0.3 and 1 eV (Dyakonov, 2004; Cao et al., 1999).

Silicon devices are structured from regions doped with impurities to create an excess of holes (p-type) or an excess of electrons (n-type). Organic semiconductors cannot be doped in the same way. Organic semiconductors remain largely intrinsic, that is, they have relatively little charge. However, whereas there is relatively little difference in the ease of transport of electron or hole charges in silicon, organic semiconductors show a much greater difference. Thus, rather than p-type or n-type semiconductors in the case of silicon, organic semiconductors display intrinsic electron transporting or hole transporting preferences. Such preferences are important to the formation of practical devices. In addition, the speed at which charges transmit through the material, known as the *charge mobility*, is much lower than in silicon. Typical charge mobilities are 10^{-4} to 10^{-3} $cm^2V^{-1}s^{-1}$ compared to 800 (hole) to 1,200 (electron) $cm^2V^{-1}s^{-1}$ in silicon, but are generally increasing as better materials are found.

Organic semiconductor materials come in two forms: small molecules and polymers. Small molecules such as pentacene and PCBM (phenyl-C_{61}-butyric acid methylester) are often insoluble or weakly soluble. They are structurally more rigid, but can potentially form crystalline structures that can increase charge mobility. Polymers tend to have wider spectral absorption band gaps as delocalizing can occur over longer chain segments. They can be made soluble and are therefore solution processable, making them potentially compatible with high throughput roll to roll techniques. Also, polymers can be used with flexible substrates, opening up many potentially interesting applications (Woo et al., 1993). Images of a semiconducting small molecule (PCBM) and polymer (MEH-PPV) can be seen in Figure 10.16.

10.4.2 Organic Photodetection

Similarly to silicon photodiodes, organic semiconductor photodiodes consist of a p-type semiconductor and n-type semiconductor sandwiched between two electrodes.

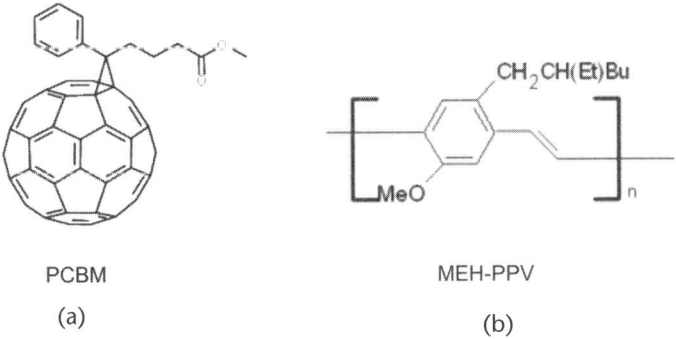

Figure 10.16 (Left) PCBM (phenyl-C_{61}-butyric acid methylester), a fullerene molecule commonly used in organic photosensors and solar cells. (Right) MEH-PPV, a common conducting polymer.

However, unlike in the silicon case, the metals must have different work functions and the nature of p-type and n-type semiconductors is very different for organic semiconductors. A more detailed description of organic photodiode theory has been given by Peumans (Peumans, 2004a, b) and Komatsu (Komatsu et al., 2004), but we will briefly introduce the concept here.

When light is absorbed in the organic semiconductor a bound exciton state is formed, as described previously. This binding energy significantly inhibits exciton dissociation into separate electrons and holes, which is required to generate a photocurrent. Thus, to aid dissociation, an energy incentive must be given for one of the charges to jump to a new material, thus splitting the exciton. This can be achieved by sandwiching electron and hole conducting materials. In the case of small molecule detectors this will be a bilayer. However, in the case of polymer devices, the mixing of two polymers leads to a phase separation (Jones, 1995), resulting in a much larger surface area of contact between the two semiconductors (known as a bulk heterojunction), greatly improving the efficiency of the device.

The mechanism of converting light into electric current is described in Figure 10.17. The main steps are as follows.

1. Light (photons) passes through a transparent electrode and is absorbed by the active layer (mainly by the polymer), leading to the formation of an excited state (an exciton), for example a bound electron-hole pair; the electron is in the LUMO and the hole in the HOMO of the polymer.

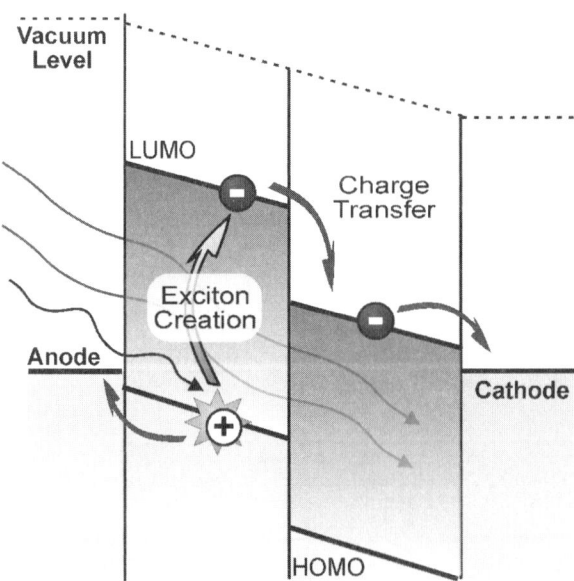

Figure 10.17 Schematic diagram of the working principle for polymer/polymer or polymer/fullerene bulk heterojunction solar cells. Procedure 1: Absorption of light by the polymer and formation of exciton. Procedure 2: Exciton diffusion and charge dissociation at the interface between the electron-donating polymer and the electron-accepting material. Procedure 3: Charge transport and/or recombination. Procedure 4: Collection of charges by the two electrodes.

10.4 Phototransduction with Organic Semiconductor Devices

Figure 10.18 Working structure of organic photodiodes. The layer stack. Initially there is a dielectric substrate for support (e.g., glass), then a conductive anode (e.g., indium tin oxide), then the active semiconductor layer (e.g., P3HT: PCBM blend), and the cathode (e.g., aluminum).

2. The exciton diffuses to the region where exciton dissociation occurs, for instance, to the interface of the electron-donating polymer that has high ionization potential and the electron accepting material that has a high electron affinity. The exciton then dissociates into free charges. The electron and hole are transferred to the electron donating and accepting materials, respectively. The driving force for charge separation and transfer, in this case, is the difference in electron affinity or in LUMO levels between two materials at the interface.
3. Charges are transported under an electric field. These charges may recombine when an electron and a hole are close enough during the transport.
4. Charges are collected by the two electrodes.

The physical device structure is realised in Figure 10.18. Indium tin oxide (ITO) acts as a transparent anode with a PEDOT covering layer to enhance the surface smoothness and improve the work function. The active layer absorbs light, generates the electron hole pairs and transports the separated charge to the electrodes. The aluminum cathode also acts as a mirror to increase the effective optical thickness of the device.

10.4.3 Organic Photodiode Structure

Organic photodiode technology can benefit greatly from the advances in organic LEDs. The device design and structure are generally similar (see Figure 10.18), as are many of the materials and fabrication techniques. Thus, the commercial drivers for organic LED displays (ease of preparation, low cost, low weight, and flexibility of the devices) can make commercial fabrication or organic photodetection systems viable.

Much of the work to the present date on organic photoelectric conversion has focused on energy generation rather than photodetection. This is because the possibility of cheap solar cell fabrication could be a great boost to energy production. There are a number of different types of conjugated polymer-based solar cell: polymer/polymer blends (Breeze et al., 2004; Krueger et al., 2004), polymer/polymer double layer (Granstrom et al., 1999), polymer/fullerene double layer blends (Pettersson et al., 1999; Roman et al., 1998), polymers blended with soluble fullerene derivatives (polymer/fullerene bulk heterojunction) (Padinger et al., 2003; Mozer et al., 2005), and polymer/inorganic hydride (Huynh et al., 2002; Ravirajan et al., 2005) solar cells. Among polymer solar cells, polymer/fullerene

Table 10.1 PCBM (Phenyl-C_{61}-Butyric Acid Methylester)

Detector Color	Electron Transporter	Hole Transporter
Blue	PCBM	BFB
Blue–Green	PCBM	F8T2
White	PCBM	P3HT

A soluble c60-based molecule, is used as the main electron transporter and three different polymers are used as hole transporters. BFB and F8T2 are polymers derived from the fluorine moiety, whereas P3HT is derived from a thiophene moiety.

bulk heterojunction solar cells are the most promising because they have the highest power conversion efficiency and easiest fabrication. The difference between solar cells and photodetectors is subtle. In terms of device structure, they are almost the same. Electrically however, the bias will be different, and optically, photodetectors will tend to concentrate on certain wavelengths, whereas solar cells attempt to be broadband.

Typical materials to create organic photodetectors can be seen in Table 10.1. Generally PCBM is used as an electron-conducting layer and a conducting polymer as the hole transporting layer. The PCBM has a rather large band gap. Thus the polymers are responsible for light absorption and, hence, chromatic response.

10.4.4 Organic Photodiode Electronic Characteristics

For imaging purposes, the important characteristics of organic photodiodes can be given by their photoresponse characteristics, their spectral capture efficiency, and their device performance characteristics.

10.4.4.1 Photocurrent and Efficiency

When an organic photodetector is illuminated, with zero bias (short circuit condition), current flowing in the circuit is pure photoinduced current, namely the short circuit current. With increasing applied forward bias, the combined photo- and electric field–induced current (injection current) competes with the photoinduced current and at a certain applied voltage, the two currents cancel each other. This voltage is known as the open-circuit voltage, V_{oc}. The band gap between the LUMO of the electron-accepting molecule and the HOMO of the electron-donating polymer has a major effect on V_{oc}. (Brabec, 2001; Gadisa et al., 2004) while the working function of electrodes has only a weak influence on V_{oc} (Brabec et al., 2002). Due to the complex physical processes involved, a lot of factors affect J_{sc}, such as mobility of charge carriers, the molecular weight of the active materials (Schilinsky et al., 2005), morphology of the active layer (van Duren et al., 2004; Hoppe et al., 2005; Shaheen et al., 2001), and the electrodes used (Brabec et al., 2002).

There are two modes of operation. At 0 or slightly forward bias, the device will work in solar cell mode. Without external bias, incoming light will generate a photocurrent which can be used to power a circuit. Charge accumulation at the load resistance of the circuit will drive the diode into the forward bias region. In the case of photodetection, readout circuitry generally puts the photodiode into

10.4 Phototransduction with Organic Semiconductor Devices

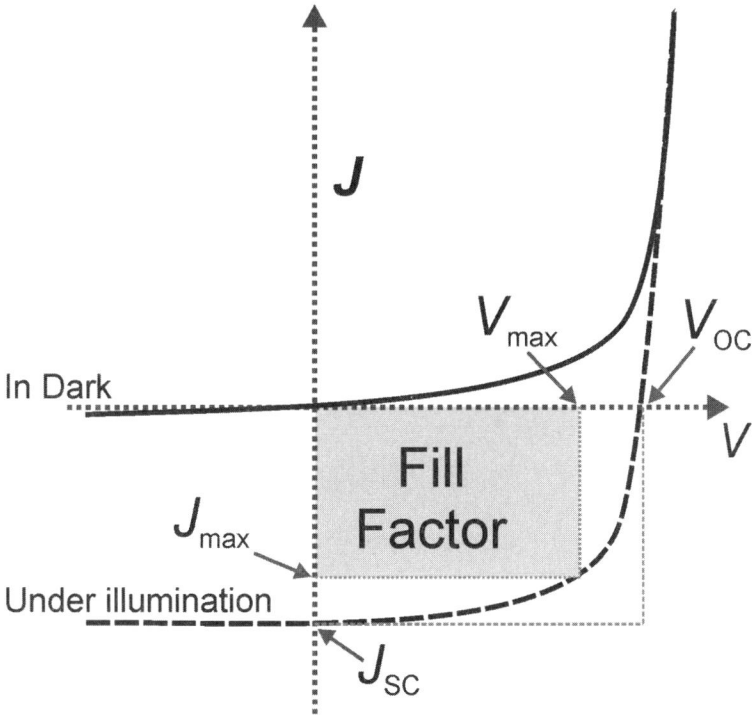

Figure 10.19 Typical J–V characteristic of a solar cell. The region to the right of the 0 bias point is the solar cell region, while the region to the left is the photodetection region.

reverse bias. In both cases, the photocurrent will give an indication of the light intensity.

The typical J–V curve for a diode in light and dark can be seen in Figure 10.19. Under illumination, the ratio of the maximum of the product of J and V to the products of J_{sc} and V_{oc} is referred to as the fill-factor, and reflects the charge transport properties, and the electronic quality of the diode:

$$FF = \frac{(J \cdot V)_{\max}}{J_{SC} \cdot V_{OC}} \qquad (10.6)$$

The internal photocurrent conversion efficiency (IPCE) is used to determine the efficiency of the device. The IPCE is defined as the ratio between the total number of collected electrons in an electric circuit and the total number of photons incident to a photodetector. Another related measure is the external quantum efficiency (EQE), which takes absorption, reflection, and light coupling losses into account. It can be determined by measuring the photoinduced current of a photodetector under illumination from an incident monochromic light under short circuit conditions and is given by:

$$EQE(\lambda) = 1{,}240 \frac{J_{sc}}{\lambda \cdot L_0} \qquad (10.7)$$

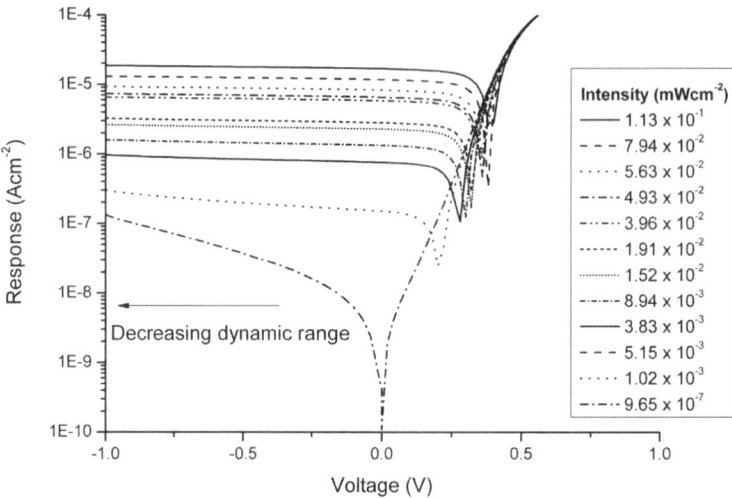

Figure 10.20 The current voltage characteristics of a P3HT:PCBM organic photodiode under varying intensity of illumination.

where J_{sc} is the photoinduced photocurrent density in units of $\mu A/cm^2$ under short circuit conditions, and λ and L_0 are wavelength and power density of the incident light, in units of nm and W/m², respectively.

Figure 10.20 shows the response characteristics of a high-quality ITO/PEDOT/P3HT:PCBM/LiF/Al heterojunction device, with a peak quantum efficiency of around 70%.

On a linear plot, the device exhibits diode-like characteristics. However, the logarithmic plot shows an increase in dark current with reverse bias. This is ex-

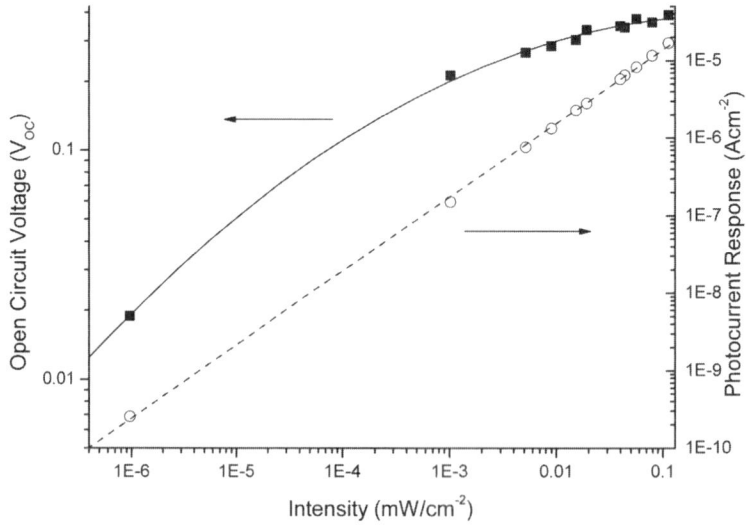

Figure 10.21 Variation of open circuit voltage and photocurrent with light intensity.

pressed in the diminishing dynamic range between the highest and lowest intensity response. The photocurrent response at zero bias displays a very linear dependence on light intensity, as can be seen in Figure 10.21. The open circuit voltage V_{OC} in contrast is strongly nonlinear. Thus, it is the linear response of the photocurrent characteristics over a very wide intensity range that makes these devices highly desirable for use as photodetectors.

10.4.4.2 The Equivalent Circuit and Shunt Resistance

A diode cannot just have good quantum efficiency at the required wavelengths to function as required. The response of the device will be heavily dependent on how it interacts with the readout circuitry. As photodiodes will be generally under reverse bias, properties such as the leakage current are almost equally as important. Figure 10.22 shows the equivalent circuit for an organic photodiode. This simple model consists of a diode, a light-modulated shunt resistance, sheet resistance, capacitance, and a light-modulated current source.

Physically R_{sheet} represents the interface resistance, R_{shunt} represents the shunt resistance, and I_{photo} represents the photocurrent. There is also a capacitance and a diode representing behavior under forward bias. In practice, each of these components is strongly correlated to the fabrication protocol, quality of material constituents, and device dimensions.

Typically R_{sheet} is between 100Ω and $1~k\Omega$ and is independent of illumination, field strength, device depth, or area. I_{photo} represents the photocurrent produced by exciton photogeneration, dissociation, and collection. Optimal efficiencies depend on device depth, but in general this component is dependent on illumination intensity and illumination area. There is also a slight field dependence as exciton dissociation is more efficient under an applied electric field (Peumans et al., 2004). R_{shunt} is the effective resistance to charge flow in the reverse direction. Under illumination there is an increased space charge and this causes the resistance to decrease. The device dimensions also have a direct influence. In addition, there may be a field dependence. The capacitance of the device is determined by the dielectric properties of the material constituents and dimensions. The forward

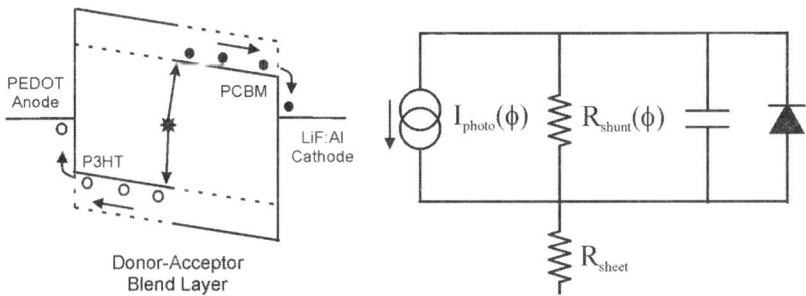

Figure 10.22 Simple energy level model for an organic photodiode based on a PCBM/P3HT blend, an ITO/PEDOT anode, and an LiF/Al cathode. Dotted lines in the band diagram represent the fact that both blend components potentially have contact with the cathode and anode. I_{photo} and R_{shunt} are both dependent on the light intensity ϕ.

bias characteristics will be determined by device dimensions and the nature of the interface, that is, whether or not there are energetic barriers to charge injection.

Figure 10.23 shows how the diode response behaves under reverse bias. As the bias increases, the leakage current across the shunt resistance increases. The response then splits into three regions: a linear region where the photocurrent greatly exceeds the leakage. a saturation region where the photocurrent is significantly less than the leakage, and a nonlinear region, where the photocurrent and leakage current are comparable. The sensitivity of the photodiode to light intensity is therefore greatly affected by the reverse bias across it.

The leakage current is the result of leakage over the shunt resistance of the photodetector. The shunt resistance is not fixed and is affected by temperature, light absorption, and field. This is because of the intrinsic nature of organic semiconductor materials. Thus, any increase in charge carriers can greatly increase conductivity. Figure 10.24 illustrates the relationship in a fabricated PCBM:P3HT device. The shunt resistance varies between 16 MΩcm^{-2} in the dark and 15 kΩcm^{-2} at an illumination intensity of 1 mWcm^{-2} × 10^{-4}mWcm^{-2}, and agrees with previous studies (Dyakonov, 2004; Riedel et al., 2004). The intensity dependence of the shunt resistance at low electric fields appears to be controlled by the following relationship:

$$R_{shunt} = \frac{1}{W \cdot L} \cdot \frac{R_{max}}{1 + m\phi^{0.9}} \qquad (10.8)$$

where W and L are the width and length of the photodiode, respectively, R_{max} is the shunt resistance in the dark, ϕ is the light intensity, and m gives the gradient

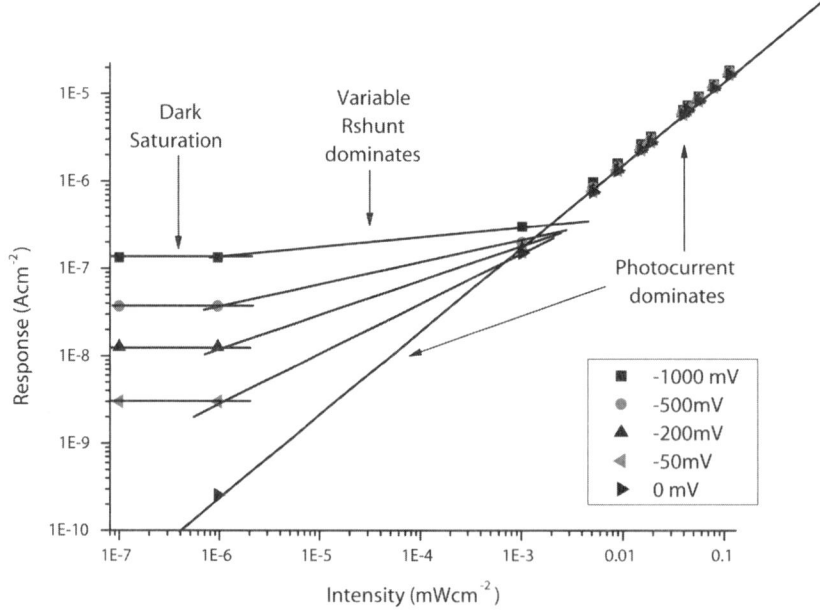

Figure 10.23 Nonlinear variations in photocurrent with photodiode bias.

10.4 Phototransduction with Organic Semiconductor Devices

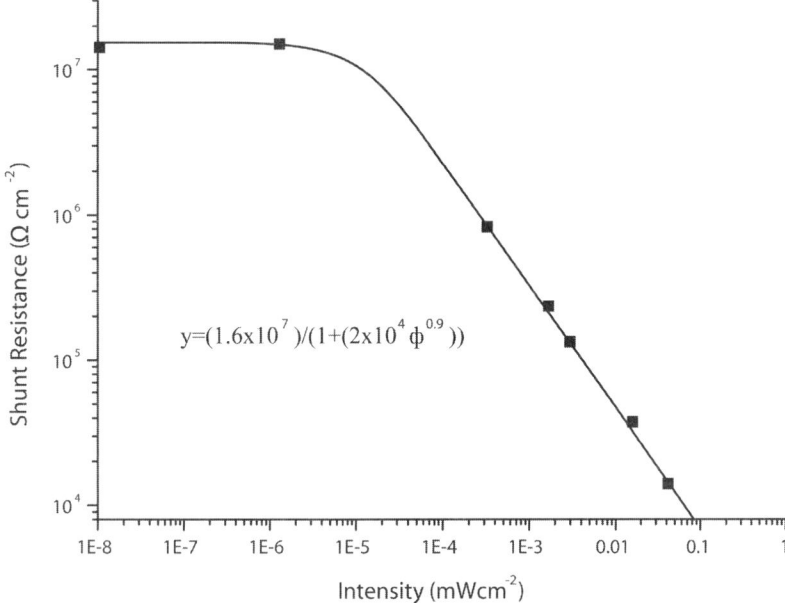

Figure 10.24 The variation in organic photodiode shunt resistance with light intensity for our P3HT:PCBM device.

of the almost linear intensity resistance relationship. This accords with an increase in carrier density under illumination, compelling higher leakage currents for any imposed bias. R_{max} and m are material and device architecture dependent

The effective response of the organic photodiode can be described in terms of three parallel current sources:

$$J = J_{Diode}(V) - J_{leakage}(\phi, V) - J_{photo}(\phi) \quad (10.9)$$

J_{Diode} describes the current voltage relationship for the diode. Its behavior depends on whether the contacts are ohmic or not. If there are Schottky barriers, these will be the limiting factor (Kwok) to the current, which will have an exponential relationship with the applied field of the form:

$$J_{Diode} = A^* T^2 \exp\left(\frac{-q\phi_b}{kT}\right) \left[\exp\left(\frac{qV}{nkT}\right) - 1\right] \quad (10.10)$$

where J_{Diode} is the current density, A^* is the Richardson constant, ϕ_b is the barrier potential, and V is the applied voltage. If instead there are ohmic contacts, then there will be a space charged limited current (Rose, 1955) given by:

$$J_{Diode} = \frac{9}{8}\varepsilon\varepsilon_0\mu_0 \exp\left(\frac{V^2}{d^3}\right) \quad (10.11)$$

where d is the film thickness. This equation is altered by Poole Frenkel field dependence at medium fields, trap filling effects, and at low fields follows a more ohmic dependence (Rose, 1955). These relationships become important when the diode moves into forward bias but do not apply at negative bias. The photocurrent output is simply given by:

$$J_{Photo} = \alpha(\lambda) \cdot \phi \tag{10.12}$$

Here α is the photoresponse in amps per watt of illumination. This is strongly dependent on wavelength and the external quantum efficiency. ϕ is the light intensity. This equation corroborates empirically with the results shown in Figure 10.25.

Finally the leakage current is simply given by an ohmic current voltage relationship involving the shunt resistance:

$$J_{leak} = \frac{V_{bias}}{R_{shunt}(\phi)} = V_{bias} \cdot \left(\frac{1 + m\phi^{0.9}}{R_{max}} \right) \tag{10.13}$$

As the core materials in organic photodetectors are intrinsic semiconductors, the base conductivity will be affected by current, field, and light intensity. This will need to be taken into account in final device structures.

As organic semiconductors can be considered instrinsic systems, their current response will be strongly influenced by anything which inserts charge into the system. More importantly, for a given bias voltage, the resistivity of the channel will change according to light intensity, temperature, and bias itself. This can be

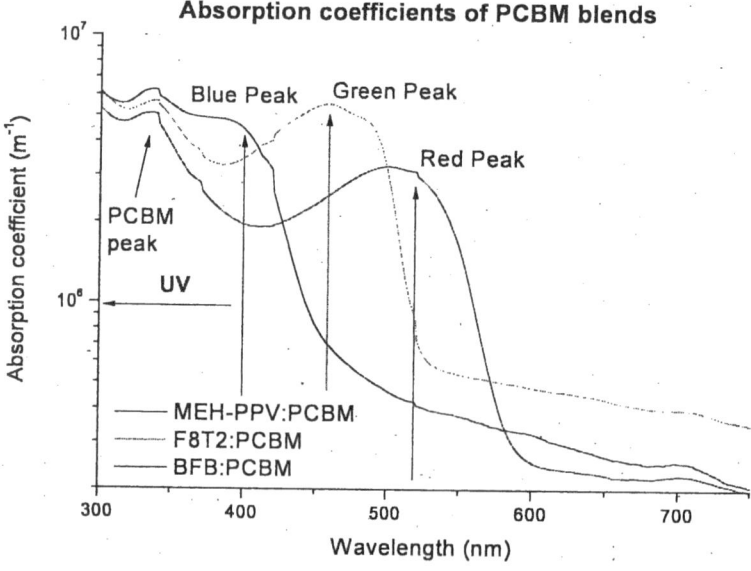

Figure 10.25 Spectral response for the differing polymer components. PCBM is the electron transporter that has a band gap in the UV region. Thus, chromatic variance is generally achieved via the polymers. There is residual subband gap absorption due to stimulation of trapping and impurity states.

borne out by the equations above. In Figure 10.20, it is shown that the best dynamic range can be achieved at zero bias. This is because leakage current will be zero for zero bias. In practice, however, circuit readout systems tend to have a few volts of reverse bias that can reduce the ultimate sensitivity of the system. In the fabrication section we discuss a new fabrication method to decrease this leakage and in the next chapter we also look at circuit methods for achieving a very small bias.

10.4.4.3 Spectral Response Characteristics

The optical band gap of silicon is 1.1 eV, which means that silicon will generally absorb light wavelengths from the near infrared through to the ultraviolet. There is a variance of absorption between the wavelengths. Thus it is possible to vary spectral sensitivity by changing the depth of the p–n junction. Absorption filters are, however, more commonly used.

In contrast, organic semiconductors have optical band gaps in the visible range. Thus, by tuning the band gap for a desired wavelength range, it is possible to obtain chromatic separation between photodetectors. The absorption coefficients for different organic blend materials can be seen in Figure 10.25.

The color response of these photodiodes is not RGB but white, blue–green, blue. This is closer to the color scheme in the human eye. Simple image processing can easily convert these responses into the RGB or YUV color schemes typically used for image processing. Such detection systems are capable of detecting many decades of intensity difference, which can therefore easily achieve the 8-bit dynamic ranges in present imaging systems.

10.4.5 Fabrication

The field of organic semiconductor optoelectronics is attracting increasing attention for many application sectors including displays, lighting, electronics, photodetection, solar energy conversion, and communications (DeGrans and Schubert, 2004). A strong attraction lies in the potential to form physically flexible, plastic devices. In all cases a low-cost, reliable fabrication method is required for these devices to be able to compete with the entrenched silicon and other inorganic semiconductor device technologies. The ideal organic layer deposition technique should include the following: compatibility with high throughput (e.g., reel to reel) processing, ability to create multilayer and multielement structures, and compatibility with atmospheric processing conditions (i.e., not requiring costly vacuum environments). The manufacture of high-quality devices also requires deposition techniques that can produce uniform, substantially defect-free layers of micron and sub-micron thicknesses and control their location on the substrate with high spatial resolution.

Many deposition techniques have been developed to achieve the specifications above. These include inkjet and ink dispenser printing for multicolor light emitting diodes (DeGrans and Schubert, 2004), doctor blade deposition (Padinger et al., 2002), and gravure printing (Makela, 2001) techniques for polymer transistor fabrication. However, these techniques involve depositions of solutions, and are not suitable for multilayer deposition. The solvent used for one polymer can also dissolve or partially dissolve previously deposited polymer layers.

10.4.5.1 Contact Printing

An exciting new technique of depositing solid-state films of conducting polymers is a stamping method known as heat and plasticizer assisted micro contact printing (H&mPA-μCP) (Chen et al., 2007). This recent development can be used to address many problems outlined above. It provides a processing method for organic semiconductor devices that can allow straightforward fabrication of single or multilayer patterned structures of commonly used solution processible materials. The process can be used with both water and organic-solvent soluble materials. Additionally it can demonstrate clear advantages relative to current methods of organic semiconductor device fabrication.

PDMS stamps are the bedrock of microcontact printing techniques. Primarily, PDMS stamps are in a semiliquid state at room temperature as the glass transition temperature of PDMS is $-120°C$. This makes it is easy to make conformal contact with flat substrates. Secondly, its surface is hydrophobic, and can be converted temporally into a hydrophilic form after oxygen or air plasma treatment. This results from a very thin (2 nm) and rigid SiOx layer forming on the surface of the stamp (Olah et al., 2005). This very thin layer is, however, unstable. Small oligomer molecules from the PDMS bulk will soon diffuse to the surface, which then reverts to its hydrophobic form. A film deposited on the surface of a PDMS stamp, therefore, gradually reduces its adhesion to the stamp because of the diffusion of these oligomer molecules to the interface. It can then be readily transferred

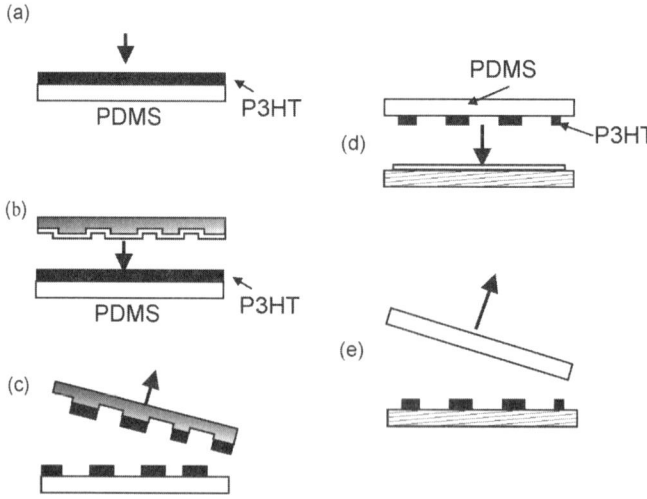

Figure 10.26 The process of patterning P3HT on a PDMS stamp. (a) Deposit active polymer (e.g., P3HT) onto a flat PDMS stamp. (b) Press patterning stamp onto the deposited polymer. (c) Remove the patterning stamp to reveal the desired pattern. (d) Press the flat stamp with patterned polymer onto the destination substrate. (e) Remove the flat PDMS stamp.

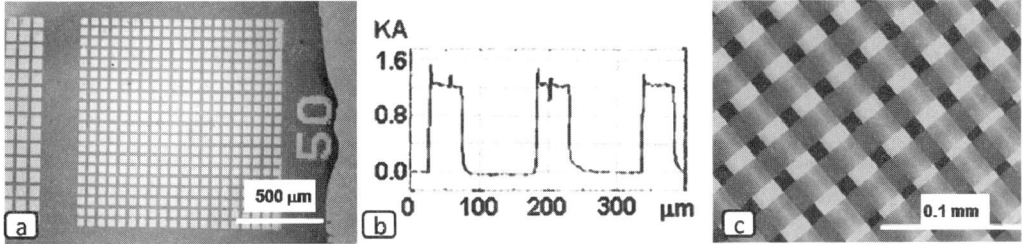

Figure 10.27 (a) The optical image of patterned P3HT:PCBM on a glass substrate. The area of open small squares in (a) is 100 mm² × 100 mm² with a 50 nm rim. (b) The thickness of patterned film in (c). (c) Double printed strips of P3HT:PCBM crossing each other.

to a target substrate. The H&mPA-μCP process can be divided into 4 steps, as shown in Figure 10.26.

Step (1): Spin coat the polymer layer on a flat PDMS stamp that has been pretreated with oxygen plasma.
Step (2): Treat a patterning stamp with oxygen plasma and attach it to the polymer layer on the flat PDMS stamp.
Step (3): Remove the patterning stamp to leave a patterned polymer layer on the flat PDMS stamp.
Step (4): Attach the patterning stamp to the destination substrate and then remove it to leave a patterned polymer layer.

The technique above has to be fine-tuned with plasticizer and heating steps as outlined in (Chen et al., 2007), but can produce impressive results. Figure 10.27 shows an optical image of printed patterned P3HT:PCBM on a glass substrate. Figure 10.27(a) shows that it is possible to print a high-quality pattern over a large area with high-quality (b). In addition, Figure 10.27(c) shows that it is possible to double print layers of P3HT on top of each other. The light red color in the optical imaging is from the P3HT:PCBM (1:1) blend film, and the light blue from the glass substrate. The optical images clearly show the fidelity of the pattern.

The total area of printed polymer is determined by the stamp size. In Figure 10.28 the printed polymer area is 1.2 cm² × 1.2 cm². For larger areas the thickness of the PDMS stamp has to be increased or else a backplane has to be supplied to keep uniformity. The smallest P3HT:PCBM lateral patterning dimension is limited by the mask resolution, which is 50 μm in the example above. Submicron dimension lateral patterns should be available using this technique with better mask technology, though it is not known at present how far into the nanoscale this technique will remain valid.

Polymer film thickness and quality are essential parameters for functional polymers, such as conducting polymers used as light emitting diodes, photodiodes, and transistors. Very thin but high-quality polymer films are desirable for polymer electronics possessing lower, driving voltage and less leakage current properties.

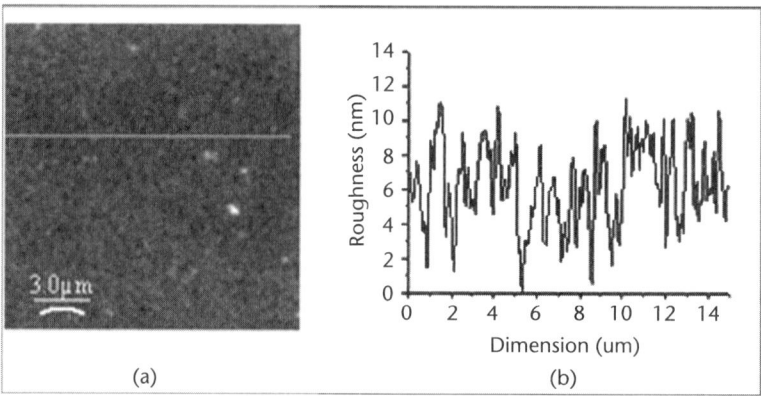

Figure 10.28 The AFM topography (a) and height profiles (b) of triple printed trilayer P3HT:PCBM/P3HT/PEDOT-PSS on ITO/substrate. The surface of the printed films is smooth. The average RMS roughness of the printed film is around 2.9 nm (8 nm peak to peak). No pin holes are observed on these printed polymer films. This is also confirmed by the later prepared polymer device, which has significantly lower leakage current.

Using this technique it is possible to contact print smooth polymer films from 15 to 220 nm.

The ability to deposit high-quality multilayer structures is essential for optimizing polymer electronics performance and efficiency (Chen et al., 2000). As stressed previously, electronic factors such as shunt resistance are very important to any final device. These properties are derived partially from the inherent quality of the materials, but also in the device structure. A major advantage with the printing technique is the ability to improve such engineering properties through the fabrication of multilayer devices, previously not possible with spin coating techniques.

10.4.5.2 Printing on CMOS

In intelligent CMOS imaging chips, there is competition between space allocated for processing and that allocated for photodetection. It is hard to grow good-quality crystalline silicon on top of amorphous silicon oxides and as a result CMOS designers only have a two-dimensional plane with which to construct both detection and processing devices. As described above, if the photodetectors are small, then their ultimate signal-to-noise ratio will also be small. If they are big, then the pixel density of the imager will be small. As a result, three-dimensional integration of photodetectors on the surface of CMOS chips would greatly improve this situation. Then the limiting factor would be purely circuit size rather than photodiode size.

However, given the ability to postfabricate organic semiconductor devices on top of silicon, it is possible to place the photosensors on top of the CMOS chip and leave the valuable silicon real estate purely for processing. Techniques such as the printing technique discussed above are now beginning to make it possible to print active layers on the surface of CMOS chips, and thus make novel photosensor devices. The scheme can be seen in Figure 10.29. Bondpad openings can be used

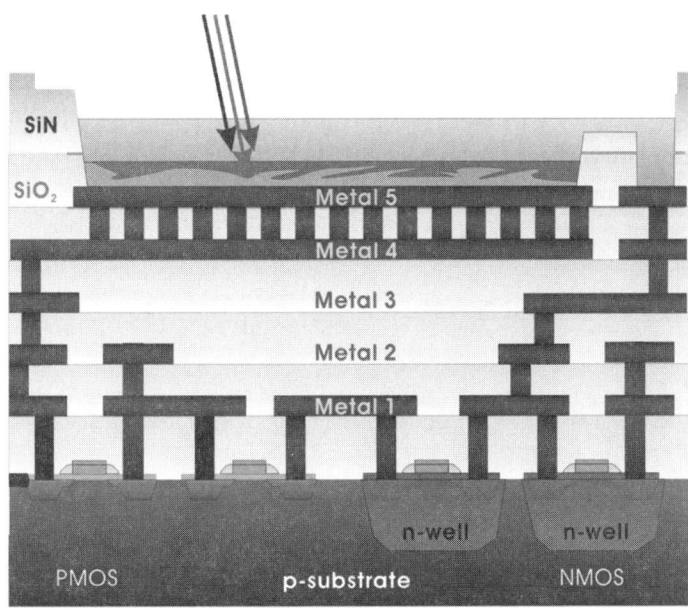

Figure 10.29 The structure of a CMOS/organic hybrid imaging chip. Metal 5 acts as a bondpad and a base metal layer for the photosensor device. Active polymer layers are then deposited on top, followed by a final transparent metal layer.

to deposit organic material with via interconnects to make connections with the transistors below.

Vertically integrated photodiodes deposited on a CMOS imaging device can occupy large portions of the surface. Without the real estate constraints of transistor-level photodiodes, their size may be increased by a factor of four or more. This will significantly increase the photodiode pixel fill-factor to above 90%. Additionally, if organic photodiodes can be deposited into specific pixel well locations and different materials are used corresponding to red, green, and blue spectral absorption, then wavelength filtering will not be required. Finally, as the photodiode sits on top of the dielectric stack, there will be no interference effects on the absorption spectrum. All this will significantly increase the light EQE of surface deposited photodiodes in comparison to transistor-level diodes.

10.5 Conclusions

In this chapter we show how the architecture of the eye allows for detection of single photons while in comparison commercial imaging cameras tend to have minimum sensitivity in the millions of photons. New developments in organic semiconductor imaging could allow for three-dimensional imaging systems to develop, which would improve intelligent imaging systems that have processing in the imaging plane.

The really interesting question for the research community would be: can we develop a photosensor with light adaptivity and internal gain similar to that of rods and cones? Even in the human eye rods and cones function at different ends of the intensity scale, so a further question would be: would it be possible to integrate a matrix of normal photodiodes with avalanche photodiodes in order to have a camera which can function both at low and high light intensities?

References

Adams, J., Parulski, K. and Spaulding, K. (1998) "Color processing in digital cameras." *Micro, IEEE*, 18, 20-30.

Bartek, M., et al. (1994) "An integrated silicon colour sensor using selective epitaxial growth." *Sensors and Actuators A: Physical*, 41, 123-128.

Baylor, D. A. (1987) "Photoreceptor signals and vision. *Investigative Ophthalmology and Visual Science*, 28, 34-49.

Blanksby, A. J. and Loinaz, M. J. (2000) "Performance analysis of a color CMOS photogate image sensor." *Electron Devices, IEEE Transactions on*, 47, 55-64.

Blanksby, et al. (1997) "Noise performance of a color CMOS photogate image sensor." *Electron Devices Meeting, 1997. Technical Digest., International*.

Brabec, C. J., et al. (2002) "Effect of LiF/metal electrodes on the performance of plastic solar cells." *Applied Physics Letters*, 80, 1288.

Brabec, C. J. C., et al. (2001) "Origin of the open circuit voltage of plastic solar cells." *Advanced Functional Materials*, 11, 374-380.

Breeze, A. J., et al. (2004) "Improving power efficiencies in polymer—Polymer blend photovoltaics." *Solar Energy Materials and Solar Cells The Development of Organic and Polymer Photovoltaics*, 83, 263-271.

Burns, M. E. and Arshavsky, V. Y. (2005) "Beyond counting photons: trials and trends in vertebrate visual transduction." *Neuron*, 48, 387-401.

Cao, Y., et al. (1999) "Improved quantum efficiency for electroluminescence in semiconducting polymers." *Nature*, 397, 414-417.

Chang, S. C., et al. (1999) "Multicolor organic light-emitting diodes processed by hybrid inkjet printing," *Advanced Materials*, 11, 734.

Chen, L., Degenaar, P. and Bradley, D. (2007) A method for patterning organic semiconductor materials. Patent pending.

Chen, L., et al. (2000) *Advanced Materials*, 12, 1367.

Chouikha, M. B., et al. (1998) "Color detection using buried triple pn junction structure implemented in BiCMOS process." *Electronics Letters*, 34, 120-122.

Degrans, B. J. and Schubert, U. S. (2004) "Inkjet printing of well-defined polymer dots and arrays," *Langmuir*, 20, 7789.

Dyakonov, V. (2004) "Mechanisms controlling the efficiency of polymer solar cells." *Applied Physics A: Materials Science and Processing*, 79, 21-25.

Fain, G. L., et al. (2001) "Adaptation in vertebrate photoreceptors." *Physiological Reviews*, 81, 117-151.

Filipek, S., et al. (2003) "G protein-coupled receptor rhodopsin: a prospectus." *Annual Review of Physiology* 65, 851-879.

Findlater, K. M., et al. (2003) "A CMOS image sensor with a double-junction active pixel." *Electron Devices, IEEE Transactions on*, 50, 32-42.

Gadisa, A., et al. (2004) "Correlation between oxidation potential and open-circuit voltage of composite solar cells based on blends of polythiophenes/fullerene derivative." *APL*, 84, 1609.

Granstrom, M., et al. (1999) "High efficiency polymer photodiodes." *Proceedings of the 1999 International Conference on Science and Technology of Synthetic Metals (ICSM-98), Jul 12-Jul 18 1998*, 102, 957-958.

Hamer, R. D. (2000) "Computational analysis of vertebrate phototransduction: combined quantitative and qualitative modelling of dark- and light-adapted responses in amphibian rods." *Visual Neuroscience*, 17, 679-699.

Hamer, R. D., et al. (2005) "Toward a unified model of vertebrate rod phototransduction." *Visual Neuroscience*, 22, 417-436.

Hamer, R. D., et al. (2003) "Multiple steps of phosphorylation of activated rhodopsin can account for the reproducibility of vertebrate rod single-photon responses." *The Journal of General Physiology*, 122, 419-444.

Hoppe, H., et al. (2005) "Kelvin probe force microscopy study on conjugated polymer/fullerene bulk heterojunction organic solar cells." *Nano Letters*, 5, 269-274.

Huynh, W. U., Dittmer, J. J. and Alivisatos, A. P. (2002) "Hybrid nanorod-polymer solar cells." *Science*, 295, 2425-2427.

Jones, R., (1995) "A superficial look at polymers." *Physics World*, 8, 47-51.

Kaupp, U. B. and Seifert, R. (2002) "Cyclic nucleotide-gated ion channels." *Physiological Reviews*, 82, 769-824.

Kolb, H., Fernandez, E. and Nelson, R. Webvision—The Organization of the Retina and Visual System. http://webvision.med.utah.edu/.

Komatsu, T., et al. (2004) "Photoresponse studies of bulk heterojunction organic photodiodes" *Japanese Journal Of Applied Physics Part 2-Letters and Express Letters*, 43 (11a), L1439-L1441 Nov 1.

Krueger, H., et al. (2004) "Some new electron-affine polymers for organic photovoltaics." *Organic Photovoltaics IV*, Aug 7-8 2003, San Diego, CA.

Kwok, K. N. (2002) *Complete Guide to Semiconductor Devices*, Wiley-IEEE Press.

Lamb, T. D. (1994) "Stochastic simulation of activation in the G-protein cascade of phototransduction." *Biphysical Journal*, 67, 1439-1454.

Lamb, T. D. and Pough, E. N. (1992) "G-protein cascades: gain and kinetics." *Trans. Neur. Sci*, 15, 291-297.

Lu, G. N., et al. (1996) "Colour detection using a buried double p-n junction structure implemented in the CMOS process." *Electronics Letters*, 32, 594-596.

Magnan, P. (2003) "Detection of visible photons in CCD and CMOS: a comparative view." *NuclearInstrum. and Methods in Physics Research A*, 504, 199.

Montag, E. (2007) course no. 1050-724: "Vision and Psychophysics." http://www.cis.rit.edu/people/faculty/montag/vandplite/course.html.

Mozer, A. J., et al. (2005) "Charge transport and recombination in bulk heterojunction solar cells studied by the photoinduced charge extraction in linearly increasing voltage technique." *Applied Physics Letters*, 86, 112104-112101.

Nikolic, K. and Forshaw, M. (2005) "New imaging devices and image processing challenges. *Lecture notes: 'From nanostructures to nanosensing applications,'* " *Amsterdam and Bologna, IOS Press and Societa Italiana di Fisica*, 263-318.

Nikonov, S., Lamb, T. D. and Pugh, E. N. (2000) "The role of steady phosphodiesterase activity in the kinetics and sensitivity of the light-adapted salamander rod photoresponse." *Journal of General Physiology*, 116, 795-824.

Olah, A., Hillborg, H. and Vancso, G. J. (2005) "Hydrophdoic recovery of UV/ozone treated poly (dimethylsiloxane): adhesion studies by contact mechanisms and mechanism of surface modification," *Applied Surface Science*, 239, 410.

Padinger, F., Rittberger, R. S. and Sariciftci, N. S. (2003) "Effects of postproduction treatment on plastic solar cells." *Advanced Functional Materials*, 13, 85–88.

Pettersson, L. A. A., Roman, L. S. and Inganäs, O. (1999) "Modeling photocurrent action spectra of photovoltaic devices based on organic thin films." *Journal of Applied Physics*, 86, 487.

Peumans, P., Yakimov, A. and Forrest, S.R. (2004a) "Small molecular weight organic thin-film photodetectors and solar cells." *Journal of Applied Physics* 95 (5), 2938–2938.

Pudas, M., et. al., (2005) "Gravure printing of conductive particulate polymer inks on flexible substrates," *Progress in Organic Coatings*, 54(4), pp. 310–316.

Poenar, D. P., French, P. J. and Wolffenbuttel, R. F. (1997) "Colour sensors based on active interference filters using silicon-compatible materials." *Sensors and Actuators A: Physical*, 62, 513–523.

Pye, D. (2003) "Amazing octave." *Physics World*, 56.

Ravirajan, P., et al. (2005) "Efficient charge collection in hybrid polymer/TiO/sub 2/ solar cells using poly(ethylenedioxythiophene)/polystyrene sulphonate as hole collector." *Applied Physics Letters*, 86, 143101–143101.

Riedel, I., et al. (2004) "Polymer solar cells with novel fullerence-based acceptor." *Thin Solid Films*, 451, 43–47.

Roman, L. S., et al. (1998) "High quantum efficiency polythiophene." *Advanced Materials*, 10, 774–777.

Rose, A. (1955) "Space-charge-limited currents in solids." *Physical Review*, 97, 1538–1544.

Schilinsky, P., et al. (2005) "Influence of the molecular weight of poly(3-hexylthiophene) on the performance of bulk heterojunction solar cells." *Chemistry of Materials*, 17, 2175–2180.

Shaheen, S. E., et al. (2001) "2.5% efficient organic plastic solar cells." *APL*, 78, 841.

Simpson, M. L., et al. (1999) "Application specific spectral response with CMOS compatible photodiodes." *Electron Devices, IEEE Transactions on*, 46, 905–913.

Steiger, J., Heun, S. and Tallant, N. (2003) "Polymer light emitting diodes made by inkjet printing," *Journal of Imaging Science and Technology*, 47, 473.

Sze, S., M. (1981) *Physics of Semiconductor Devices,* New York, John Wiley and Sons.

Van Duren, J. K. J., et al. (2004) "Relating the morphology of poly(p-phenylene vinylene)/methanofullerene blends to solar-cell performance." *Advanced Functional Materials*, 14, 425–434.

Waser, R. (2003) "Sensor arrays and imaging systems." *Nanoelectronics and Information Technology*, ed. R. Waser, Wiley-VCH 2003, 813–819.

Wong, H.-S. (1996) "Technology and device scaling considerations for CMOS imagers." *IEEE Transactions on Electron Devices*, 43, 2131.

Woo, H. S., et al. (1993) "Optical spectra and excitations in phenylene vinylene oligomers." *Synthetic Metals*, 59, 13–28.

CHAPTER 11
Analog Retinomorphic Circuitry to Perform Retinal and Retinal-Inspired Processing

Patrick Degenaar and Dylan Banks

11.1 Introduction

The eye is a biological sensory system which has independently evolved over fifty times (Hofmann et al., 2005). Of mammals with sight, ten distinct and separate vision architectures have evolved. The ability to see is a strategic evolutionary advantage, which is so important that all animals who inhabit our world in lit conditions have developed some form of sight. In some cases, in dark areas such as the deep sea, creatures have developed light emissive processes to complement rudimentary sight. Each of these vision sensors optimize, through micromutations, to work effectively within their given "boundary conditions." The trade-offs include energy cost, signal processing, image sensing and processing, information transfer, noise, and size (Darwin, 1975).

The expression "eyes like a hawk" is often used to describe individuals with particularly good sight. Hawks are birds with highly acute vision, yet have smaller eyes (so less light can enter) and a smaller visual cortex (so, in theory less processing occurs), yet they apparently have visual acuity that is much better than that of a human. Why haven't humans evolved such sight? Perhaps the answer is subtle. The visual system of birds of prey is designed to achieve certain tasks, notably detecting the motion of small objects, and high-speed imaging. In contrast, humans need to be able to operate over a wide variety of spatial scales, and within a wide range of contexts. Thus, visual systems in different animals are optimized to maximize their function given the limitations of size, and energy investment.

These trade-offs are a necessary function of optimization within all engineering development systems. Retinal signal processing operates in a wet biological technology platform. The evolved vision architecture optimizes our ability to see within this environment, both to its relative benefits and disadvantages. The human technology is warm, between $36.1°$ and $37.8°C$. At these temperatures, molecules frequently change conformation and state as low-energy polar and nonpolar (but noncovalent) bonds, break, and reform. Relative to silicon, there is high entropy within our bodies, creating an environment with increased dynamic noise. Biological signal processing has developed techniques to negate the reduced dynamic range within static environments. These techniques may be directly implemented within artificial systems, increasing information dynamic range, but trading off against temporal resolution.

Both silicon circuits and biological neurons use charge distribution and current flow to process signals, but the intrinsic resistance of a biological system is far higher than a silicon one. Neurons designed to convey electrical signals in biological technologies have an intrinsic resistivity of about 100 ohms per centimeter. The width of a neuron fiber containing axoplasm, which is loosely analogous to an electrical wire connecting several components of a system, may be 0.1 to 20 microns in diameter. For a 1-micron diameter fiber the electrical resistance per unit length is 10^{10} ohms per centimeter (Nicholls et al., 1992), causing significant inhibition to the conductance of electrical signals. Aluminum wires that connect circuits on many integrated circuits (ICs) have an intrinsic resistance of 2.4×10^{-10} ohms per centimeter, allowing conduction of electrical signals over far greater distances within an artificial technology. The human body negates this problem by transferring charge over large distances by sequential diffusion of ions over short distances (action potentials).

Inorganic semiconductor and wet biological technologies are not only significantly different in electrical terms. Structurally, wet biology can provide massive parallelism and interconnectivity in a way that is not possible in the silicon case with present technology. A typical neuron can have hundreds of individual connections, whereas silicon connectivity is at least an order of magnitude worse off. In contrast, the speed of inorganic semiconductor technology allows for structures not possible within wet biology. In the case of the human retina, information extraction is achieved through a million parallel neurons. In the case of the typical CMOS (complementary metal oxide semiconductor) or CCD (charge coupled device) imaging chip, a single or duplex high-speed serial link with a couple of power lines is more common. Thus, the purpose should not be to simply map biological processing structures directly onto silicon. Instead, it is better to understand the algorithms and their associated trade-offs and implement them taking the capabilities and drawbacks of CMOS technology into account.

In this chapter we will first summarize the human eye from an engineering perspective. Then the implementation of the processing elements will be investigated in detail. The method of phototransduction will have been investigated in the preceding chapter.

11.2 Principles of Analog Processing

Before an informed decision of which image processing pathways may best be implemented, an understanding of the environment in which they will be fabricated is necessary.

Most electronic imaging devices have similar distinguishable functions: they collect light (usually with a lens system), they may filter photons into separate wavelength regions, and they can absorb photons in an active region and can extract the intensity information in some way. Within semiconductor and organic photodetector materials, photons are converted into electrons, electrons are recorded and processed, converted into a digital signal, and output for further processing.

Power consumption is a major factor affecting the technology choices. Comparing charge coupled devices (CCDs) and CMOS cameras, a low-power, million pixel CCD may consume hundreds of mWatts, outputting an image at seven to eight frames per second (SANYO, 2004), whereas a CMOS device consuming tens of mWatts can perform a similar function at hundreds of Hz (MICRON, 2005), offering an order of magnitude improvement in power consumption (Ackland and Dickinson, 1996; Fossum, 1995). Part of this can be explained by improvements to both techniques, but CMOS image sensors hold several explicit advantages over CCD counterparts.

- *Maturity:* CCD imaging technology is 50 years old but CMOS has been the backbone technology of the microprocessor industry with huge quantities of investment.
- *Circuitry:* CMOS imaging chips can have built-in processing circuitry, whereas CCD chips can have only the most basic circuitry.
- *Readout:* CCD readout is limited to a row raster, whereas CMOS allows for many interesting possibilities such as asynchronous address.

Most commercial CMOS image sensors are concerned with imitating the CCD readout method and packing high pixel densities to achieve megapixel resolutions in a small pixel area. However it is possible to implement highly efficient image processing in the imaging plane. The parallel nature of such distributed amplification and processing circuitry facilitates high-speed processing with extremely low-power consumption. Although it is not always necessary to operate at high frame rates (25 to 100 Hz is fine for most visual stimulus), it can have advantages for high-speed imaging applications. A standard CCD device with external processing software will scale image processing time directly with the number of pixels, as pixel information is output sequentially. With a distributed processing system this computation takes place at the focal plane and in real time, significantly reducing the post-IC processing required. Bus speeds will still limit data communication between the imaging device and off-chip processing units such as a reconfigurable logic or DSP device, but the ability to perform image-processing computation without scaling time constraints compels CMOS as an attractive option for fabrication.

Significant differences exist between wet chemical technology used by biological systems and the inorganic technology platform of silicon. Neverless, there are mathematical algorithms which are applicable to both. Therefore it is neither desirable, nor feasible to exactly replicate biological neurons in silicon. It is, however, possible to mimic the basic functionality of neuronal behavior such as depolarization, hyperpolarization, excitation and inhibition, and action potentials. By combining these complex excitatory and inhibitory neuronal feedback systems may be implemented within CMOS.

Neurons perform basic mathematical functions: they add, subtract, divide, perform exponentials, inhibit, and excite and perform low pass filtering. All of these functions are achievable with MOSFET transistor circuits if arrayed in the correct fashion. There are CMOS characteristics that are analogous to neuronal information transfer and mathematical computation. The next step towards developing a neuromorphic CMOS methodology is the understanding of interneuron computation and the high-level functions they perform within the visual system—essentially investigating how neurons combine and process input signals so that we may see.

Transistors can be operated in two modes: the voltage switch (digital) and the voltage-controlled transconductor (analog). Transistors in both forms of operation can achieve complex mathematical function. In much of today's computational machines the accuracy of digital logic is preferred. However, analog logic can be compact and consume very low power. It is these characteristics that will be investigated in this chapter.

11.2.1 The Metal Oxide Semiconductor Field Effect Transistor

The metal oxide semiconductor field effect transistor (MOSFET) is the most significant large-scale integrated circuit device in terms of cost and mass production. It is a four terminal structure with a source, drain, gate, and bulk. The standard nMOS device (Figure 11.1) has a p-type semiconductor substrate (where dopants create an excess of holes) and n-type implants (where dopants create an excess of electrons). These materials are used to form the source and drain terminals. Together these layers act to form p–n diode junctions between the substrate, and the relative implants. An oxide layer between the gate and channel gives high gate input impedance, and in an ideal mode acts as a capacitor dielectric allowing charge buildup on the gate to influence the effective channel conductivity, controlling the channel drain source current (I_{ds}).

Carriers of one polarity predominantly transport current through a MOSFET: electrons in an nMOS device, and holes in a pMOS device. MOSFETs are available in two complementary forms: a p-channel (pMOS)-type, where a zero gate voltage enhances channel conductivity; and an n-channel variety (nMOS), where a positive gate voltage enhances channel conductivity.

Within a given CMOS process, engineers can only change the channel length and width characteristics of the transistors. To affect other characteristics, such as doping and gate oxide thickness, the designer has to choose from different preset foundry processes. These basic device parameters will affect the channel impedance and transistor transconductance characteristics. Transconductance is the ratio of drain current to gate voltage change over an arbitrarily small interval.

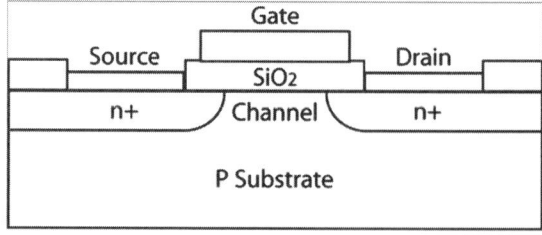

Figure 11.1 The field effect transistor, showing the channel, source and drain terminals, and the gate. The silicon dioxide layer induces charge collection on the gate, which affects the effective channel width.

11.2.1.1 Transistor Operation

The threshold voltage of a MOSFET is the gate voltage which creates a charge balance of electrons or holes within a depletion region in the substrate. At this point in an nMOS device, the concentration of holes in the gate is equal to the concentration of electrons in the substrate. The positive threshold voltage on the gate compels negative electrons within the substrate to move toward the gate side of the channel to equal the opposing charge within the gate, reducing the resistance of channel current flow through the channel laterally.

If the gate voltage is below the threshold voltage, the transistor is "off." Ideally there is no drift current through the channel except for a leakage current. If the gate voltage is larger than the threshold voltage, the transistor is on, increasing the electron density in the channel and, consequently, current flow.

As the voltage between the drain and the source (V_{ds}) increases, the gate to drain (channel) voltage decreases. At pinch off, the channel close to the gate–drain junction becomes completely depleted of charge carriers. Electrons travelling through the pinched off region are velocity saturated such that, if the drain to gate voltage increases above the pinch off voltage, the charge carrier concentration within the channel remains constant. Within this active region, the drain current is largely independent on the V_{ds} as the drain to source voltage effect saturates.

11.2.1.2 nMOS and pMOS Devices

In an n-channel (nMOS) device, positive gate voltages induce negative charge build-up in the channel, enhancing the drain to source current by reducing channel resistance (inversion). In a p-channel (pMOS) device, relative negative (below V_{dd}) gate voltages induce positive charge build-up in the channel enhancing the drain–source current. These operate broadly as the antithesis of the nMOS device, again with three regions of inversion but with different transconductance characteristics.

11.2.1.3 Transconductance Characteristics

The channel charge density is proportional to the difference between the threshold voltage and the gate voltage (effective gate–source voltage V_{eff}).

$$V_{eff} - V_{gate-source} \quad V_{threshold} \tag{11.1}$$

Conductance is a measure of the flow of current through two points across an applied voltage and is the reciprocal of resistance. Transconductance (g_m) is, more generally, described as a ratio between voltage and current changes (11.2). This can be described mathematically as:

$$g_m = \frac{\Delta I}{\Delta V} \tag{11.2}$$

where ΔI is the small change in current and ΔV is the small change in voltage.

The transfer conductance (transconductance, g_m) of a transistor is the function of gate voltage on drain current (11.3).

$$g_m = \frac{2I_{ds}}{V_{eff}} \quad (11.3)$$

where I_{ds} is the drain current and V_{eff} is the effective gate–source voltage.

11.2.1.4 Inversion Characteristics

As the gate to source voltage increases, channel inversion occurs gradually and is described by three regions: strong, moderate, and weak inversion. Where the effective voltage (V_{eff}) is larger than 150 mVolts, the transistor operates in strong inversion. Here, transconductance varies as a square law; at approximately 100 mVolts below V_{eff} is the weak inversion region where transconductance varies as an exponential; between these regions moderate inversion is observed with varying transconductance characteristics between the square law of strong inversion and the exponential of weak inversion. These regions can be used to apply mathematical functions on the voltage and current passing through them (Figure 11.2).

Weak Inversion

$$I_{ds} = \beta(2n)\phi_t^2(n-1)\left(\exp\frac{V_{gs}-V_t}{n\phi_t}\right)\left(1 - \exp\frac{-V_{ds}}{\phi_t}\right) \quad (11.4)$$

Moderate Inversion

$$I_{ds} = \beta\phi_t^2\left((V_{gs}-V_t)V_{ds} - \frac{1}{2}\alpha V_{ds}^2\right) \quad (11.5)$$

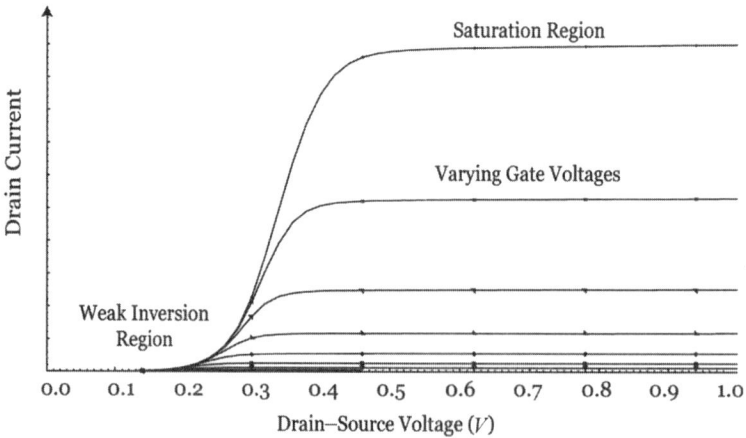

Figure 11.2 Simulation of drain current against drain–source voltage, with different gate voltages. Plot shows regions of weak, strong, and saturation.

Strong Inversion

$$I_{ds} = \frac{\beta}{2\alpha} \phi_t^2 \left(V_{gs} - V_t\right)^2$$

where

$$\beta = \frac{W}{L} \mu C'_{ox} \tag{11.6}$$

$$\phi_T = \frac{KT}{q} \tag{11.7}$$

$$\alpha = 1 + \frac{\gamma}{2\sqrt{V_{sb} + \phi_f + \phi_o}} \tag{11.8}$$

W is the channel width
L is the MOSFET channel length
C'_{ox} is the gate oxide capacitance per unit area
K is Boltzmann's constant
T is absolute temperature
q is the electronic charge
n is the weak inversion slope factor
V_{to} is the threshold voltage
V_o is the Early voltage
V_p is the pinch off voltage
V_{sb} is the potential difference between the source and the bulk
V_{db} is the potential difference between the drain and the bulk
V_{sb} is the potential difference between the source and the bulk
γ is the bulk threshold parameter
ϕ_f is the Fermi potential
ϕ_0 is the semiconductor surface potential

11.2.1.5 MOSFET Weak Inversion and Biological Gap Junctions

When the gate voltage is below the threshold voltage and the semiconductor surface is in weak inversion, a transistor is within the subthreshold region of operation (Vittoz and Fellrath, 1977) where MOSFET transistor current flows largely by diffusion alone. The diffusion characteristics of this region are analogous to the operation of a biological gap junction within a neuron, where current through the channel is linearly proportional to the difference in carrier concentrations across it. Transistors in weak inversion typically operate between 1 and 3 volts supply voltage, and at between femto and 100 nanoamps: given these power conditions they are particularly useful for micropower IC design.

11.2.2 Analog vs Digital Methodologies

Analog circuits use a continuously variable signal to process information, in some proportional relationship between the physics of the transistors and the voltage or current representing the signal. In contrast, in digital electronics signals take only one of two different levels. Most of the circuit operations that can be performed on an analog signal such as amplification, filtering, or limiting can also be duplicated in the digital domain. But which is the best to use? The answer comes from an in-depth analysis of power consumption, technology, die space, and noise.

Analog circuits are more susceptible to noise and device mismatch than their digital equivalents, as a small signal variation can have a nonlinear effect. In contrast, digital signals take only one of two different values; thus a voltage disturbance would have to be about one-half the magnitude of the digital signal to cause an error. Precision is another consideration. Fundamental physical limits such as shot and thermal noise limit the resolution of analog signals. In digital electronics, specific precision is obtained by using more or less digits to represent the signal. This is practically limited by the performance of the analog-to-digital converter, since digital operations can usually be performed without loss of precision.

However, for many operations, analog circuitry can perform the required processing at lower power. Especially when working in the weak inversion region of the transistor operation, ultralow-power operation can be acheived. Such low power must be traded with precision. But for operations where high precision is not required, it can be beneficial. The human eye works with such inherent noise limiting processes and still achieves a stunning result with an extremely low-power budget.

It is possible to mix analog and digital to get the best of both worlds, but digital clocks can strongly interfere with analog circuitry. Thus, asynchronous solutions, which have a broader power density spectrum, will cause much fewer problems to neighboring analog circuitry. The reality of the situation is that a mixture of the two techniques is probably the most efficient means of computation. Indeed, this mixed approach is undertaken by the human computational system: the brain.

11.3 Photo Electric Transduction

Detecting light intensity and converting it to an electrical voltage or current (signal), and processing the signals from an array of detectors, are the primary tasks of vision chips. This section gives a basic overview of some of the circuits that may be implemented within CMOS to derive a voltage or current signal from photodiodes.

Photodiodes operate as a voltage and current source in response to incident light between the wavelengths 200 to 1,100 nm. The current through a photodiode is approximately linear with incident power for a given spectral range within the silicon absorption spectrum. This current output is linear over several orders of magnitude, varying between the noise floor (100 fAmps) and the maximum output (depending on the pixel size, 10 nAmp or more). The voltage output, however, changes logarithmically with incident light power allowing significant dynamic range compression.

11.3.1 Logarithmic Sensors

Some basic photodetection circuits are shown in Figure 11.3. Figure 11.3(a) shows a photodiode in a simple logarithmic conversion circuit, and in a cascaded configuration in Figure 11.3(b). As the input photocurrent is usually very small and falls within the subthreshold region of an MOS diode, the current–voltage relationship is determined by Equations (11.9) to (11.11).

$$I \propto \beta \exp^{\frac{V}{\phi_t}\frac{1}{n}} \tag{11.9}$$

$$I \propto \beta \exp^{\frac{V}{\phi_t}\frac{1}{n^2+n}} \tag{11.10}$$

$$I \propto \beta \exp^{\frac{V}{\phi_t}\frac{1}{n^3+n^2+n}} \tag{11.11}$$

where:

$$\beta = \frac{W}{L}\mu C'_{ox} \tag{11.12}$$

$$\phi_T = \frac{KT}{q} \tag{11.13}$$

and:

I = input photocurrent
n = subthreshold slope factor
V = output voltage

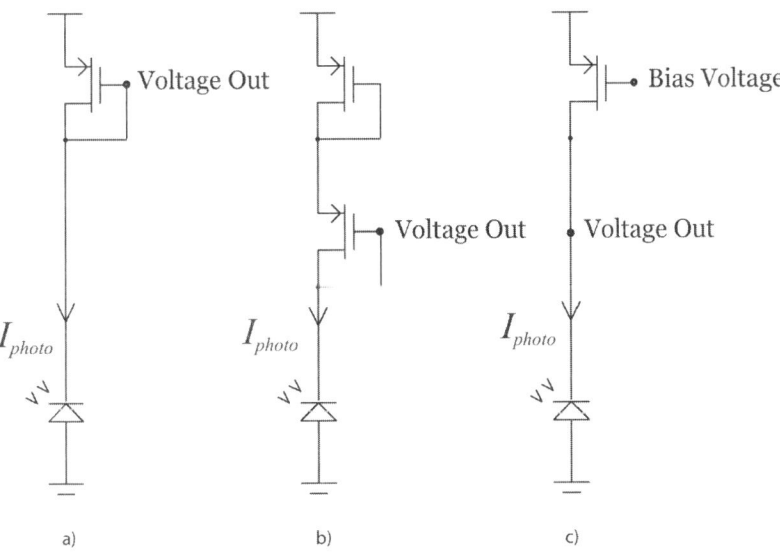

Figure 11.3 Image showing logarithmic current to voltage conversion photosensors (a) A diode connected pMOS MOSFET; (b) A cascoded diode connected MOSFET; (c) A pMOS MOSFET in the source follower configuration, acting as a bias controlled resistor.

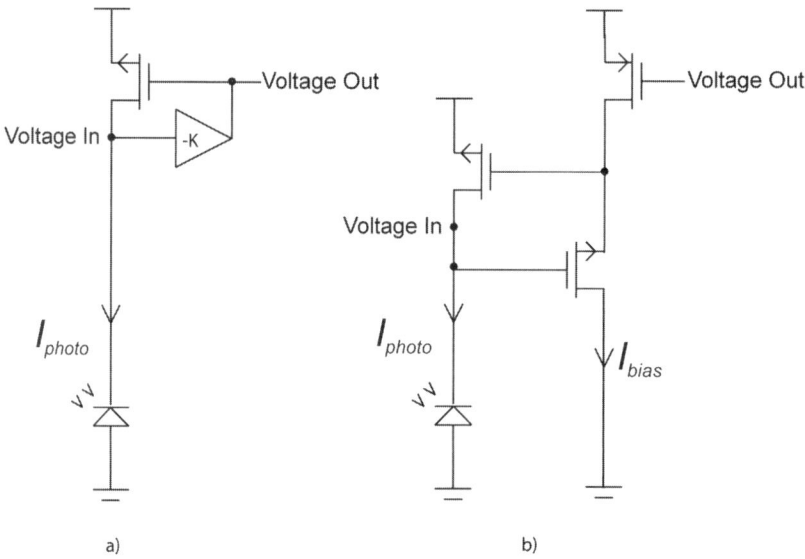

Figure 11.4 The direct current (DC) operating point of these photodetection circuits is determined by the amplifier and the input photocurrent. If an inverter configuration with the transition region at $V_{dd}/2$ is used, such as in image (b) (right), the Voltage In will settle at $V_{dd}/2$. If a high gain differential amplifier is used with its positive input connected to a reference voltage, Voltage In will be set at that reference voltage.

11.3.2 Feedback Buffers

A limitation to the photodetection circuits described in Figure 11.3 is the capacitive load at the input of the circuit. This capacitance reduces the bandwidth of temporal signals, particularly when the current is low. A method to avoid this is to use feedback compensation to reduce the input impedance, as shown in Figure 11.4. These systems clamp the photodiode input voltage, however, Voltage Out is also dependent on input photocurrent (I_{photo}). With a constant voltage clamp, Voltage Out will have a logarithmic function, with I_{photo}. As such the circuit maintains its characteristics as a logarithmic compressor. The input transistor Miller capacitance and input resistance can be reduced by adding a cascaded transistor into the amplifier path, further improving the characteristics of these circuit designs.

11.3.3 Integration-Based Photodetection Circuits

Most commercially available imaging devices use some form of charge integration combined with sample and hold circuitry to transduce the photocurrent into a voltage. The basic diagram for this circuit is shown in Figure 11.5. The advantages of this configuration are its linear transfer characteristics, and controllable integration time, which also serve to remove high-frequency noise components of the input signal. Capacitive mismatch is significantly less than transistor mismatch and photodiode mismatch is low (typically less than 2%) because it has a relatively

11.3 Photo Electric Transduction

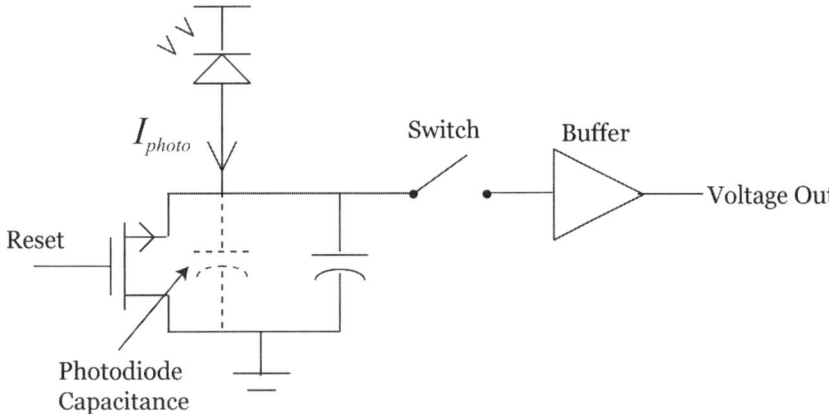

Figure 11.5 Initially the reset transistor is on (high V) and the voltage at the input node to the buffer is set to the reset value. The reset voltage is grounded and the photocurrent (I_{photo}) charges up the input capacitance, and is read when the switch is closed. After reading, the switch is opened and the reset voltage raised to V_{dd}.

large area. As such, this configuration has significantly less mismatch than those described above.

Sample and hold systems have many advantages, as outlined above. However, they have difficulty in implementing real-time parallel algorithms. Thus, they are generally not used in retinomorphic imaging chips.

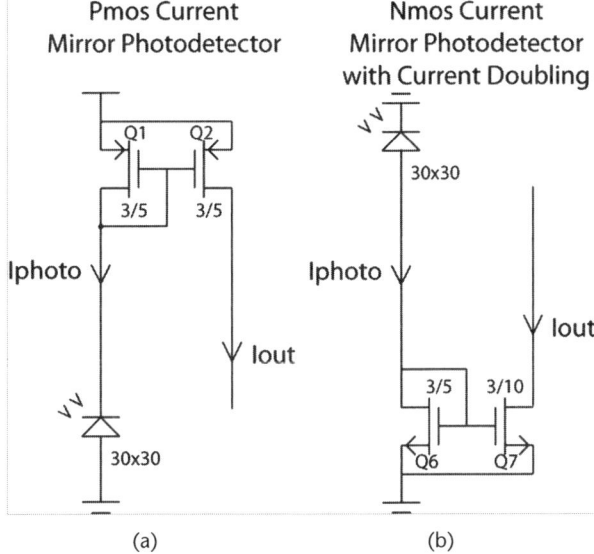

Figure 11.6 Schematic of a pMOS (a) current mirror and an nMOS (b) current mirror.

11.3.4 Photocurrent Current-Mode Readout

The simplest method of detecting the photocurrent in continuous time is to use a current mirror setup (Figure 11.6).

Using a current mirror to read the photodiode output, and to buffer the current for further processing, allows linear current mode image computations. By maintaining the dynamic range of the photodiode input over the full bandwidth of photocurrent, this technique does not logarithmically compress the input current (Gruev et al., 2004), unless the gate voltage between the transistors is amplified by a factor K, creating a logarithmic current mode sensor (Tanzer et al., 2004). This technique can output 5 orders of current magnitude corresponding to over a 16-bit dynamic range. This is the maximum photocurrent dynamic range that may be obtained from the photodiode circuitry described in this section.

11.4 Retinimorphic Circuit Processing

As discussed in Chapter 2, a very simplified model for retinal spatial processing may be found in spatial convolutions. That is, the retinal circuitry may be modelled as convolving the signals on each of the photoreceptors with a spatial template. One simplistic model for that might be a spatial Laplacian kernel. The architectural advantage of doing this processing in the visual plane is that despite individual processing components being slow, the distributed parallel nature of the processing architecture allows video rate response speeds at very low power.

To extract an image from a CMOS chip, photodetection circuits need to be combined with photodiodes in an array. This can be either a hexagonal pattern similar to the cone structure within the fovea, or in a square grid array more typical with conventional imaging systems. Each photodetection sensor element is called a pixel (taken from picture element). This needs to be small, to maximize resolution and reduce aliasing problems (Yang et al., 1998; Yang et al., 1999). There are two basic systems of pixel operation: continuous time sampling, for which the photocurrent is read as needed; and integration sampling, where the photocurrent collects on a capacitor until it is read.

Retinomorphic imaging arrays have parallel image processing circuitry distributed with the pixel photodiodes. Each pixel-level processing circuit is fairly simplistic and usually expressed in terms of current summation or subtraction circuits. However, the resultant convolution algorithm can form powerful Laplacian and/or Sobel filters.

$$R = i_0 + \sum F_{x,y} i_{x,y} \qquad (11.14)$$

where

R is the response to the convolution
i_0 is the current of the central pixel
$i_{x,y}$ is the photoresponse of a pixel at x, y
$F_{x,y}$ is the algebraic function performed on $i_{x,y}$

11.4 Retinimorphic Circuit Processing

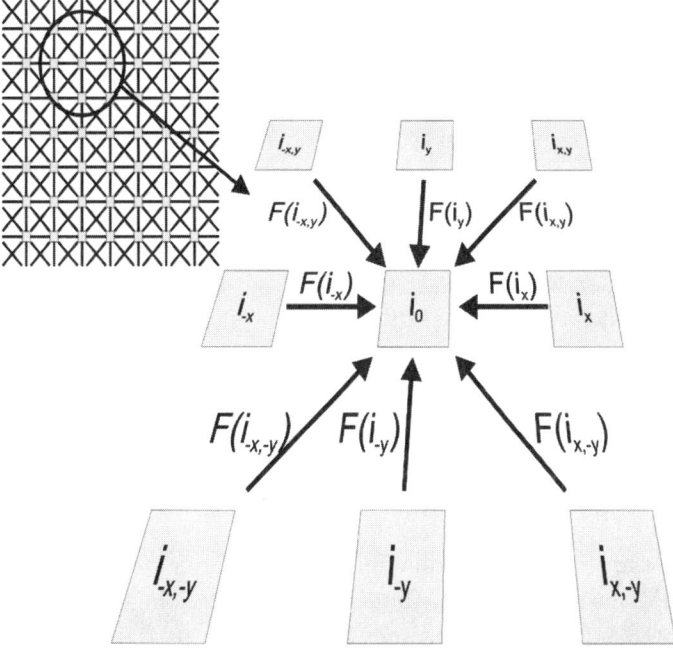

Figure 11.7 Filter matrix of a standard two-dimensional image processing algorithm. The imaging array has a network of interconnected photoreceptor points. For each photoreceptor, a function can be performed that consists of adding or subtracting the "influencing current."

On-chip signal processing of pixel sensors is commonplace within the camera-on-a-chip field (Fossum, 1997; Fossum, 1995). A direct result of the neuromorphic development instigated by Carver Mead (Mead, 1989; Mead and Mahowald, 1988) is distributed, parallel spatial processing circuitry, which is ideally suited to real-time operation. The voltage mode resistive network (Mead and Mahowald, 1988) forms the basis of many two-dimensional filter designs, implementing a smoothing array over the photodiode circuitry and enhancing image contrasts.

11.4.1 Voltage Mode Resistive Networks

Resistive networks in CMOS are constructed using either passive elements such as resistors and capacitors, which do not require power to operate, or active powered elements such as the transistor. CMOS passive resistive elements include polysilicon strips, diffusion resistors, and well resistors.

One of the first-described resistive networks was the Horizontal REsistor (HRES) (Mead, 1989). This one-dimensional array of transistors has developed into a compact single transistor diffuse network, implemented by Andreas Andreou and Kwabena Boahen (Andreou A. and K., 1995; Boahen and Andreou, 1992). With an array of 210 x 230 pixels, using on average 12 transistors per pixel, this is one of the largest retinomorphic vision chips. Voltage-controlled transistors act as nonlinear resistors interconnecting the pixel nodes. The result is a diffuse network of exponentially decreasing node voltages, forming a lowpass filter described by Equation 11.15 and shown in Figure 11.8.

$$I = I_0 \exp^{\frac{V_{bias}}{nV_T}} \tanh\frac{V1 - V2}{2} \qquad (11.15)$$

This circuit has been extended to include a two-layer diffusive network. The upper layer corresponds to horizontal cells in retina and the lower layer to cones, where nMOS transistors model chemical synapses.

The equivalent lateral spread of horizontal cells can be implemented over a two-dimensional area of pixels. A resistive grid of pixel nodes interconnect with voltage-controlled transistors as with one-dimensional networks. Each node within the grid receives an input current and distributes an output current to its neighbors. This output is read at the node voltages. With equal resistance, an exponential decrease in current will be observed with lateral spread, exhibiting a diffusion gradient similar to that observed within the horizontal cells of the retina.

One of the first silicon retina chips was implemented by Carver Mead (Mead, 1989) and Misha Mahowald (Mahowald, 1994) (see Figure 11.9). The chip mimics rods and horizontal and bipolar cells of the vertebrate retina. Where the rods detect light, the horizontal cells average the output of the rods spatially and temporally and the bipolar cells detect the difference between the averaged output of the horizontal cells and the input. Rods are implemented using parasitic phototransistors and a logarithmic current to voltage converter.

Several versions of two-dimensional chips have used the circuit shown in Figure 11.8. Andreou and Boahen's two-dimensional extensions use a hexagonal network, with six neighborhood connections in a similar arrangement to that of Figure 11.9 (Andreou et al., 1991; Andreou A. and K., 1995; Boahen, 1996). The largest chip occupies an area of 9.5 mm^2 × 9.3 mm^2, implemented in a 1.2 micron CMOS process. Used under the conditions described in the paper the chip dissipates at 50 mW. One of the major issues for discerning a power consumption figure of

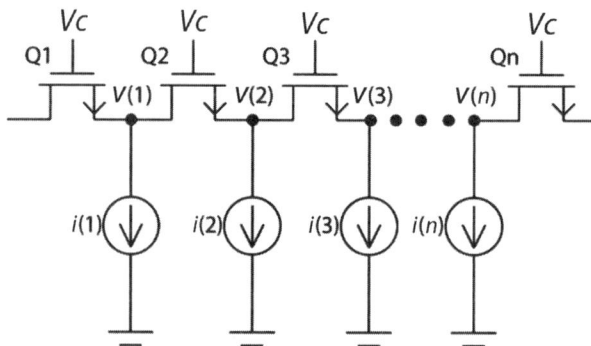

Figure 11.8 Andreou and Boahen's one-dimensional resistive network, where each globally controlled gate voltage tunes the channel resistance, implementing an exponential lateral decay on each current input.

11.4 Retinimorphic Circuit Processing

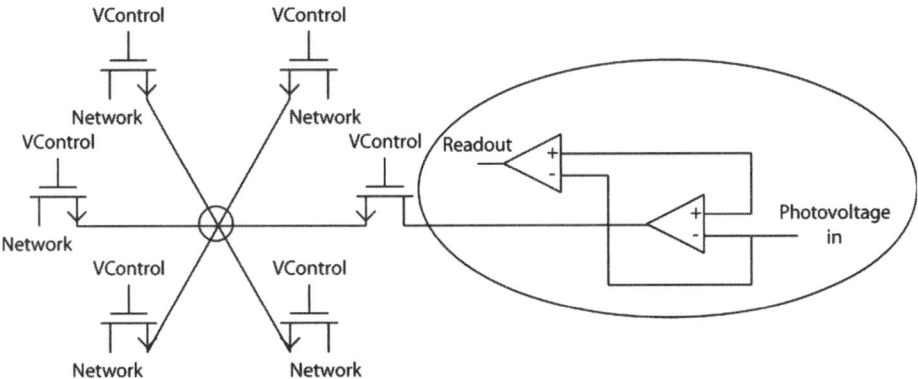

Figure 11.9 Schematic arrangement of Carver Mead and Misha Mahowald's silicon retina (1994). The two-dimensional network is implemented on a hexagonal array, with a logarithmic photovoltage input and voltage output.

merit for imaging chips is that different descriptions are often used under different power conditions, with different lighting and with different tests.

The spatial image Gaussian filter designed by Kobayashi et al. (Kobayashi et al., 1995; Kobayashi et al. 1991; Kobayashi et al., 1990) uses a hexagonal resistive network. It uses negative resistors implemented using negative impedance converters (NIC) to approximate the Gaussian function. The chip is a 45×40 array of photodetectors and resistive grid on a 7.9 mm $\times$ 9.2 mm chip using a 2-micron CMOS process.

Tetsuya Yagi and Seiji Kameda (Yagi et al., 1998; Kameda and Yagi, 2003; Kameda and Yagi, 2000; Kameda and Yagi, 2001; Kameda and Yagi, 2002), have implemented a difference of Gaussian representation by subtracting two resistive networks with different full-width half-maximums, FWHM (explained above), with impressive results.

11.4.1.1 Limitations with This Approach

Kernel distributions of voltage mode resistive networks depend on the current passing through each node. Photodiode current varies by more than four orders of magnitude, between hundreds of fAmps and tens of nAmps. The result is a spatial distribution network which is light intensity dependent. This is not like the horizontal cell network in the retina, which has fixed spatial distributions regulated by synapses and graded potentials. Voltage mode resistive network methodologies cannot implement color opponent signals in a kernel convolution, as these require fixed receptive fields.

11.4.2 Current Mode Approaches to Receptive Field Convolution

The concept of current mode image processing was first envisaged by Vivian Ward in the mid-1990s. Ward and Syrzycki describe vision sensors based on the receptive field concept (Ward, 1995; Ward et al., 1993), where current mode circuits handle positive and negative currents read directly from photodetectors. An important facet of their work is the principle of taking a photocurrent and using copies of it to convolve receptive fields.

In 2001 Etienne-Cummings et al. developed the MIMD chip architecture, providing five spatially processed images in parallel with programmable kernel sizes and configurations. This system uses a single processing core with digitally programmed current-mode analog computation to convolve the pixel arrays. There are some significant advantages to this innovative approach. Primarily the resolution of the photodiode array can be increased because the processing circuitry is no longer distributed within the pixels but separate and singular. Viktor Gruev and Ralph Etienne-Cummings extended the technique in 2002 with an improved reconfigurable current scaling system (Figure 11.10). This system samples the photodiode with an active pixel sensor (APS) scheme, the resulting voltage is read from the pixel array and converted back to a current within the processing unit. This is then scaled and outputted off-chip for further processing with a reconfigurable logic unit.

By implementing the highly saturating filters in a reconfigurable system, on-chip edge detection is improved over the voltage mode resistive networks described previously. However, the kernel-processing unit is not distributed; it is conjoined to the output readout circuitry, creating a sequential bottleneck of information transfer. Therefore, as the array size increases the frame rate of image capture will have to decrease. Each output information unit needs further processing so its address must be sequential or registered. Without further processing on-chip, this technique is bound to a scanning output, and cannot use any of the other output

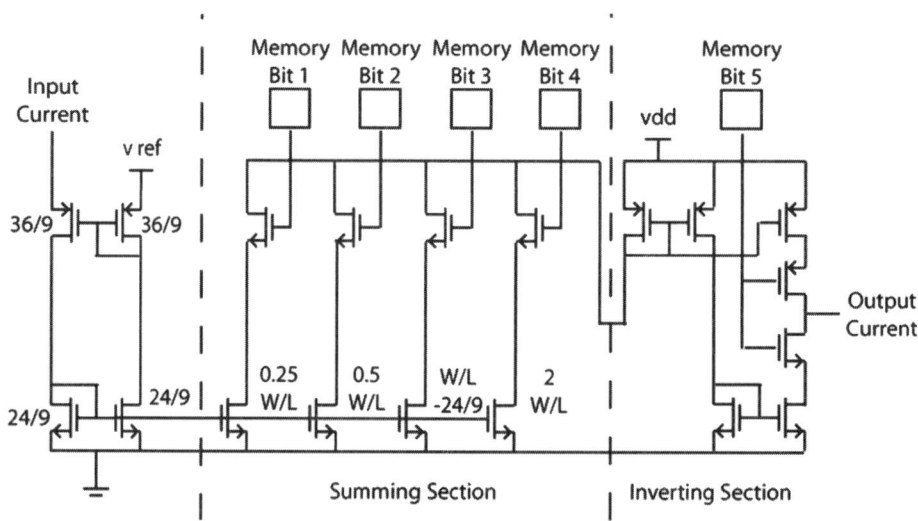

Figure 11.10 This schematic shows the circuit layout of Gruev and Etienne-Cummings' virtual ground and scaling circuit. The virtual ground on the left-hand side implements a positive feedback loop into the input. The scaling circuit is a series of scaled current mirrors (center), and the inverting section is on the right.

11.4 Retinimorphic Circuit Processing

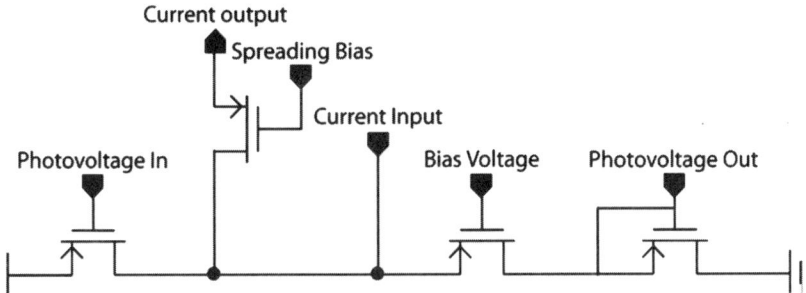

Figure 11.11 Schematic of a single component of the basic horizontal cell. Lateral spreading occurs through the network input and output. Note this system has a global bias control and no control over its output resistance.

systems described later. To implement current subtraction on-chip and output a true convolution, the same problems encountered by Ward in (Ward, 1995) become apparent. There are several ways around this problem which shall be discussed later.

11.4.2.1 Improved Horizontal Cell Circuitry

Traditionally, retinal horizontal cells have been implemented with resistive networks. These are not effective at creating flexible or precise kernels, and are intensity dependent. The eye does not implement a difference of Gaussian receptive fields at the horizontal cell level of the retina; rather, this happens in combination with the bipolar, amacrine and ganglion cells. The purpose of the horizontal cells is to control photoreceptor sensitivity and image smoothing. It is possible to redesign the basic resistive network to mimic better its functionality within the eye. Figure 11.11 shows a single component of the basic resistive network and is identical to a single transistor pictured in Figure 11.8 but with the bias and input circuitry attached.

Figure 11.12 shows a schematic of an individual component of an improved horizontal cell network with self-biasing output inhibition. To achieve this, the input voltage is put through a level shifter and used to inversely bias two output transistors acting as resistors. As such, when the input current (light intensity) is large the output resistors close off, as shown in Figure 11.13. Here the numbers 1 to 4 correspond to cell progression from the photodiode current source. When the light intensity is low, the output resistors open and the current is allowed to spread.

11.4.2.2 Novel Bipolar Circuitry

By processing information in voltage mode, the majority of retinomorphic CMOS imagers restrict the on-chip computational power. Bipolar fields form an important component of retinal processing, where graded potentials rely on positive and neg-

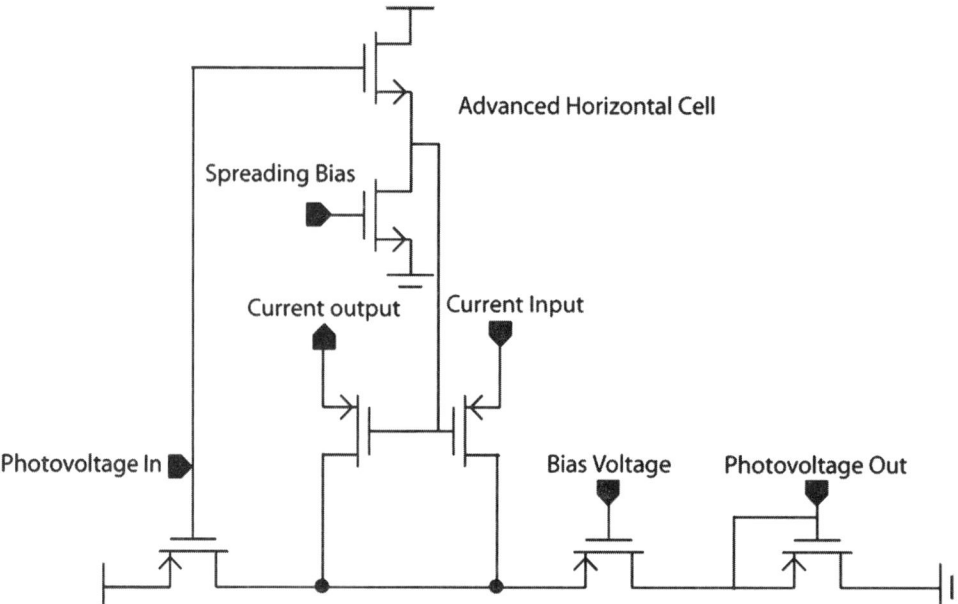

Figure 11.12 Schematic of the improved horizontal cell with local inhibition control and controlled output resistance.

ative potential differences across the neuronal membrane to enact stimulation and convolve dendrite arrays. There are, however, difficulties in trying to implement positive and negative systems in CMOS. Biological systems negate this problem by raising the zero point to a higher effective potential as a result of inhibitory inputs. However, voltage mode implementations of bipolar retinal pathways have proved cumbersome and often oscillatory. A more faithful representation of the bipolar potential differences within the retina is achieved through the use of current mode subtraction circuitry, where bidirectional circuitry can fix a zero point for computation.

11.4.2.3 Bidirectional Current Mode Processing

The human brain uses bidirectional circuitry to control graded action potentials. The flow of ions through a membrane is analogous to the sum of positive and

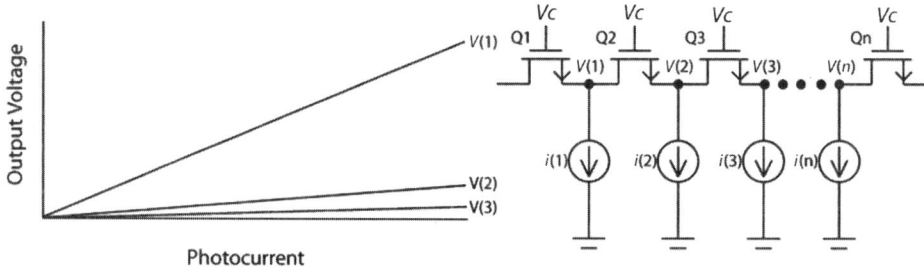

Figure 11.13 Plot showing current spreading at high light levels. Spreading between cells is minimized to enhance edge detection.

negative currents at a node. Neuronal graded action potentials play a central role in retinal processing between the photoreceptors and the ganglion cells, and retinomorphic IC designs should be analogous to biological computation. The benefits of doing so are lower power consumption of image processing convolutions and increased computational ability. These techniques have been employed within two recent publications (Banks et al., 2005; Banks et al., 2007) where novel biologically inspired current mode circuitry has been used to implement image processing methodologies.

11.4.2.4 Dealing with Multiple High Impedance Processing Channels

A significant problem with bidirectional circuitry is described perfectly by a standard two transistor comparator (Figure 11.14). When two opposing currents are connected, one of the buffer transistors will saturate to the larger current and force the voltage to ground or supply. As such, this mode cannot be used for current subtraction where either current input may be larger in magnitude.

The way to negate this problem is to use bidirectional circuitry. Essentially this form of circuitry requires two outputs: either two current outputs, or one current output and one voltage output, illustrated in the system-level block diagram (Figure 11.15). Figure 11.16 shows a schematic of a two-channel system.

This four quadrant current subtraction circuit uses single current mirrors connected to each other and then input into a low impedance node, one for each input node. Thus, in operation, the losing node will saturate and no current will be output, however, the winning node will draw or push current through its corresponding low impedance node, allowing an accurate measure of current magnitude.

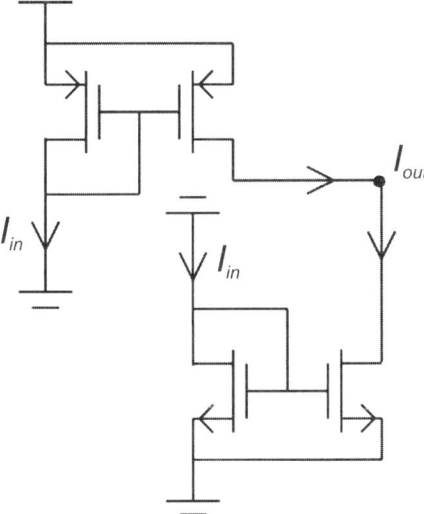

Figure 11.14 Circuit schematic of a basic current comparator. The two opposing currents are buffered and connected, the largest (magnitude) of the two currents will saturate the opposing transistor, and the voltage will switch to either ground if the pMOS is largest or supply if the nMOS is largest.

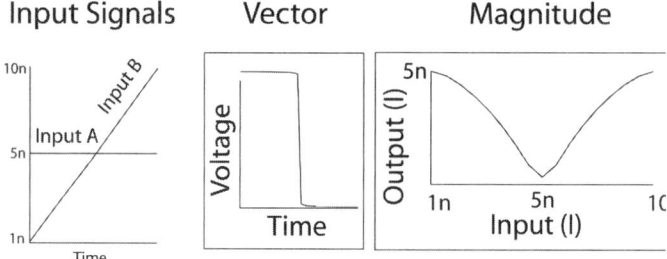

Figure 11.15 System-level design flow of vector magnitude design implementation plotting against time. Simulation results of two inputs.

The opponent output of this circuit is not suitable for all situations. Specifically, it requires two-channel processing in subsequent processing components, which may be impractical. Another solution to the two-channel output is to use a single bidirectional output configuration, shown in Figure 11.17 and simulated in Figure 11.18.

This circuit works well at all current levels except those where the comparator is not fully switched to ground or the supply voltage. Under these conditions the circuit drains large quantities of power, and does not perform a current subtraction.

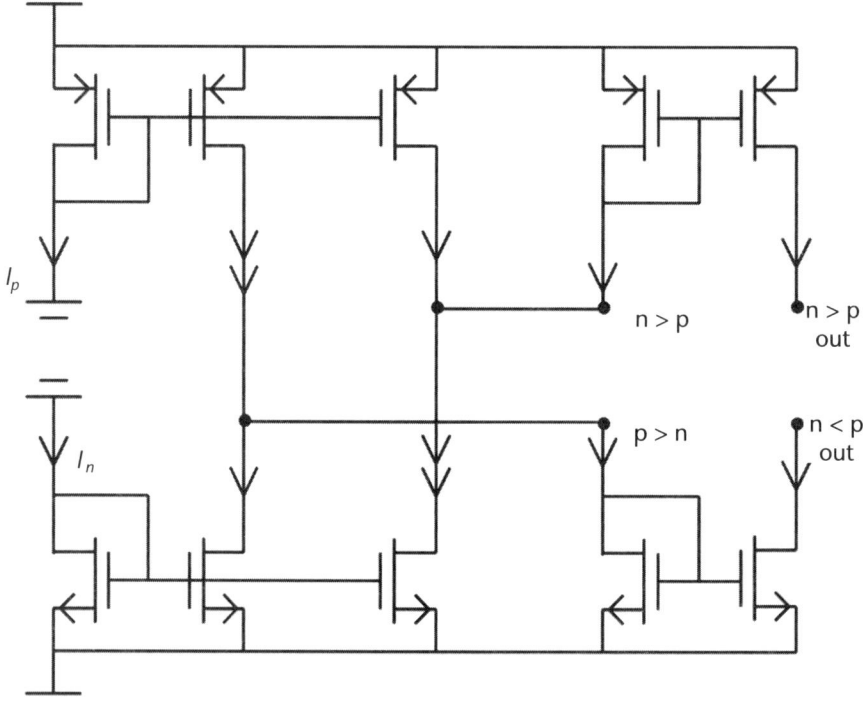

Figure 11.16 Schematic of a basic bidirectional subtraction circuit with two channel current output.

11.4 Retinimorphic Circuit Processing

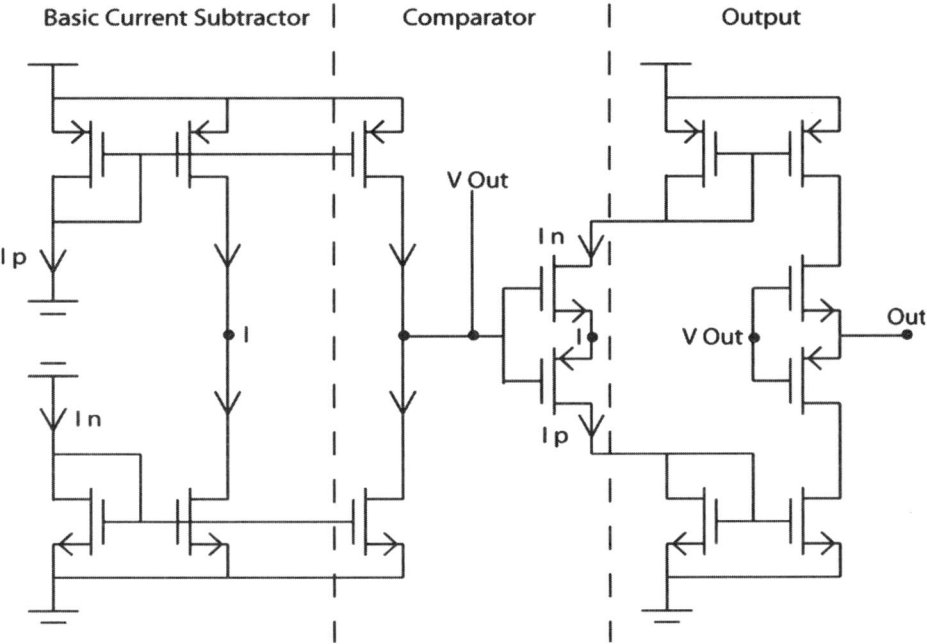

Figure 11.17 Schematic of bidirectional current subtraction circuit using a current comparator switch, which can also be used to control subsequent bidirectional units.

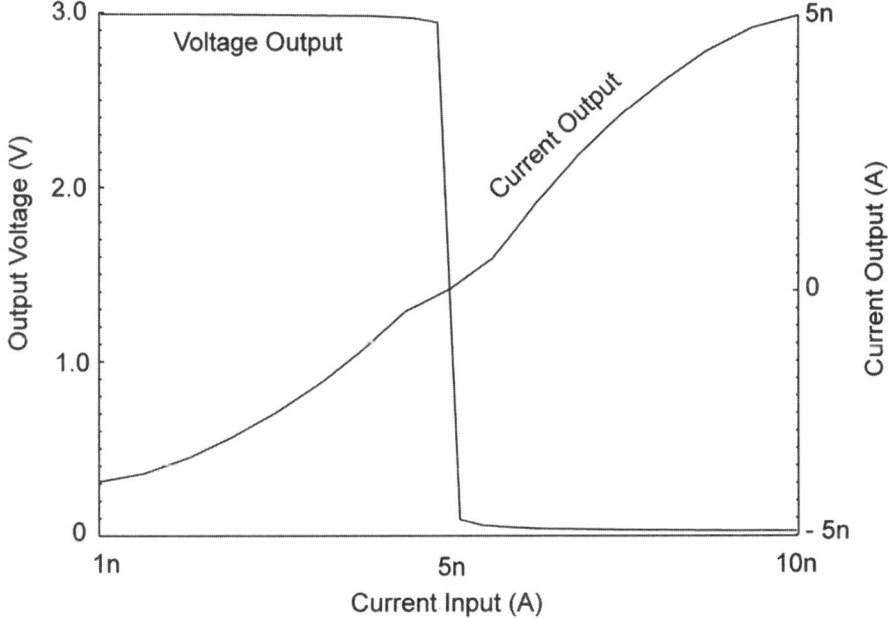

Figure 11.18 Simulation of bidirectional current simulation showing voltage switching and current output of the vector magnitude circuit with cascaded comparator.

11.4.2.5 The Current Comparator

One of the significant issues with current subtraction circuitry is the effectiveness of the current comparator used. Figure 11.19 shows three variations of basic current comparators that are simulated in Figure 11.20. This illustrates the effect of higher output impedance on current mirror comparator switching. The output impedance of the cascaded and Wilson current mirror is larger than the output impedance of the standard current mirror. The output impedance of the connected nMOS transistor determines the output impedance of the single current mirror. The largest effect on these is channel length modulation generated by the Early effect. The Wilson current mirror isolates the output voltage from the output transistor another nMOS device. This significantly increases output impedance of both the cascaded and Wilson current comparators, as shown in the simulation results (Figure 11.20).

Figure 11.21 shows the schematic and simulation results from a cascaded current comparator implemented within a standard 0.35 micron process. The transistors within the comparator are large to minimize mismatch errors. The time

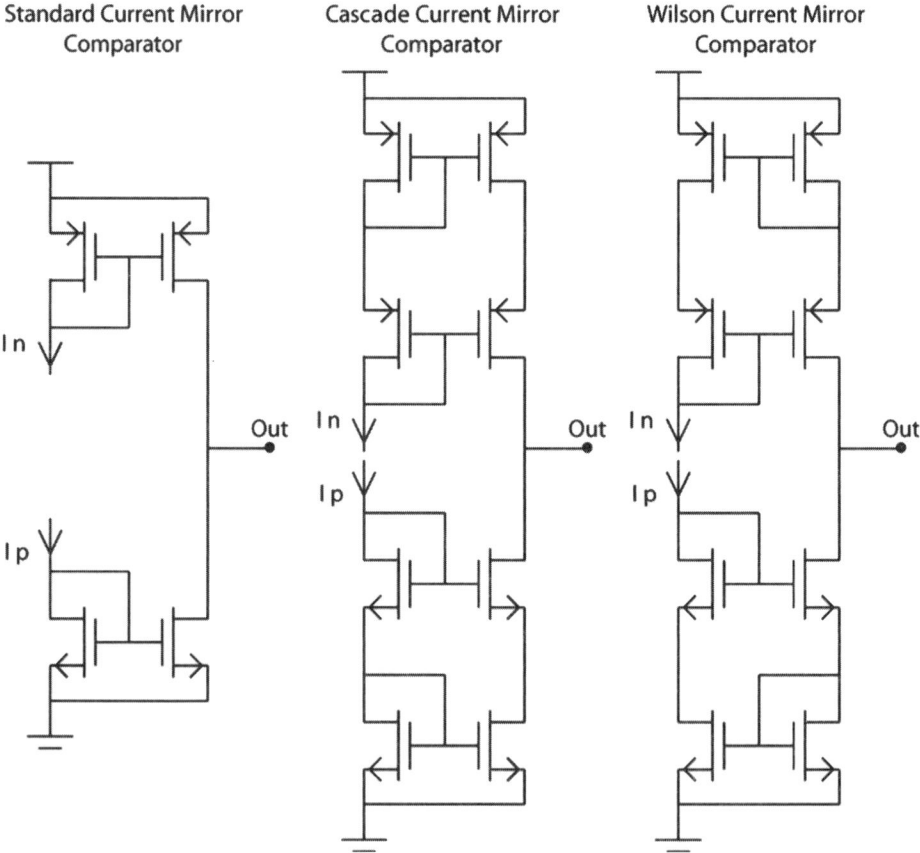

Figure 11.19 Schematics showing three main current comparator techniques, with a standard current mirror comparator (left) and a cascaded version (center) with significantly higher output impedance, and a Wilson current comparator based on the Wilson current mirror configuration.

11.4 Retinimorphic Circuit Processing

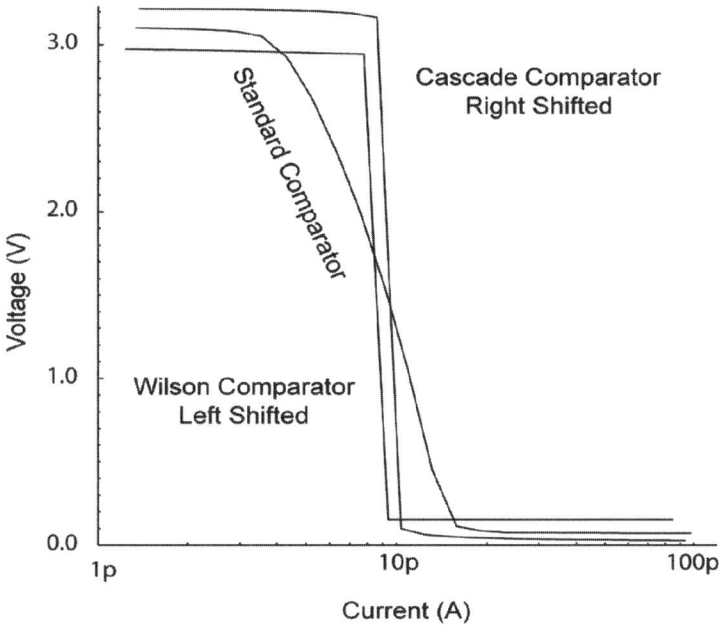

Figure 11.20 Simulations of a standard current mirror comparator: the plot shows voltage switching of a standard current mirror comparator, a Wilson current mirror comparator, and the cascade comparator.

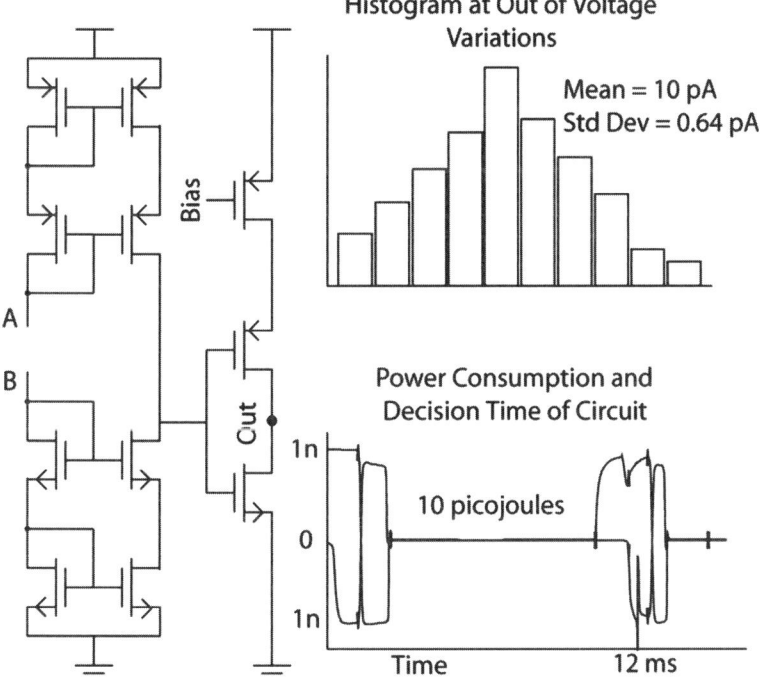

Figure 11.21 The schematic (left) shows a cascaded current comparator. Top right is a Monte Carlo evaluation of the current comparator in a 0.35 micron CMOS process, and bottom right is the total power consumption of the comparator per decision (excluding input currents).

taken for a comparator decision is dependent on the amount of current allowed through the inverter. At ultralow current levels where only picoamps are allowed through the inverter, the switching time occurs in milliseconds. At larger values, the switching time dramatically reduces and at 300 picowatts the decision time is approximately 20 microseconds. Importantly in ultralow power configuration, this comparator consumes less than 10 picojoules per decision, ensuring it is extremely low power.

11.4.3 Reconfigurable Fields

The processing schematic pictured in Figure 11.22 allows bidirectional current flow, and consists of three sections: a bidirectional input stage, a scaling circuit, and an output subtraction stage. The circuitry hard-wires into the pixel four separate reconfigurable image-processing kernels using 2-bit global control: a Laplacian, a vertical and horizontal Sobel filter, and a smoothing array. Devices Q1 and Q2 form a low impedance continuous time photocurrent detector circuit, which is inverted by Q6 and Q7 and scaled by the weighting section, devices Q3 to Q5 and Q8 to Q9. These transistors are hard-wired into kernel configurations and output to the subtraction circuit, creating a reconfigurable bidirectional current image unit, performing real-time convolution. The current output of photodiodes varies between 100 femtoamps and 10 nanoamps. Within this region, all analog devices will be within the weak inversion (WI) region, and exhibiting exponential transconductance. The kernel size is restricted by metal interconnect capacity of the CMOS process, and filter weighting is limited by the scaling of current mirrors.

It is difficult to implement a large variation of programmable analog computations as the circuitry tends to be fixed and thus adding programmability adds space. However, it is nevertheless possible to implement specific circuitry that is

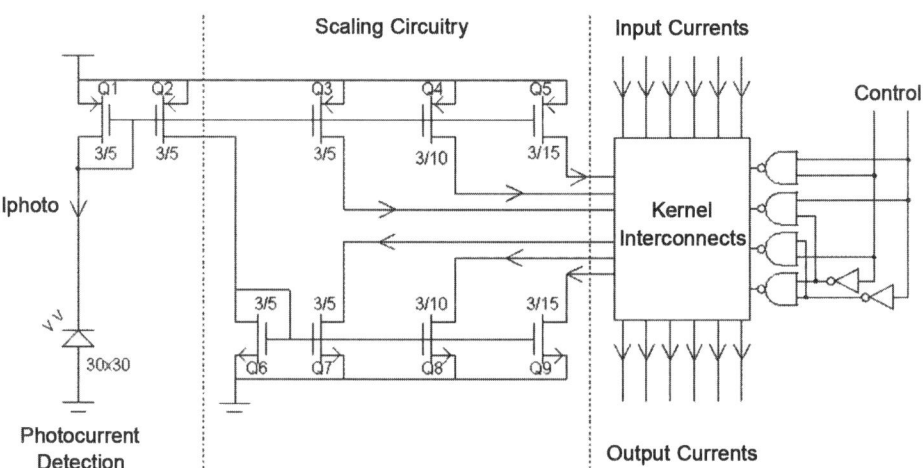

Figure 11.22 Schematic of reconfigurable current scaling units, showing photodiode input and buffer, current inverter, and scaling circuitry with 1, 2, and 4x scaling transistors, routing unit, and two-channel global switching unit.

11.4 Retinimorphic Circuit Processing

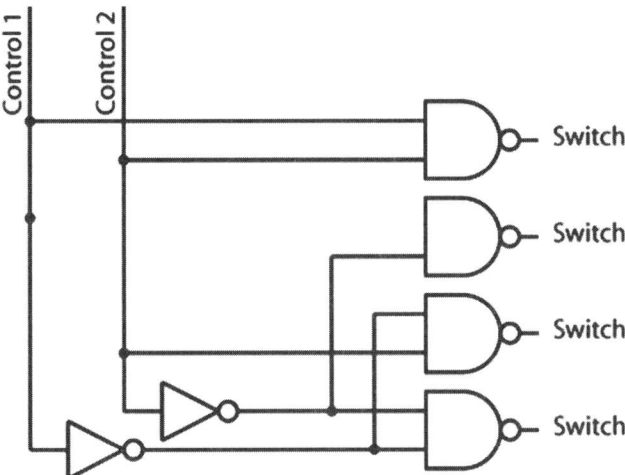

Figure 11.23 Schematic of a basic 4-bit digital switch used to globally control reconfigurable fields.

useful, such as Sobel and Laplacian filters. In such a situation, a globally controlled digital switching system, such as the simple 4-bit schematic shown in Figure 11.23, can be used.

The output of this field can be seen in Figure 11.24 below.

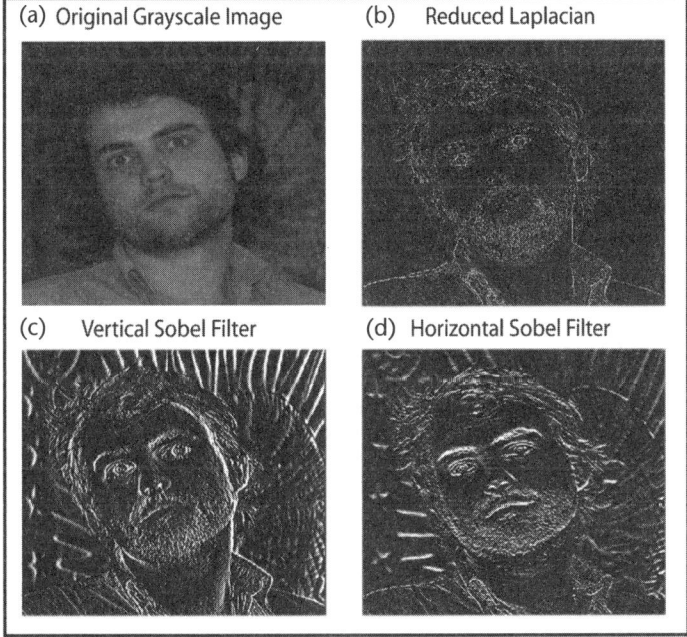

Figure 11.24 The simulated output of the reconfigurable fields. (Top left) The original image. (Top right) A reduced Laplacian which is inspired by the horizontal and bipolar cell processing in the retina. (Bottom) Directional Sobel filters that indicate lines in the vertical and horizontal planes.

11.4.4 Intelligent Ganglion Cells

In order to communicate with off-chip logic, the imaging chip needs to output digital-like information. Inorganic silicon photodiodes are capable of up to 5 orders of magnitude of dynamic range, but are usually only implemented with an 8-bit dynamic range. This can be a problem for scenes with large variations in image intensity.

Rather than raster scanning out an 8-, 10-, 12-, or 16-bit digital signal, the voltage or current representation of light or feature intensity can be converted into a frequency of spiking signals. In the human nervous system, binary encoding is not possible, thus spike rate coding is used. That is, information is contained in the rate of action potentials from a specific location. It is, therefore, also possible to use a spike rate encoding in retinomorphic vision chips. The advantage with frequency encoding is that the dynamic range can be quickly modified via the sampling time. Additionally, as pixels are spiking, they can asynchronously highlight important features to subsequent processing units without a requirement of a global raster. This can have important memory space advantages in sparse encoded systems.

Early work by Delbrück and Mead (Mead, 1989) led to an adaptive photoreceptor which could detect the contrast regardless of overall light intensity. This chip became the mainstay of the neuromorphic vision community. A current-feedback technique proposed by Culurciello *et al.* can be seen in the gray area of Figure 11.25. This technique greatly reduces the energy cost to develop the spike, while still maintaining an acceptable speed. The photocurrent acts to discharge the capacitor, causing *v_photo* to fall, and thresholded by the Q6 to Q7 inverter providing current feedback to provide positive feedback to increase speed and reduce power at the crossover point. Once a spike is induced, it dissociates the current path by switches Q2 to Q3. The photoreceptor is reset by the AER acknowledge after off-chip transmission of the previous event. The spiking rate is therefore directly correlated to the photocurrent and thus the light intensity. The inhibiting NOR gate serves to disconnect the AER handshake to inactive spike generators in pixel groups.

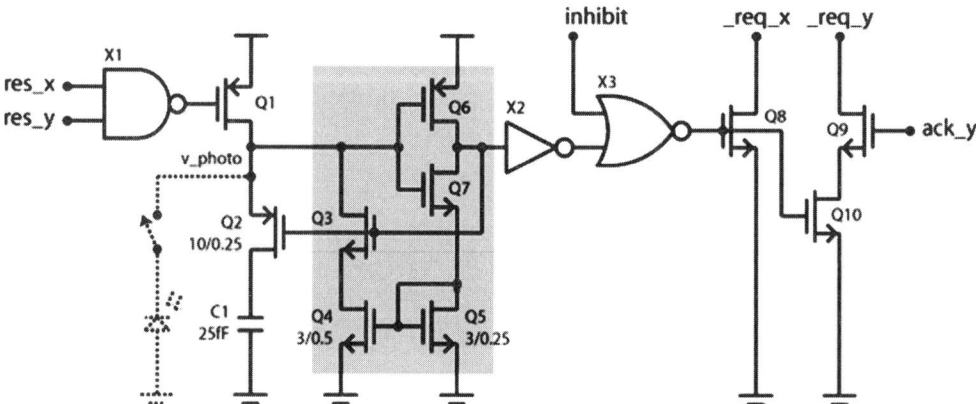

Figure 11.25 The spiking photoreceptor circuit (shaded area illustrating the current-mode feedback).

11.4.4.1 ON–OFF Ganglion Cells

The major drawback of any integrating system is that the frequency response is low for low light intensities. Here again we can learn from nature by implementing complementary ON and OFF channels. Using ON–OFF opponency, where ON-cells spike at high frequency at high light levels, and OFF-cells spike at high frequency at low light levels, there will always be an adequate frequency response, even at low light levels.

The current is buffered and mirrored to create ON and OFF channels. In the OFF channel, the current is inverted, such that low photocurrents create high off-currents and vice versa. The two channels then compete by integrating their currents into voltages through capacitance. This voltage is released in the form of a spike once a trigger threshold has been surpassed and the charge collected is reset. To reduce redundancy, only the first spike, whether ON or OFF, is released and both channels are reset. A complementary output is sent to indicate an ON or OFF spike. Hysteresis is added to stop the circuit oscillating between ON and OFF when the light intensity is close to the threshold between light and dark channels.

The circuit schematic can be seen in Figure 11.26. A simple current mirror is used to copy the photocurrent (iphoto) to two separate branches. One copy is fed into a predefined current sink (ibias) which effectively produces the difference of these two currents (ibias-iphoto.) This in turn is mirrored to obtain the OFF current. The bias current is chosen such that at maximum light intensity the OFF current is zero.

The ON and OFF currents then are used to create an increasing voltage, by means of integrating these into the parasitic capacitance of their respective nodes. High-gain digital buffers (NOT gates) or the circuit configuration used in Figure 11.25 is used to threshold detect and the first channel (ON or OFF) to reach threshold is collected through a logic OR gate. An additional output is provided to specify whether the response is ON or OFF by using an RS flip-flop to determine which channel is dominant. Hysteretic feedback is provided to the digital

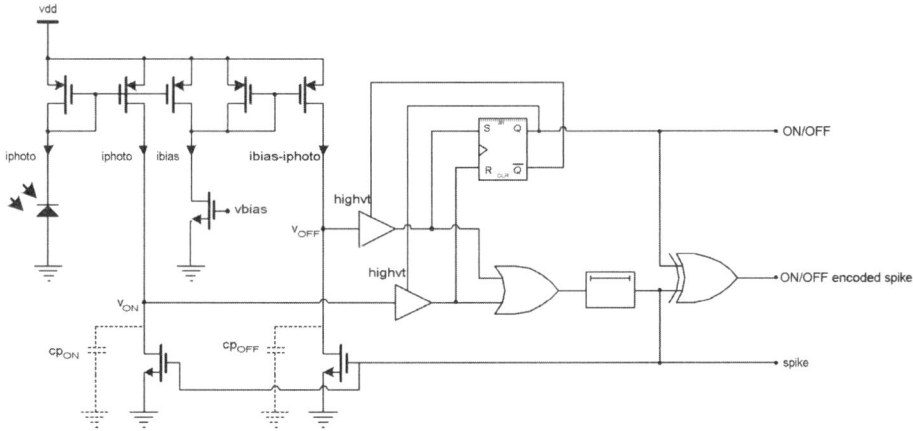

Figure 11.26 Basic circuit topology of the ON–OFF spiking photoreceptor circuit.

buffers to provide a 10% to 20% lag on channel selection changeover to prevent rapid channel toggling when the ON and OFF responses are comparable. The two outputs (SPIKE and ON/OFF) are also combined into a single signal by using a logic XOR operation.

Simulation results for the individual ON and OFF channels can be seen in Figure 11.27. The slow responses can be seen for the ON channel at low light intensities, and in the OFF channel at high light intensities. The spike interval at the maximum firing rate is 3 μs, corresponding to over 500 kHz in response when considering that the data is effectively compressed by half. 500 kHz is sufficient to provide a 16-bit dynamic range at 10 Hz refresh. This maximum firing rate is limited by the combined parasitic capacitance at the integrating node.

11.4.4.2 Pulse Width Encoding

Pulse width encoding can be used to add additional information to spikes. ON–OFF encoding can, for example, be described in a single pulse width rather than

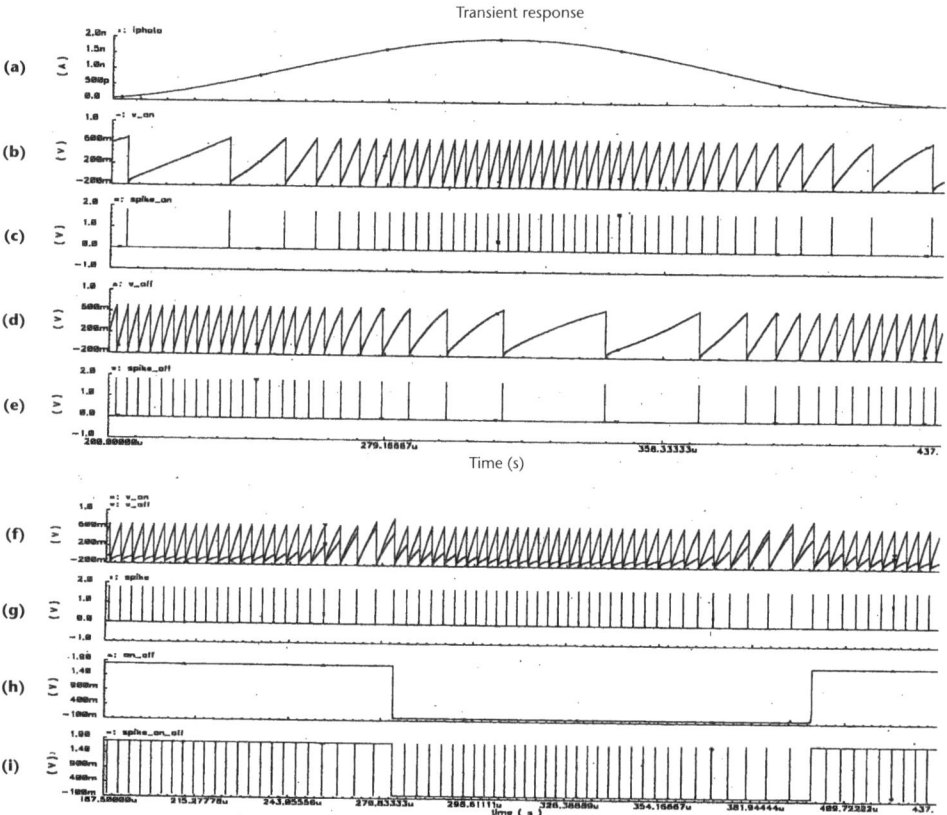

Figure 11.27 Simulated results for the individual ON and OFF channel spike generators. The waveforms shown from top to bottom: (a) photocurrent; (b) ON channel charging response; (c) ON channel spike output; (d) OFF channel charging response; (e) OFF channel spike output; (f) competing ON–OFF charging response; (g) spike output; (h) ON–OFF channel selection; (i) combined SPIKE and ON–OFF channel encoded output.

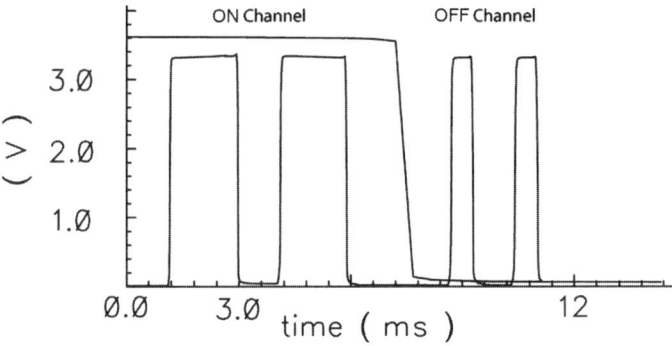

Figure 11.28 This simulation result shows two output plots. The vector output plot that switches from V_{dd} to V_{ss} controls the pulse width of the spiking output channel. When the vector output is high the spiking output is wider than when it is low. This technique allows dual signed data to be sent down a single channel.

from two separate output lines. In this case, the vector output of the bipolar cell dictates the pulse width of the spiking ganglion cell output (Figure 11.28).

Within each of these circuits current mode feedback is analogous to the excitatory and inhibitory stages of depolarization and hyperpolarization, and also to the function of neurotransmitters within synapses. Using these techniques, accurate silicon neurons may be fabricated and used as part of an accurate artificial neuron (Banks et al., 2007).

11.5 Address Event Representation

Traditional image extraction from CMOS and CCD imaging chips is performed by raster scanning the pixel matrix and extracting the information. This method scans each pixel of the imaging array in turn, converts the electronic representation of light intensity into a digital signal, and forwards the information to the subsequent communication bus. The light intensity is binary encoded, and the pixel positional information is contained within its sequential place in the outbound bit stream. The information is then fed down a single high-speed serial connection to the subsequent processing, storage, and display. This method is highly effective at extracting high-quality large image matrices. There are some variations, such as interlaced raster scanning (common in YUV color extraction), but, ultimately, image extraction is similar in the broad range of commercial devices.

In contrast, mammalian vision systems do not have a fast serial bus. In the human eye, information from over 130 million photoreceptors is compressed and passed through the 1 million parallel ganglion cells of the optic nerve (Kolb et al.). Given that each ganglion has a maximum burst frequency of 1 KHz, the total burst bandwidth is 1 GHz. This is still not enough to encode the information from all 100 million cells in real time, hence the retinal processing, and the foveation scheme, which will be described in the next section. Encoding along these parallel ganglion cells is frequency based and burst based, with perhaps some correlated interganglion spike timing encoding.

Retinomorphic imaging chips can learn from this scheme. Thus, rather than sending binary encoded blocks of feature intensity representation, the pixels can output a series of event spikes. Intensity content can then be translated from the frequency of spikes over a given sample period, or the time between two spikes or time to first spike:

- *Rate coding* is carried out by counting the frequency of events in a specific time period. In effect this form is averaging the redundant information over a set of spikes.
- *Spike coding* times the period between spikes and thus does use redundant information.
- *Time to first spike*, sets a starting trigger, and measures the time taken to achieve the first spike.

In biology, neurons have a massive parallel connectivity. Thus, hundreds of event pathways can be connected directly to their destination location. The receiver then "knows" both the timing information and the origin of the spikes. In CMOS, it is difficult to achieve more than a handful of interconnects. However, those fewer interconnects can be much faster, allowing for massive multiplexing. In order for the destination to know the origin of an individual event, it can be tagged with the address of the origin location. Hence, the title for this information scheme is *address event representation* (AER).

The AER protocol was first developed by Mahowald (Mahowald, 1992). In addition to the intelligent imaging systems outlined above, this protocol has been used to design high dynamic range imaging chips (Culurciello et al., 2003), auditory chips, and neural network processing chips. The key difference in AER from raster scanned methods is that the information is asynchronous. As a result simple but powerful algorithms can be implemented without significant additional energy expenditure. Image thresholding, for example, can be performed by simply adding a global time cutoff to a time to first spike system.

11.5.1 The Arbitration Tree

For imaging arrays, adding an address tagging scheme at the pixel level would require considerable silicon real estate. In addition, whereas a scanning system sequentially extracts information from a given matrix, AER is, by its nature, stochastic. As a result, there is a chance that collisions can occur, distorting the information at the multiplexing stage.

To counter this, arbitration tree schemes have been developed that try to order the extraction of information and tag extracted events with the address of their origin. Figure 11.29 describes the process involved in the information extraction from a one-dimensional array of spiking addresses. When a specific address location desires to output an event, its output request line will go high. This will then activate an arbitration unit that will determine whether to transmit the request to the next level. The determination is a first past the postlogic process. The arbiter unit will only allow transmission if no prior transmission has occurred from the other line.

11.5 Address Event Representation

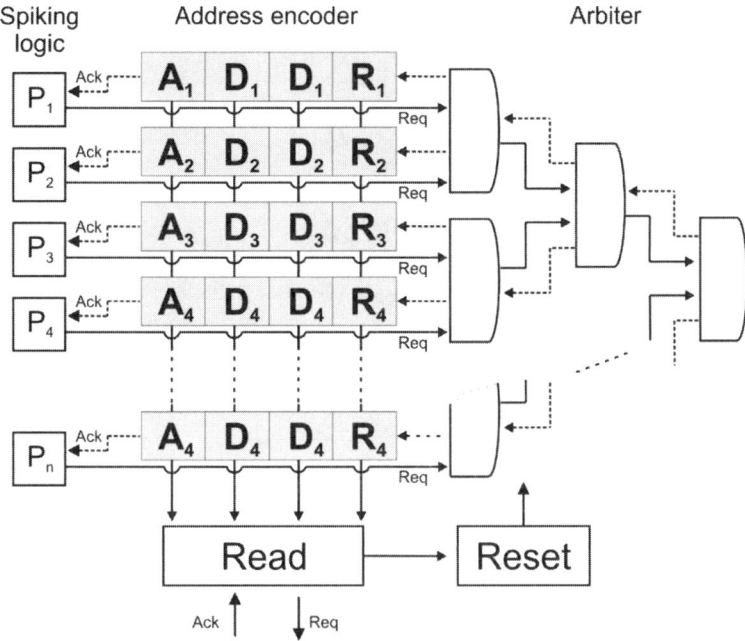

Figure 11.29 The arbitration scheme for a one-dimensional array of spiking addresses. When an address location sends a request, it passes the request to an arbitration tree that will arbitrate between competing signals. The arbitration tree will then send the winning signal backwards to the address encoder and a signal to the read block to request a read and reset.

This arbitration then cascades through a tree until the final unit is reached. When this occurs, an acknowledge will be sent back to the winning unit and then to the address encoder. A signal is also sent to the read unit to request an output. When this is acknowledged, the read unit will read out the highlighted address and send a signal to the reset unit that will reset the arbiter. An acknowledge is also sent back to the requesting pixel to reset the spiking process.

Two-dimensional systems can extract the spike timing of a single unit in a row, or the entire row. In the case of extraction of a single unit, the method is known as bit-AER. Often, logic circuits will convert the bit event from parallel to serial form to be compatible with serial communication protocols such as USB. In this case, it is known as bit-serial.

A schematic for the bit-AER scheme can be seen in Figure 11.30. When a pixel spikes, it checks if the row is already active and then outputs a request if free. This request is then transmitted along the row request to the arbiter and progresses through the arbitration tree. If successful, a signal is passed to activate the X-encoder and the column arbiter. When the column arbiter has chosen the spiking unit to be output, it activates the Y-encoder and sends a signal to the read block to transmit the event. On acknowledge, a signal is sent to the reset block to reset the arbiter and the pixel that produced the event.

The maximum throughput of the bit-AER will either be limited by the capacitive-resistive limit of the row and column lines or the arbitration time. Either way, the maximum throughput of the bit-AER is determined by the size of

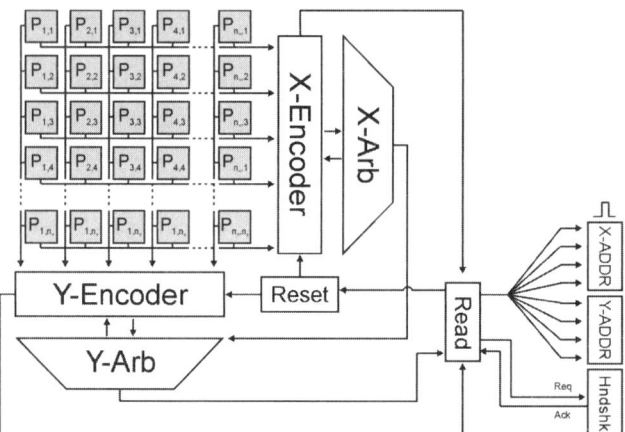

Figure 11.30 two-dimension bit coded address event arbitration. The spike from a pixel passes along a row and requests attention from the arbiter. On successful routing through the arbitration tree, the arbiter activates the X-encoder, and sends a signal to the column arbitration to arbitrate the columns. The output is then a spike consisting of the X- and Y-addresses.

the address matrix $n_x n_y$ and the arbitration unit time t_{unit}. As the matrix increases in size the maximum possible throughput roughly reduces in a logarithmic step function:

$$\eta_{arb} \propto \frac{1}{\log_2 \sqrt{n_x n_y}} \quad (11.16)$$

This can be seen schematically in Figure 11.31 below.

A modification to this theme is the word-serial scheme (Boahen, 2004a; Boahen, 2004b; Boahen, 2004c). In the word-serial scheme, when an event gets through the row arbitration scheme, all the column information is read out from that row rather than the individual pixel. Subsequent nonevents can be ignored. The advantage of the word-serial scheme is that it increases bandwidth. This, however, comes at a cost. AER systems are particularly useful for non-linear imaging arrays, such as foveated chips. The word-serial system, however, is more suited to uniform matrix arrays.

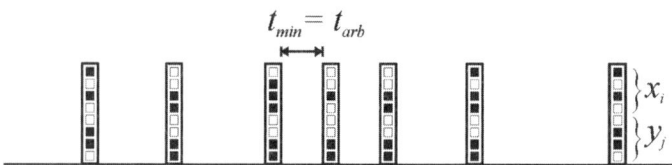

Figure 11.31 The rate of output events from addresses (x_i, y_j) will vary according to the signal, but the maximum rate is determined by the minimum time interval t_{min}, which is determined by the arbitration delay.

11.5 Address Event Representation

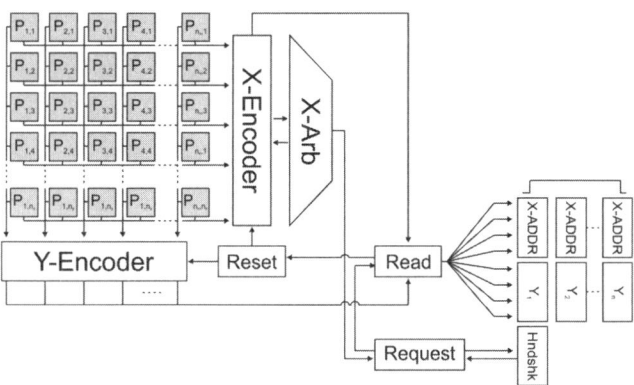

Figure 11.32 The scheme for word-arbitration. A given address requesting an attention sends the signal to the X-arbiter, which performs the row arbitration as described previously. It then sends a signal to the Y-encoder and the read. No arbitration is required for the Y-encoder. The output is then a compilation of pulses from the row.

Figure 11.32 shows the scheme for the word-AER system. When an address produces an event, it does not check whether the row is free, but signals a request to the arbiter seeking attention. The row arbiter carries out its arbitration tree as described before. On completion, a signal is sent back to the X-encoder and to the read. The X-encoder reads and buffers the signals from the pixels. The read unit then reads the row and the Y-encoder submits addresses for the lines that have addresses that have produced events. The reset then resets the X-arbiter and the pixels on the row which has fired.

The output information stream is described in more detail in Figure 11.33. If the Y-encoder in Figure 11.32 simply reads in the pixels with events from a given row, then its output into a shift register would be as shown in (a). It is, however, conceivable to remove the noneventing columns as shown in (b). This is desirable both in terms of information compression and bandwidth, and because the output

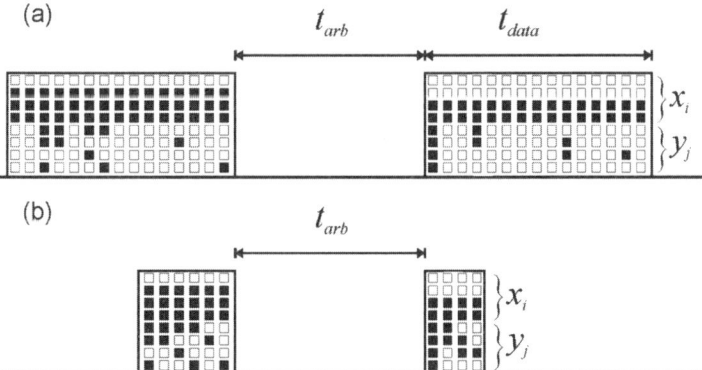

Figure 11.33 Events are output in row blocks with active addresses (x_i, y_j). (a) In this case there is no removal of the columns that have no events. (b) Removal of addresses without events.

is still a stream of event locations, whereas the scheme of (a) is a hybrid between AER and a traditional row raster scan.

11.5.2 Collisions

The major drawback in the AER method is the time gap between spikes, which does not exist in the raster method. This time gap means that spikes attempting to output during that dead time have to be queued. If, however, the spiking rate exceeds the overall extraction rate, there is a danger that the output information will be distorted.

To date, most retinomorphic imaging chips have been fabricated with small photoreceptor densities (e.g., 4.8K pixels (Andreou and Boahen, 1994). However, commercially useful vision systems require VGA resolutions (307 kpixels) or above. As the pixel densities scale up, then so too does the arbitration tree, and the capacitance of the output line. For an output line with a bandwidth of 100 Mhz, if the delay in the arbitration tree is 100 μs, the effective bandwidth would be only 10 kHz!

Collisions distort the output matrix in two ways:

- If colliding events are dropped then the output value will not match the attempted natural event generation value of the pixels
- If the colliding events are queued, then the queue time will lead to a distortion in the inter-event timing compared to the natural event generation frequency.

In practice, a mixture of the two types of distortion will occur. When a pixel or row is in the process of outputting its data, other pixels that have generated events will be effectively queued. This queuing process distorts their event timing output. Additionally, once they have generated an event, their event generator can no longer function, so any additional events that could have been generated while in the queue have to compete with any other pixels that also want to request the bus.

For rate coding, the distortion will increase massively with increasing matrix size as the maximum throughput decreases and the event output increases. In spike-coded AER and time to first spike AER, fewer generated events will in theory allow more throughput, but collisions will be more catastrophic to the information output.

11.5.3 Sparse Coding

The AER user community is fully aware of the collision problems. Thus, the dominant argument for the use of address event systems is for sparsely coded data. The human retina performs spatial and motion processing and outputs the resultant compressed data down the parvocellular and magnocellular pathways, respectively. The resulting image thus has a small percentage of the output of the original image.

11.5 Address Event Representation

If an image Matrix $I(m, n)$ is converted into spike form, then the total number of spikes to be extracted is given by:

$$\eta_I \Delta t = \sum_x \sum_y I_{ij} \cdot f(I) \qquad (11.17)$$

where η_I is the spiking rate and I_{ij} is the pixel intensity at a given position. For a given convolution process, the total number of spikes will be reduced according to that convolution. Examples are given in Figure 11.34, below:

In the case of the figures above, for a 256×256 image the spiking intensities would be:

Image	Percent of (a)	Spikes
(a)	100%	1.10×10^7
(b)	30.96%	3.41×10^6
(c)	7.88%	8.69×10^5
(d)	2.11%	2.33×10^5

Thus, if we wish to compare the output of the AER with different output percentages, we can simply scale an output image within a matrix.

11.5.4 Collision Reduction

The argument in favor of address event systems is that their purpose is not to extract whole images but to output the salient features, such as edges of a particular image. Thus, perhaps only 10% of a given image matrix needs to be output within a given time frame.

Queuing systems can also help (Culurciello et al., 2003), as can reducing the total number of spikes by using a time to first spike protocol rather than rate coding. The word-AER is also a useful method for increasing bandwidth. However, for salient feature extraction systems, often only a small proportion of a large array may need to be extracted for each frame. The foveation scheme described in the next section may also be useful in reducing information content.

Ultimately, the AER scheme will probably remain in use for low-bandwidth systems only, that is, small sensor arrays rather than large ones. The scaling of transistor technology is making individual transistors smaller and faster, but the capacitance of the output lines is actually increasing. This capacitance is the basis

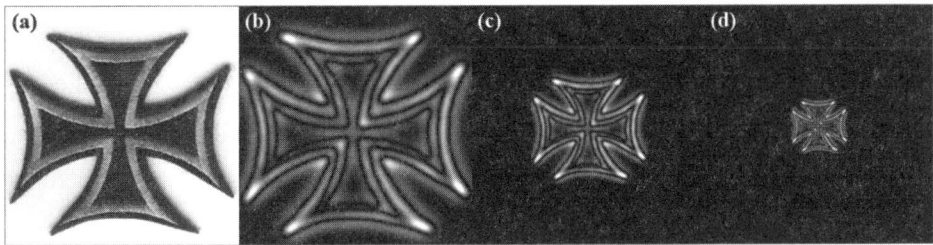

Figure 11.34 (a) An original image with a roughly even histogram of intensities. (b) A Laplacian convolution output of (a). (c and d) Size reductions of the Laplacian output to reduce spiking output.

for an RC time constant which does not scale favorably with transistor downsizing, and is thus fast becoming the limiting factor in the speed of most electronic circuits, and especially asynchronous circuits.

11.6 Adaptive Foveation

Foveation provides the advantage of being able to trade spatial resolution for temporal resolution in a limited bandwidth regime. Previous work (Azadmehr M. et al., 2005; Bernardino A. et al., 2002; Etienne-Cummings R. et al., 2000; Sandini G. et al., 2002; Wodnicki R. et al., 1997) concentrated on mimicking biology, developing foveated vision chips by physically defining a foveal region of increased resolution at the center of the CMOS imager. The biological eye has exceptionally good optomechanics. This enables the eye to be moved around the visual field at high speeds. In contrast, even with modern MEMS technologies, it is challenging to position silicon-based cameras with as many degrees of freedom at such high speed and accuracy, in addition to being compact. In silicon, however, it is possible to electronically define and reposition the fovea; this may be seen as an analogy to foveation through movement that is encountered in biology, but does not require sensor motion. Thus, a good approach to artificial foveation is electronic rather than physical (Constandinou et al., 2006).

Adaptable foveation allows for an area to be specified as either a high spatial resolution fovea or high temporal resolution periphery. The scheme developed in this paper acts not only to designate a spot which is to act as a fovea, but to be able to dynamically adjust the size of that region at will. The broadening or thinning of the fovea spot is adaptable to imaging requirements and requires only a simple coordinate in feedback.

The basic principle is to group pixels into photoreceptor clusters and to take intensity information on an asynchronous event timing basis rather than through

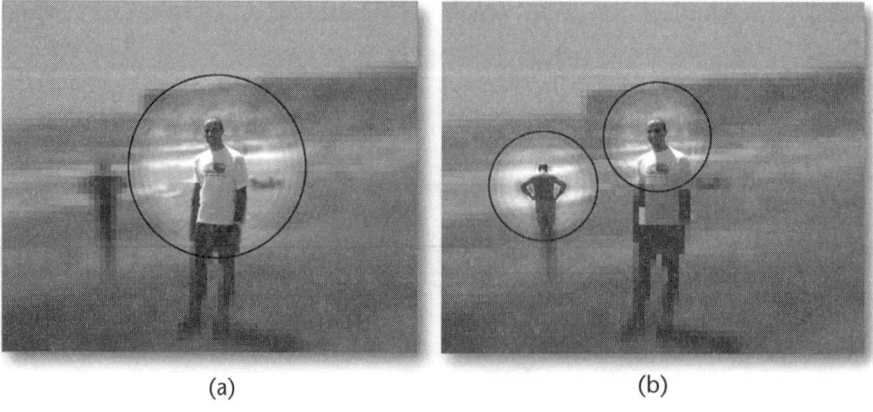

Figure 11.35 The effect of foveation on imaging. Shown is: (on the left) imaging with a single large fovea and (on the right) imaging with two foveal regions (Constandinou, et al., 2006).

raster scanning. The photoreceptor clusters can either let all their constituents send out spatial information or integrate their response into a single combined output. A global current spreading network between photoreceptor clusters can be used to determine which clusters relative to an initial position get defined as fovea or periphery.

11.6.1 System Algorithm

This scheme proposes to combine pixels into repeat unit structures of 3×3 pixels. The 9 pixels in these unit structures can act as either individual photoreceptors or as a singular compound photosensor. When combined, the photocurrents are added for better light sensitivity, and the output can be 9 times faster at the cost of 9 times less spatial resolution. In this way the unit structures can be defined as foveal (individual sensors) or peripheral (singular sensors). The scheme is visualized in Figure 11.36 below.

To achieve the foveal and peripheral definition, a current spreading and thresholding technique is used. Photoreceptor groups within the array can be addressed using (x, y) coordinates. On addressing a specific group, a foveal current is sourced to that group; subsequently being attenuated and spread to neighboring photoreceptor groups. Thus, there is a decay function from the designated fovea to the farthest periphery. Given that the fovea is not predetermined, each point will feed back to its surrounding photoreceptor groups, as in Figure 11.37. A thresholding function can then be used to differentiate between the fovea and the periphery.

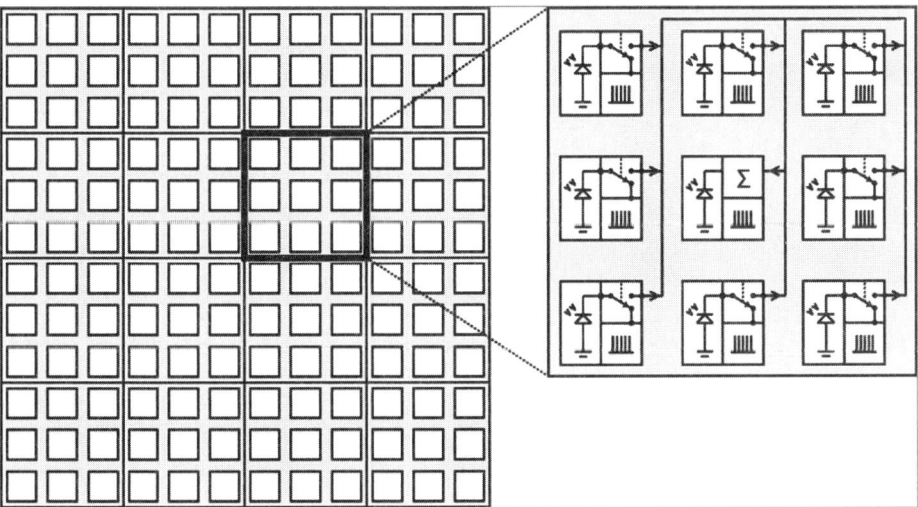

Figure 11.36 The scheme of how the photoreceptor groups can be reconfigured to operate as either foveal or periphery units.

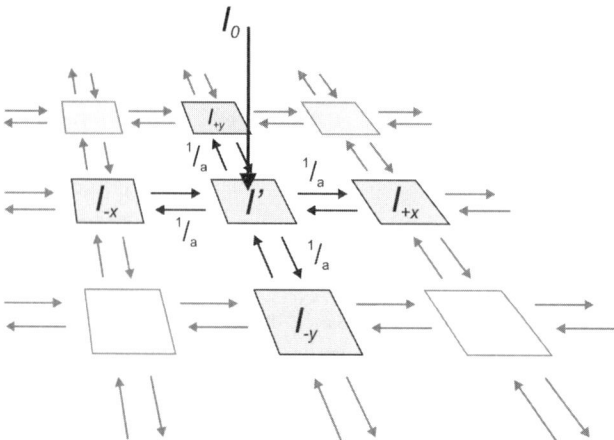

Figure 11.37 Foveal current spreading technique.

Mathematically, the foveal decay current can be described in terms of a recursive spreading function, where each component's feed forward influence from its preceding neighbor is:

$$I' = I_{+x+y} \sum_{n=0}^{\infty} \frac{(n+1)}{a^{(2n+1)}} + I_{-x+y} \sum_{n=0}^{\infty} \frac{(n+1)}{a^{(2n+1)}}$$

$$+ I_{+x-y} \sum_{n=0}^{\infty} \frac{(n+1)}{a^{(2n+1)}} + I_{-x-y} \sum_{n=0}^{\infty} \frac{(n+1)}{a^{(2n+1)}}$$

where I_{xy} is the interactive influence from one of the neighboring points in the grid and a is the attenuation factor. For the array to be stable at the central point the attenuation factor must be greater than 4. These function relationships can be solved using iterative recursion in standard mathematical software. Results of these simulations can be seen in Figure 11.38, where an initial attenuation factor of 4 was used at the fovea, followed by an attenuation factor of 6 between all other connections.

The foveal current spreading technique simulation can be seen for a 27 × 48 pixel array that has been sent for fabrication and also for a 320 × 200 pixel array.

For the 27 × 48 pixel array the propagation forms a readily tunable percentage of the array. However, this spreading function forms a small percentage of a 320 × 200 array. If desired the attenuation factor can be reduced to enhance spreading. In order to be fully scalable to a mega-pixel array, a logarithmic output function may be required to allow broadly tunable spreading of the fovea.

11.6.2 Circuit Implementation

The foveal control circuit, as previously explained, sources a current to a specific location spreading through a network of interconnecting pixel groups. The circuit used to achieve this functionality is illustrated in Figure 11.39.

11.6 Adaptive Foveation

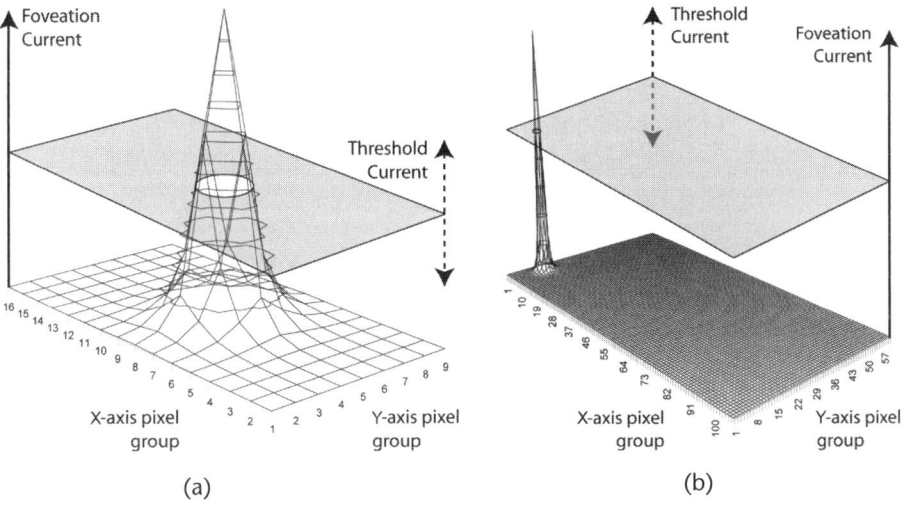

Figure 11.38 The spreading function of the foveal current. Left, the spread for a 27 × 48 chip. Right, the spread for a 320 × 200 array.

In this scheme, v_fov is used to set the seeding foveal current, defining the initial current level, and therefore defining the fovea shape. The x_fov and y_fov inputs provide the row and column coordinates that act to select the foveal centroid. The foveal current is subsequently attenuated by a factor of 2 and distributed to each of the neighboring pixels via simple mirror Q7 to Q10. Furthermore, the received currents from adjacent cells also act to influence a cell's effective foveal current contribution through the received i_sum, which is attenuated by a factor of 6. For pixel groups situated away from the specified foveal center, this influence

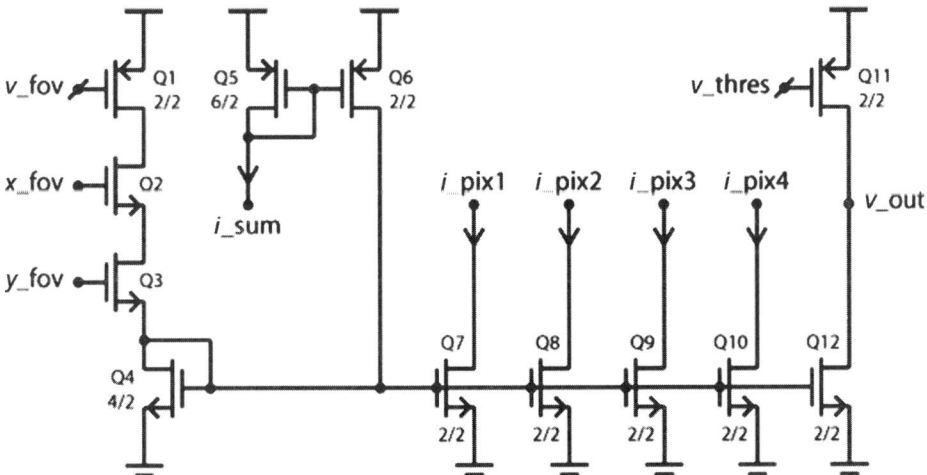

Figure 11.39 The implementation of the current spreading network determining the foveation state.

provides the only source of current, as the initial foveal current is supplied to only a single pixel group. This effective current is then thresholded using a simple current comparator and implemented using an opposing CMOS pair Q11 and Q12. This operates on the principle that the effective foveal current will be sunk from a current source Q11, set by v_thres, acting to drive either Q11 or Q12 into its ohmic region. This will therefore result in a voltage swing at node v_out, providing the foveal control signal.

This circuit is akin to previous pseudoresistive networks (Andreou A. and K., 1995; Vittoz E. A., 1997) except that it is in current mode and has some advantages for this structure. Reducing the mismatch on this circuit is not critical as the specific shape of the fovea is not significant. The overhead this circuit therefore places on the fill-factor is subsequently not great.

The surface fill-factor is very important to the utility of this structure. Foveation is desirable in order to add utility, but not at the expense of scalability, for example, to a megapixel resolution. To provide a proof of principle, a 27×48 pixel (9×16 photoreceptor cluster) array has been designed and submitted for fabrication in a standard 0.25 μm CMOS process. The tessellating pixel-group layout is shown in Figure 11.40, below.

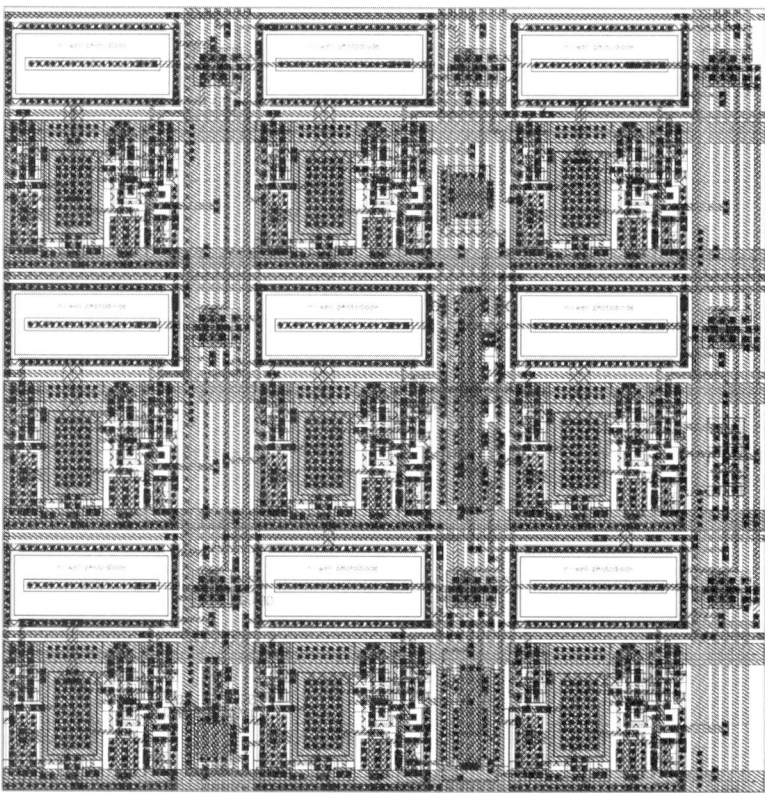

Figure 11.40 The 3×3 pixel-group layout implemented in a 0.25 μm CMOS process. The implemented surface fill-factor of the photodiodes was 22.4%, but this is without major optimization. Fill-factors of 30% or higher should be possible within the same overall cell size.

11.6 Adaptive Foveation

Each individual photoreceptor within the 3 × 3 pixel-group has a dimension of 25 × 25 microns, including the foveation overhead. This compares favorably with commercial imaging array dimensions, allowing for implementation of a megapixel resolution at 25 mm × 25 mm.

The readout system being not of standard matrix dimensions and readout functionality requires the use of AER readout, described previously. The main functional difference in this case from those previously is that in this case, several address events are inhibited from signalling. This will occur when pixel groups act as peripheral pixels, sending their information as a single address event instead of 90 distinct address events. The AER scheme can be seen in Figure 11.41.

11.6.3 The Future

In principle this foveation technique is not purely limited to a single foveating point. Multiple foveating points could be implemented, posing some interesting information processing possibilities. In the human eye, the physical limitations of our biological components have limited our visual architecture to a single foveating point. If a silicon system is not limited in this way, is there an advantage in having multiple foveating points? Could a system with specific fovea for different tasks be

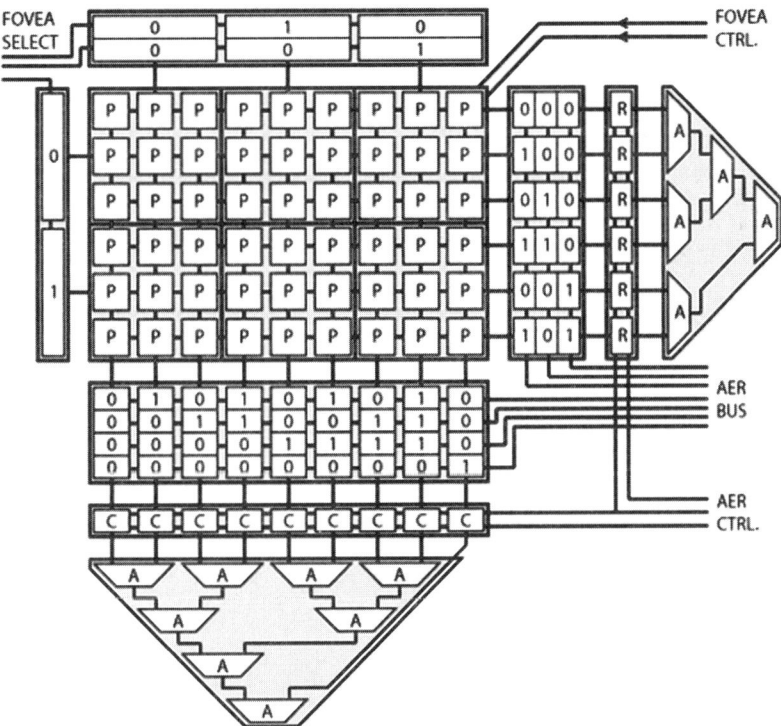

Figure 11.41 System architecture of the adaptable foveating vision chip. Sub-blocks illustrated include: P = Pixel, R = Row Latches, C = Column Latches, A = Arbiters, and 0/1 = Address Encoders or Decoders.

developed? For example, a fovea for reading text and another for face recognition. These questions as yet remain unanswered.

11.7 Conclusions

In the world of Moores law and reconfigurable digital logic and high-power computer processors, it can seem that analog electronics is somewhat old fashioned. Certainly, initial competitive comparisons would favor multi-GHz digital processors over their KHz speed silicon or biological neurons. However, the fastest digital computational engines are hard pressed to achieve the visual pattern recognition capabilities of the human visual system. Furthermore, the human visual system has a power consumption of 4W compared to the hundreds of watts used in high-power digital systems. Clearly, for applications requiring portability, low-power algorithms are required, yet not at the expense of the mathematical requirements.

In Chapter 12, we consider a generic analog platform for spatiotemporal processing, and show how this may be designed to mimic V1-type anisotropic (orientation-selective) responses. This complements the design strategy in this chapter, which has focused mainly on retinomorphic spatial and temporal processing.

In later chapters, digital parallel processing concepts will be described, which can also provide lower power consumptions. Their advantage is that by using reconfigurable digital logic, they retain an element of programmability. However, in this chapter the algorithms described investigate ultra-low-power analog configurations that can be implemented directly into the CMOS imaging chip. Furthermore, if organic/CMOS configurations, as outlined in the previous chapter, can be used, the valuable silicon real estate underneath such pixels can be put to best use by performing analog algorithms

Ultimately, the analog and digital worlds must meet, and we have investigated methods of their interface in this chapter. Traditional CMOS and CCD chips tend to raster scan their outputs. This is fine for taking high-quality images, but there may be some advantage in looking at asynchronous approaches, not least because it may be easier to perform certain types of algorithms on event-based information upstream from the imaging array.

The neuromorphic field is over 20 years old, but has remained young as a result of being at the periphery of the imaging world. As transistor sizes move ever deeper into the nanoscale, aspects of this type of intelligent imaging will develop increasing interest from a community obsessed with mobile technologies running on limited battery supply.

References

Ackland, B. and Dickinson, A. (1996) "Camera on a chip." *Solid-State Circuits Conference, 1996. Digest of Technical Papers. 43rd ISSCC, 1996 IEEE International*, 412, 22–25.

Andreou, A. and Boahen, K. (1995.) "A 590,000 transistor 48,000 pixel, contrast sensitive, edge enhancing, CMOS imager-silicon retina." *Proceedings of the 16th Conference Advanced Research in VLSI*, 225–240.

Andreou, A. G. and Boahen, K. A. (1994) "A 48,000 pixel, 590,000 transistor silicon retina in current-mode subthreshold CMOS." *Circuits and Systems, 1994. Proceedings of the 37th Midwest Symposium on Circuits and Systems*, 102–197.

Andreou, A. G., et al. (1991) "Current-mode subthreshold MOS circuits for analog VLSI neural systems." *Neural Networks, IEEE Transactions on*, 2, 205–213.

Azadmehr, M., Abrahamsen, J. P. and P, H. (2005) "A Foveated AER imager chip." *IEEE International Symposium on Circuits and Systems*, 2751–2754.

Banks, D. J., Dagenaar, P. and Toumazou, C. (2005) "A colour and intensity contrast segmentation algorithm for curent mode pixel distributed edge detection." *Eurosensors XIX*. Barcelona.

Banks, D. J., Degenaar, P., Toumazou, C. (2007) "Low-power pulse-width-modulated neuromorphic spiking circuit allowing signed double byte data transfer along a single channel." *Electron Lett*, 43, 704–709.

Bernardino A., Santos-Victor J. and G., S. (2002) "Foveated active tracking with redundant 2D motion parameters." *Robotics and Autonomous Systems*, 3, 205–221.

Boahen, K. and Andreou, A. G. (1992) "A contrast sensitive silicon retina with reciprocal synapses." *Advances in Neural Information Processing*, 4, 764–772.

Boahen, K. A. (1996) "A retinomorphic vision system." *Micro, IEEE*, 16, 30–39.

Boahen, K. A. (2004a) "A burst-mode word-serial address-event link-I: transmitter design." *Circuits and Systems I: Regular Papers, IEEE Transactions on [see also Circuits and Systems I: Fundamental Theory and Applications, IEEE Transactions on]*, 51, 1269–1280.

Boahen, K. A. (2004b) "A burst-mode word-serial address-event link-II: receiver design." *Circuits and Systems I: Regular Papers, IEEE Transactions on [see also Circuits and Systems I: Fundamental Theory and Applications, IEEE Transactions on]*, 51, 1281–1291.

Boahen, K. A. (2004c) "A burst-mode word-serial address-event link-III: analysis and test results." *Circuits and Systems I: Regular Papers, IEEE Transactions on [see also Circuits and Systems I: Fundamental Theory and Applications, IEEE Transactions on]*, 51, 1292–1300.

Constandinou, T., Degenaar, P. and Toumazou, C. (2006) "An adaptable foveating vision chip." *Proceedings of the IEEE International Symposium on Circuits and Systems, Kos*, 3566–3569.

Culurciello, E., Etienne-Cummings, R. and Boahen, K. A. (2003) "A biomorphic digital image sensor." *Solid-State Circuits, IEEE Journal of*, 38, 281–294.

Darwin, C. (1975) "Difficulties on theory." *On the Origin of Species a Facsimile of the First Edition*. Cambridge MA, Harvard University Press.

Etienne-Cummings R., Van Der Spiegel J., Mueller P. and Zhang, M.-Z. (2000) "A foveated silicon retina for 2D tracking." *IEEE Transactions on Circuits and Systems II*, 47, 504–517.

Fossum, E. R. (1995) "CMOS image sensors: electronic camera on a chip." *Electron Devices Meeting*, 17–25.

Fossum, E. R. (1997) "CMOS image sensors: electronic camera-on-a-chip." *Electron Devices, IEEE Transactions on*, 44, 1689–1698.

Gruev, V., Etienne-Cummings, R. and Horiuchi, T. (2004) "Linear current mode imager with low fix pattern noise." *Circuits and Systems, 2004. ISCAS '04, Proceedings of the 2004 International Symposium*, 4.

Hofmann, O., et al. (2005) "Thin-film organic photodiodes as integrated detectors for microscale chemiluminescence assays." *Sensors and Actuators B: Chemical*, 106, 878–884.

Kameda, S. and Yagi, T. (2001) "An analog vision chip applicable to real-time image processings in indoor illumination." *Neural Networks, 2000. IJCNN 2000, Proceedings of the IEEE-INNS-ENNS International Joint Conference on*.

Kameda, S. and Yagi, T. (2001) "A silicon retina calculating high-precision spatial and temporal derivatives." *Neural Networks, 2001. IJCNN '01. Proceedings of the International Joint Conference on*, 1, 201–205.

Kameda, S. and Yagi, T. (2002) "Calculating direction of motion with sustained and transient responses of silicon retina." *SICE 2002. Proceedings of the 41st SICE Annual Conference*, 3, 1853–1858.

Kameda, S. and Yagi, T. (2003) "An analog VLSI chip emulating sustained and transient response channels of the vertebrate retina." *Neural Networks, IEEE Transactions on*, 14, 1405–1412.

Kobayashi, H., Matsumoto, T. and Sanekata, J. (1995) "Two-dimensional spatio-temporal dynamics of analog image processing neural networks." *Neural Networks, IEEE Transactions on*, 6, 1148–1164.

Kobayashi, H., White, J. L. and Abidi, A. A. (1990) "An analog CMOS network for Gaussian convolution with embedded image sensing." *Solid-State Circuits Conference, 1990. Digest of Technical Papers. 37th ISSCC, 1990 IEEE International*, 300, 216–217.

Kobayashi, H., White, J. L. and Abidi, A. A. (1991) "An active resistor network for Gaussian filtering of images." *Solid-State Circuits, IEEE Journal of*, 26, 738–748.

Kolb, H., Fernandez, E. and Nelson, R. Webvision—The Organization of the Retina and Visual System. http://webvision.med.utah.edu/.

Mahowald, M. (1994) "Analog VLSI chip for stereocorrespondence." *Circuits and Systems, 1994. ISCAS '94, 1994 IEEE International Symposium on*, 6, 347–350.

Mahowald, M. A. (1992.) VLSI Analogs of Neuronal Visual Processing: A Synthesis of Form and Function. *PhD thesis, Dept. of Computation and Neural Systems, California Institute of Technology, Pasadena, CA*.

Mead, C. A. (1989) *Analog VLSI and Neural Systems*, New York, Addison-Wesley.

Mead, C. A. and Mahowald, M. A. (1988) "A silicon model of early visual processing." *Neural Networks*, 1, 91–97.

MICRON "Megapixel CMOS Active Pixel Digital Image Sensor."http://www.micron.com/products/imaging/products/megapixel.html.

Nicholls, J. G., Martin, A. R. and Wallace, B. G. (1992) *From Neuron to the Brain*, Sunderland MA, Sinauer Associates, Inc.

Sandini G., Santos-Victor, J., Paidia, T. and Burton, F. (2002) "OMNIVIEWS: direct omnidirectional imaging based on a retina-like sensor." *Proceedings of IEEE Sensors*, 1, 27–30.

Sanyo (2004) Thinnest and Lowest Power Consuming Megapixel CCD. http://www.sanyo.co.jp/koho/hypertext4-eng/0403news-e/0302-e.html#3.

Tanzer, M., Graupner, A. and Schuffny, R. (2004) "Design and evaluation of current-mode image sensors in CMOS-technology." *Circuits and Systems II: Express Briefs, IEEE Transactions on [see also Circuits and Systems II: Analog and Digital Signal Processing, IEEE Transactions on]*, 51, 566–570.

Vittoz, E. and Fellrath, J. (1977) "CMOS analog integrated circuits based on weak inversion operations." *Solid-State Circuits, IEEE Journal of*, 12, 224–231.

Vittoz, E. A. (1997) "Pseudo-resistive networks and their application to analog collective computation." *Proceedings of MicroNeuro*, 163–172.

Ward, V. (1995) "VLSI implememtation of receptive fields with current-mode signal processing for smart vision sensors." *Analog Integrated Circuits and Signal Processing*, 7, 167–179.

Ward, V., Syrzycki, M. and Chapman, G. (1993) "CMOS photodetector with built-in light adaptation mechanism." *Microelectronics Journal*, 24, 547–553.

Wodnicki R., Roberts, G. W. and Levine, M. D. (1997) "A log-polar image sensor fabricated in a standard 1.2um ASIC CMOS process." *IEEE Journal of Solid-State Circuits*, 32, 1274–1277.

Yagi, T., Hayashida, Y. and Kameda, S. (Eds.) (1998) *An Analog VLSI Which Emulates Biological Vision*, New York, Addison Wellesley.

Yang, D. X. D., Fowler, B. and El Gamal, A. (1999) "A Nyquist-rate pixel-level ADC for CMOS image sensors." *Solid-State Circuits, IEEE Journal of*, 34, 348–356.

Yang, D. X. D., Fowler, B. and Gamal, A. E. (1998) A Nyquist rate pixel level ADC for CMOS image sensors. *Custom Integrated Circuits Conference, 1998. Proceedings of the IEEE 1998*, 348–356.

CHAPTER 12
Analog V1 Platforms
Henry M. D. Ip, Emmanuel M. Drakakis, and Anil Anthony Bharath

12.1 Analog Processing: Obsolete?

It is surely provocative to suggest that analog computational cores have anything at all to offer in the digital world of today. Yet, digital circuitry is built using devices that rely on the same fundamental physics as those of analog circuitry. Despite the ease and flexibility of programmable digital devices, and the precision of computation attainable with digital systems in general, the modern electronic design engineer has surely been faced with the interesting realization that a digital implementation of, say, a standard eighth-order Butterworth filter is vastly more power-hungry than a traditional analog version of the same, a consideration that becomes increasingly important as power consumption becomes an ever-more important design constraint.

In short, the main requirements that determine the choice of hardware platform for a specific application tend to be (a) the ease of implementation to meet desired performance characteristics; (b) the ability to accommodate changes in design specifications; (c) costs incurred in either design, manufacture, or maintenance; and (d) power consumption. Thus, for an application that has a design requirement of high-precision computation, but is relatively unconstrained by power costs, a digital platform is almost always the method of favor, particularly in view of the flexibility offered by programmable systems. However, where a device must be power efficient, in order to be mobile, or is required to perform front-end processing that is not particularly demanding in terms of precision, then an analog or mixed-mode device might be a valid choice.

For convenience, Table 12.1 presents a tabular summary of previous "vision chips" with same design details and power consumptions.

Another interesting and obvious characteristic that is offered by networks of *analog devices* is the ability to exploit the physics and temporal dynamics of one or several devices connected together in a spatial arrangement to reproduce the dynamics of groups of neurons; see, for example, the retinal network model of Chapter 2. From the perspective of transforming moving visual input into a "V1-like" representation domain, a mixed model of processing, in which a small amount of programmability encapsulates an analog computational core, is of great potential interest.

A well-known mixed-mode platform for computation that has been applied to the area of vision is the Cellular Neural Network (CNN). This flexible, programmable device uses a spatial network of identical units in order to perform spatial filtering in an extremely power-efficient manner.

Table 12.1 A Comparison of Some Vision Chips

Ref	Institution	Year	Technology	Total Power	Transistor Count	Cell Size	Total Size in Sq Pixels	Receptors	Description
[3]	UMIST, U.K.	2005	0.6 μm	40 mW	128/cell	98.6 × 98.6 μm²	(21 × 21)	Yes	Switch current-based programmable processor. 1.1 GIPS for programmable SIMD core. 2.5 MHz clock. 2.5% error in image filtering.
[4]	Univ. Utah, U.S.A.	2005	0.5 μm²	140 uW		100 × 100 μm²	2.24 × 2.24 mm² (17 × 17)	Yes	One-dimensional motion detectors arranged radially. Delay based motion computation architecture. Temporal bandpass filtering is integrated into the photoreceptors.
[5]	CRST, Italy	2005	0.35 μm	6 mW	30/cell		8.7 mm² (32 × 32)	Yes	3 × 3 edge detection, motion detection, sensor amplification, and dynamic-range boosting.
[6]	Univ. Arizona, U.S.A.	2005	1.6/2.0 μm	500/650/ 200 μW	53/35/29/cell	29,950/20,233/ 17,620 μm²	10.2 mm² (12 × 40) 4.41 mm² (24 × 1) 4.41 mm² (5 × 13)	Yes	A comparison of three implemented one-dimensional motion architectures.
[7]	Univ. Helsinki, Finland	2004	0.25 μm	192 mW	7,092/cell		10 mm² (2 × 72)	No	Digital multipliers. 1.027 million transistors. Analog polynomial weights up to 3rd order. On chip D/A conversion.

12.1 Analog Processing: Obsolete?

Ref	Institution	Year	Technology	Total Power	Transistor Count	Cell Size	Total Size in Sq Pixels	Receptors	Description
[8]	IMSE-CNM-CSIC, Spain	2004	0.35 μm	<2.95 W	3,748,170	73.3 × 75.7 μm^2	(128 × 128) 11.9 × 12.2 mm^2	Yes	3.75 million transistors. 330 GOPS, 82.5 GOPS/W. 0.6 μs time constant. 8-bit analog accuracy. 120 MBytes/s I/O. Dynamic power consumption with idle blocks switching off when not needed.
[9]	Univ. New York, U.S.A.	2004	0.8 μm	9.68 mW	203/cell	170 × 170 μm^2	2.7 mm^2 × 2.95 mm^2 (11 × 11)	Yes	CNN-inspired design. Dual analog network for motion computation and handling motion discontinuity respectively. Read out circuit capable of 1,000 fps at 11 × 11.
[10]	Lodz Univ., Poland	2003	0.8 μm	82 mW	207/cell	132.8 μm^2 × 374.5 μm^2	4.81 mm^2 × 4.3 mm^2 (300)	No	Analog continuous-time processing. Weighted order statistics nonlinear filtering. Analog input sampling.
[11]	IMSE-CNM-CSIC, Spain	2003	0.5 μm	300 mW			9.27 mm^2 × 8.45 mm^2 (32 × 32)	No	Two-layered CNN architecture. Current conveyor analog processing. Single transistor synapse. OTA offset calibration. Analog switch current memories. 10 MB/s I/O 8-bit weight accuracy and 7 to 8-bit image sampling resolution. 100 ns, 470 GOPS.

Table 12.1 Continued

Ref	Institution	Year	Technology	Total Power	Transistor Count	Cell Size	Total Size in Sq Pixels	Receptors	Description
[12]	Univ. Veszperm, Hungary	2003		3W				No	Xilinx Virtex-3000–based FPGA CNNUM emulation.
[13]	CEM Switzerland	2003	0.5 µm	303 mW		69 µm² × 69 µm²	99.7 mm² (128 × 128)	Yes	Carries out magnitude and direction spatial gradient computation. 120 dB DR. 2 ms time constant. 50 fps minimum depending on lighting (integration receptor takes time).
[14]	Univ. Cagliari-Piazza, Italy	2002	0.5 µm	50 mW static	84/cell	80 µm² × 80 µm²	86 mm² (100 × 100)	Yes	1,000 fps, 850k transistors. Performs spatial processing to extract magnitude and direction of spatial gradients.
[15]	Nat. Taiwan Univ.	2002	0.5 µm	30 mW		124 µm² × 124 µm²	5.1 mm² × 5.1 mm² (32 × 32)	Yes	Motion chip based on solving the optical flow equation. Nonlinear resistive grid architecture for continuity of motion. Output writing > 80 fps.
[16]	Johns Hopkins Univ., U.S.A.	2001	1.2 µm	1 mW	65k	45.6 µm² × 45 µm²	(80 × 78) 22 mm²	Yes	Single processing element: digitally programmable analog multipliers for spatiotemporal convolution. Processing electronics routed away from photoreceptor array. 12.4 GOPS/mW at 9.6 kfps. Up to 11 × 11 mask size.
[17]	Univ. Science & Tech., Hong Kong	2000	1.2 µm	<200 µW	52/cell	132 µm² × 108 µm²	2.2 mm² × 2.2 mm² (12 × 14)	Yes	Current mode processing. Two layer CNN Gabor filter implementation.

12.1 Analog Processing: Obsolete?

Ref	Institution	Year	Technology	Total Power	Transistor Count	Cell Size	Total Size in Sq Pixels	Receptors	Description
[18]	Johns Hopkins Univ., U.S.A.	1999	1.2 μm		88/cell		2.2 mm² × 2.2 mm² (12 × 10)	Yes	CNN-inspired analog processing core using CMOS pseudoresistors for diffusion.
[19]	U.C. Berkeley, U.S.A.	1998	0.8 μm	0.3W		232 μm² × 263.5 μm²	(16 × 16) 5.5 mm² × 4.7 mm²	No	Analog I/O. 5000 analog multipliers. 90 ns time constant. 1 billion operations/sec excluding I/O.
[20]	IMSE-CNM-CSIC, Spain	1997	0.8 μm	1.1W		190 μm² × 190 μm²	30 mm² (20 × 22)	Yes	Current conveyer-based analog processing. On-chip DACs for weight adjustments. 10 MHz I/O rate.
[21]	Univ. De Granada, Spain	1997	1.2 μm	0.64 mW			(8 × 8)	Yes	0.4 μs time constant. Differential pair-based integrators. Gilbert multipliers. 150 μs time constant.
[22]	Univ. Illinois, U.S.A.	1997	2 μm	350 μW/cell		110 μm² × 220 μm²	2.3 mm² × 2.3 mm² (1 × 9 and 5 × 5)	Yes	Motion chip to solve the optical flow equation. Photoreceptors packed away from processor circuitry.
[23]	Inst. Ricerca Science & Tech., Italy	1995	1.2 μm	12 mW		48 μm² × 48 μm²	13 mm² (64 × 64)	Yes	Opamp and switched capacitor-based motion chip.

12.2 The Cellular Neural Network

The cellular neural network paradigm was first introduced by Chua and Yang in 1988 [24]. Over the years, CNNs have attracted much attention, establishing themselves as a distinct subject area in the circuits community. The study of CNNs requires tools of nonlinear control, considerations of system stability, linear filters, analog VLSI design principles, and more; this makes their study both challenging and highly interesting. Applications of CNNs span several areas, including the modelling of biological systems, control of robotic systems, computer vision, and general artificial intelligence. The implementations of CNNs include not only their "native" mixed-mode (so-called "analogic") VLSI systems, but also of the CNN universal-machine (CNNUM) analog computers, and CNN-inspired application-specific integrated circuits (ASICs). For the core of a CNN processor, analog continuous-time or discrete data processing could be employed via a wide range of strategies.

Even classical CNNs might be seen as fitting well with the goal of creating generic platforms for vision. Indeed, the use of CNNs as linear image filtering tools has been discussed extensively in [1], where a general, recursive, two-dimensional spatial transfer function design strategy was proposed using standard tools, such as the Z-transform.

12.3 The Linear CNN

A linear two-dimensional CNN[1] with time-invariant templates consists of a rectangular array of interconnected cells. The state-space of a general cell at location $\vec{k} = (k_x, k_y) \in \mathbb{Z}^2$ is determined by both the inputs to the cells over a neighborhood $B_r^2 = (a, b)$ such at $|a|, |b| \leq r$ for integers a and b, *and* the states of the other cells in the neighborhood. Accordingly, we write

$$c\dot{s}(\vec{k},t) = -\frac{1}{R}s(\vec{k},t) + \sum_{\vec{\xi} \in B_r^2} b_0(\vec{\xi})e(\vec{k}+\vec{\xi},t) - \sum_{\vec{\xi} \in B_r^2} a_0(\vec{\xi})s(\vec{k}+\vec{\xi},t) \quad (12.1)$$

where $\vec{k} = (k_x, k_y) \in \mathbb{Z}^2$ and $\vec{\xi} = (\xi_x, \xi_y) \in \mathbb{Z}^2$ are spatial indexing vectors in the two-dimensional (x, y) plane. Functions $a_0(\vec{\xi})$ and $b_0(\vec{\xi})$ are real functions defined on the discretized of discretized space that govern how the states and inputs of neighboring cells affect the dynamics of the current cell, and R is a constant defined in the original CNN literature that carries the uni of ohms.

For simplicity, a one-dimensional illustration of linear CNNs is shown in Figure 12.1. Considering (12.1), and assuming steady-state operation of a *stable* network, that is, $\dot{s}(\vec{k}) = 0$, we have:

$$s(\vec{k}) = R \sum_{\vec{\xi} \in B_r^2} b_0(\vec{\xi})e(\vec{k}+\vec{\xi}) - R \sum_{\vec{\xi} \in B_r^2} a_0(\vec{\xi})s(\vec{k}+\vec{\xi}) \quad (12.2)$$

---

1. Unless otherwise stated, we refer specifically to a single–layer linear CNN [25].

12.3 The Linear CNN

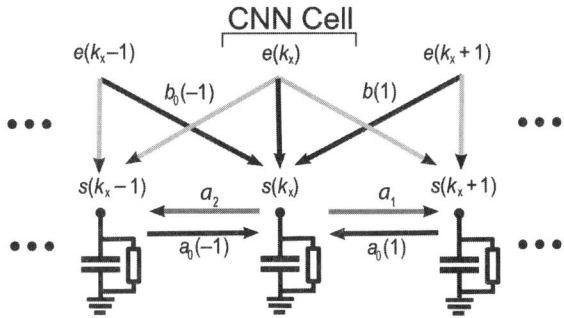

Figure 12.1 Architecture schematic of a one-dimensional CNN.

Equation (12.2) is in the form of an infinite impulse response (IIR) filter. The output at each spatial location is determined by weighted sums of the outputs and inputs at neighboring spatial locations (Figure 12.3). The coupling coefficients $a_0(\vec{\xi})$ and $b_0(\vec{\xi})$ form matrices $\mathbf{A} = a(\vec{\xi})_{\vec{\xi} \in B_r^2}$ and $\mathbf{B} = a(\vec{\xi})_{\vec{\xi} \in B_r^2}$ that are known as CNN *cloning templates*, and are analogous to the convolution and feedback kernels used for describing spatial IIR filtering. See Figures 12.2 and 12.3 for an illustration of the distinction between FIR and IIR filters for the two-dimensional case.

Recognition of the steady-state nature of the CNN, as defined by (12.2), has led to the application of CNNs in their steady-state regime for implementing quite general spatial filtering. In order to maximize the frames-per-second throughput in vision chips, the capacitance at each node is minimized to achieve the smallest time constant for settlement of the output image. Conversely, Shi [2] actually exploited the continuous-time dynamics embodied by (12.1) in the form of the the "CNN Filtering Array" to perform spatially discrete, continuous-time, spatiotemporal filtering, an approach that was extended to encompass a more general class of CNNs by Ip *et al.* [26].

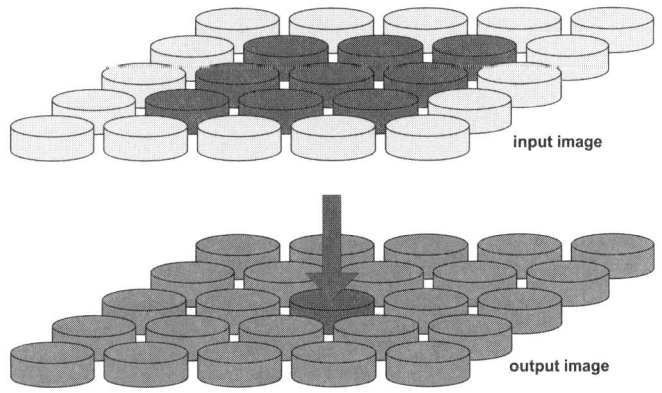

Figure 12.2 Architectural schematic of FIR filtering.

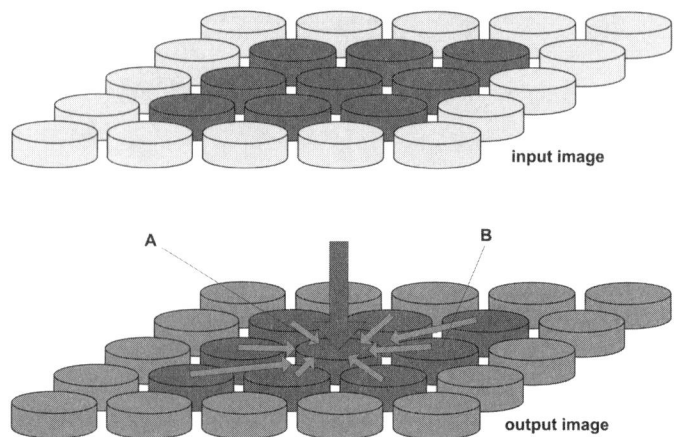

Figure 12.3 Architectural schematic of IIR filtering.

Equation (12.1) alone completely characterizes the dynamics of each CNN cell (assuming spatial invariance). In order to further understand the filtering function, we express Equation (12.1) in the spatiotemporal Z-Laplace domain:[2]

$$H(z_x, z_y, p_t) = \frac{\sum_{\vec{\xi} \in B_r^2} b_0(\vec{\xi})}{\sum_{\vec{\xi} \in B_r^2} a_0(\vec{\xi}) + cRp_t} \qquad (12.3)$$

where (z_x, z_y) are Z-domain variables corresponding to the discretized spatial indices (k_x, k_y), and p_t is a Laplace variable corresponding to time. Equation (12.3) is obtained by taking the Z-transform across discrete spatial indexes of Equation (12.1) and, subsequently, a Laplace transform on the temporal variable. We shall adopt the notation of transfer functions in preference to the differential equation or the cloning template representations.

12.4 CNNs and Mixed Domain Spatiotemporal Transfer Functions

In this section, we explore the spatiotemporal filtering characteristics of the original Chua-Yang CNNs (CYCNNs) operating in its linear region. Emphasis will be placed on analysis with spatiotemporal transfer functions (STTFs).

Let us first consider a simple CNN structure with each cell connected to its four immediate neighbors only [27], operating in its linear region (or equivalently, without applying the static nonlinearity at the output of each cell) and implemented by the use of ideal operational transconductance amplifiers (OTAs). The circuit of a single CNN cell is shown in Figure 12.4, with $s(k_x, k_y)$ and $e(k_x, k_y)$ representing the output and input voltage at location (k_x, k_y), respectively. The

---

2. We define this domain to be the equivalent of the Laplace domain for two-dimensional, spatially sampled, continuous-time functions.

12.4 CNNs and Mixed Domain Spatiotemporal Transfer Functions

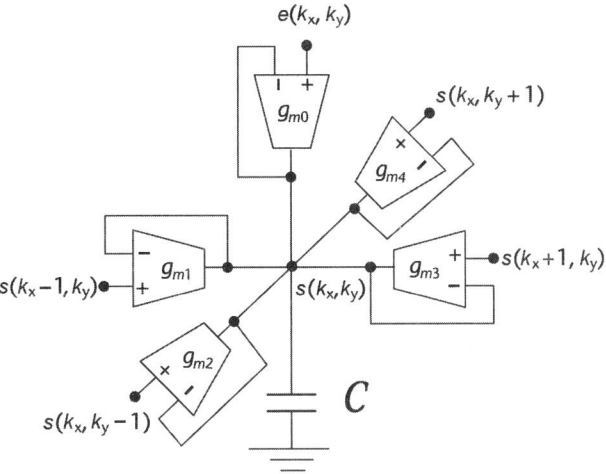

Figure 12.4 Linear CNN implementation by means of transconductors.

gain (transconductance) of each OTA is denoted as g_{mi}, for $i \in \{0,\ldots,4\}$. This CNN has a trivial input template and a feedback cloning template given by:

$$A_T = \begin{bmatrix} 0 & g_{m4} & 0 \\ g_{m1} & -\sum_{i=0}^{4} g_{mi} & g_{m3} \\ 0 & g_{m2} & 0 \end{bmatrix} \qquad (12.4)$$

Applying Kirchhoff's current law (KCL) at $s(k_x, k_y)$ and assuming steady-state operation in time, and subsequently applying the two-dimensional Z-transform with an assumption of spatial invariance, yields:

$$\frac{S_{z_x,z_y}}{E_{z_x,z_y}} = \frac{g_{m0}}{\sum_{i=0}^{4} g_{mi} - g_{m1}z_x^{-1} - g_{m2}z_y^{-1} - g_{m3}z_x^{+1} - g_{m4}z_y^{+1}} \qquad (12.5)$$

The quantities z_x and z_y correspond, respectively, to the x and y variables in the spatial Z-domain, whereas S_{z_x,z_y} (i.e., $Z\{s(k_x,k_y)\}$) and E_{z_x,z_y} (i.e., $Z\{e(k_x,k_y)\}$) denote the Z-transformed output and input, respectively. In general, a CNN operating in its linear region implements rational discrete spatial transfer functions [1]. This becomes clearer if we consider the "diffusion"[3] from the neighboring output nodes $s(k_x - \xi_{x0}, k_y - \xi_{y0})$ at relative location (ξ_{x0}, ξ_{y0}) in addition to the four

---

3. The term *diffusion* is commonly used to describe symmetrical flow across a medium. Here, we adopt the term to denote a directional flow of information from one node to another.

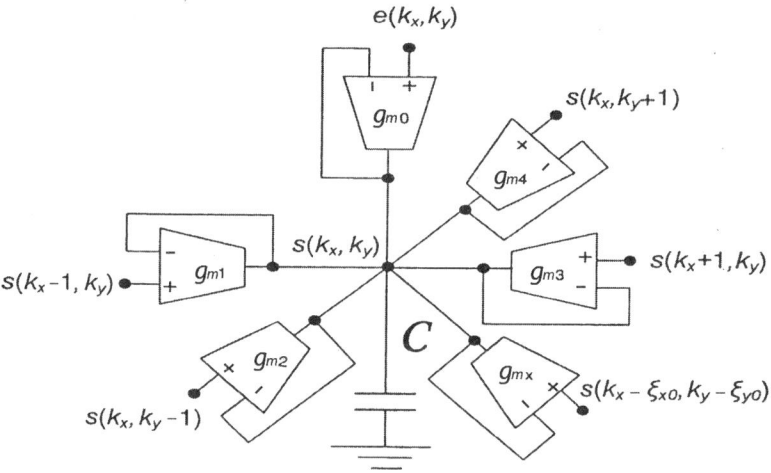

Figure 12.5 "Diffusion" from node $s(k_x - \xi_{x0}, k_y - \xi_{y0})$ in a linear CNN.

neighbors in Figure 12.4. The network with the additional diffusion is shown in Figure 12.5. Considering the steady-state KCL, after $s(k_x, k_y)$ settles over time,

$$g_{m0}e(k_x,k_y) + g_{m1}s(k_x - 1, k_y) + g_{m2}s(k_x, k_y - 1) + g_{m3}s(k_x + 1, k_y)$$
$$+ g_{m4}s(k_x, k_y + 1) + g_{mx}s(k_x - \xi_{x0}, k_y - \xi_{y0}) = \left(\sum_{i=0}^{4} g_{mi} + g_{mx}\right)s(k_x, k_y)$$
(12.6)

which leads to:

$$\frac{S_{z_x, z_y}}{E_{z_x, z_y}} =$$
$$\frac{g_{m0}}{\left(\sum_{i=0}^{4} g_{mi} + g_{mx}\right) - g_{m1}z_x^{-1} - g_{m2}z_y^{-1} - g_{m3}z_x^{+1} - g_{m4}z_y^{+1} - g_{mx}z_x^{-\xi_{x0}}z_y^{-\xi_{y0}}}$$
(12.7)

Hence, the diffusion from node $s(k_x - \xi_{x0}, k_y - \xi_{y0})$ creates terms of the form $z_x^{-n}z_y^{-m}$ in the denominator of the transfer function. Similarly, when "diffusing" from an additional input node $e(k_x - \xi_{x0}, k_y - \xi_{y0})$, the extra term $z_x^{-\xi_{x0}}z_y^{-\xi_{y0}}$ is created in the numerator. In other words:

> *the diffusion location and strength in the CNN templates is directly related to the form of the polynomials comprising the spatial rational transfer function of the network* [25].

Note that the above elementary analysis considers only the spatial dynamics and assumes static input with respect to time. When temporal dynamics are included (with time varying inputs/outputs as $e(k_x,k_y,t)$ and $s(k_x,k_y,t)$, respectively) and the Laplace transform is applied upon the Z-transformed KCL equation at $s(k_x,k_y,t)$ in Figure 12.4, we derive the following spatiotemporal transfer function,

$$\frac{S_{z_x,z_y,p_t}}{E_{z_x,z_y,p_t}} = \frac{g_{m0}}{\left(\sum_{i=1}^{4} g_{mi}\right) - g_{m1}z_x^{-1} - g_{m2}z_y^{-1} - g_{m3}z_x^{+1} - g_{m4}z_y^{+1} + cp_t} \quad (12.8)$$

In Equation (12.8), p_t denotes the temporal Laplace variable. Under certain conditions on its coefficients, this transfer function realizes an efficient velocity selective filter [27]. In the next section, we introduce extra novel connections to the network described in Figure 12.4 and show how this allows us to implement a large range of general spatiotemporal filters.

12.5 Networks with Temporal Derivative Diffusion

In this section, we explore the temporal derivative coupling between cells of the original Chua-Yang (CYCNN)[4] operating in their linear region [1]. This results in a network that realizes general, rational, nonseparable spatiotemporal transfer functions (STTFs).

Notice that the grounded capacitor in Figures 12.4 and 12.5 is included in the original CNN topology [24]. How would the network dynamics change if capacitive connections existed between neighboring cells? Indeed, more generally, how would temporal derivative diffusion from neighbors alter the spatiotemporal response properties of the network?

Consider a general lth-order derivative diffusion scheme from the general location $(k_x - \xi_{x1}, k_y - \xi_{y1})$, as shown in Figure 12.6. The "derivative transconductor" (conceptually speaking) provides a current output proportional to the input voltage's lth time derivative, that is, $i_{\xi_{x1},\xi_{y1},t} = T_{\xi_{x1},\xi_{y1}} \frac{d^l}{dt^l} s(k_x - \xi_{x1}, k_y - \xi_{y1}, t)$. The quantity $T_{\xi_{x1},\xi_{y1}}$ should be treated as a general coefficient of appropriate dimensions resembling the role of a capacitor. The KCL equation at node (k_x, k_y) with voltage input and output $e(k_x, k_y, t)$ and $s(k_x, k_y, t)$, respectively, is now given by:

$$g_{m0}e(k_x,k_y,t) + g_{m1}s(k_x-1,k_y,t) + g_{m2}s(k_x,k_y-1,t) + \\ g_{m3}s(k_x+1,k_y,t) + g_{m4}s(k_x,k_y+1,t) + \\ T_{\xi_{x1},\xi_{y1}}\frac{d^l}{dt^l}s(k_x-\xi_{x1},k_y-\xi_{y1},t) = c\dot{s}(k_x,k_y,t) + \left(\sum_{i=1}^{4}g_{mi}\right)s(k_x,k_y,t) \quad (12.9)$$

---

4. Unless specified otherwise, we refer to a *single* CNN layered structure. A discussion concerning multi-layered CYCNNs is elaborated in Section 12.6.2.

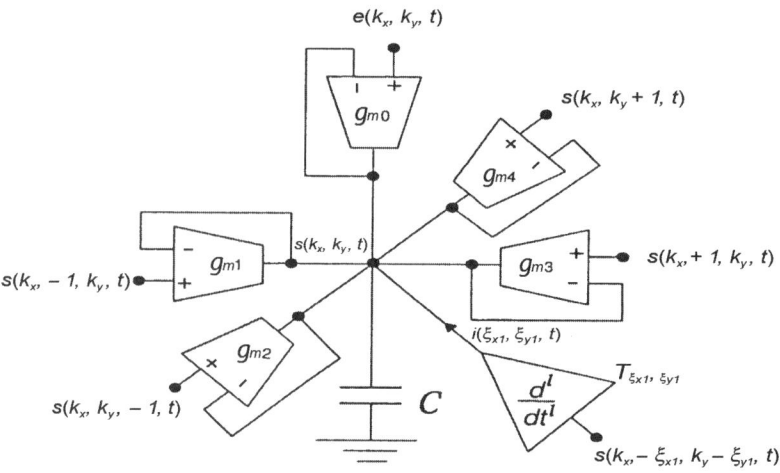

Figure 12.6 CNN with temporal derivative diffusion from node $s(k_x - \xi_{x0}, k_y - \xi_{y0})$.

Taking the Z-transform in space and assuming spatial invariance:

$$g_{m0}E_{z_x,z_y,t} + S_{z_x,z_y,t}(g_{m1}z_x^{-1} + g_{m2}z_y^{-1} + g_{m3}z_x^{+1} + g_{m4}z_y^{+1})$$
$$+ z_x^{-\xi_{y1}}z_y^{-\xi_{y1}}T_{\xi_{x1},\xi_{y1}}\frac{d^l}{dt^l}S_{z_x,z_y,t} = c\dot{S}_{z_x,z_y,t} + \left(\sum_{i=1}^{4}g_{mi}\right)S_{z_x,z_y,t} \quad (12.10)$$

Subsequently, taking the Laplace transform of Equation (12.10) [2], yields:

$$g_{m0}E_{z_x,z_y,p_t} + S_{z_x,z_y,p_t}(g_{m1}z_x^{-1} + g_{m2}z_y^{-1} + g_{m3}z_x^{+1} + g_{m4}z_y^{+1})$$
$$+ z_x^{-\xi_{x1}}z_y^{-\xi_{y1}}T_{\xi_{x1},\xi_{y1}}p_t^l S_{z_x,z_y,p_t} = cp_t S_{z_x,z_y,p_t} + \left(\sum_{i=1}^{4}g_{mi}\right)S_{z_x,z_y,p_t} \quad (12.11)$$

Rearranging (12.11), a spatiotemporal transfer function in the mixed temporal-Laplace/spatial Z-domain emerges:

$$\frac{S_{z_x,z_y,p_t}}{E_{z_x,z_y,p_t}} = g_{m0}\left[g_{m0} + \sum_{i=1}^{4}g_{mi} - g_{m1}z_x^{-1} - g_{m2}z_y^{-1} - g_{m3}z_y^{+1} - \right.$$
$$\left. g_{m4}z_x^{+1} - T_{\xi_{x1},\xi_{y1}}z_x^{-\xi_{x1}}z_y^{-\xi_{y1}}p_t^l + cp_t\right]^{-1} \quad (12.12)$$

Now, it should be clear that an lth derivative diffusion from a general node at $(k_x - \xi_{x1}, k_y - \xi_{y1})$ creates the additional term $z_x^{-\xi_{x1}}z_y^{-\xi_{y1}}p_t^l$ in the original spatiotemporal transfer function of (12.8). If we diffuse the temporal derivative from an input location $(k_x - \xi_{x1}, k_y - \xi_{y1})$, the extra term $z_x^{-\xi_{x1}}z_y^{-\xi_{y1}}p_t^l$ will appear in the numerator of the transfer function. Likewise, diffusing from an output allows the extra term to be created in the denominator as in (12.12). To obtain a term consisting of the Laplace variable p_t^l only, we simply have to introduce a temporal derivative diffusion from the ground node.

12.5 Networks with Temporal Derivative Diffusion

The above analysis reveals that

the extra temporal derivative diffusion introduced to linear CNNs, as illustrated in Figure 12.6, enables the implementation of a general mixed-domain spatiotemporal rational transfer function of the form

$$\frac{S_{z_x,z_y,p_t}}{E_{z_x,z_y,p_t}} = \frac{\sum_{l\in 1...D}\sum_{\vec{\xi}\in B_r^2} b'_l(\vec{\xi}) z_x^{\xi_x} z_y^{\xi_y} p_t^l}{\sum_{l\in 1...D}\sum_{\vec{\xi}\in B_r^2} a'_l(\vec{\xi}) z_x^{\xi_x} z_y^{\xi_y} p_t^l} \quad (12.13)$$

with a one-to-one correspondence of network connections and transfer function polynomial terms.

The region of support B_r^2 [2] is defined in two-dimensional space as:

$$B_r^2 = \{(x,y) \in N^2 | x|, |y| \leq r\} \quad (12.14)$$

The quantities $a'_l(\vec{\xi})$ and $b'_l(\vec{\xi})$ are, in general, complex constants for $l \in Z_+$, though we only consider real values in this current presentation. The differential equation corresponding to (12.14) is given by:

$$c\dot{s}(\vec{k},t) = -\frac{1}{R}s(\vec{k},t) + \sum_{\vec{\xi}\in B_r^2} a_0(\vec{\xi})s(\vec{k}+\vec{\xi},t) + \sum_{\vec{\xi}\in B_r^2} b_0(\vec{\xi})e(\vec{k}+\vec{\xi},t)$$
$$+ \sum_{l\in 1...D}\left[\sum_{\vec{\xi}\in B_r^2} a_l(\vec{\xi})\frac{d^l s(\vec{k}+\vec{\xi},t)}{dt^l} + \sum_{\vec{\xi}\in B_r^2} b_l(\vec{\xi})\frac{d^l e(\vec{k}+\vec{\xi},t)}{dt^l}\right] \quad (12.15)$$

with parameters related to (12.13) as follows:

$$a'_0(0,0) = 1/R, \text{ otherwise } a'_0(\vec{\xi}) = -a_0(\vec{\xi}) \quad (12.16)$$

$$a'_1(0,0) = c, \text{ otherwise } a'_l(\vec{\xi}) = -a_1(\vec{\xi}) \text{ for } l \geq 1 \quad (12.17)$$

$$b'_l(\vec{\xi}) = b_l(\vec{\xi}) \quad (12.18)$$

where R is treated as a real constant. From this differential equation, we can now define network cloning templates. Immediately we recognize that Equation (12.15), neglecting the expression in the third line, corresponds to the dynamics of the original CYCNN operating in its linear region (linear CNN). Therefore, we have CNN feedback and feed forward templates being $\mathbf{A} = a_0(\vec{\xi})_{\vec{\xi}\in B_r^2}$ and $\mathbf{B} = b_0(\vec{\xi})_{\vec{\xi}\in B_r^2}$, respectively. We can define extra cloning templates for the time-derivative terms from (12.15), $\mathbf{A}_l = a_l(\vec{\xi})_{\vec{\xi}\in B_r^2}$ and $\mathbf{B}_l = b_l(\vec{\xi})_{\vec{\xi}\in B_r^2}$, respectively, as the lth-order derivative feedback and feed forward cloning templates.

Note that each term in (differential) (12.15) corresponds to one term in the transfer function (12.13), and also to one physical network connection (one coefficient in the templates). This facilitates the interpretation of the filtering function directly from the network topology.

The general spatiotemporal transfer function, representing linear CNNs without the extra time derivative diffusion, is given by

$$\frac{S_{z_x,z_y,p_t}}{E_{z_x,z_y,p_t}} = \frac{\sum_{\vec{\xi} \in B_r^2} b_0(\vec{\xi}) z_x^{\xi_x} z_y^{\xi_y}}{\sum_{\vec{\xi} \in B_r^2} a_0(\vec{\xi}) z_x^{\xi_x} z_y^{\xi_y} + c p_t} \quad (12.19)$$

Comparing (12.13) and (12.19) reveals that the extra temporal derivative diffusion connections to linear CNNs enables a richer class of spatiotemporal characteristics to be realized.

12.5.1 Stability

An arbitrary, stable, rational spatiotemporal transfer function, either in the Z- or Laplace domain, can be converted to the form described by Equation (12.13) by applying the appropriate bilinear transform (or its inverse) to the spatial (given a continuous prototype) or temporal (given a discrete prototype) variable(s) [28].

Note that no restrictions are imposed on the signs of $\vec{\xi} = (\xi_x, \xi_y)$ in Equation (12.15). This corresponds to the current cell taking information from neighbors of all directions just like the CYCNN. From a filtering point of view, we have general support (recurring in both causal and anti-causal directions) discrete systems that are not necessarily recursible [29]. Despite nonrecursible structures not being suitable for direct implementation on delay-based digital systems, analog VLSI arrays such as the CNNUM [11] are particularly suited for the implementation of general support dynamics. Preliminary results from ongoing research on analog VLSI implementations of the network proposed here are reported in [26]. The general support nature of CNNs results in a rather slack stability criterion that allows a greater freedom for pole locations than purely causal or purely anticausal systems. This facilitates a convenient three-dimensional pole placement scheme for filter synthesis [30], which cannot be carried out for designing stable causal/anticausal filters for delay implementation in digital systems. Specifically, we treat the denominator of Equation (12.13) as a polynomial of the temporal Laplace variable p_t and solve for the poles in terms of (z_x, z_y). Equation (12.13) describes a system with spatial general support and causal time. The stability of such systems are guaranteed if all poles fall outside the following space [31]:

$$|z_x| = 1 \cap |z_y| = 1 \cap Re\{p_t\} \geq 0 \quad (12.20)$$

12.6 A Signal Flow Graph–Based Implementation

i.e. for stability we have no poles lying on the dual unit circles and on the right half Laplace plane.

12.6 A Signal Flow Graph–Based Implementation

As hinted at the beginning of this chapter, analog VLSI systems are particularly suited for implementing spatiotemporal filtering networks. In analog VLSI, linear signal processing operations are often realized with continuous time integrators as a fundamental building block [32,33]. Here we propose a transformation, based on signal flow graph manipulation, for implementing the suggested network dynamics with spatially coupled continuous-time integrators. With the aid of the SFG, we can also examine how our proposed time-derivative diffusion connections on linear CNNs compares with the spatiotemporal dynamics realized by multilayer CNNs (MLCNNs).

12.6.1 Continuous Time Signal Flow Graphs

Consider the differential equation describing the network dynamics, Equation (12.15), and let L and D be the respective degrees of the temporal Laplace polynomials in the denominator and numerator of the transfer function. We also make the reasonable assumption that $L \geq D$. Consider the Laplace transform of Equation (12.15):

$$p_t S(\vec{k}, p_t) = -\frac{1}{RC} S(\vec{k}, p_t) + \frac{1}{c} \sum_{\vec{\xi} \in B_r^2} a_0(\vec{\xi}) S(\vec{k} + \vec{\xi}, p_t)$$
$$+ \frac{1}{c} \sum_{\vec{\xi} \in B_r^2} b_0(\vec{\xi}) E(\vec{k} + \vec{\xi}, p_t) + \sum_{l \in 1 \ldots L} \sum_{\vec{\xi} \in B_r^2} a_l(\vec{\xi}) p_t^l S(\vec{k} + \vec{\xi}, p_t)$$
$$+ \sum_{l \in 1 \ldots D} \sum_{\vec{\xi} \in B_r^2} b_l(\vec{\xi}) p_t^l E(\vec{k} + \vec{\xi}, p_t) \quad (12.21)$$

Setting $A = a_L(0,0) \neq 0$, yields:

$$p_t S(\vec{k}, p_t) = -\frac{1}{RC} S(\vec{k}, p_t) + \frac{1}{c} \sum_{\vec{\xi} \in B_r^2} a_0(\vec{\xi}) S(\vec{k} + \vec{\xi}, p_t)$$
$$+ \frac{1}{c} \sum_{\vec{\xi} \in B_r^2} b_0(\vec{\xi}) E(\vec{k} + \vec{\xi}, p_t) + A p_t^L S(\vec{k}, p_t)$$
$$+ \sum_{\vec{\xi} \neq (0,0) \in B_r^2} a_L(\vec{\xi}) p_t^L S(\vec{k} + \vec{\xi}, p_t)$$
$$+ \sum_{l \in 1 \ldots L-1} \sum_{\vec{\xi} \in B_r^2} a_l(\vec{\xi}) p_t^l S(\vec{k} + \vec{\xi}, p_t)$$
$$+ \sum_{l \in 1 \ldots D} \sum_{\vec{\xi} \in B_r^2} b_l(\vec{\xi}) p_t^l E(\vec{k} + \vec{\xi}, p_t) \quad (12.22)$$

In other words, we assume that at least one of the highest-power p_t^L terms in the denominator is free of the Z-domain variables, z_x and z_y, in its coefficient. This is a reasonable assumption, since any terms in the form $p_t^L z_x^{-a} z_y^{-b}$ can have the Z-domain variables eliminated by multiplying the whole transfer function by $z_x^{+a} z_y^{+b} / z_x^{+a} z_y^{+b}$. However, this affects the stability criterion of the filter [31]. In the case for which an unstable transfer function results after the multiplication with $z_x^{+a} z_y^{+b} / z_x^{+a} z_y^{+b}$, the filter is not suitable for the integrator implementation outlined here.

An integral equation in the Laplace domain can be obtained after dividing Equation (12.22) by $A p_t^L$ and rearranging:

$$S(\vec{k}, p_t) = \frac{1}{RcAp_t^L} S_{\vec{k}, p_t} - \frac{1}{c} \sum_{\vec{\xi} \in B_r^2} \frac{a_0(\vec{\xi}) S(\vec{k}+\vec{\xi}, p_t)}{Ap_t^L} \frac{1}{c} \sum_{\vec{\xi} \in B_r^2} \frac{b_0(\vec{\xi}) E(\vec{k}+\vec{\xi}, p_t)}{Ap_t^L}$$
$$+ \frac{1}{Ap_t^{L-1}} S(\vec{k}, p_t) - \sum_{l \in 1 \ldots L-1} \sum_{\vec{\xi} \in B_r^2} \frac{a_l(\vec{\xi}) S(\vec{k}+\vec{\xi}, p_t)}{Ap_t^{L-l}}$$
$$- \sum_{l \in 1 \ldots D} \sum_{\vec{\xi} \in B_r^2} \frac{b_l(\vec{\xi}) E(\vec{k}+\vec{\xi}, p_t)}{Ap_t^{L-l}}$$
$$- \sum_{\vec{\xi} \in B_r^2, \neq \vec{0}} \frac{a_L(\vec{\xi}) S(\vec{k}+\vec{\xi}, p_t)}{A} \qquad (12.23)$$

Equation (12.23) governs the temporal dynamics at each cell, with input/outputs of neighboring cells as inputs to the current cell. For clarity, we use $S_{\vec{\xi}+\vec{k}}^{L-l}$ to denote the $S(\vec{\xi}+\vec{k}, p_t)$ terms in (12.23) with a coefficient involving the term $1/p_t^{L-l}$. Similarly, for neighboring inputs that are also connected to the current cell, we have $E_{\vec{\xi}+\vec{k}}^{L-l}$. Hence, (12.23) becomes:

$$S(\vec{k}, p_t) = \frac{1}{RcAp_t^L} S_{\vec{k}}^L - \frac{1}{c} \sum_{\vec{\xi} \in B_r^2} \frac{a_0(\vec{\xi}) S_{\vec{k}+\vec{\xi}}^L}{Ap_t^L} - \frac{1}{c} \sum_{\vec{\xi} \in B_r^2} \frac{b_0(\vec{\xi}) E_{\vec{k}+\vec{\xi}}^L}{Ap_t^L}$$
$$+ \frac{1}{Ap_t^{L-1}} S_{\vec{k}}^{L-1} - \sum_{l \in 1 \ldots L-1} \sum_{\vec{\xi} \in B_r^2} \frac{a_l(\vec{\xi}) S_{\vec{k}+\vec{\xi}}^{L-l}}{Ap_t^{L-l}}$$
$$- \sum_{l \in 1 \ldots D} \sum_{\vec{\xi} \in B_r^2} \frac{b_l(\vec{\xi}) E_{\vec{k}+\vec{\xi}}^{L-l}}{Ap_t^{L-l}} - \sum_{\vec{\xi} \in B_r^2, \neq \vec{0}} \frac{a_L(\vec{\xi}) S_{\vec{k}+\vec{\xi}}}{A} \qquad (12.24)$$

It is now possible to construct a signal flow graph from (12.24), based on integrators only. A simplified SFG omitting, for the sake of clarity, the detailed "gain" terms is shown in Figure 12.7. The precise mapping between the integral (12.24)

12.6 A Signal Flow Graph–Based Implementation

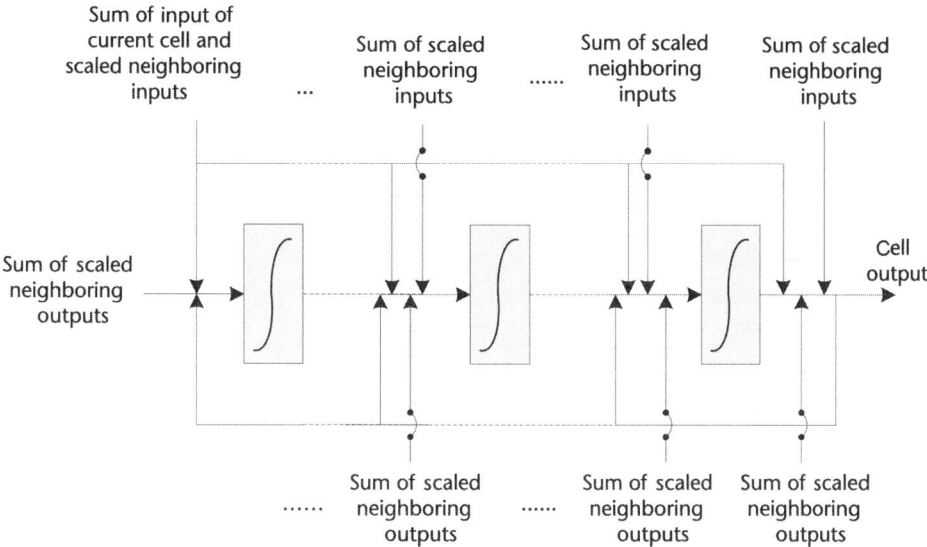

Figure 12.7 A schematic of the SFG implementation of the proposed spatiotemporal filtering network. Detailed correlation with SFG gain terms and STTF coefficients can be found in [34].

and SFG gains has already been reported in [34]. The input of the integrators originates not just from the current cell's input/output but also the input/output of neighboring cells. Therefore, the SFG of our modified CNN presented here is spatially coupled. Note that the SFG presented in Figure 12.7 is general. For any $s^{L-l}_{\vec{k}+\vec{\xi}}$ with $\vec{\xi} = \vec{0}$, we have local feedback in the SFG. Also, if there exist terms in the form of $s^{L-l}_{\vec{k}+\vec{\xi}}$ with $L - l = 0$, the output becomes:

$$S_{\vec{k},p_t} = \text{integrated terms} - \sum_{\vec{\xi} \neq \vec{0} \in B_r^2} \frac{a_L(\vec{\xi}) S_{\vec{k}+\vec{\xi}}}{A} \tag{12.25}$$

This corresponds to the presence of a nonconstant coefficient of the highest power p_t term in the denominator of (12.13).

A state-space representation can be obtained by assigning a state variable to the outputs of each one of the integrators in Figure 12.7. Equation (12.25) suggests that the output of the system is not equal to the state itself, but rather to the difference of the state of the corresponding cell and its "inputs" (outputs of neighboring cells). Bearing in mind that the neighboring output terms $S_{\vec{k}+\vec{\xi}}$ in (12.25) must also satisfy the same dynamics, (12.25) has, effectively, to be satisfied for each network cell.

When the constraints embodied by (12.25) are imposed [30], these extra algebraic constraints *do not* represent impractical restrictions to the network, but are instead satisfied in real time by the physics of circuit components and fundamental electrical laws. Ongoing work on physical network realizations is reported in a recent publication [26].

12.6.2 On SFG Relations with the MLCNN

Multilayered CNNs (MLCNNs) have been suggested in the earlier CNN literature, and defined as a state-space (i.e., coupled first-order differential equations) formalism in [24]. This enables a direct translation from state-space to an SFG with each first-order equation represented as a single integrator in the SFG (given that we remove the output nonlinearity). The general SFG for an MLCNN cell is shown in Figure 12.8. Despite its rich dynamics for system modelling [35], limited published literature on exploiting it for linear spatiotemporal filtering exists.

To reveal the spatiotemporal filtering capability of MLCNNs in the context of STTFs, we consider a two layer *linearized* MLCNN with feedforward interlayer interactions only [24]:

$$c_1 \dot{x}_1(\vec{k}) = -R_1^{-1} x_1(\vec{k}) + I_1 + \sum_{\vec{\xi} \in B_r^2} a_0^{11}(\vec{\xi}) x_1(\vec{k} - \vec{\xi}) + \sum_{\vec{\xi} \in B_r^2} b_0^{11}(\vec{\xi}) u_1(\vec{k} - \vec{\xi}) \quad (12.26)$$

$$c_2 \dot{x}_2(\vec{k}) = -R_2^{-1} x_2(\vec{k}) + I_2 + \sum_{\vec{\xi} \in B_r^2} a_0^{21}(\vec{\xi}) x_1(\vec{k} - \vec{\xi}) + \sum_{\vec{\xi} \in B_r^2} a_0^{22}(\vec{\xi}) x_2(\vec{k} - \vec{\xi})$$
$$+ \sum_{\vec{\xi} \in B_r^2} b_0^{21}(\vec{\xi}) u_1(\vec{k} - \vec{\xi}) + \sum_{\vec{\xi} \in B_r^2} b_0^{22}(\vec{\xi}) u_2(\vec{k} - \vec{\xi}) \quad (12.27)$$

where x_1 and x_2 denote the state variable of the first and second layer, respectively, u_1 and u_2 denote the respective input of the first and second layer, coefficients $a_0^{11}(\vec{\xi}), b_0^{11}(\vec{\xi}), a_0^{21}(\vec{\xi}), a_0^{22}(\vec{\xi}), b_0^{21}(\vec{\xi})$, and $b_0^{22}(\vec{\xi})$ originate from the intercell/layer

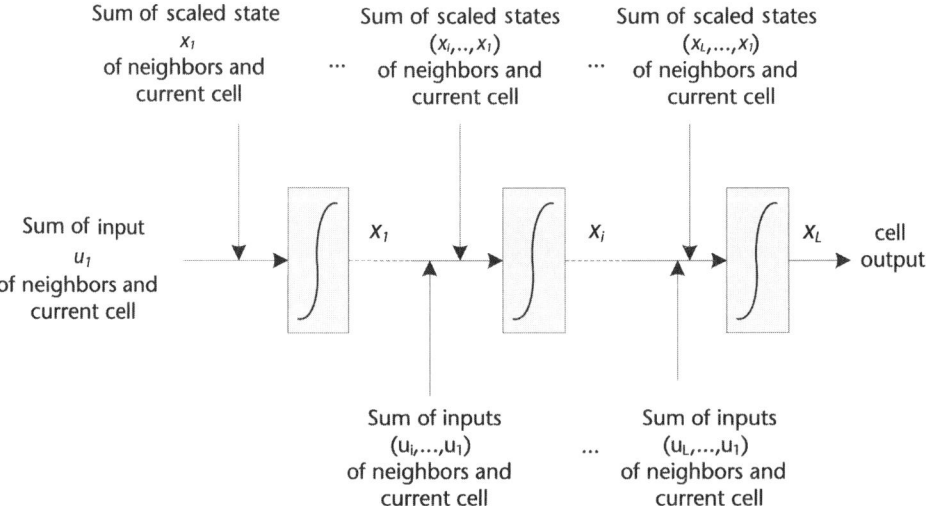

Figure 12.8 A schematic of the SFG implementation of the original MLCNN [24] (feed forward interlayer interaction only). Symbols $x_1, \ldots, x_{L-l}, \ldots, x_L$ are used to denote the multiple states at a single cell with each integrator implementing a single first-order state-space equation.

12.6 A Signal Flow Graph–Based Implementation

coupling templates. Assuming a single input layer u_2 with $u_1 = 0$ and taking $x_2 = y$ as the output, we can derive the STTF by taking the Laplace and Z-transforms of Equations (12.26) and (12.27), and eliminating x_1:

$$\frac{Y}{U_2} = \frac{[A_{21}B_{11} + B_{22}(R_1^{-1} - A_{11})] + B_{22}c_1 p_t}{c_1 c_2 p_t^2 + [c_2(R_1^{-1} - A_{11}) + c_1(R_2^{-1} - A_{22})]p_t + (R_2^{-1} - A_{22})(R_1^{-1} - A_{11})} \quad (12.28)$$

The above spatiotemporal fourier transform (STTF) explicitly confirms that the MLCNN is indeed suitable for spatiotemporal filtering tasks with higher-order (≥ 1) temporal dynamics. Equation (12.28) is similar to the form of the general STTF of networks with second-order temporal derivative diffusion, which is, from Equation (12.13), given by

$$\frac{Y}{U} = \frac{\sum_{l \in \{1,2\}} \sum_{\vec{\xi} \in B_r^2} b_l'(\vec{\xi}) z_x^{\xi_x} z_y^{\xi_y} p_t^l}{\sum_{l \in \{1,2\}} \sum_{\vec{\xi} \in B_r^2} a_l'(\vec{\xi}) z_x^{\xi_x} z_y^{\xi_y} p_t^l} \quad (12.29)$$

However, in order for the two-layer CNN to implement the arbitrary second-order (temporal) STTF realizable by the second-order temporal derivative diffusion network of Equation (12.29), the following constraints need to be satisfied:

$$\sum_{\vec{\xi} \in B_r^2} a_0'(\vec{\xi}) z_x^{\xi_x} z_y^{\xi_y} = (R_2^{-1} - A_{22})(R_1^{-1} - A_{11}) \quad (12.30)$$

$$\sum_{\vec{\xi} \in B_r^2} a_1'(\vec{\xi}) z_x^{\xi_x} z_y^{\xi_y} = c_2(R_1^{-1} - A_{11}) + c_1(R_2^{-1} - A_{22}) \quad (12.31)$$

$$\sum_{\vec{\xi} \in B_r^2} a_2'(\vec{\xi}) z_x^{\xi_x} z_y^{\xi_y} = c_2 c_1 \quad (12.32)$$

$$\sum_{\vec{\xi} \in B_r^2} b_0'(\vec{\xi}) z_x^{\xi_x} z_y^{\xi_y} = A_{21}B_{11} + B_{22}(R_1^{-1} - A_{11}) \quad (12.33)$$

$$\sum_{\vec{\xi} \in B_r^2} b_1'(\vec{\xi}) z_x^{\xi_x} z_y^{\xi_y} = B_{22} c_1 \quad (12.34)$$

$$\sum_{\vec{\xi} \in B_r^2} b_2'(\vec{\xi}) z_x^{\xi_x} z_y^{\xi_y} = 0 \quad (12.35)$$

Equations (12.30) to (12.32) and (12.33) to (12.35) originated from, respectively, comparing the denominators and numerators of (12.28) and (12.29). They suggest limitations on the spatiotemporal filtering function of a two-layer CNN compared with a second-order temporal derivative diffusion network in the sense that the form of STTFs implementable by the feedforward interlayer MLCNN is relatively restricted.

(a) The non-p_t polynomial in the denominator should be expressed as a factorizable polynomial according to Equation (12.30) for variables A_{11} and A_{22}. In the case where either A_{11} or A_{22} is a constant, Equations (12.30) to (12.31) suggest that $\sum_{\vec{\xi} \in B_r^2} a'_0(\vec{\xi}) z_x^{\xi_x} z_y^{\xi_y}$ and $\sum_{\vec{\xi} \in B_r^2} a'_1(\vec{\xi}) z_x^{\xi_x} z_y^{\xi_y}$ have to be in the same form.
Note that this restriction on the STTF of the two-layer CNN originates from the original MLCNN definition [24], according to which, for each state x_i, the state dynamics differential equation allows coupling only from states with x_j, $1 \leq j \leq i^5$, that is, only feed forward interlayer coupling is permitted. If one allows arbitrary state coupling across the layers (feedback as well as feed forward) and further allows input coupling to any desirable layer, the aforementioned restriction does not apply to STTFs of MLCNNs any more. Indeed, such an extension to the MLCNN exists, but only limited to system modelling applications [35] rather than in the context of linear spatiotemporal filtering, which is the focus of this chapter.

(b) Equation (12.32) restricts the coefficient of p_t^2 in the denominator to be a constant instead of a bivariate polynomial of (z_x, x_y) promised by Equation (12.13), that is, the STTF implementable by MLCNNs does not include the subset of STTFs that leads to "algebraic constraint implementation" mentioned at the end of Section 12.6.1.

(c) The numerator is limited to first order in (12.35). For a general M-layer MLCNN, the numerator of its STTF is always of $(M-1)$th order in p_t and the coefficient of p_t^M is constant. The STTF of one example given in the next section does not have the above restriction, and so it is also implementable with a two-layer CNN. However, to the best of our knowledge, the specific example has not been reported in literature to be linked with MLCNN theory. Despite MLCNNs being capable of implementing general STTFs with higher-order temporal dynamics given some limitations, the temporal derivative diffusion network provides a "one-to-one" relationship of network elements to STTF terms. For (12.30) to (12.35), MLCNNs generally do not "create" an STTF term with a single template value. Moreover, the authors bear the view that the "adding one circuit element at a time" derivation, presented earlier, of a single-layer network/differential equation, enables a more intuitive and intimate bond between the filtering (STTF) and the practical network structure described by the original CNN and time-derivative template, compared to the notion of a multiple differential equation state-space of MLCNNs.

---

5. Equation (12.26) is independent of x_2.

12.7 Examples

In this section, we provide two examples of spatiotemporal responses implementable by CNNs with additional time-derivative diffusion connections. Our first example involves the design of a specific highpass three-dimensional spatiotemporal filter with a cone shaped pass/stop-band boundary. In a recent publication [34], we included a first-order velocity selective filter [28, 36] as an example. In the second example, we consider the implementation of a spatiotemporal transfer function derived through analytical modelling of recursive visual cortical neural interactions. We thus show that the temporal-derivative enhanced CNN (TDCNN) can be used to approximate STTFs that describe both the spatial and temporal characteristics of classical, dynamical receptive fields of simple cell neurons.

12.7.1 A Spatiotemporal Cone Filter

In this example, we apply an empirical method to design a spatiotemporal "cone" filter. We start with a two-dimensional spatial filter and by transforming its coefficients into temporal frequency–dependent terms, a temporal frequency–dependent spatial filter is created. Cone filters are three-dimensional highpass filters with their passband and stop-band separated by two tip-to-tip cones. An ideal cone response is illustrated in Figure 12.9(a). Due to the high computational cost of implementing three-dimensional convolutions, recursive cone filter approximations have drawn considerable attention from the research community [28, 29, 37, 38]. Here, we show how approximate cone-like responses can be produced by CNNs with our additional time-derivative diffusion connections.

Consider a two-dimensional CNN filter [25] with two-dimensional Z-domain transfer function:

$$H_L(z_x, z_y) = \frac{1}{1 + 4a_c - a_c(z_x^{+1} + z_x^{-1} + z_y^{+1} + z_y^{-1})} \qquad (12.36)$$

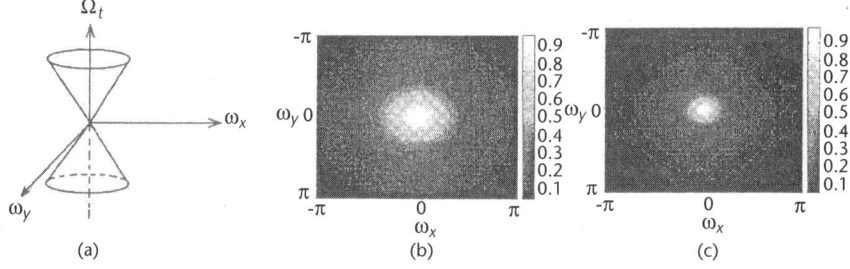

Figure 12.9 (a) Ideal cone filter response. The amplitude of the magnitude response equals to unity inside the cone and zero outside. (b) Amplitude response of (12.36) with $a_c = 1$. (c) Amplitude response of (12.36) with $a_c = 5$.

and frequency response given by:

$$H_L(z_x, z_y) = \frac{1}{1 + 2a_c(1 - \cos\omega_x) + 2a_c(1 - \cos\omega_y)} \quad (12.37)$$

The magnitude response of (12.37), as the parameter $a_c \in R$ is varied, is shown in Figure 12.9(b, c). Two things can be noticed. First, the pass- and stopband boundary (for simplicity, we hereafter refer to this as the "passband boundary") is approximately circular. Second, the passband boundary decreases in "radius" as a_c increases. Observe the ideal spatiotemporal cone characteristic in Figure 12.9(a) and assume a slice parallel to the (ω_x, ω_y) axis: the passband boundary is circular. As we move along the Ω_t axis, the boundary increases in radius with the absolute value of Ω_t. Therefore, we aim to express a_c in (12.37) as a function of Ω_t such that an increase in $|\Omega_t|$ causes a decrease in a_c and hence an increase in radius of the pass-band boundary of the spatial filter defined by (12.36). In this way, the desired temporal frequency–dependent spatial filter response can be synthesized. A reasonable choice is:

$$a_c = \frac{a'_c}{p_t} = \frac{a'_c}{j\Omega_t} \quad (12.38)$$

Substituting (12.38) into (12.36) and rearranging, we have:

$$H'_L(z_x, z_y) = \frac{p_t}{p_t + 4a'_c - a'_c(z_x^{+1} + z_x^{-1} + z_y^{+1} + z_y^{-1})} \quad (12.39)$$

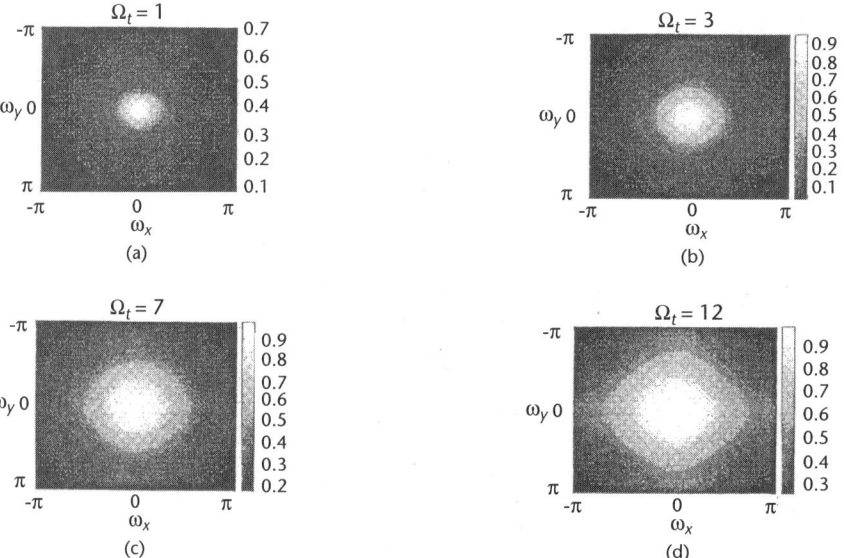

Figure 12.10 Magnitude response "slices" parallel to the $\Omega_t = 0$ plane of a cone approximation implemented with a CNN with temporal derivative diffusion.

12.7 Examples

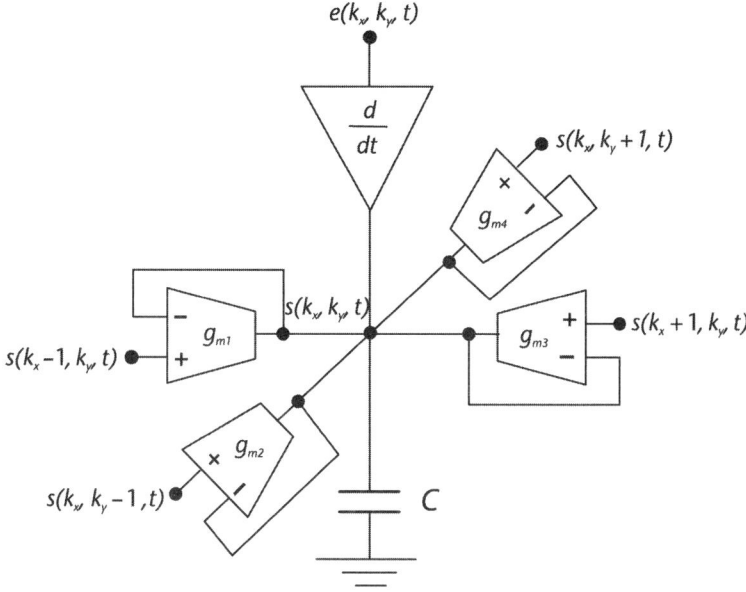

Figure 12.11 Network for cone filter approximation.

To ensure filter stability [2, 31], we introduce an extra constant term τ_c in the denominator:

$$H'_L(z_x, z_y) = \frac{p_t}{p_t + 4a'_c - a'_c(z_x^{+1} + z_x^{-1} + z_y^{+1} + z_y^{-1}) + \tau_c} \quad (12.40)$$

The corresponding network and SFG are shown in Figures 12.11 and 12.12, respectively. As seen in Figure 12.11, a temporal derivative diffusion from the input node is needed. The response of (12.40) as we move along the Ω_t axis is shown in Figure 12.10, with an approximate "cone" characteristic (since the amplitude

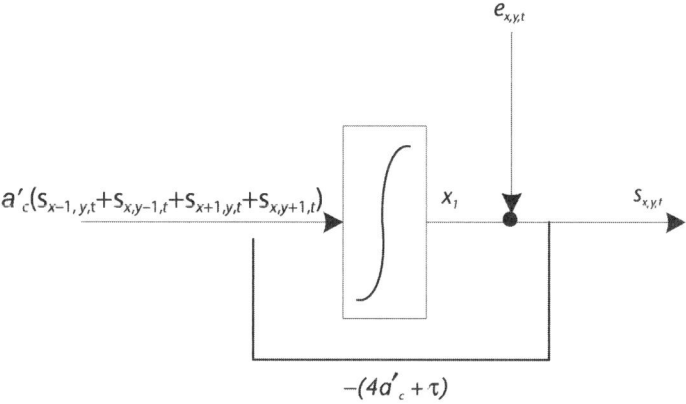

Figure 12.12 SFG for the implementation of a cone filtering network.

response is an even function of Ω_t, we only show responses for positive Ω_t values). Therefore, by introducing the appropriate temporal derivative connections and utilizing the concept of temporal frequency–dependent spatial filters, we tailor the desired temporal characteristics to a spatial filter defined by (12.36), and consequently obtain a cone filter approximation.

Note that (12.40) can be re-expressed as the sum of a constant term and a linear CNN STTF with a nonconstant numerator independent of p_t. This suggests an alternative implementation of the cone approximation based on linear CNNs and some postprocessing (which has not been reported elsewhere in CNN literature). However, the network based on temporal derivative diffusion provides a simpler approach with only a single numerator term corresponding to a single element in the input time-derivative template.

Simulation results are shown in Figures 12.13 to 12.15. Spatiotemporal sinusoidal signals with varying spatial and temporal frequencies are applied to a 30×30 network and the steady-state output is presented in Figures 12.14 and 12.15. Figure 12.13 shows a typical output at a particular time instant. Figure 12.14 shows the magnitude of the output spatiotemporal sine wave when the temporal frequency of the input is varied while the spatial frequency is fixed

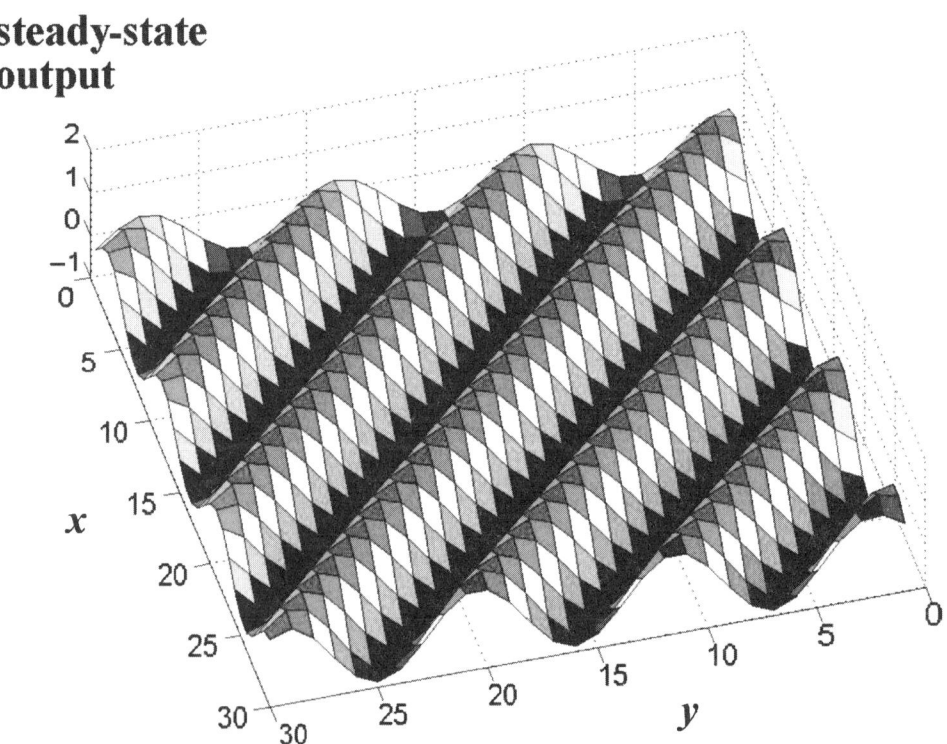

Figure 12.13 A steady-state output snapshot at a particular time instance (after the temporal transient had died down) when an input at frequency $(\omega_x, \omega_y, \Omega_t) = (0.7, 0.7, 10)$ is applied to the cone filtering network.

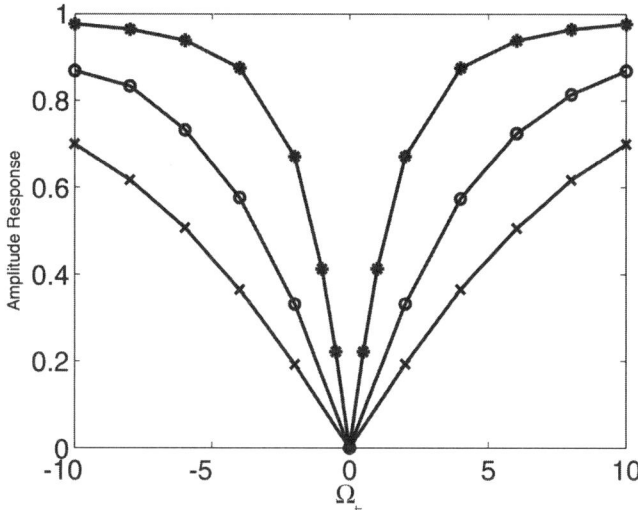

Figure 12.14 Simulation results for a 30 × 30 network approximating the cone response. Spatiotemporal sine waves of different spatial and temporal frequencies are applied to the network and the output is observed for each input spatiotemporal sine wave. The vertical axis shows the output sinusoidal amplitude of the middle pixel at location (15, 15). The horizontal axis shows temporal frequency of input. Results for three input spatial frequencies are shown. Symbols $(*, o, x)$ are for spatial frequencies $(\omega_x, \omega_y) = (0.35, -0.35)$, $(0.7, 0.7)$, and $(-1, -1)$ respectively.

at three different points. As we move away from the frequency domain origin by varying the spatial frequency from $(0.35, -0.35)$ to $(-1, -1)$, we observe that an input of higher temporal frequency is required to give the same output level as inputs with lower spatial frequencies. This confirms the cone filtering characteristic. Figure 12.15 shows the output signal at various spatial locations when a pulse of height 100 (relative units) and width of 0.01 seconds is applied at cell (15, 15) of the 30 × 30 network, the spatiotemporal output thus approximates the impulse response of our cone filter. As seen from Figure 12.15, the response drops quickly

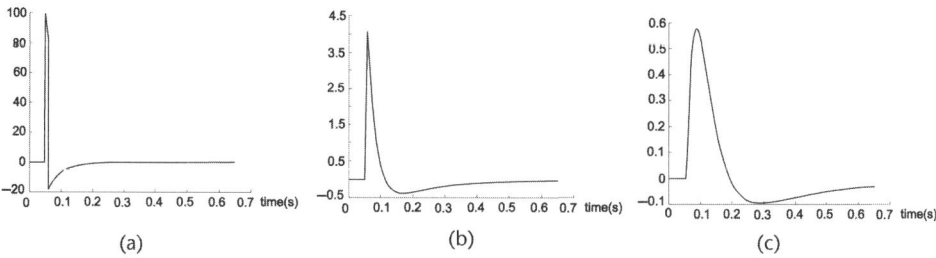

Figure 12.15 (a) Output signal at spatial location (15,15) when an impulse is applied to a 30 × 30 cone network at location (15,15); (b) Identical output signals observed at spatial locations (15,16), (16,15), (15,14), and (14,15) when an impulse is applied to a 30 × 30 cone network at location (15,15); (c) Identical output signals observed at spatial locations (14,16), (16,16), (16,14), and (14,14) when an impulse is applied to a 30 × 30 cone network at location (15,15).

across time in cell (15, 15) (where the "impulse" is applied) and the output amplitude decays abruptly across space. This suggests that the "preferred input" of the network is a fast flashing dot occupying a small space, which confirms the highpass characteristic of the filter. Furthermore, note that the temporal characteristic (time constant) changes across space across from Figure 12.15(a) to Figure 12.15(c), that is, the response takes a longer time to decay in Figure 12.15(c) than it takes in Figure 12.15(a). This confirms the space-time nonseparable characteristic of our cone filtering network.

12.7.2 Visual Cortical Receptive Field Modelling

Mineiro and Zipser proposed a second-order (temporal) model for simple-cell spatiotemporal receptive fields found in the visual cortex in [39]. Their analysis took into account recurrent interactions in the cortex resulting in a recursive filtering structure similar to (12.13). Their receptive field in the frequency domain is given by:

$$H_v(\omega_x, \omega_y) = \frac{\hat{T}_{\Omega_t}[\hat{X}^1_{\omega_x,\omega_y}(1 + j\Omega_t - \hat{W}^{22}_{\omega_x,\omega_y}) + \hat{W}^{22}_{\omega_x,\omega_y}\hat{X}^2_{\omega_x,\omega_y}]}{\hat{W}^{12}_{\omega_x,\omega_y}\hat{W}^{21}_{\omega_x,\omega_y} - (1 + j\Omega_t - \hat{W}^{11}_{\omega_x,\omega_y})(1 + j\Omega_t - \hat{W}^{22}_{\omega_x,\omega_y})} \quad (12.41)$$

where $\hat{T}_{\Omega_t}$ denotes the separable part of the temporal profile of the receptive field, and:

$$\hat{W}^{12}_{\omega_x,\omega_y} = \hat{W}^{21}_{\omega_x,\omega_y} = -Be^{-\frac{\sigma_m^2(\omega_x^2 + \omega_y^2)}{4\pi}} \quad (12.42)$$

$$\hat{W}^{11}_{\omega_x,\omega_y} = -j\frac{v_0 \tau}{2\pi}(\omega_x + \omega_y)e^{-\frac{\sigma_m^2(\omega_x^2 + \omega_y^2)}{4\pi}} = -\hat{W}^{22}_{\omega_x,\omega_y} \quad (12.43)$$

$$\hat{X}^1_{\omega_x,\omega_y} = \frac{-jC}{2}\left(e^{j\phi}e^{-\frac{\sigma_x^2(\omega_x - \omega_{x0})^2 + \sigma_y^2(\omega_y - \omega_{y0})^2}{4\pi^2}} + e^{-j\phi}e^{-\frac{\sigma_x^2(\omega_x + \omega_{x0})^2 + \sigma_y^2(\omega_y + \omega_{y0})^2}{4\pi^2}}\right) \quad (12.44)$$

$$\hat{X}^2_{\omega_x,\omega_y} = \frac{-jC}{2}\left(e^{j\phi}e^{-\frac{\sigma_x^2(\omega_x + \omega_{x0})^2 + \sigma_y^2(\omega_y + \omega_{y0})^2}{4\pi^2}} + e^{-j\phi}e^{-\frac{\sigma_x^2(\omega_x - \omega_{x0})^2 + \sigma_y^2(\omega_y - \omega_{y0})^2}{4\pi^2}}\right) \quad (12.45)$$

where B, C, v_0, τ, ω_{x0}, ω_{y0}, σ_m, σ_x, ϕ, and σ_y are real constants. Note that the original form of Equation (12.41) in [39] only has one spatial dimension. We have extended the STTF from [39] to two spatial dimensions. Equations (12.44) and (12.45) are produced by taking the Fourier Transform of a two-dimensional Gaussian function modulated by two-dimensional sinusoids with a phase shift of ϕ. As seen from (12.42) to (12.45), we have Gaussian, derivative of Gaussian, and Gabor-like functions of the spatial frequency variables. In [30], we analyzed STTFs in terms of sinusoidal functions of the spatial frequencies. These can be used to approximate (12.42) to (12.45). Since we restrict ourselves to filters with real coefficients in this chapter, we only consider the denominator of (12.41) involving the spatial terms of (12.42) and (12.43). The Gabor functions from (12.44) and

(12.45), which lead to complex coefficients in the numerator of the STTF, are not considered here. Complex spatiotemporal filtering with two layers of CNN involving our additional time-derivative connections is discussed in Section 12.8, and in more detail in [40]. For particular parameter values, we can use the following approximation for (12.42) and (12.43) in the range of $(\omega_x, \omega_y) \in (-\pi, \pi] \times (-\pi, \pi]$:

$$\hat{W}^{12}_{\omega_x,\omega_y} = \hat{W}^{21}_{\omega_x,\omega_y} \simeq -\frac{B}{4}[2 + \Gamma_{1,0} + \Gamma_{0,1}] \quad (12.46)$$

for $\sigma_m = \sqrt{2}$.

$$\hat{W}^{11}_{\omega_x,\omega_y} = -\hat{W}^{22}_{\omega_x,\omega_y} \simeq -jA(\Delta_{1,0} + \Delta_{0,1}), A = \frac{3.04 v_0 \tau}{2\pi} \quad (12.47)$$

for $\sigma_m = \sqrt{2}$.

The abbreviation of spatial sinusoidal functions in (12.46) and (12.47) is given by

$$\Gamma_{\xi_x, \xi_y} = \cos(\xi_x \omega_x + \xi_y \omega_y) \quad (12.48)$$

and

$$\Delta_{\xi_x, \xi_y} = \sin(\xi_x \omega_x + \xi_y \omega_y) \quad (12.49)$$

The denominator of (12.41) can now be approximated by:

$$\frac{B^2}{16}(5 + 4\Gamma_{1,0} + 4\Gamma_{0,1} + \Gamma_{1,-1} + \Gamma_{1,1} + \frac{\Gamma_{2,0}}{2} + \frac{\Gamma_{0,2}}{2}) \\ - A^2(1 - \frac{\Gamma_{2,0}}{2} - \frac{\Gamma_{0,2}}{2} + \Gamma_{1,-1} - \Gamma_{1,1}) - 1 + \Omega_t^2 - 2j\Omega_t \quad (12.50)$$

We evaluate the magnitude of Equation (12.50), which is the sum of squares of its real R_v and imaginary parts I_v:

$$R_v = \left[\frac{B^2}{16}(5 + 4\Gamma_{1,0} + 4\Gamma_{0,1} + \Gamma_{1,-1} + \Gamma_{1,1} + \frac{\Gamma_{2,0}}{2} + \frac{\Gamma_{0,2}}{2}) \\ - A^2\left(1 - \frac{\Gamma_{2,0}}{2} - \frac{\Gamma_{0,2}}{2} + \Gamma_{1,-1} - \Gamma_{1,1}\right) - 1 + \Omega_t^2\right]^2 \quad (12.51)$$

with

$$I_v = 4\Omega_t^2 \quad (12.52)$$

The visualization of (12.51) as Ω_t is varied, is shown in Figure 12.16. We see the spatial response is symmetrical about the line $\Omega_t = 0$ and no spatiotemporal orientation is observed. This is expected as both (12.51) and (12.52) are even functions of Ω_t; this is also true for the original (12.42) to (12.43). Therefore,

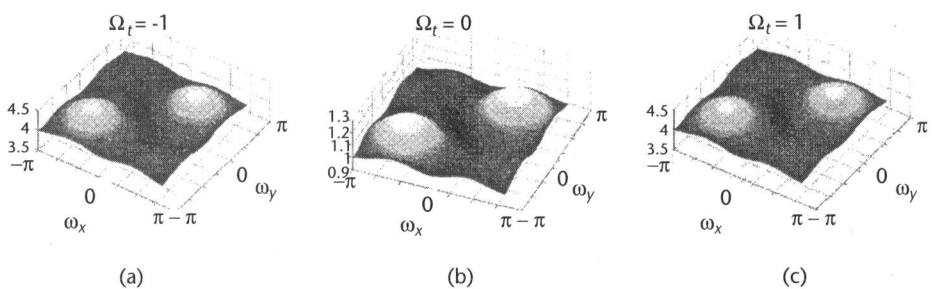

Figure 12.16 *Rv* as an even function of temporal frequency. Visualization of (12.51) shows identical patterns at $\Omega_t = 1$ (a) and $\Omega_t = -1$ (b).

we conclude that any velocity selectivity due to spatiotemporal orientation in the frequency domain [41] in (12.41) is purely due to the numerator terms. As for the STTF, we use the relation

$$z_x^{\xi_x} = e^{j\xi_x \omega_x} \text{ and } z_y^{\xi_y} = e^{j\xi_y \omega_y} \tag{12.53}$$

on the unit circles and Euler's theorem and write (12.46) and (12.47) in terms of variables z_x and z_y. The denominator of (12.41) can be approximated by the frequency response of the following STTF:

$$\left[\frac{5B^2}{16} - 1 - A^2 + \frac{B^2}{8}(z_x + z_x^{-1} + z_y + z_y^{-1}) \right.$$
$$+ \left(\frac{B^2}{32} - \frac{A^2}{2} \right) \cdot (z_x z_y + z_x^{-1} z_y^{-1} + z_x z_y^{-1} + z_x^{-1} z_y)$$
$$\left. + \left(\frac{B^2}{64} + \frac{A^2}{2} \right) \cdot (z_x^2 + z_y^2 + z_x^{-2} + z_y^{-2}) - 2p_t - p_t^2 \right]^{-1} \tag{12.54}$$

This corresponds to a network with first- and second-order temporal derivative diffusion to ground.

12.8 Modeling of Complex Cell Receptive Fields

As discussed in Chapter 5, a practical model for implementing complex cell receptive fields is obtained by combining the results of a series of neurons displaying spatially overlapping (approximately) receptive fields, but of different spatial phases. A bandpass spatiotemporal filter may be used for representing the linear portion of such a complex cell.

Following [40], such a filter, in complex form, may be designed using the placement of spatial frequency–dependent temporal poles, known as the SDTP placement technique, creating a lowpass complex filter prototype. A three-dimensional version of the SDTP technique was suggested by Ip [30]. A frequency shifting approach may then be applied to create a complex bandpass filter prototype. Ip

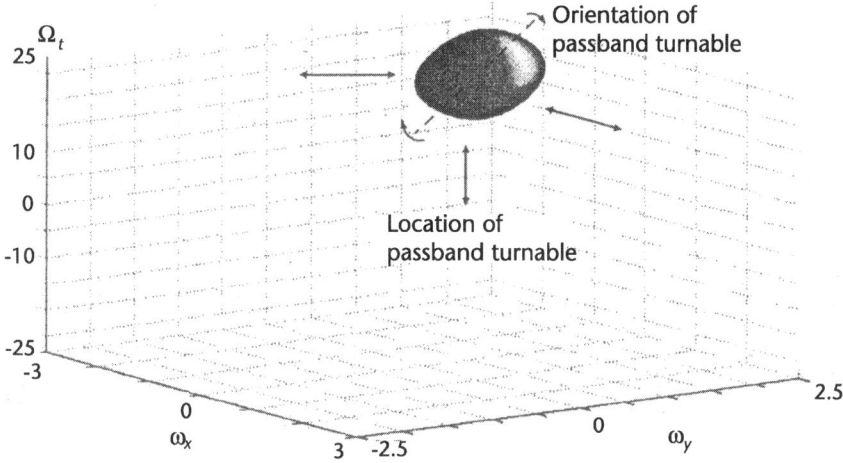

Figure 12.17 Indicative–3 dB isosurface of the magnitude response of a two-layered complex bandpass TDCNN. Both the orientation of the passband volume and the location of the center of the passband are tunable.

suggested an SFG implementation for complex bandpass filters based on a two-layer TDCNN. The output from one layer produces a symmetric receptive field response, while that of the other layer generates the antisymmetric response. Despite the apparent cell complexity, an advantage of the two-layer approach is that the two-layered version of the complex-valued bandpass filter requires interaction with only the 4 nearest cells, thereby simplifying the routing required.

A three-dimensional isosurface of such a complex bandpass filter with tunable characteristics is shown in Figure 12.17. Further details are to be found in [40], in which detailed simulation results and SFG implementations for both real and complex bandpass filters are discussed at length.

12.9 Summary and Conclusions

The demands placed on VLSI systems in order to reproduce responses of the complexity of visual neurons is very high. The most popular approaches tend to rely on static models of processing, which are best implemented using FIR complex digital filter design paradigms. However, for reproducing the variety of spatiotemporal responses seen in V1 neurons, a more efficient approach might seem to be recursive networks, among which the analog networks seem most natural, exploiting the temporal properties of VLSI components.

A new, modified, general form of linear CNNs, targeting spatiotemporal filtering applications, was introduced by incorporating novel temporal derivative "diffusion" between neighboring CNN cells. This results in a convenient one-to-one relationship of three-dimensional rational spatiotemporal transfer function terms with network connections. It is hoped that this has provided insight into the richness of mixed-domain spatiotemporal transfer functions that are realizable with such network topologies, and the appropriateness of such topologies for generating linear response to stimuli that are similar to some spatiotemporal models of receptive field properties found in the primary visual cortex.

Recent work has shown how spatiotemporal, V1 receptive field outputs, when grouped appropriately, permit the emergence of V5-like receptive fields, also known to be quite velocity selective. More recent work has shown that certain aspects of human vision may be explained by such models [42]. Thus, it is almost certain that the use of spatiotemporal receptive field modelling and implementation, of the form enabled by the approaches described in this chapter, ultimately provide a means of reverse engineering the motion pathways of the visual system more efficiently than is currently routinely feasible. This opens up the prospect of improving the computation of new motion-based features, such as those of motion saliency as already discussed in Chapter 9, and as will be addressed in Chapter 14.

Acknowledgments

Parts of the contents of this Chapter have appeared in:

- H.M. Ip, E.M. Drakakis and A.A. Bharath, "On Analogue Networks and Mixed-Domain Spatiotemporal Frequency Response" in *IEEE Transactions on Circuits and Systems, Part I: Regular Papers*, February 2008, pages 264–285, 2008.
- H.M. Ip, E.M. Drakakis and A.A. Bharath, "Synthesis of Nonseparable 3-D Spatiotemporal Bandpass Filters on Analogue Networks," in *Transactions on Circuit Systems, Part I: Regular Papers*, February 2008, pages 286–298, 2008.

References

[1] Crounse, K., Wee, C., and Chua, L. O., "Methods for image processing and pattern formation in cellular neural networks: a tutorial," *IEEE Journal on Circuits and Systems*, Vol. 42, no. 10, October 1995, pp. 583–601.

[2] Shi, B. E., Roska, T., and Chua, L. O., "Design of linear cellular neural networks for motion sensitive filtering," *IEEE Transactions on Circuits and Systems II: Analog and Digital Signal Processing*, Vol. 40, no. 5, May 1993, pp. 320–331.

[3] Dudek, P. and Hicks, P. J., "A general-purpose processor-per-pixel analog SIMD vision chip," *IEEE Transactions on Circuits and Systems I: Regular Papers*, Vol. 52, no. 1, January 2005, pp. 13–20.

[4] Harrison, R. R., "A biologically inspired analog ic for visual collision detection," *IEEE Transactions on Circuits and Systems I: Regular Papers*, Vol. 52, no. 11, November 2005, pp. 2308–2318.

[5] Massari, N., et al., "A cmos image sensor with programmable pixel-level analog processing," *IEEE Transactions on Neural Networks*, Vol. 16, no. 6, November 2005, pp. 1673–1684.

[6] Giggins, C. M., Pant, V., and Dutschmann, R., "Analog VLSI implementation of spatio-temporal frequency tuned visual motion algorithms," *IEEE Transactions on Circuits and Systems I: Regular Papers*, Vol. 52, March 2005, no. 3, pp. 489–502.

[7] Laiho, M., et al., "A mixed-mode polynomial cellular array processor hardware realization," *IEEE Transactions on Circuits and Systems I: Regular Papers*, Vol. 51, no. 2, February 2004, pp. 286–297.

[8] Cembrano, G. L., et al., "A 1000 FPS at 128 × 128 vision processor with 8-bit digitized I/O," *IEEE Journal of Solid-State Circuits*, Vol. 39, no. 7, July 2004, pp. 1044–1055.

[9] Stocker, A. A., "Analog VLSI focal-plane array with dynamic connections for the estimation of piecewise-smooth optical flow," *IEEE Transactions on Circuits and Systems I: Regular Papers*, Vol. 51, May 2004, no. 5, pp. 963–973.

[10] Kowalski, J., "0.8 µm cmos implementation of weighted-order statistic image filter based on cellular neural network architecture," *IEEE Transactions on Neural Networks*, Vol. 14, no. 5, September 2003, pp. 1366–1374.

[11] Galan, R. C., et al., "A bio-inspired two-layer mixed-signal flexible programmable chip for early vision," *IEEE Transactions on Neural Networks*, Vol. 14, no. 5, September 2003, pp. 1313–1336.

[12] Nagy, Z. and Szolgay, P., "Configurable multilayer CNN-UM emulator on FPGA," *IEEE Transactions on Circuits and Systems I: Fundamental Theory and Applications*, Vol. 50, no. 6, June 2003, pp. 774–778.

[13] Ruedi, P., et al., "A 128×128 pixel 120dB dynamic-range vision sensor chip for image contrast and orientation extraction," *IEEE Journal of Solid-State Circuits*, Vol. 38, no. 12, December 2003, pp. 2325–2333.

[14] Barbaro, M., et al., "A 100×100 pixel silicon retina for gradient extraction with steering filter capabilities and temporal output coding," *IEEE Journal of Solid-State Circuits*, Vol. 37, no. 2, February 2002, pp. 160–172.

[15] Lei, M. and Chiueh, T., "An analog motion field detection chip for image segmentation," *IEEE Transactions on Circuits and Systems for Video Technology*, Vol. 12, no. 5, May 2002, pp. 299–308.

[16] Etienne-Cummings, R., Kalayjian, Z. K., and Cai, D., "A programmable focal-plane mimd image processor chip," *IEEE Journal of Solid-State Circuits*, Vol. 36, no. 1, January 2001, pp. 64–73.

[17] Shi, B. E., "A low-power orientation-selective vision sensor," *IEEE Transactions on Circuits and Systems II: Analog and Digital Signal Processing*, Vol. 47, no. 5, May 2000, pp. 435–440.

[18] Cauwenberghs, G. and Waskiewicz, J., "Focal-plane analog VLSI cellular implementation of the boundary contour system," *IEEE Transactions on Circuits and Systems I: Fundamental Theory and Applications*, Vol. 46, no. 2, February 1999, pp. 327–334.

[19] Cruz, J. M. and Chua, L. O., "A 16×16 cellular neural network universal chip: the first complete single-chip dynamic computer array with distributed memory and with gray-scale input-output," *Analog Integrated Circuits and Signal Processing*, Vol. 15, no. 3, March 1998, pp. 227–237.

[20] Dominguez-Castro, R., et al., "A 0.8-µm cmos two-dimensional programmable mixed-signal focal-plane array processor with on-chip binary imaging and instructions storage," *IEEE Journal of Solid-State Circuits*, Vol. 32, no. 7, July 1997, pp. 1013–1026.

[21] Anguita, M., et al., "A low-power cmos implementation of programmable CNN's with embedded photosensors," *IEEE Transactions on Neural Networks*, Vol. 44, no. 2, February 1997, pp. 149–153.

[22] Etienne-Cummings, R., Spiegel, J. V. D., and Mueller, P., "A focal plane visual motion measurement sensor," *IEEE Transactions on Circuits and Systems - I: Fundamental Theory and Applications*, Vol. 44, no. 1, January 1997, pp. 55–66.

[23] Simoni, A., et al., "A single-chip optical sensor with analog memory for motion detection," *IEEE Journal of Solid-State Circuits*, Vol. 30, no. 7, July 1995, pp. 800–806.

[24] Chua, L. O. and Yang, L., "Cellular neural networks: theory and applications," *IEEE Transactions on Circuits and Systems*, Vol. 35, October 1988, pp. 1257-1290.

[25] Shi, B. E. and Chua, L. O., "Resistive grid image filtering: inputs/output analysis via the CNN framework," *IEEE Transactions on Circuits and Systems I: Fundamental Theory and Applications*, Vol. 39, no. 7, July 1992, pp. 531-548.

[26] Ip, H. M. D., Drakakis, E. M., and Bharath, A. A., "Towards analog VLSI arrays for nonseparable 3d spatiotemporal filtering," in *Proceedings of the IEEE International Workshop on Cellular Neural Networks and Its Applications (CNNA'06)*, Istanbul, Turkey, August 2006, pp. 1-6.

[27] Torralba, A. B. and Herault, J., "An efficient neuromorphic analog network for motion estimation," *IEEE Transactions on Circuits and Systems I: Special Issue on Bio-Inspired Processors and CNNs for Vision*, Vol. 46, no. 2, February 1999, pp. 269-280.

[28] Zhang, Y. and Bruton, L. T., "Applications of 3D LCR networks in the design of 3-D recursive filters for processing image sequences," *IEEE Journal on Circuits and Systems for Video Technology*, Vol. 4, no. 4, August 1994, pp. 369-382.

[29] Mutluay, H. E. and Fahmy, M. M., "Analysis of N-D general-support filters," *IEEE Transactions on Acoustics, Speech, and Signal Processing*, Vol. 33, no. 4, August 1985, pp. 972-982.

[30] Ip, H. M. D., Drakakis, E. M., and Bharath, A. A., "Synthesis of nonseparable 3D spatiotemporal bandpass filters on analog networks," *IEEE Transactions on Circuits and Systems I*, vol. 55, no. 1, 2008, pp. 286-298.

[31] Justice, J. H. and Shanks, J. L., "Stability criterion for N-dimensional digital filters," *IEEE Transactions on Automatic Control*, Vol. 18, no. 3, 1973, pp. 284-286.

[32] Johns, D. and Martin, K., *Analog intergrated circuits design*. Wiley Text Books, 1996.

[33] Huelsman, L. P., *Active and passive analog filter design, an introduction*. New York, McGraw-Hill, 1993.

[34] Ip, H. M. D., Drakakis, E. M., and Bharath, A. A., "Analog networks for mixed domain spatiotemporal filtering," in *Proceedings of the IEEE International Symposium on Circuits and Systems 2005 (ISCAS'05)*, Kobe, Japan, May 2005, pp. 3918-3921.

[35] Balya, D., et al., "Implementing the multilayer retinal model on the complex-cell CNN-UM chip prototype," *International Journal of Bifurcation and Chaos*, Vol. 14, no. 2, February 2004, pp. 427-451.

[36] Bertschmann, R. K., Bertley, N. R., and Bruton, L. T., "A 3-D integrator-differentiator double-loop (IDD) filter for raster-scan video processing," in *Proceedings of the IEEE International Symposium on Circuits and Systems 1995 (ISCAS'95)*, Seattle, WA, April 1995, pp. 470-473.

[37] Pitas, J. K. and Venetsanopoulos, A. N., "The use of symmetries in the design of multidimensional digital filters," *IEEE Journal of Circuits and Systems*, Vol. 33, no. 9, September 1986, pp. 863-873.

[38] Bolle, M., "A closed form design method for recursive 3-d cone filters," in *Proceedings of the IEEE International Symposium on Acoustics, Speech, and Signal Processing 1994*, Vol. 6, Adelaide, Australia, 1994, pp. VI/141-VI/144.

[39] Mineiro, P. and Zipser, D., "Analysis of direction selectivity arising from recurrent cortical interactions," *Neural Computation*, February 1998, pp. 353-371.

[40] Ip, H. M. D., "Bioinspired analog networks for spatiotemporal filtering: Theory and analog VLSI implementations," Ph.D. dissertation, Imperial College London, 2007.

[41] Jähne, B., HauBecker, H., and Geibler P. (Eds.), *Handbook of computer vision and application, volume 2: signal processing and pattern recognition*. Academic Press, 1999.

[42] Rust, N. C., et al., "How MT cells analyze the motion of visual patterns," *Nature Neuroscience*, Vol. 9, November 2006, pp. 1421-1431.

CHAPTER 13
From Algorithms to Hardware Implementation

Christos-Savvas Bouganis, Suhaib A. Fahmy, and Peter Y. K. Cheung

13.1 Introduction

This chapter addresses the process of mapping algorithms into hardware. Two case studies are considered. The first case is focused on the mapping of a novel, biologically inspired, complex steerable wavelet implementation onto reconfigurable logic. This wavelet construction performs a sub-band decomposition of images into orientation and scale-selective channels that can then be used for analysis purposes. Although there are numerous decompositions that satisfy such a requirement, the proposed construction utilizes polar separable functions, so that the orientation selectivity can be specified independently of radial frequency (or scale selectivity). The wavelet construction is of special interest because it is biologically inspired, and it requires a small number of kernels for its realization, which leads to an efficient hardware design.

The second case of interest is focused on a face authentication algorithm and its acceleration through hardware. The algorithm under investigation is the trace transform. The trace transform has been recently proposed for the face authentication task, achieving one of the best authentication rates [17]. One of the main drawbacks of the transform is its computational complexity, which prevents its usage in a real-life environment. However, the algorithm's inherent parallelism can be exploited in a hardware implementation leading to the algorithm's acceleration.

The target hardware platform is a field-programmable gate array (FPGA). FPGAs are programmable devices that offer fine-grain parallelism and reconfigurability. The chapter discusses the mapping of the above wavelet construction and the trace transform into FPGA, and demonstrates that a considerable speedup factor is achieved compared to software implementations. This chapter focuses on a detailed description of the implementation of the designs in hardware, alongside a discussion on the various decisions during the design process. Moreover, results regarding the speed of the proposed design compared with software implementations, and the error that is inserted to the system due to fixed point representations, are discussed. The chapter concludes by demonstrating the applicability of the design.

13.2 Field Programmable Gate Arrays

Field programmable gate arrays (FPGAs) were first introduced in 1984. FPGAs are semiconductor devices that can be programmed to perform an operation after the manufacturing process. This is achieved by containing logic and interconnection components that are programmable. The programmable logic can be configured to perform a series of functions like AND, OR, XOR, and other mathematical functions, where the programmable interconnections are responsible for connecting the different components of an FPGA together, leading to realizations of more complex functions.

The programmable components make up the building blocks of an FPGA. Each building block usually contains a look-up table (LUT) with four inputs, which can be configured to perform any mathematical function that requires up to four inputs and one output, a flip-flop, which can store one bit of information, and some other support logic. Support logic is used to enhance some of the most common tasks the device should perform. Figure 13.1 shows an overview of such a building block. The programmable interconnections connect the different components together for implementation of more complex functions. These interconnections are programmable, allowing a flexibility in the connection of the programmable components.

When FPGAs were first introduced, they used to have only programmable logic and interconnections. However, the requirement for more powerful devices and the vendors' target to take part of the DSP market pushed the FPGA vendors to include extra components in their devices. These components are embedded multipliers, that can operate general multiplication more efficiently than mapping it to reconfigurable logic, and embedded random access memory (RAM) modules, for storing information inside the device. Nowadays, in the high-end FPGAs, the embedded multipliers have been replaced by DPS blocks that can perform logical operations too, and some devices also contain a hard-core processor.

This fine-grain parallelism and the reconfigurability properties make FPGAs a good candidate for acceleration of many applications. By appropriately dividing the problem into different blocks that can be executed in parallel and mapping them into an FPGA, a considerable amount of acceleration can be achieved. Initially, FPGAs were treated by the designers as devices for fast prototyping of applications by taking advantage of their reconfigurability properties. However, the increasing cost of ASIC manufacturing has led the field to adopt the FPGA as the target platform in many situations.

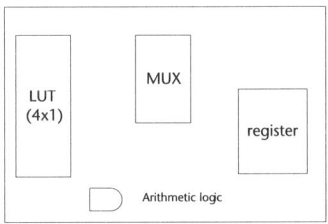

Figure 13.1 Programmable component.

13.2.1 Circuit Design

Many programming languages exist that can be used for circuit design. One of the most common is VHDL (VHSIC[1] Hardware Description Language). It is a register transfer level (RTL) language, where the designer has to specify the operations that should happen in the device in every clock cycle.

The need for a language that would permit the designer to design circuits faster and without a deep knowledge of hardware directed the hardware field in the creation of higher-level languages. An example of such languages is the Handel-C from Celoxica [4]. Handel-C [16] is based on the syntax of ANSI[2] C with additional extensions in order to take advantage of the specific characteristics of the hardware. It is independent of the targeting platform which makes the design easily transferable to other hardware platforms. Recently, higher-level languages have appeared, like SystemC [20]. SystemC is a set of routines that allows the simulation of concurrent processes and also mapping of the design into hardware. It is a transaction-level language, which means that the language focuses more on the transactions that have to be performed and less on the implementation issues.

13.2.2 Design Process

The steps that a hardware designer has to follow in designing a hardware device are briefly described in this section. The first step is the analysis of the problem and the decision of the architecture of the system. Then, a description of the system using VHDL or a higher-level language is performed. The design is tested using a set of input data (test-vectors) that should test all the corner cases[3] in the design. The next step is to map the design onto the FPGA, where the different components/modules in the design are mapped to the resources of the FPGA. The final step is the so-called place and route step, where the components are actually placed on specific places in the FPGA (place) and are connected together (route).

13.3 Mapping Two-Dimensional Filters onto FPGAs

The need for more flexibility and fast prototyping of signal processing algorithms has led the FPGA community to investigate tools for easy mapping of signal processing algorithms onto FPGAs. One approach is to provide the designers with building blocks that are common in DSP applications [14]. Another approach is to provide tools that allow the engineers to describe their design in a high-level language [2].

The applications of wavelets to signal and image compression are well researched [7, 10, 19]. The work we shall describe in the next section focuses on providing a detailed example based around the implementation of the specific

---

1. Very-high speed integrated circuit.
2. American National Standards Institute.
3. The term "corner cases" is used to indicate extreme cases that may not arise often during an application that should be tested to test the robustness of the system.

steerable wavelet pyramid described in Chapter 5, including performance and precision trade-offs that must be made in order to target a specific FPGA platform.

13.4 Implementation of Complex Wavelet Pyramid on FPGA

The complex steerable wavelet design discussed in Chapter 5 may be implemented on an FPGA device for speed of processing. The original design should be tailored in order to ensure that it is "hardware friendly" by targeting a minimum number of distinct and symmetric kernels. The first decision that should be made is which part of the algorithm will be mapped to hardware and which to software. The part of the algorithm that is mapped onto hardware should be fixed, that is, it has a very small probability to change, and the operations involved should be hardware friendly. The part of the algorithm that should be flexible is mapped onto software. Thus, in the current problem, "high-level" decisions that make use of the resulting images from the application of the steerable pyramid on the input image, that is, corner detection, are mapped to be executed by the host central processing unit (CPU), where the FPGA is responsible for performing the convolution of the input image with the kernels in order to construct the pyramid structure.

The target board that is used for implementation is the RC1000-PP from Celoxica. It is a PCI[4] bus plug-in card for PCs with a Virtex V1000 FPGA and four memory banks of 2 MBytes each. All four memory banks are accessible by both the FPGA and any device on the PCI bus. However, at any time instance, only one device can access a memory bank. In this work, Handel-C is used as the main tool to describe the steerable complex wavelet pyramid on hardware.

The available device (V1000) has limited hardware resources, which does not allow the realization of all levels of the pyramid in the device at the same time. Thus, only part of the pyramid is accelerated in hardware at any given time, and the full decomposition is realized through reuse of the same hardware.

13.4.1 FPGA Design

Software realizations of algorithms make use of floating point operations. However, a block in FPGA that can perform double precision arithmetic is expensive in terms of area, speed, and power. Thus, it is a common design technique to perform the operations using fixed point arithmetic. That is, for each variable in the design, a number of bits are allocated to represent this number where the position of the decimal point is fixed. This introduces errors in the algorithm, and further analysis on the performance of the algorithm is required in order to assess the impact of the introduced errors. This word-length optimization is an active area of research in high-level synthesis, which focuses on assigning optimum word lengths in the variables of the design in order to bound the error at the output of the device [1,3].

The quantization of the variables in the proposed design is as follows: eight bits are used to represent a pixel in the image and ten bits are used to represent the coefficients of each filter and also the output of each convolution. The impact to

---
4. A peripheral component interconnect (PCI) bus is a path between the components in a computer.

the final accuracy of the algorithm by selecting these numbers of bits to represent the variables is discussed in Section 13.4.3. In order for the decomposition to be performed as fast as possible, the whole design is pipelined to produce three convolution results per clock cycle.

The kernels reported in Chapter 5 have a size of 15×15 pixels. This would require in the current design 225×3 general purpose multipliers, which cannot be accommodated by the available device. Thus, the kernels have been truncated to sizes of 7×7 pixels, implying that only 49×3 multipliers are required. The difference in the frequency domain between the original and truncated version of the kernels is illustrated in Figure 13.2. The figure shows that only a small degradation in the frequency response takes place, permitting us to perform this truncation. However, this truncation depends on the specific application and cannot always be applied without significant degradation of the output. Moreover, Figure 13.2 also depicts the effect of quantizing the filter coefficients of the truncated version using 10-bits precision. Only a small further degradation on the frequency response takes place.

Figure 13.3(a) shows an overview of the design. The pixels are stored in a raster scan in sets of four in the memory allowing the FPGA to fetch four pixels per read cycle.

The *ManageBuffer* process is responsible for buffering the data to provide a region of the image to the next process in the design. By applying such a design technique we make use of the locality of the two-dimensional convolution operator and reuse the data that has been fetched from the external memories in the previous step. Thus, using the above design methodology and by requiring one pixel only from the external RAM, a new window for convolution is constructed at every clock cycle. The drawback of such a design is that a number of lines of the image should be stored internally in first-in first-out (FIFO) blocks in the device. However, our targeted device has embedded RAM blocks. Thus, the FIFOs are mapped to block RAMs in the FPGA for a more efficient use of resources.

The next process, the *ProcessWindow*, performs the convolution between a window in the image and the appropriate masks. It contains three programmable processes *FilterBankA (FBA)*, *FilterBankB (FBB)*, and *FilterBankC (FBC)*, each one of which can apply three different filters by loading a specific set of coefficients. A shift register and a RAM to store the coefficients are selected to form the appropriate masks for each level of the pyramid. In each of these programmable blocks, the multiplications between the pixel values and the coefficients are performed in parallel, requiring one clock cycle. The resulting values are added together using a pipelined adder tree, achieving one result of the convolution per clock cycle.

It should be noted that the current hardware design has not been optimized with respect to the symmetric properties of the kernels, but rather it gives a general methodology of mapping such designs into hardware. The final results are concatenated and stored in the external RAM. Figure 13.3(b) shows a detailed diagram of the *FilterBank* process. Moreover, the result from the last filter, which represents the input image for the next level of the pyramid, is decimated, saturated in the range [0,255] and stored in the external memory by the *NextLevelImage* process.

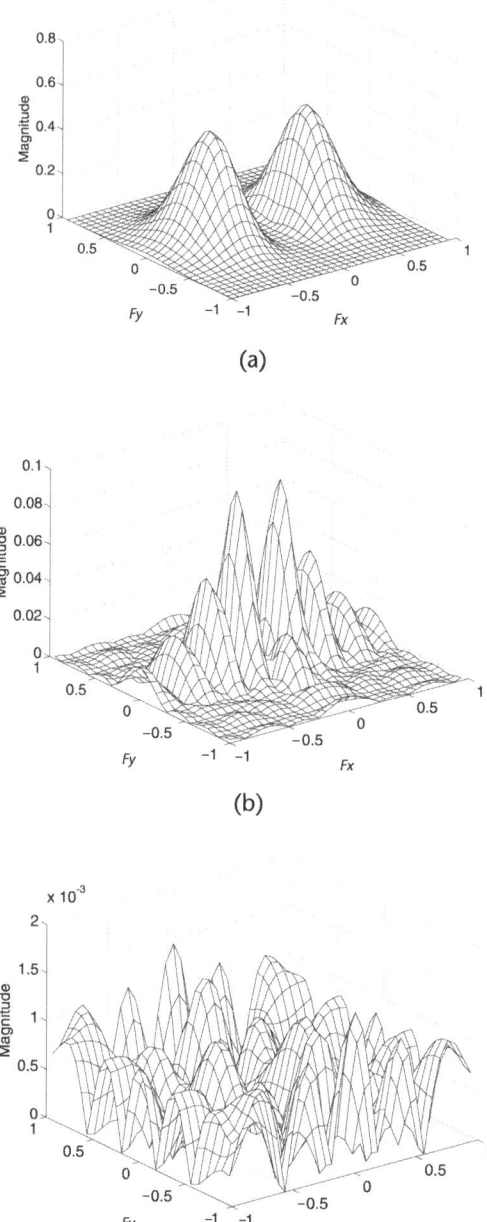

Figure 13.2 (a) Frequency response of one of the filters. (b) The absolute difference in the frequency response between the original filter and its truncated to 7×7 pixel size version. (c) The absolute difference in the frequency response between the truncated version of the pixel and the same size filter where the coefficients have been quantized using 10-bits precision. Note the different scales on the vertical axes.

13.4 Implementation of Complex Wavelet Pyramid on FPGA

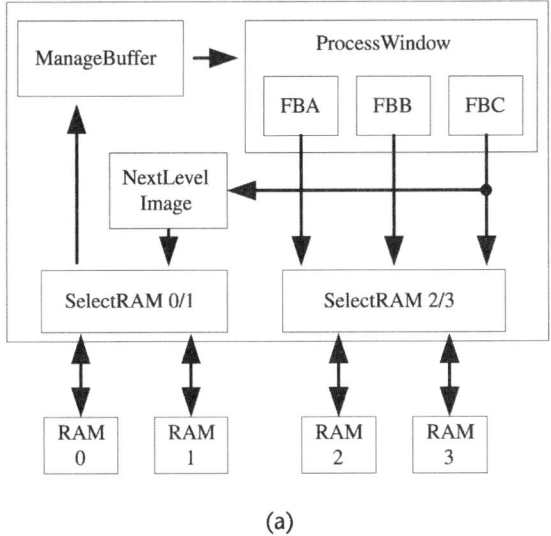

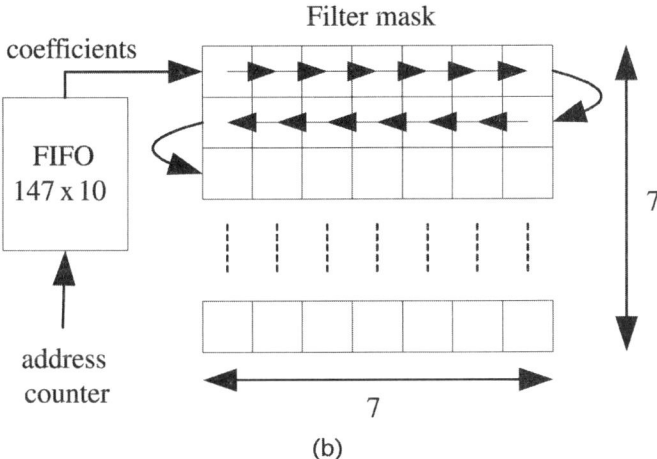

Figure 13.3 (a) The top-level diagram of the design. (b) The *FilterBank* process. The FIFO contains the coefficients for the three kernels that are realized in the filter bank.

13.4.2 Host Control

The CPU controls the operation of the FPGA by a handshake protocol. Due to the associated latency of each transfer through the PCI bus, the data are transferred using direct memory access (DMA) between the CPU and the board [6]. In order to speed up the process, the decomposition of the image and the transfer of the data to/from the host are performed in parallel. The following scheme is used. Out of the four memory banks, the first two are used to store the new frame that is sent by the host for processing, the previous frame that is being processed by the FPGA, and the decimated images that are used for the different levels of the pyramid. The other two banks are used by the FPGA to store the result of the convolutions. The handshake protocol operates as follows. When a new frame is available, the CPU sends the data to RAM 0 or 1 and signals to the FPGA that a new frame is waiting for processing. In the meantime, the FPGA processes the previous frame from RAM 1 or 0, respectively. The results from the convolutions are stored in RAMs 2 and 3. The output data is distributed between RAMs 2 and 3 such that while the FPGA writes the results from a convolution to one RAM the CPU performs a DMA transfer to the already calculated results from the other RAM. The distribution of the results is necessary, since the design should be able to handle images with size 640 by 480 pixels.

13.4.3 Implementation Analysis

Experiments are performed to investigate the impact of the number of bits that is used to represent the kernel coefficients (N_c) and the bits that are used to represent the result of the convolution (N_o) to the filter responses. The mean square error of the estimation of each filter response between full precision and fixed point for each combination of N_c and N_o is estimated using the Lena image shown in Figure 13.4. Figure 13.5 shows the average mean square error over all filters using the same combination of N_c and N_o. In our design, N_o is set to 10 in order to be able to store the results of three parallel convolutions by performing only one 32-bit access to the external memory. From the figure it can be concluded that the number of bits used for the coefficients has a small effect on the error of the response compared with the number of bits used to represent the result of the

Figure 13.4 The Lena image used for testing.

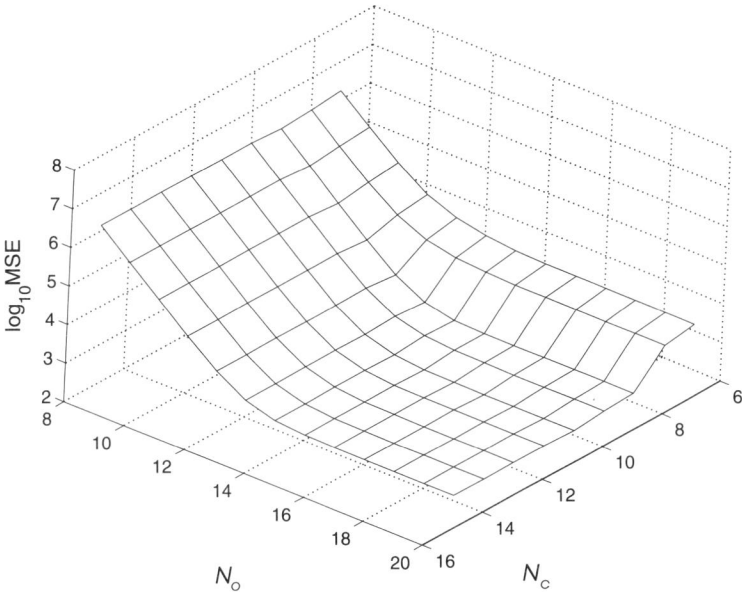

Figure 13.5 Mean square error in the filters' response using the Lena image between fixed-point and floating-point arithmetic.

filters. Also, it should be mentioned that the error in the filter responses increases after changing levels since the decimated result of the lowpass channel is reused for the next bandpass decomposition. A more thorough investigation of the injected error in the design by quantizing the variables can be found in [1,3].

Using the Xilinx ISE 6.1 software, we place and route the design. The overall design uses 12,286 slices out of the 12,288 slices of V1000. Due to the large size of the design compared with the available space in Virtex V1000, the optimum clock rate cannot be achieved. The synthesis results of the design using Xilinx ISE 6.1 gives 99% usage of slices. It is clear that there is not enough available space for optimum routing, which reduces the optimal clock frequency. Also, the design has used 15 out of the 32 available block RAMs. Furthermore, the slow RAMs that are available on the board reduce the effective speed of the design. Due to the nature of the decomposition, the data that is generated corresponds to an equivalent size of 14.6 times the input image. This amount of data cannot be stored in the internal memory of the FPGA and should be transferred to the external RAMs. Thus, the available bandwidth to the external memories reduces the effective speed of the current design. A rate of 16.6 MHz was achieved giving 13.1 frames per second in video graphics array (VGA) resolution (640 × 480).

13.4.4 Performance Analysis

Experiments were performed to compare the speed of the new design with a software implementation. A decomposition with four orientations and four levels is performed on two test images. The first test image is the Lena image with size

256×256, shown in Figure 13.4, where the second test image is the Boats image with size 512×512, shown in Figure 13.6.

Table 13.1 shows a summary of the results. The first row of the table corresponds to a machine with dual hyperthreading xeons at 2.66 GHz with 2 GB of RAM. The software runs under MATLAB using the *Intel SIMD Integrated Performance Primitives* libraries, which also takes advantage of multiprocessors. The second row corresponds to a similar machine but without hyperthreading technology. The software version of the design was implemented using single, dual, and quad threads. The RC1000-PP board is placed on a Dual Pentium III machine at 450 MHz and 512 MB of RAM. The result for the software is the average over 40 frames, where for the hardware the result is the average of 4,000 frames. The timing for the FPGA includes the DMA transfers. In both cases the required time to read the data from the hard disk is excluded. The speedup factor is calculated with respect to the best performance of the software implementation in each row. It can be seen that an average improvement of 2.6 times in the speed can be achieved. Moreover, by placing the design in an XC2V6000 we can investigate how fast the current design can be clocked without any restrictions from the size of the FPGA device or by the timing constraints of the external memories. The synthesis tool showed that the design can be clocked up to 50 MHz giving an average *speedup factor* of 8 compared with the software implementation.

Figure 13.6 The Boats image used for testing.

Table 13.1 Comparison Results in Speed Between Software and Hardware Implementations

Image Size	HT	1 Thr.	2 Thr.	4 Thr.	FPGA	Speedup
Lena	Yes	0.06035s	0.04897s	0.05088s	0.0170s	2.88
(256 × 256)	No	0.05953s	0.04737s	—		2.78
Boats	Yes	0.21074s	0.21281s	0.16113s	0.0653s	2.46
(512 × 512)	No	0.20720s	0.15724s	—		2.40

Using an XC2V6000 device a speedup factor of 8 is achieved.

13.4.4.1 Corner Detection

Further experiments are performed to assess the performance of the design to real-life situations. The application under consideration is corner detection using the algorithm described in Chapter 5. We investigate how precisely the corner of a structure in the image is detected given the limited number of bits that is used to represent the coefficients and the response of the filters. Figure 13.7 shows the performance of the above design compared with a software implementation. The image at the top is the result of the corner detection when the whole algorithm is implemented in software. The image at the bottom is the result of the corner detection when the decomposition of the image is performed in FPGA. It can be seen that most of the features have been detected correctly except for 8 mismatches. Further investigation revealed that by assigning 16 bits to represent the output of the filters we obtain zero mismatches.

13.4.5 Conclusions

This section presented a mapping of the novel steerable pyramid, discussed in Chapter 5, for image decomposition and feature detection into reconfigurable hardware. The section illustrates the various steps in the design process of mapping a steerable pyramid into hardware and highlights issues that should be considered in such mappings, such as memory management, limited resources, and so on. A detailed investigation of the impact of the quantization of the variables to the filter responses is carried out and potential problems in the design of such multilevel transforms are highlighted. Due to the nature of the algorithm, a huge amount of data is produced and can be stored only in the external RAMs. The current design is limited by the available bandwidth to the external memories.

13.5 Hardware Implementation of the Trace Transform

13.5.1 Introduction to the Trace Transform

The trace transform (see also Chapter 9) was recently introduced by Kadyrov and Petrou [11-13, 15]. They have demonstrated that the trace transform is a powerful tool, showing excellent results in image database search, industrial token registration, activity monitoring [12], character recognition, and face authentication [17, 18].

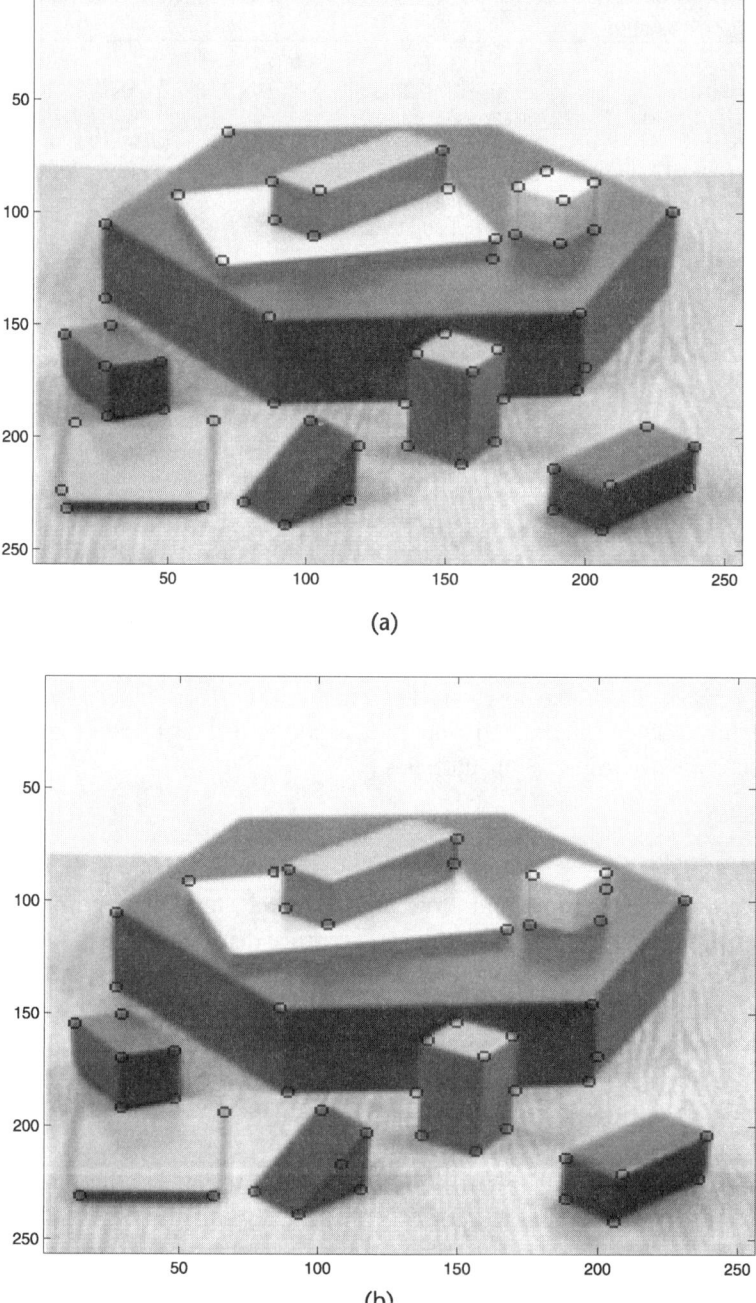

Figure 13.7 (a) The result of the corner detection algorithm described in Chapter 5 implemented in software. (b) The same result using the hardware implementation of the same algorithm.

13.5 Hardware Implementation of the Trace Transform

The proposed transform produces a redundant representation of an image, from which features can be extracted. The strength of the transform lies in its ability to extract features that are robust to affine transformations, occlusion, and even nonlinear deformations. In fact, careful selection of the transform functionals, one of the parameters of the transform, can allow not just robustness, but also recovery of the transformation coefficients, which would allow an affine-transformed image to be returned to its original state [13,15]. One of the primary obstacles to further investigation and adoption of the algorithm has been its computational complexity.

FPGAs provide an ideal platform for accelerating the trace transform through exploitation of the inherent parallelism that exists within the algorithm. Hardware implementations of algorithms can process multiple parallel computations, which leads to considerable acceleration of the algorithm. Furthermore, the flexibility of the FPGA platform suits the generic nature of the transform, since alternative functionals can be swapped in and out with ease due to the ability to change the configuration of an FPGA device on the fly. A possible acceleration of the trace transform would allow a further investigation of the transform in terms of the used functionals and applications [8].

The trace transform of an image is a transformation from the spatial domain to a domain with parameters ϕ and p. Each (ϕ,p) point in the transformed domain is the result of applying a defined functional on the intensity function of a line that crosses the original image tangential to an angle ϕ and at a distance p from the center. This is shown in Figure 13.8. The resultant representation is an $n_\phi \times n_p$ image where n_ϕ is the number of angles considered and n_p is the maximum number of lines across the image for each angle. An image can be traced with any number of functionals, each producing a corresponding trace. Any functional can be used to reduce each trace line to a single value. The simplest functional, the sum of pixel intensities along the lines, yields the radon transform [5] of an image.

It may be easier to consider drawing multiple parallel lines across an image. This can be done for any angle. Then a functional can be applied to the pixel intensities in each line to yield a value for the (ϕ,p) point in the new domain. How the corresponding line pixels are selected is decided by the specific implementation. Standard interpolation methods, including nearest-neighbor and bicubic interpolation, can be used. A functional maps a vector to a single value. The type of functional can be one of many that have been proposed in [12], or can take any

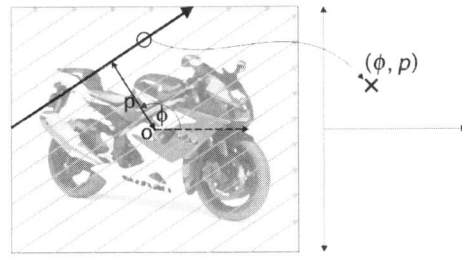

Figure 13.8 Mapping of an image to the new space.

other form that maps a vector to a single value. This might include sum, median, mode, sum of differences, root mean squared, and so on. Table 13.2 shows the 22 functionals proposed by the trace transform authors for face authentication [18]. These were chosen for their strength in texture classification and their robustness to different types of image transformation. By inspecting the functionals in Table 13.2, it is clear that some functionals are computationally more intensive than others, and there is a large variation in their structure. Accelerating these functionals in hardware can yield a significant performance boost.

In the case where the image is to be mapped to a single feature value, further steps are needed; these are shown in Figure 13.9. First, a "diametrical" functional (D) is applied to the columns of the trace image, reducing the image to a single vector. Finally, a "circus" functional (C) is applied to this vector to yield a single value feature. By combining a number of functionals at each stage, numerous features can be extracted. The focus in this chapter is solely on the first step of the trace transform, that is, mapping the image to the trace domain, since this is where

Table 13.2 The Trace Functionals T

No.	Functional	Details
1	$T(f(t)) = \int_0^\infty f(t)dt$	Radon transform
2	$T(f(t)) = \left[\int_0^\infty |f(t)|^{\frac{1}{2}}dt\right]^2$	
3	$T(f(t)) = \left[\int_0^\infty |f(t)|^4 dt\right]^{\frac{1}{4}}$	
4	$T(f(t)) = \int_0^\infty |f(t)'|dt$	$f(t)' = (t_2 - t_1), (t_3 - t_2), \ldots, (t_n - t_{n-1})$
5	$T(f(t)) = \text{median}_t\{f(t), |f(t)|\}$	Weighted median
6	$T(f(t)) = \text{median}_t\{f(t), |f(t)'|\}$	
7	$T(f(t)) = \left[\int_0^{n/2} |\mathcal{F}\{f(t)\}(t)|^4 dt\right]^{\frac{1}{4}}$	$\mathcal{F}$ means taking the discrete fourier transform
8	$T(f(t)) = \int_0^\infty \left|\frac{d}{dt}\mathcal{M}\{f(t)\}\right| dt$	$\mathcal{M}$ means taking the median over a length 3 window, and $\frac{d}{dt}$ means taking the difference of successive samples
9	$T(f(t)) = \int_0^\infty rf(t)dt$	$r = |l - c|; l = 1, 2, \ldots, n; c = \text{median}_l\{l, f(t)\}$
10	$T(f(t)) = \text{median}_t\left\{\sqrt{rf(t)}, |f(t)'|^{\frac{1}{2}}\right\}$	
11	$T(f(t)) = \int_0^\infty r^2 f(t)dt$	
12	$T(f(t)) = \int_{c*}^\infty \sqrt{rf(t)}dt$	$c*$ signifies the nearest integer to c
13	$T(f(t)) = \int_{c*}^\infty rf(t)dt$	
14	$T(f(t)) = \int_{c*}^\infty r^2 f(t)dt$	
15	$T(f(t)) = \text{median}_{t*}\{f(t^*), |f(t^*)|^{\frac{1}{2}}\}$	$f(t^*) = \{f(t_{c*}), f(t_{c*+1}), \ldots, f(t_n)\}$
16	$T(f(t)) = \text{median}_{t*}\{rf(t^*), |f(t^*)|^{\frac{1}{2}}\}$	$l = c^*, c^* + 1, \ldots, n; c = \text{median}_l\{l, |f(t)|^{\frac{1}{2}}\}$
17	$T(f(t)) = \left|\int_{c*}^\infty e^{i4\log(r)}\sqrt{rf(t)}dt\right|$	
18	$T(f(t)) = \left|\int_{c*}^\infty e^{i3\log(r)} f(t)dt\right|$	
19	$T(f(t)) = \left|\int_{c*}^\infty e^{i5\log(r)} rf(t)dt\right|$	
20	$T(f(t)) = \int_c^\infty \sqrt{rf(t)}dt$	$r = |l - c|; l = 1, 2, \ldots, n;$
21	$T(f(t)) = \int_c^\infty rf(t)dt$	$c = \frac{1}{S}\int_0^\infty l|f(t)|dt; S = \int_0^\infty |f(t)|dt$
22	$T(f(t)) = \int_c^\infty r^2 f(t)dt$	

13.5 Hardware Implementation of the Trace Transform

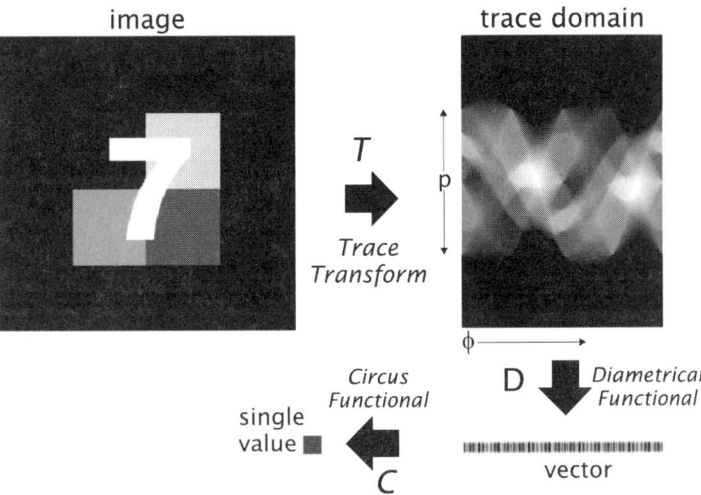

Figure 13.9 An image, its trace, and the subsequent steps of feature extraction.

most of the computational complexity lies; the diametrical and circus functionals are applied fewer times.

13.5.2 Computational Complexity

There are a number of parameters of the transform that can be adjusted according to the requirements of the specific application. The values of these parameters determine the computational complexity of the implementation. First, the number of angles to consider, n_ϕ, can be adjusted. In the extreme case, one may consider all lines that intersect the image at any angle down to 1 degree increments or lower. It is, however, possible to consider larger increments, and this reduces the computational complexity, though the image is sampled more coarsely. It is also possible to vary the distance between successive lines, N/n_p, for an image of size $N \times N$ pixels. Again, in the extreme case, this could be a single pixel increment from one line to the next. Thus for a $N \times N$ image, there would be a maximum of $\sqrt{2}N$ lines for each angle. It is further possible to vary the sampling granularity, n_t, along the lines, with the extreme case being to read every pixel, resulting in a maximum of $\sqrt{2}N$ pixels per line. It is important to note, however, that these parameters will affect the performance of the algorithm, and an in-depth study is needed to determine the trade-offs involved.

The parameters that are used in the computational complexity discussion are shown in Table 13.3. Let us consider first the access of the pixels in the image. For each of n_ϕ angles, a number of C_ϕ accesses (operations) are required. Second, for each of n_T functionals, we are required to process n_t points on each of the n_p lines for each of the n_ϕ angles. If C_T denotes the number of operations that are required on average per functional per pixel, we require a total of $n_\phi n_p n_t n_T C_T$ operations for the trace. Therefore the total computational complexity is given by $n_\phi C_\phi + n_\phi n_p n_t n_T C_T$.

Table 13.3 Trace Transform Parameters

Parameter	Explanation
n_ϕ	the number of angles to consider
n_p	the number of distances (inverse of the interline distance)
n_t	the number of points to consider along each trace line
n_T	the number of trace functionals
C_ϕ	the operations required for trace line address generation
$C_T = \frac{1}{N}(C_{t1} + C_{t2} + \cdots + C_{tN})$	the average operations per pixel for each functional

For an $N \times N$ image, common values for these parameters are $n_\phi = 180$, $n_p \simeq N$, and $n_t \simeq N$. This yields the total computation complexity to be $180C_\phi + 180N^2 n_T C_T$. Considering that a standard system may include 8 to 10 functionals, the high computational complexity is clear. The required time to perform C_ϕ and C_T operations depends on the implementation platform. In custom-designed hardware, we may be able to perform many of these operations in parallel. Thus, the parallelization of angles, line accesses, and functionals has the potential to reduce the required time to perform these operations, resulting in a faster system, capable of running in real time.

13.5.3 Full Trace Transform System

13.5.3.1 Acceleration Methods

To run any sort of computational process on a hardware system, the instructions involved must be broken into portions that are sufficiently simple to be run by the target hardware. Consider a multiplication: on a simple arithmetic logic unit (ALU), which only includes addition capability, the multiplication must be broken into its constituent additions in order to run on the ALU. Standard programming languages make this easier by allowing a coder to specify complex instructions, while the compiler breaks this up into "machine code." Typically, a general purpose processor will have a core processing unit that is capable of computing a fixed variation of small operations at very high speed. While the number of computations per second may be very high, the number of complete operations per second is significantly reduced. Hardware acceleration allows the system designer to break free from the limitations of this computational core. By implementing a core that is custom designed to implement specific operations, and by facilitating parallel computation of independent data, the core can run slower than a general purpose processor but still have significantly greater processing throughput.[5]

The key to accelerating the trace transform in hardware is to exploit the inherent parallelism of the algorithm. Where a typical software implementation would contain loops that iterate through various values of ϕ and p, it is possible in hardware to compute these results in parallel. The main bound in hardware would be the area used by an implementation and the available memory bandwidth. So, as

---

5. Throughput refers to the real overall processing speed of a system, and is typically measured in full data units completed per second. For image and video processing, the typical measure would be the number of frames processed per second.

long as there is area to spare and data is efficiently passed around, more operations can be performed in parallel. This can potentially achieve a large speedup of the algorithm.

Further acceleration in hardware can be achieved through "pipelining." Under this technique, complex operations are broken down into simpler constituent stages that take place in successive clock cycles. Though this increases the latency of the overall operation, since what used to take one cycle now takes many, it increases the throughput of the operation, that is the number of operations per unit of time. The key is that each of these stages can process different stages of data in parallel, and at the same time the required time for each stage is reduced. Thus, the design can now be clocked faster.

13.5.3.2 Target Board

The target board for the current implementation is the Celoxica RC300 development board. The board features a plethora of peripherals and connectivity options. Of concern here are the USB connection to a host computer and the four on-board ZBT SRAMs. These RAMs can be read from in a single clock cycle (provided the read address has been provided in the previous cycle). Each of these RAMs holds 8 MB of data composed of 32-bit words. The board hosts a Xilinx Virtex II 6000 FPGA. The FPGA provides ample logic capacity as well as 144 hard wired multipliers and 144 18 Kb BlockRAMs. The system is designed and implemented using Celoxica Handel-C [4], a high-level hardware description language.

13.5.3.3 System Overview

The overall system is constructed as follows. A USB camera is connected to a host PC, which captures the image data, applies some preprocessing, and then stores successive frames in the board SRAMs. The hardware system reads each of these frames in turn, computing the multiple traces of the source image in parallel, before storing the results in a different board SRAM. These are read back by the PC and used for further processing.

The hardware implementation presented here consists of a few simple blocks. The overall architecture is shown in Figure 13.10. The top-level control block oversees the control of other blocks. The initialization block manages the configuration of the functional blocks. The rotation block takes the input image and

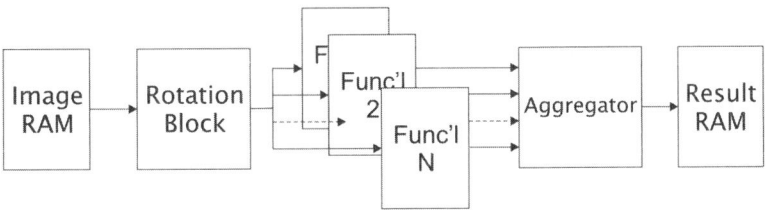

Figure 13.10 Architecture overview.

outputs the raster scan of the input image rotated by an angle. Each functional block reads the rotated image and performs the functional calculation for each line of the image. Note that each functional block is run in parallel. Finally, the aggregator polls each of the functional blocks in turn and stores the results for each in the on-board SRAM.

Before the system begins processing, it must be initialized. This is done over the universal serial bus (USB) interface to the computer. Once initialization is complete, the computer sends one image at a time (via USB) to the on-board RAMs, which are double buffered[6] to increase throughput. The input to the system is a preprocessed image consisting of the four concatenated primary orthogonal rotations stored in on-board RAM. This RAM is accessed out of order by the rotation block, which takes a rotation angle and produces a raster scan of the input rotated by the specified angle. The rotation block does this for a full 360 degree revolution in steps of $\delta\theta$ degree, defined in this implementation as $2°$.

Each of the rotated images is then passed through to the functional blocks, which compute the relevant results in parallel. The functional blocks keep track of the beginnings and ends of rows and also use the mask (which is passed with the image from the external RAM) to determine which pixels are to be incorporated. At the end of each image row, the results are passed to the aggregator, which proceeds to output them serially to another board RAM. This external RAM is read (again double buffered) by the USB connection to the PC, where results are sorted and accumulated for further processing.

13.5.3.4 Top-Level Control

The top-level control block oversees the other blocks, sending the first rotation angle to the rotation block. When the last result for each rotation is ready, the most latent functional sends a signal, informing the top-level control block to initiate the next rotation, and so on. Once all angles are completed, the full result can be read from the output RAM.

13.5.3.5 Rotation Block

Conceptually, the first step in the process should be a line tracer that samples pixel values in the original image that fall along the designated trace line. Hence, this functional unit should be able to produce a set of addresses from which to read pixel values in the original image, given ϕ and p as inputs. However, given that the image is to be traced at all values of those inputs (within specified sampling parameters), the process can be simplified. Rather than trace the image in such a manner, the image itself can be rotated, then sampled in horizontal pixel rows. This would simplify the required circuitry by applying the trigonometric rotation

---

6. Double buffering is a technique where one module writes to a RAM while another module reads from another RAM. When both operations are finished, the first module writes to the second RAM, while the second module reads from the first RAM.

functions once over the whole image, and allowing the subsequent blocks to have clearly ordered inputs.

An important caveat must be stated here: it is assumed that the input image is masked and when rotated will not suffer from detrimental cropping. Any masked object that falls within the bounds of a circle with diameter N will satisfy this condition. It is important to note that the trace transform only performs well when the subject of interest is masked; background noise is highly detrimental to its performance. Hence this assumption should be valid for a trace transform implementation in any domain as long as the objects are suitably sized.

The line tracer thus takes an angle as its input and produces a rotated version of the original image at its output. To simplify the system further, and negate the need for an image buffer inside the system, the line tracer is configured to produce the output in raster scan format. Thus it reads from the source image out of order. This also precludes the need for any data addresses to pass through the system, since the address structure is inherent in the data.

Implementing the algorithm in hardware means it is possible to consider performing multiple rotations in parallel. However, given that the image is stored in on-board SRAM, duplicates of the input image in separate RAMs would be required in order to achieve this, as well as further rotation blocks running in parallel. Even so, it is possible to quadruple performance without increasing the area. Consider that a rotation by any multiple of 90 degrees is simply a rearrangement of addressing, which can be implemented in software or through minimal hardware. It is then possible to deduce that any rotation above 90 degrees can be implemented as a trivial rotation by a multiple of 90, followed by a further rotation by the remaining angle. To clarify this point, a rotation by 132 degrees is the same as a 90-degree rotated image being rotated through 42 degrees. Furthermore, given that there is ample word length in the external SRAMs, the four orthogonal base rotations (0°, 90°, 180°, and 270°) can be stored together in a single word. Four rotations can now be computed in parallel by using the line tracer to rotate the image in the SRAM, then splicing the resultant word. This effectively turns a single-ported RAM into a four-ported one.

The rotation block is implemented as follows: an on-board SRAM contains the source image. This source image is stored as all four primary rotations and their respective masks concatenated, as shown in Figure 13.11. Since the on-board RAMs available have a 36-bit word length, this is a perfect fit.

The block takes an angle as its input. This is used to address lookup tables (in on-chip block RAMs) containing sine and cosine values, which are used to implement the standard Cartesian rotation equations:

$$x' = x \cos\theta - y \sin\theta; \quad y' = x \sin\theta + y \cos\theta \qquad (13.1)$$

These equations all use simple 8-bit fixed point word lengths. Nearest-neighbor approximation is used to avoid more complex circuitry; bicubic or biquadratic approximation are more computationally intensive. In the current implementation, the image resolution has been doubled in each dimension compared with the initial software version to make up for any errors that might arise due to this approxi-

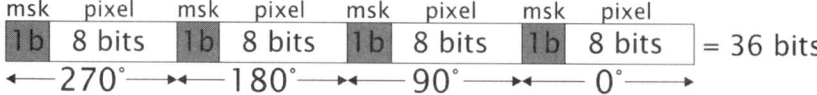

Figure 13.11 Image RAM word.

mation. The resultant output of the block is a raster scan of the rotated image still in its 36-bit concatenated form. This is fed to the functional blocks, all running in parallel, where the computation occurs.

13.5.3.6 Functional Blocks

Each functional block can implement a set of functionals described in Table 13.2 (or other sets). The specific selection of a functional is achieved through the initialization phase. The functional blocks follow a standard design. They await a "start" signal to indicate the first pixel of a rotation is arriving before beginning to compute the results. The functional block splices the incoming signal into the four constituent rotations and masks. The mask corresponding to each pixel determines whether it is considered in the calculation; masked-out pixels are simply ignored in the calculation. Since the image size is fixed, each functional block keeps track of the current row number and position within the row in the image to avoid the use of many control signals. When the end of each row is reached, it stores the results for that row in an output buffer and sends a "new result" signal to the aggregator.

The resulting word length depends on the functional, though all results can be individually scaled if necessary. Note that each of the functional blocks is duplicated four times, in order to perform each of the four orthogonal rotations in parallel.

The flexibility of the functionals is gained by using the on-chip block RAMs as lookups for some basic computations. This is discussed in Section 13.5.4

13.5.3.7 Initialization

Before the system can begin processing the input images, all internal signals are initialized to prevent the propagation of undefined values. Further, the running parameters (like number of angles, lines, and pixel spacing) are initialized. Finally, the configuration data for each of the functional units is loaded into the respective functional blocks. A typical functional block will contain 2 or 3 lookup RAMs. These are used to implement simple arithmetic functions such as x, x^2, $\cos(x)$, and so on. For each functional block, the RAMs must be loaded with the relevant data to compute the correct functional. Furthermore, some functionals have optional processing datapaths that can be either enabled or disabled.

Given that there may be 8 to 10 functionals to compute, a large number of RAMs have to be initialized (discussed in Section 13.5.3.6). To do this from one single unit would be slow and inefficient in terms of routing. As a result, the intialization phase consists of the following steps: first, the initialization settings

13.5 Hardware Implementation of the Trace Transform 387

are read from the host PC over the USB connection, and stored as-is in the on-board RAM. Then a distributor block takes these values and places them onto an initialization bus. Finally, a dedicated unit beside each functional block reads from this bus, ignoring initialization data that is not meant for it, and using those which are correctly labelled to initialize its settings.

The initialization data is formatted as shown in Figure 13.12 (bear in mind that the USB interface only transports a single byte at a time, so some data is split).

13.5.4 Flexible Functionals for Exploration

To extend the system completed thus far, and enable it to be used for exploring the functional space, it becomes necessary to develop some sort of reconfigurable platform for experimenting with various functionals. Initially, the trace transform proposers worked with 22 functionals for face authentication, as defined in Table 13.2. These functionals have not been fully evaluated against each other for their relative performance in the recognition task, nor compared with other possibilities. Indeed, the computational complexity of the trace transform would mean that a thorough investigation of a large functional space would be time consuming.

Through the use of available hardware building blocks, it is possible to design a small number of flexible functionals that can compute the majority of those initially proposed alongside a large number of additional functionals. The key step is to group the existing functionals into families of similarly defined functionals and then use the power of hardware reconfiguration to allow a single generalized functional block to compute a large number of possible functionals.

The resources that most assist in this endeavor are the on-chip block RAMs that pervade the fabric of modern FPGAs. These are small random access memories (RAMs) that exist in the FPGA fabric, which can be used in multiple width and depth configurations. The primary task of benefit here is their use as lookup RAMs.

```
initialization data size (byte 1 of 3)
initialization data size (byte 2 of 3)
initialization data size (byte 3 of 3)
functional number
blockram number //255 = config register code
data 1
data 2
data 3
...
functional number
blockram number //255 = config register code
data 1
data 2
data 3
...
```

Figure 13.12 Functional block lookup initialization data.

A lookup RAM typically contains some function precomputed for the value of each input address. As an example, if a lookup RAM is to be used to compute sin(x), then the contents of the RAM would contain the precomputed value of sin(address), where "address" is the address of the word. Thus when a value x is used to address the RAM, the output is equal to sin(x).

The strength of this scheme comes from the fact that these RAMs can be configured at runtime, precluding the need to resynthesize the design, which is time consuming. Add to this the fact that a wide variety of functions can be precomputed in this way, and it is clear that a simple functional with these RAMs incorporated permits significant flexibility. Just some of the possible arithmetic functions that can be computed in such a manner include x, x^2, $\sqrt{x}$, $\ln(x)$, and $\sin(x)$, among a large array of others. The only limitation of this scheme is that the input value must be bound. The larger the range of a value, the more bits are required, and this can increase the resources used dramatically. In the case of the trace transform system, as well as many image processing applications, image samples are typically constrained to 8-bit word length.

13.5.4.1 Type A Functional Block

This functional block is able to compute functionals 1, 2, and 4 from the original list of 22. A block diagram is shown in Figure 13.13. Note that each circuit element takes a single cycle to run and that the dashed parts are optional, determined by the configuration register. "D" is a single cycle delay register.

The block takes an input pixel then applies a function, l_1, to it. Optionally, function l_2 is applied to a one cycle delayed version of the input pixel and the absolute difference is taken: the actual datapath is decided by the configuration register. Each of the resultant values is then summed and at the end of the row, the final result is optionally squared.

13.5.4.2 Type B Functional Block

This functional block, as illustrated in Figure 13.14, is more complex than Type A; it implements functionals 9, 11, 12, 13, and 14 from Table 13.2, which depend on the weighted median calculation. The weighted median is implemented using an

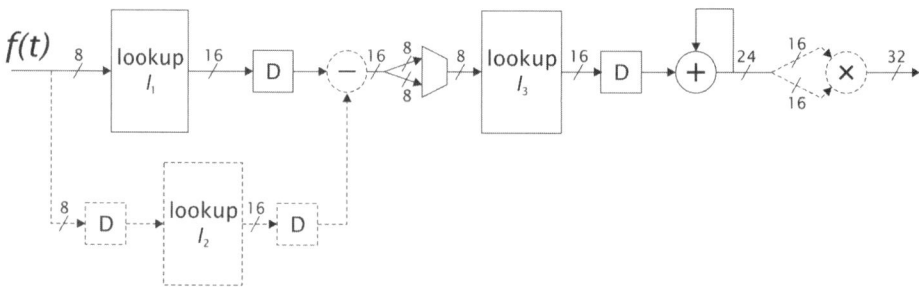

Figure 13.13 Type A functional block.

efficient hardware block that can compute medians for very large windows using a cumulative histogram [9]. Since the median block can only return a result after a whole row has been processed, the current implementation uses the result from the previous row in each calculation. Lookup l_1 offers flexibility in modifying the skew of the median calculation with the default being $\sqrt{\cdot}$.

Here, the functional computes the intermediate values c and r as described in row 9 of Table 13.2. Note that the value c is only updated once per row (shown shaded in the figure). Function l_2 is then applied to r and l_3 to $f(t)$. These results are then multiplied before being summed in an accumulator.

13.5.4.3 Type C Functional Block

This functional block implements functionals 20, 21, and 22 of the initial list. A block diagram is shown in Figure 13.15. The system follows similar design to the Type B block, except that c is computed as described in row 21 of Table 13.2. Note that the values c and S are only updated once per row.

13.5.5 Functional Coverage

These three functional types cover all 11 functionals required by the shape trace transform for face authentication [18]. Furthermore, by using alternative lookup functions in the block RAMs, it is possible to add further functionals.

13.5.6 Performance and Area Results

Before discussing the results of implementing the trace transform in hardware, an overview of the operating parameters is needed. Recall that the source image is stored in an off-chip SRAM; these SRAMs can be read pipelined in a single cycle. A single 256 × 256 image contains 65,536 pixels, and so it would take an equal number of cycles to read. As such, a new rotation is complete every 65,536 cycles plus a few cycles used to fill and flush the pipeline. With a rotation angle incremented by 2° each time, 180 rotations are needed for a full cycle. Since 4 rotations are computed in parallel, the actual number of rotations is 45. Hence,

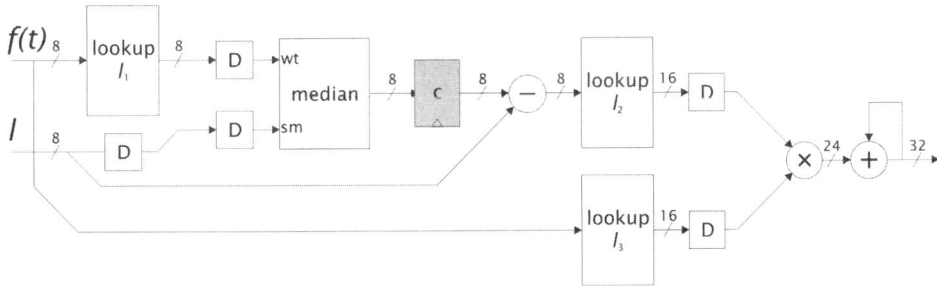

Figure 13.14 Type B functional block.

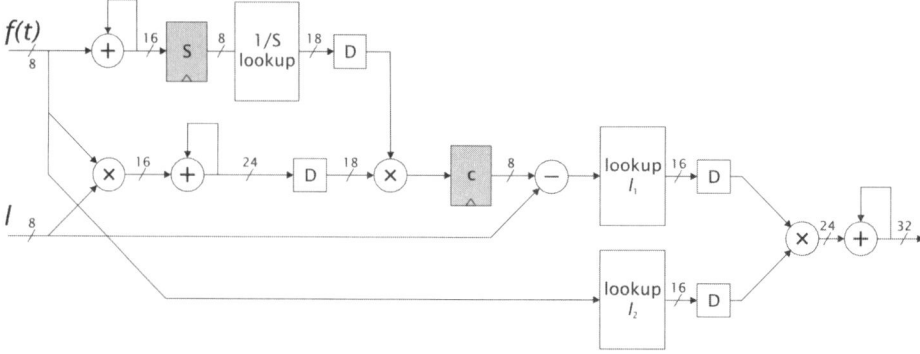

Figure 13.15 Type C Functional Block.

assuming a rotation latency of 65,560 cycles, to include the margins mentioned above, a full set of rotations is complete in just under 3 million clock cycles. Each of the functionals instantiated runs in parallel, and computes results as the pixels stream through, so they do not add to the cycle length.

The tool that synthesizes the design gives final area and speed results that indicate the efficiency of the design. The area of a design is typically measured in terms of logic usage. On a Xilinx FPGA, the most basic unit is a slice which includes two four-input lookup tables (LUTs) and some other logic. The FPGA fabric also includes block RAMs and embedded 18×18-bit multipliers. These are the three common types of available resources in a modern FPGA, and usage figures for each give a measure for area.

Table 13.4 shows the area results for each of the units in the design. All units were successfully synthesized to run at 79 MHz. Since each full trace takes 3 million cycles, this means the full system can process 256×256 pixel images, producing traces at 26 frames per second; this satisfies the real-time requirement of the system. Note that multiple traces are produced in parallel per image. There is sufficient time in the system to implement 36 functionals. However, the resource availability on the current target device, a Xilinx Virtex-II XC2V6000, would limit the number of functionals that can be placed simultaneously in the device. The actual number depends on which types are chosen.

Table 13.5 shows the speedup in computing a single functional using the discussed hardware implementation. As a software reference an optimized MATLAB code is used that is running on a Pentium 4 at 2.2 GHz with 1 GB of mem-

Table 13.4 Synthesis Results

Unit	Slices	Embedded Multipliers	block RAMs
Framework	1,300	4	2
Type A Functional	800	4	6
Type B Functional	17,500	4	22
Type C Functional	1,300	8	6
Total Available	33,792	144	144

Table 13.5 Running Times and Speedup

Functional Type	Software (ms)	Hardware (ms)	Speedup
Type A Functional	1,400	38.5	36×
Type B Functional	3,900	38.5	101×
Type C Functional	2,600	38.5	67×

ory. In all the cases, the hardware design outperforms the software version by a considerable margin. It is important to note that these numbers are for a single functional. In the hardware implementation, additional functionals are computed in parallel, resulting in an even greater performance boost. Consider a system with three functionals—one of each type. The software implementation would take 7.9 seconds to compute the three traces, whereas the hardware system would still take 38.5 ms, a speedup of over 200 times. Increasing the number of functionals further, to the 11 implemented out of Table 13.2, a software version takes 31 seconds, while the hardware implementation would still take the same time as implementing one functional. This results in a speedup of over 800 times. This would require a larger FPGA device, however, higher capacities are common in the latest generation of FPGAs. The more functionals, the greater the speedup.

13.5.7 Conclusions

This section presents a mapping of the trace transform to reconfigurable logic. The section demonstrates how a considerable speedup can be achieved by exploiting the fine-grain parallelism of FPGA devices. The proposed system is able to realize a large set of functionals by using only a small set of building blocks. The reconfigurability property of these blocks allows the mapping of many different functionals in each block. A comparison with a software version of the algorithm demonstrates the scalability of the hardware design and the considerable speedup gain when the number of functionals increases. Finally, important hardware techniques like pipelining and double buffering that boost the performance of the system are described and explained.

13.6 Summary

This chapter demonstrates that a considerable acceleration can be achieved by placing algorithms into hardware. The first part of the chapter focuses on the mapping of the steerable pyramid discussed in Chapter 5 into an FPGA using an XC2V6000 device, where a speedup factor of 8 is achieved with a minimum impact in the accuracy of the algorithm. The second part of the chapter focuses on the mapping of a face authentication algorithm, the trace transform (discussed further in Chapter 6) to hardware. The computational complexity of the algorithm does not permit its use in real-time scenarios. This chapter demonstrates how the inherent parallelism of the algorithm can be exploited in a hardware implementation achieving a considerable speedup. Detailed techniques of hardware mapping have been reported in both algorithms and two of the most common issues in a hardware implementation, accuracy and memory bandwidth, have been discussed.

References

[1] Bouganis, C. S., Constantinides, G. A., and Cheung, P. Y. K., "Heterogeneity exploration for multiple 2D filter designs." In *International Conference on Field Programmable Logic and Applications*, 2005, pp. 263-268.

[2] Bellows, P. and Hutchings, B., "Designing run-time reconfigurable systems with JHDL." *Journal of VLSI Signal Processing Systems for Signal, Image, and Video Technology*, Vol. 28(1-2), May/June 2001, pp. 29-45.

[3] Constantinides, G. A., Cheung, P. Y. K., and Luk, W., "Wordlength optimization for linear digital signal processing." *IEEE Transactions on Computer-Aided Design of Integrated Circuits and Systems*, Vol. 22(10), October 2003, pp. 1432-1442.

[4] http://www.celoxica.com/, 2006.

[5] Deans, S. R., *The Radon Transform and Some of Its Applications*. John Wiley and Sons, 1983.

[6] http://www.celoxica.com/techlib/files/cel-w0307171jkx-33.pdf, 2006.

[7] Freeman, W. T. and Adelson, E. H., "The design and use of steerable filters." *IEEE Transactions on Pattern Analysis and Machine Inteligence*, Vol. 13(9), 1991, pp. 891-906.

[8] Fahmy, S. A., et al., "Efficient realtime FPGA implementation of the trace transform." In *International Conference on Field Programmable Logic and Applications*, 2006, pp. 1-6.

[9] Fahmy, S. A., Cheung, P. Y. K., and Luk, W., "Novel FPGA-based implementation of median and weighted median filters for image processing." In *International Conference on Field Programmable Logic and Applications*, 2005, pp. 142-147.

[10] Kingsbury, N., "Image processing with complex wavelets." *Philosophical Transactions Of The Royal Society Of London - Series A*, Vol. 357(1760), September 1999, pp. 2543-2560.

[11] Kadyrov, A. and Petrou, M., "The trace transform as a tool to invariant feature construction." In *Fourteenth International Conference on Pattern Recognition, Proceedings*, Vol. 2, 1998, pp. 1037-1039.

[12] Kadyrov, A. and Petrou, M., "The trace transform and its applications." *IEEE Transactions on Pattern Analysis and Machine Intelligence*, Vol. 23(8), 2001, pp. 811-828.

[13] Kadyrov, A. and Petrou, M., "Affine parameter estimation from the trace transform." In *16th International Conference on Pattern Recognition, Proceedings*, Vol. 2, 2002, pp. 798-801.

[14] Nibouche, M., et al., "An FPGA-based wavelet transforms coprocessor." In *IEEE International Conference on Image Processing*, vol. 3, 2001, pp. 194-197.

[15] Petrou, M. and Kadyrov, A., "Affine invariant features from the trace transform." *IEEE Transactions on Pattern Analysis and Machine Intelligence*, Vol. 26(1), 2004, pp. 30-44.

[16] Page, I. and Luk, W., in "Compiling occam into FPGAs," *FPGAs*, pages 271-283. Abingdon EE&CS Books, Abingdon, UK, 1991.

[17] Srisuk, S., et al., "Face authentication using the trace transform." In *IEEE Computer Society Conference on Computer Vision and Pattern Recognition, Proceedings*, Vol. 1, 2003, pages I-305-I-312.

[18] Srisuk, S., et al., "A face authentication system using the trace transform." *Pattern Analysis and Applications*, Vol. 8(1-2), 2005, 50-61.

[19] Sendur, L. and Selesnick, I. W., "Bivariate shrinkage functions for wavelet-based denoising exploiting interscale dependency." *IEEE Transactions on Signal Processing*, Vol. 50(11), 2002, pp. 2744-2756.

[20] http://www.systemc.org/, 2006.

CHAPTER 14
Real-Time Spatiotemporal Saliency
Yang Liu, Christos-Savvas Bouganis, and Peter Y. K. Cheung

14.1 Introduction

In complex dynamic environments, the overwhelming amount of data prevents detailed visual analysis over the entire scene. Certain regions of a scene are attentively distinctive and create some form of immediate significant visual arousal. Such regions are referred to as being "salient." Humans have the unmatched ability to rapidly find "salient" regions. For example, humans are especially responsive to the color red; among several competing hypotheses are that this response might have evolved, or in some cases might be learned, due to the association of red with danger (blood), or of the association with food (ripe fruit!). Of course, in daily life, our attention is not influenced by color alone. Other factors that contribute to visual saliency include intensity and motion. In this chapter, we describe a spatiotemporal saliency framework [14] that emulates the attentive analysis function of the human visual system. More complex and high-level scene analysis can therefore focus on the most salient regions detected by the framework. Immediate applications of the spatiotemporal framework include surveillance, compression, and prediction of viewers' attention in advertisement.

Saliency of a still image is referred to as spatial saliency. Itti et al. [8] proposed a biologically plausible model, which combines intensity, color, and orientation information in order to generate a saliency map. Kadir et al. [9] resort to information theory and use entropy as the saliency measure. Oliva et al. [15] analyzed the global distribution of low-level spatial features to render a saliency map.

Saliency in video refers to spatiotemporal saliency, which requires the incorporation of the temporal information in addition to the spatial information. Laptev et al. [13] proposed the extension of the Harris corner detector [4] from the spatial domain to the spatiotemporal domain. Saliency occurs where the image values have significant local variations in both space and time. Zhong et al. [18] divided the video into equal-length segments and classified the extracted features into prototypes. The segments of the video that could not be matched with any of the computed prototypes were considered as salient. Similarly, Boiman and Irani [2] compared two-dimensional and three-dimensional video patches with a training database to detect abnormal actions present in the video. In terms of storage requirement, both types of these spatiotemporal algorithms [2, 13, 18] required the access of a stack of video frames and they were not suitable for applications using streaming video input. Zhai et al. [17] constructed both spatial and temporal saliency maps, which were subsequently fused into the overall spatiotemporal saliency map.

Research on human perception and memory suggests that the human visual system effectively approximates optimal statistical inference and correctly combines new data with an accurate probabilistic model of the environment [5]. For example, imagine a situation where we are crossing a busy street. Before we cross the street, we observe the oncoming cars for a period of time. This effectively builds up statistical models on the movements of the cars in our brain. Future locations of the cars can be inferred according to the models. Hence, the decision of when it is safe to cross the road is determined by such motion models. Based on such statistical models, Itti et al. [7] proposed to compute 72 feature maps first and model the data from each feature map at location $[x,y]$ at time t as Poisson distributions, which can reasonably describe the behavior of neurons in the visual system, notably the cortical pyramidal cell conditional firing statistics. Finally, the Kullback-Leibler divergence, a probability distribution distance measure, was used to quantify the differences between a priori and a posteriori distributions over models. In this way, the spatiotemporal saliency, which the authors called "surprise," was obtained.

In this chapter, we describe a saliency framework [14] that also uses a probability distribution distance measure to quantify saliency. In contrast to [7], this framework concentrates on the motion of spatially salient features that contradict a given motion prediction model. Moreover, the framework has the ability to update the model parameters with new measurements of the environment, which is in line with the current research on human perception and memory. From the storage requirement point of view, the described framework requires only two full frames to be stored, which makes it a good candidate for real-time applications.

14.2 The Framework Overview

We begin the description of the framework by introducing the underlying idea that governs the saliency detection. The idea is that it uses the unpredictability in the motion of spatial features as a measure of spatiotemporal saliency. In other words, when the motion path of a spatial feature in the image is predictable, the feature is classified as nonsalient, otherwise it is classified as salient. The above definition of spatiotemporal saliency distinguishes this framework from the previous approaches, since it is based on the prediction of the feature's motion. As a result, this approach can be thought of as working at a higher conceptual level than previous approaches, since statistical inference is taken into account. Based on the above new definition of spatiotemporal saliency, a generic framework is conceived that can be built up from numerous established techniques. It is an unsupervised process due to its low-level nature, and it operates without any knowledge of the shape or size of the object(s) that may be of interest.

The overview of the framework is shown in Figure 14.1, where two-dimensional features refer to spatial features within a video frame. They are detected in the initialization step of the algorithm. As time elapses, some existing two-dimensional features may exit the field of view of the camera, hence the scene, while some new two-dimensional feature may enter the scene. Therefore, it is necessary to refresh the two-dimensional features every n frames to reflect the up-to-date presence of two-dimensional features, where n is set according to the prior knowledge of the scene's activity. These two-dimensional features are the cues to be both tracked and predicted between consecutive frames and the discrepancies between

14.2 The Framework Overview

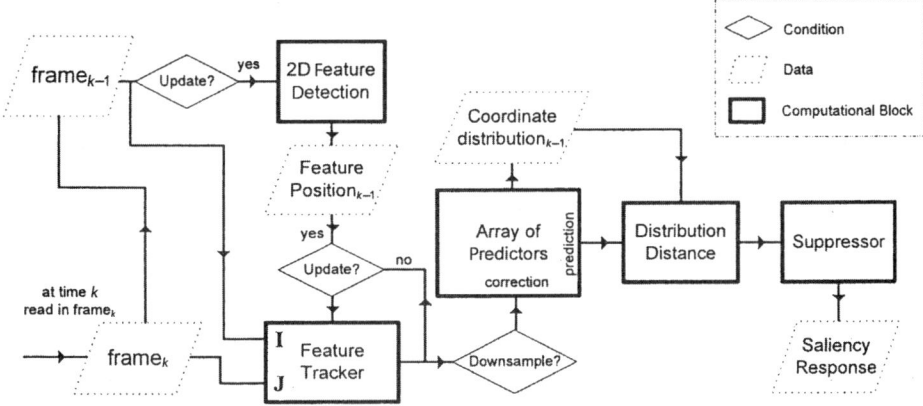

Figure 14.1 Overview of the proposed spatiotemporal saliency framework.

the tracking and predicting results contribute to the spatiotemporal saliency map. Features can be obtained by any spatial saliency algorithm. The justification of involving a spatial saliency algorithm is that in order for the features to be qualified as spatiotemporally salient, they must first be spatially salient. From those features that satisfy the spatial criterion, the temporally salient ones are screened out as spatiotemporal salient points.

The two-dimensional features are then tracked by feature trackers. Let **u** be a feature in frame **I**. The purpose of feature tracking is to find its corresponding location in frame **J**.

A number of previous locations of two-dimensional features are stored in the memory, from which the average speed of each two-dimensional feature in the past few frames can be estimated. A very small average speed implies that the current frame rate of the camera is too high to observe any significant movement of the two-dimensional feature or that the projected speed of the object to the image plane is small. This is problematic because it is difficult to determine unpredictability out of a sequence of insignificant movements. To rectify this, the coordinates of individual two-dimensional features are decimated according to the average speed of the corresponding two-dimensional feature.

Each feature is assigned a predictor and is treated separately to evaluate the unpredictability of each two-dimensional feature. The decimation process previously mentioned adapts to the average speed of the two-dimensional feature to downsample the feature coordinates, which are the measurement inputs of the predictors. Each predictor updates its model on receiving new measurements and predicts the coordinates of the feature in the upcoming frame according to a particular motion model. At time k, each feature has a set of *predicted* coordinates with different probabilities computed at time $k - 1$ by its predictor and a set of *current* coordinates each with different probabilities, given by the feature tracker. The distance between the two probability density functions of the *predicted* and *current* coordinate determines the predictability of the motion behavior of the feature in question, that is, the temporal saliency of the feature.

It is suggested in [14] that several spatiotemporally salient features close to each other should make a bigger impact on the spatiotemporal saliency responses than one single feature relatively distant from other features. This is achieved by a "suppressor" module. The suppressor promotes clusters of salient features that are excited by one another while playing down singular features whose silent neighbors inhibit their saliency response.

This framework is efficient in terms of storage cost. It needs to store only the current frame and the previous frame for two-dimensional feature tracking. The remaining storage is linearly proportional to the number of features, which is much less than the number of pixels in a frame. Nevertheless, using the appropriate predictor, statistical distributions of the motions of two-dimensional features can be recursively built over the past frames, so information from earlier frames can be implicitly taken into account. Effectively, the spatiotemporal saliency map has temporal support of more than just two frames.

14.3 Realization of the Framework

Given the framework structure in Section 14.2, algorithms that realize the functions of the blocks have to be applied. In this section, the realization of the framework is based on the assumption that features with linear motions are classified as non-salient, and features with more complicated motions are classified as salient. It should be noted that this assumption is made to provide an intuitive example of one possible realization of the framework. However, the framework itself is not restricted to these assumptions.

14.3.1 Two-Dimensional Feature Detection

The Shi and Tomasi algorithm [16] has been adopted for two-dimensional feature detection. The choice of this algorithm lies in the fact that corners can be detected with robustness and that it has been developed to maximize the quality of tracking, which is important for the subsequent steps of the spatiotemporal saliency framework. Given an image $\mathbf{I}$, $\mathbf{u}$ is a pixel with coordinates $[u_x, u_y]^T$, and $\mathbf{w}$ is the neighborhood $[u_x - w_x, u_x + w_x] \times [u_y - w_y, u_y + w_y]$ around pixel $\mathbf{u}$. The spatial gradient matrix $\mathbf{G_u}$ for pixel $\mathbf{u}$ is defined in (14.1).

$$\mathbf{G_u} = \sum_{x=u_x-w_x}^{x=u_x+w_x} \sum_{y=u_y-w_y}^{y=u_y+w_y} \begin{bmatrix} \mathbf{I}_x^2(x,y) & \mathbf{I}_x(x,y)\mathbf{I}_y(x,y) \\ \mathbf{I}_x(x,y)\mathbf{I}_y(x,y) & \mathbf{I}_y^2(x,y) \end{bmatrix} \quad (14.1)$$

where

$$\mathbf{I}_x(x,y) = \frac{\mathbf{I}(x+1,y) - \mathbf{I}(x-1,y)}{2}$$
$$\mathbf{I}_y(x,y) = \frac{\mathbf{I}(x,y+1) - \mathbf{I}(x,y-1)}{2}$$

The eigenvalues of the matrix $\mathbf{G}$ are the keys to determine how well the pixel $\mathbf{u}$ can be tracked. Shi and Tomasi prove that if the minimum eigenvalue of $\mathbf{G}$ is large enough, then **u** can be corners, salt-and-pepper textures, or any other pattern that can be tracked reliably.

Based on the eigenvalue criterion, the first step of the Shi and Tomasi algorithm is to find the minimum eigenvalue of spatial gradient matrix **G** defined in Equation (14.1) for each pixel in the image. Each pixel is associated with one corresponding eigenvalue. Pixels with eigenvalues above the predefined threshold remain and the rest are rejected. The pixels with eigenvalues that are local maxima in their respective 3×3 neighborhood are selected as features. Finally, the algorithm ensures that all the features are far enough apart from each other by removing features that are close to a feature with higher eigenvalue. Figure 14.2 shows the input image to the Shi and Tomasi algorithm, and the detected features are represented by crosses in Figure 14.3.

14.3.2 Feature Tracker

The goal of a feature tracker is to locate the corresponding features between frames. The criteria of a good feature tracker are accuracy and robustness. All the pixels in the surrounding region around the feature point in question, known as the integration window, are taken into account and analyzed for the purpose of tracking the feature. A small integration window is preferred to avoid smoothing out the details in order to achieve high accuracy. However, the robustness in handling large motion requires having a large integration window. The representation of data by a pyramidal image structure facilitates a trade-off between robustness and accuracy. Given an image, such a representation is created by reproducing a copy of the image at progressively lower resolutions. Imagine a pyramid with the image

Figure 14.2 Input image for the Shi and Tomasi algorithm.

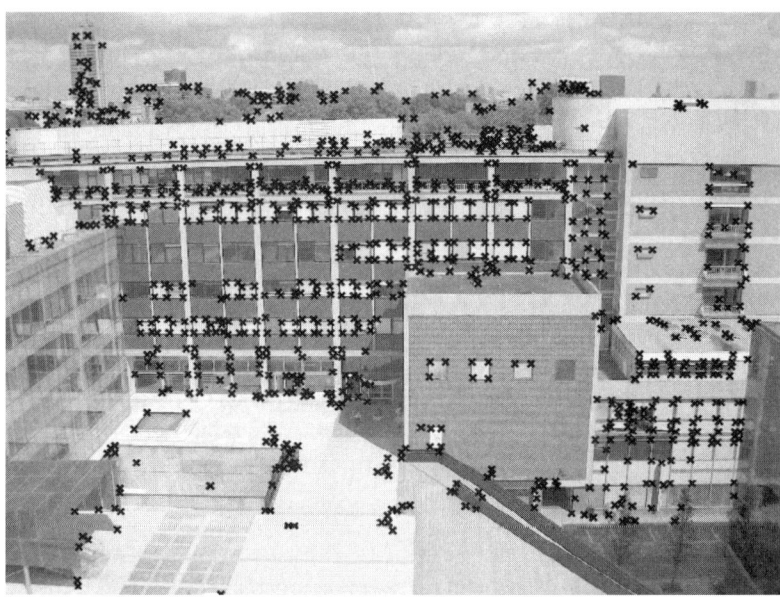

Figure 14.3 Features detected by the Shi and Tomasi algorithm. Locations of the features are indicated by the crosses.

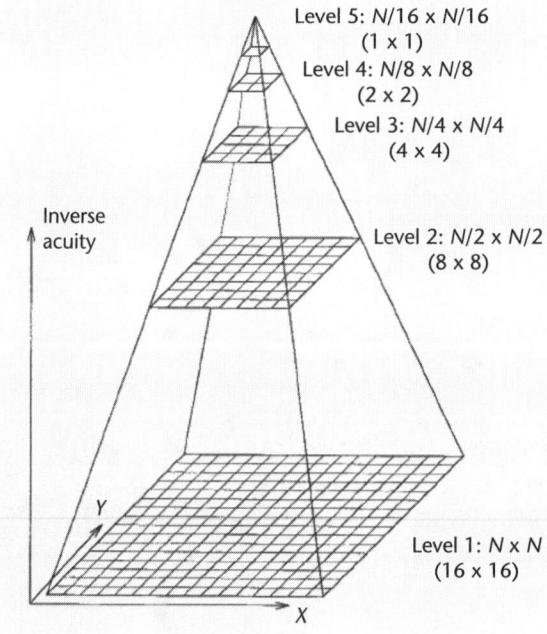

Figure 14.4 Representation of a pyramidal image data structure.

14.3 Realization of the Framework

Figure 14.5 Example of pyramidal image data structure.

at the lowest resolution at the top and the highest (original) resolution at the bottom, such as the one shown in Figure 14.4. Within the pyramid, the image at each level has the same content as the original image. Figure 14.5 shows an example of a pyramidal image data structure with 9 levels of resolution.

Such pyramidal implementation of the iterative Lucas-Kanade optical flow algorithm [3] is adopted as the feature tracker in this particular realization of the saliency framework. A feature pixel is tracked in the lowest-resolution level L_m within the pyramid and the result is propagated to the next finer level until the original resolution L_0 is reached.

At each level L, the corresponding coordinates of feature $\mathbf{u}^L$ in image $\mathbf{I}^L$ can be found by (14.2).

$$\mathbf{u}^L = [u_x^L, u_y^L] = \frac{\mathbf{u}^{L_0}}{2^L} \tag{14.2}$$

$\mathbf{u}^{L_0}$ represents the coordinates of feature $\mathbf{u}$ in the original resolution image $\mathbf{I}^{L_0}$, which lies at level 0 in the pyramid. The spatial gradient matrix $\mathbf{G}$, with respect to the integration window of $\mathbf{u}^L$, is computed by (14.3).

$$\mathbf{G}_{\mathbf{u}^L} = \sum_{x=u_x^L-w_x}^{u_x^L+w_x} \sum_{y=u_y^L-w_y}^{u_y^L+w_y} \begin{bmatrix} \mathbf{I}_x^2(x,y) & \mathbf{I}_x(x,y)\mathbf{I}_y(x,y) \\ \mathbf{I}_x(x,y)\mathbf{I}_y(x,y) & \mathbf{I}_y^2(x,y) \end{bmatrix} \tag{14.3}$$

where

$$\mathbf{I}_x(x,y) = \frac{\mathbf{I}(x+1,y) - \mathbf{I}(x-1,y)}{2}$$
$$\mathbf{I}_y(x,y) = \frac{\mathbf{I}(x,y+1) - \mathbf{I}(x,y-1)}{2}$$

Then, the feature tracking process at level L is carried out as follows.
Initialization:

$$\mathbf{g}^L = 2(\mathbf{g}^{L+1} + \mathbf{v}_K^{L+1}), \quad \text{with } \mathbf{g}^{L_m} = [0\ 0]^T \tag{14.4}$$

$$\mathbf{v}_0 = [0\ 0]^T \tag{14.5}$$

Lucas-Kanade iterations; for $k = 1$ to K, incrementing by 1:

$$\delta\mathbf{I}_k(x,y) = \mathbf{I}(x,y) - \mathbf{J}(x + g_x^L + v_x^{k-1}, y + g_y^L + v_y^{k-1})$$
$$\mathbf{b}_k = \sum_{x=u_x^L-w_x}^{u_x^L+w_x} \sum_{y=u_y^L-w_y}^{u_y^L+w_y} \begin{bmatrix} \delta\mathbf{I}_k(x,y)\mathbf{I}_x(x,y) \\ \delta\mathbf{I}_k(x,y)\mathbf{I}_y(x,y) \end{bmatrix}$$
$$\mathbf{v}_k = \mathbf{v}_{k-1} + (\mathbf{G}_{\mathbf{u}^L})^{-1}\mathbf{b}_k \tag{14.6}$$

In (14.4), $\mathbf{v}_K^{L+1}$ is the tracking result after all K iterations of the Lucas-Kanade process at level $L + 1$, which is the immediate previous level to the current level L. $\mathbf{g}$ is an accumulator of tracking results. The implication of (14.4) is that $\mathbf{g}^L$ contains the accumulative tracking result obtained from all levels that have been processed. $\mathbf{g}^L$ provides the initial "guess" for the Lucas-Kanade iteration at the current level L. At the end of the Lucas-Kanade iteration, $\mathbf{v}_K$ represents the displacement vector of the feature $\mathbf{u}$ from image $\mathbf{I}^L$ to image $\mathbf{J}^L$. $\mathbf{v}_K$ is then used to update the accumulator $\mathbf{g}$ for the next level of finer image.

After level 0 is reached, the displacement vector $\mathbf{d} = \mathbf{v}_K + \mathbf{g}^{L_0}$ at the original image resolution is found. Thus, the corresponding coordinates $\mathbf{s} = \mathbf{u} + \mathbf{d}$ in the image $\mathbf{J}$ are obtained for the feature $\mathbf{u}$ in image $\mathbf{I}$.

Figure 14.6(a) and Figure 14.6(b) show the first input images **I** and the second input image **J** for the Lucas-Kanade feature tracker, respectively. Relative to Figure 14.6(a), the car in Figure 14.6(b) moves to the right while everything else remains static. Two-dimensional features in Figure 14.6(a) are detected by the Shi and Tomasi algorithm described earlier. These same features are tracked in Figure 14.6(b) using the pyramidal Lucas-Kanade feature tracker. The tracking result is shown in Figure 14.7, where the displacement vectors for the features are represented by arrows.

(a)

(b)

Figure 14.6 The input images for the pyramidal Lucas-Kanade feature tracker. (a) is the first image **I**; (b) is the second image **J**.

Figure 14.7 The output of the pyramidal Lucas-Kanade feature tracker. Arrows indicate displacement vector.

14.3.3 Prediction

Based on the nonsaliency assumption of linearity, the Kalman filter is a good instrument in predicting motion. The Kalman filter is a recursive estimator that provides the best estimation, in a least-square error sense, of the "hidden" information of interest from noisy measurements. It was first introduced in the seminal papers [11, 12]. One of the first public known applications was made at NASA to guide the *Apollo 11* lunar module to the moon's surface. Since then, the Kalman filter has found its way to become an indispensable part of many applications ranging from signal processing to communication and control.

It is not always feasible to measure all the variables of a dynamic system. The Kalman filter provides a way to infer the missing variables, or states, from noisy measurements. It can also be used to predict future states taking into account all the measurements in the past. It must be noticed that the Kalman filter is a *recursive* filter. The model is updated when a new measurement is available. However, it does not need to store all the measurements. A formulation of the Kalman filter is shown below.

$$\begin{aligned} \mathbf{x}_{k+1} &= \mathbf{F}\mathbf{x}_k + \mathbf{w}_k \\ \mathbf{z}_k &= \mathbf{H}\mathbf{x}_k + \mathbf{v}_k \end{aligned} \quad (14.7)$$

$\mathbf{x}_k$ is the n-dimensional state vector and $\mathbf{z}_k$ is the m-dimensional measurement vector at time k. The state $\mathbf{x}_k$ evolves in time and the transition is governed by the transition matrix $\mathbf{F}$, which is $n \times n$ in size. The $m \times n$ matrix $\mathbf{H}$ is called the measurement matrix and relates the measurements to the states. The system is corrupted by two types of noise. $\mathbf{w}$ is the n-dimensional process noise vector and $\mathbf{v}$ is the m-dimensional measurement noise vector. The noise vectors are assumed to be independent Gaussian processes with zero mean and covariance matrices $\mathbf{Q}$ and $\mathbf{R}$, respectively.

The Kalman filter assumes that the probability density of the states at every time step is Gaussian and hence exactly and completely characterized by two parameters, its mean and covariance [1]. The means are represented by the state vector $\mathbf{x}_k$ and the variances are represented by the estimation error covariance $\mathbf{P}_k$. The task of a Kalman filter is to estimate or predict the state vector $\mathbf{x}_k$ from noise-perturbed measurements $\mathbf{z}_k$. At time k, given a new measurement vector $\mathbf{z}_k$, a Kalman filter computes the optimal prediction of the state vector $\hat{\mathbf{x}}_{k+1} = \hat{\mathbf{x}}_{k+1|k}$ based on known measurement vectors $\{\mathbf{z}_1, \mathbf{z}_2, \ldots, \mathbf{z}_k\}$.

The Kalman filter can be interpreted as having two stages: *measurement update* and *prediction*. In the *measurement update* stage, the parameters of the Kalman filter are updated using (14.8) to (14.10).

$$\mathbf{K}_k = \mathbf{P}_k^- \mathbf{H}^T [\mathbf{H} \mathbf{P}_k^- \mathbf{H}^T + \mathbf{R}]^{-1} \tag{14.8}$$

$$\hat{\mathbf{x}}_k = \hat{\mathbf{x}}_k^- + \mathbf{K}_k (\mathbf{z}_k - \mathbf{H} \hat{\mathbf{x}}_k^-) \tag{14.9}$$

$$\mathbf{P}_k = [\mathbf{I} - \mathbf{K}_k \mathbf{H}] \mathbf{P}_k^- [\mathbf{I} - \mathbf{K}_k \mathbf{H}]^T + \mathbf{K}_k \mathbf{R}_k \mathbf{K}_k^T \tag{14.10}$$

$\mathbf{K}_k$ is the Kalman gain, $\hat{\mathbf{x}}_k^-$ is the a priori state vector, $\hat{\mathbf{x}}_k$ is the a posteriori state vector, $\mathbf{P}_k^-$ is the a priori state estimation error covariance matrix, and $\mathbf{P}_k$ is the a posteriori state estimation error covariance matrix. The Kalman gain in Equation (14.8) is the statistically optimal gain. This Kalman gain minimizes $E\{\|\mathbf{x}_k - \hat{\mathbf{x}}_k\|^2\}$, which is the expected value of the square of the magnitude of the error in the posterior state estimation. By using the optimal gain in (14.8), the update of the state estimation error covariance matrix in (14.10) can be simplified as in (14.11).

$$\mathbf{P}_k = [\mathbf{I} - \mathbf{K}_k \mathbf{H}] \mathbf{P}_k^- \tag{14.11}$$

In the *prediction* stage, the future states are predicted by using the projection equations:

$$\text{Project state:} \quad \hat{\mathbf{x}}_k^- = \mathbf{F} \hat{\mathbf{x}}_{k-1} \tag{14.12}$$

$$\text{Project covariance:} \quad \mathbf{P}_k^- = \mathbf{F} \mathbf{P}_{k-1} \mathbf{F}^T + \mathbf{Q} \tag{14.13}$$

More details on the derivations of the Kalman filter, (14.8) to (14.13), can be found in [6]. The definitions of the symbols used in these equations are summarized as follow:

$\hat{\mathbf{x}}_k^-$: a priori state vector
$\hat{\mathbf{x}}_k$: a posteriori state vector
$\mathbf{P}_k^-$: a priori estimation error covariance
$\mathbf{P}_k$: a posteriori estimation error covariance
$\mathbf{F}$: transition matrix
$\mathbf{H}$: measurement matrix
$\mathbf{Q}$: process noise covariance
$\mathbf{R}$: measurement noise covariance

Within the spatiotemporal saliency framework, the Kalman filter serves as the predictor. The coordinates of a two-dimensional feature from the feature tracker are the measurements and the probability distributions of possible coordinates and velocities are the states. As mentioned at the beginning of this section, a *linear* motion model is assumed, thus *linear* motions are predictable and temporally nonsalient. On the other hand, *nonlinear* motions that cannot be predicted are temporally salient. The transition matrix **F**, state vector **x**, and measurement matrix **H** are set to reflect these assumptions.

$$\mathbf{F} = \begin{bmatrix} 1 & 0 & 1 & 0 \\ 0 & 1 & 0 & 1 \\ 0 & 0 & 1 & 0 \\ 0 & 0 & 0 & 1 \end{bmatrix}, \quad \mathbf{x} = \begin{bmatrix} x \\ y \\ x' \\ y' \end{bmatrix}$$

$$\mathbf{H} = \begin{bmatrix} 1 & 0 & 0 & 0 \\ 0 & 1 & 0 & 0 \end{bmatrix}$$

x and y are the means of the Gaussian probability density of the coordinates in the x-direction and y-direction, respectively. x' and y' are the means of the Gaussian probability density of the velocity components in the x and y direction, respectively. The Kalman filter noise covariance matrices **Q** and **R** are both set to be identity matrices, which implies that the position and speed variables are treated as uncorrelated, and the process noise variances and the measurement noise variances are assumed to be 1.

At time k, there is the a priori state vector $\hat{\mathbf{x}}_k^-$ and the a priori estimation error covariance $\mathbf{P}_k^-$ that are both "predicted" from the previous time step $k-1$ using (14.12) and (14.13). A measurement vector $\mathbf{z}_k$ that contains new information is available at time k. This is used to update the Kalman filter using (14.8), (14.9), and (14.11). After the update, the a posteriori state vector $\hat{\mathbf{x}}_k$ and the a posteriori estimation error covariance $\mathbf{P}_k$ are obtained. These a posteriori statistics incorporate the new measurement vector $\mathbf{z}_k$, whereas the a priori statistics are "predicted" without the knowledge of $\mathbf{z}_k$. Therefore, the distance between a priori and a posteriori statistics quantifies how accurate the prediction is.

14.3.4 Distribution Distance

Since the comparison is made between two probability distributions, a distribution distance measure is required. The Bhattacharyya distance [10] is used as the distribution distance to quantify saliency from the probability distributions constructed by the predictor. It is a symmetric measure of the distance between two probability distributions $p_p(x)$ and $p_q(x)$ and it is defined as:

$$B = -\log\left(\int_{-\infty}^{+\infty} \sqrt{p_p(x) p_q(x)} dx\right) \quad (14.14)$$

The Kalman filter, as the predictor, assumes that the probability density function of the two-dimensional feature coordinates follows a Gaussian distribution.

The Bhattacharyya distance of two univariate Gaussian distributions p and q is given by

$$B_{\text{Gaussian}}(q;p) = \frac{1}{4}\frac{(\mu_p - \mu_q)^2}{\sigma_p^2 + \sigma_q^2} + \frac{1}{2}\log\left(\frac{\sigma_p^2 + \sigma_q^2}{2\sigma_p\sigma_q}\right) \quad (14.15)$$

where μ_p and μ_q are the mean values and σ_p and σ_q are the variances of the distributions p and q, respectively. The first term in (14.15) accounts for the difference due to the mean, and the second term gives the separability caused by the variance. The Bhattacharyya distance in (14.15) for Gaussian distribution can be employed to compute the distance between the a priori and a posteriori distributions at time k. The distances are computed in x and y directions separately as in (14.16) and (14.17), then the distances in two directions are combined in (14.18).

$$B_k^x = \frac{1}{4}\frac{(\hat{x}_k^- - \hat{x}_k)^2}{(p_k^{x-})^2 + (p_k^x)^2} + \frac{1}{2}\log\left(\frac{(p_k^{x-})^2 + (p_k^x)^2}{2p_k^{x-}p_k^x}\right) \quad (14.16)$$

$$B_k^y = \frac{1}{4}\frac{(\hat{y}_k^- - \hat{y}_k)^2}{(p_k^{y-})^2 + (p_k^y)^2} + \frac{1}{2}\log\left(\frac{(p_k^{y-})^2 + (p_k^y)^2}{2p_k^{y-}p_k^y}\right) \quad (14.17)$$

$$B_k = \sqrt{(B_k^x)^2 + (B_k^y)^2} \quad (14.18)$$

where

$\hat{x}_k^-$	the x-coordinate mean in the a priori state vector $\hat{\mathbf{x}}_k^-$
$\hat{y}_k^-$	the y-coordinate mean in the a priori state vector $\hat{\mathbf{x}}_k^-$
$\hat{x}_k$	the x-coordinate mean in the a posteriori state vector $\hat{\mathbf{x}}_k$
$\hat{y}_k$	the y-coordinate mean in the a posteriori state vector $\hat{\mathbf{x}}_k$
$(p_k^{x-})^2$	the variances of x-coordinates in the a priori estimation error covariance $\mathbf{P}_k^-$
$(p_k^{y-})^2$	the variances of y-coordinates in the a priori estimation error covariance $\mathbf{P}_k^-$
$(p_k^x)^2$	the variances of x-coordinates in the a posteriori estimation error covariance $\mathbf{P}_k$
$(p_k^y)^2$	the variances of y-coordinates in the a posteriori estimation error covariance $\mathbf{P}_k$

B_k in (14.18) is used to deduce the short-term unpredictability of a two-dimensional feature. Therefore, one scalar value is generated for one two-dimensional feature at time k as its temporal saliency value. The combination of the Kalman filter and the Bhattacharyya distance to compute the temporal saliency is demonstrated with the following examples.

In Figure 14.8, the measurements of the Kalman filter are the two-dimensional feature coordinates obtained by the feature tracker. The two-dimensional feature travels from bottom left to top right approximately in a straight line at constant velocity. The a priori means of coordinates are represented in crosses in Figure 14.8.

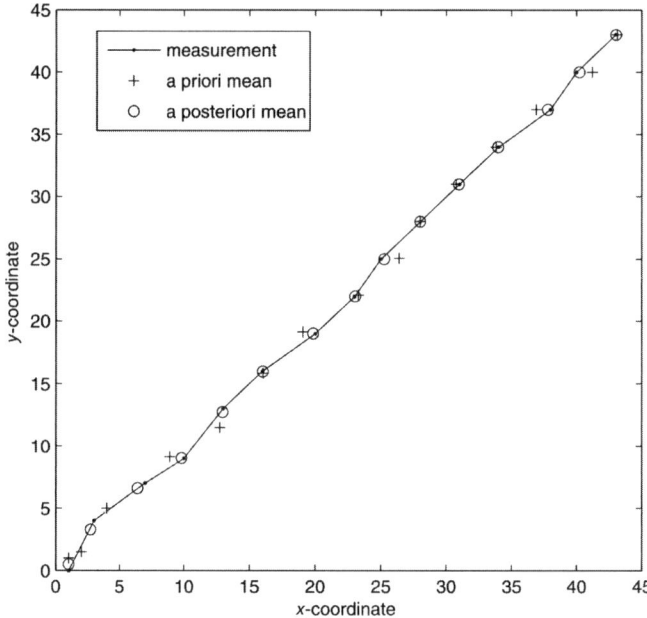

Figure 14.8 The Kalman filter variables for the two-dimensional feature in a linear motion.

It can be seen that these predicted values are close to the measurements. It should be noted that the measurements are in future time steps with respect to the a priori means. Hence, the a priori means are the predicted values by the Kalman filter. The a posteriori means of coordinates are rendered after the Kalman filter is updated with new measurements.

The corresponding Bhattacharyya distance between the a priori and the a posteriori coordinate distributions are shown in Figure 14.9. This distance is used to represent the temporal saliency of the two-dimensional feature. The x-axis of Figure 14.9 is chosen to be the same as that of Figure 14.8 in order to facilitate

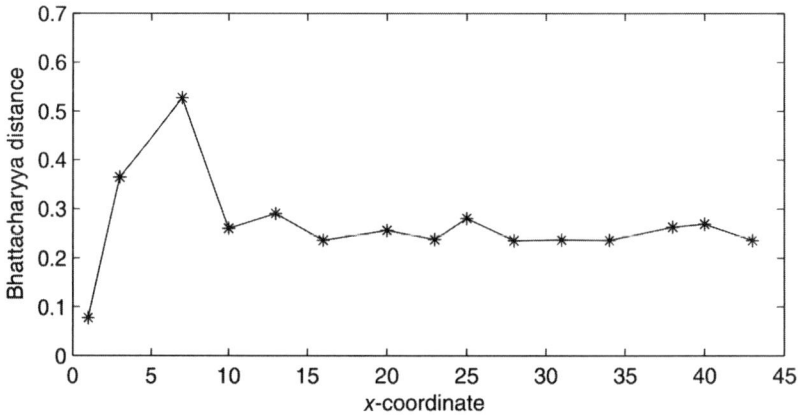

Figure 14.9 The Bhattacharyya distance for the two-dimensional feature in a linear motion.

14.3 Realization of the Framework

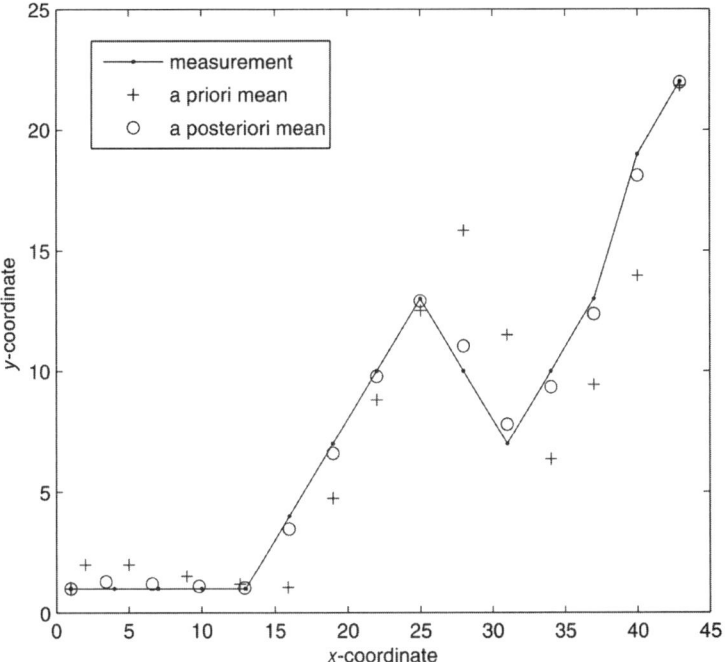

Figure 14.10 The Kalman filter variables for the two-dimensional feature in an unpredictable motion.

correspondence. The second coordinates have large Bhattacharyya distance value. This is because the Kalman filter is initialized without any knowledge of the velocity of the two-dimensional feature, and it needs a few measurements to infer these missing states.

In contrast to the last example, Figure 14.10 shows a two-dimensional feature travels in an unpredictable fashion, which changes course several times. As a result, the a priori means of coordinates are less able to predict the upcoming measurements. The Bhattacharyya distance between the a priori and the a

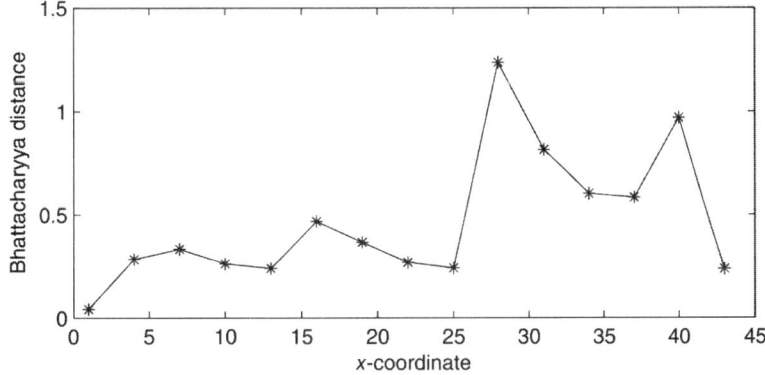

Figure 14.11 The Bhattacharyya distance for the two-dimensional feature in an unpredictable motion.

posteriori coordinate distributions is shown in Figure 14.11. The peaks of the Bhattacharyya distance can be associated with the unpredictable change of movement direction of the two-dimensional feature. Such response to unpredictability in motion demonstrates the capability of the Bhattacharyya distance to represent the temporal saliency of the two-dimensional feature.

14.3.5 Suppression

The suppressor postprocesses the preliminary saliency map rendered by the distribution distance stage. The features are subject to a threshold process and those whose distribution distance values are less than the threshold are disregarded. The neighboring pixels of each feature are then investigated. A feature only remains salient and advances into the final saliency map if there are a sufficient number of other features surviving the threshold process within the neighborhood, otherwise the feature is removed. Under this suppression scheme, only clusters of salient features remain and outliers are removed. The suppressor has the effect of disentangling noise from genuine salient movement by relying on multiple saliency responses. Figure 14.12 gives an illustration of the suppression process.

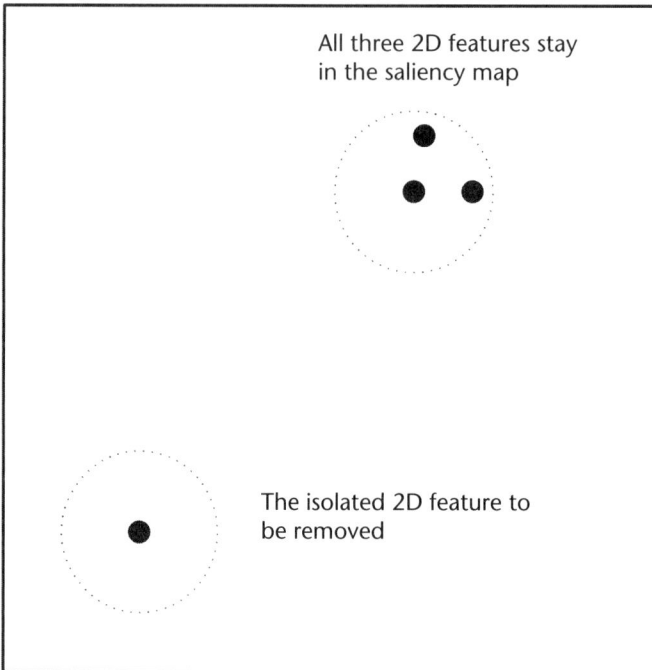

Figure 14.12 The suppression process: after the two-dimensional features are put through the threshold screening, the neighborhood of each two-dimensional feature is investigated. The black dot represents the two-dimensional feature and the dotted circle represents the radius of the investigated neighborhood. The two-dimensional feature on the bottom left does not have any other two-dimensional feature in its neighborhood. It is thus removed from the saliency map. The two-dimensional features on the top right are in close proximity to each other and hence they all remain in the saliency map.

14.4 Performance Evaluation

Experiments were conducted to demonstrate the viability and robustness of the spatiotemporal framework using the realization described in Section 14.3. In all the experiments, the maximum number of features was limited to 1,000. The suppressor threshold was 2. Each feature had to have at least one neighboring feature within a circle of three pixel radius in order to be qualified as a salient feature. The Kalman filter noise covariance matrices $\mathbf{Q}$ and $\mathbf{R}$ were both set to be identity matrices, which implies that the position and speed variables were treated as uncorrelated, and the a priori estimated process and measurement noise variances were assumed to be 1.

14.4.1 Adaptive Saliency Responses

The experiment illustrated in Figure 14.13 shows how the movement of the car triggers the adaptive spatiotemporal saliency responses. For presentation purpose, the centers of the circles are the feature coordinates that give rise to the saliency responses and the radii of the circles are determined by the Bhattacharyya distance between the a priori and a posteriori distributions.

From frames F1 to F3, the car moves at a constant velocity, therefore no saliency was detected by the model. In order to adapt to the low speed of the car, feature coordinates in frame F2 are decimated. Notice that it is the feature

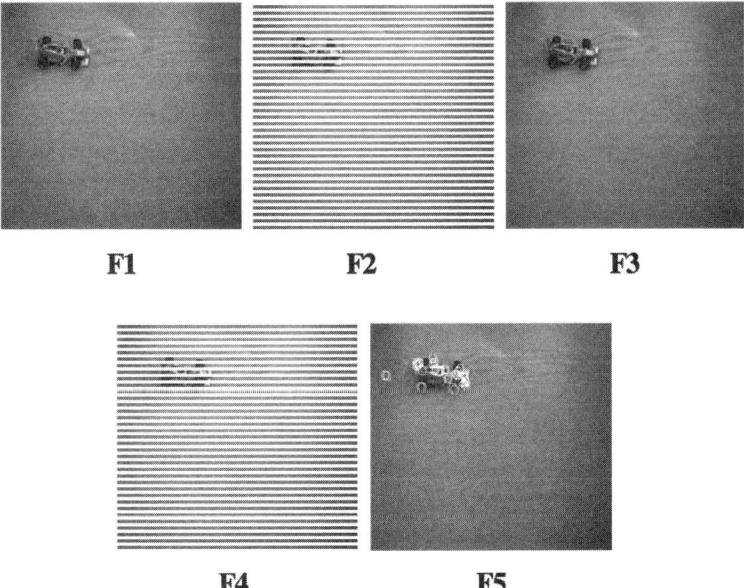

Figure 14.13 Spatiotemporal saliency results: the car moves slowly at a constant velocity before frame F3, to adapt to the slow motion, feature coordinates in one frame are decimated out of every two frames. Feature coordinates in frames F2 and F4 are decimated and disregarded by the predictor stage. The car changes velocity after frame F3. The salient responses are picked up by two-dimensional features automatically placed on the car and they are shown as circles in frame F5.

coordinates that are decimated as opposed to the frame itself. If there were two cars, a fast moving car and a slow one, the coordinates of features on the fast moving car would not be decimated whereas the coordinates of the features on the slow one would be as in this experiment. This is very beneficial, because the framework is capable of analyzing dynamic scenes, where motions of objects can differ a lot in speed.

The car changes velocity from frame F3 and the speed is still low so the feature coordinates in frame F4 are decimated. The salient responses are picked up by two-dimensional features automatically placed on the car and they are shown as circles in frame F5. The absolute velocity change of the car is small, so without the decimation, the small change would be undetected. However, relative to the slow motion of the car, the velocity change is significant and this adaptive mechanism allows us to capture it.

This experiment illustrates that the spatiotemporal framework detects motion pop-up phenomena successfully.

14.4.2 Complex Scene Saliency Analysis

This experiment demonstrates the robustness of the spatiotemporal saliency model dealing with a complex scene. In Figure 14.14, a massive number of fish flock towards one spot where food is supplied. In the region around that spot, fish movements are especially vigorous and irregular, hence these points are the most unpredictable in the scene.

Figure 14.14(a) shows spatiotemporal saliency detection results at the beginning of the video clip and only the very first two frames of the video are known to the framework. The saliency model suggests a majority of two-dimensional fea-

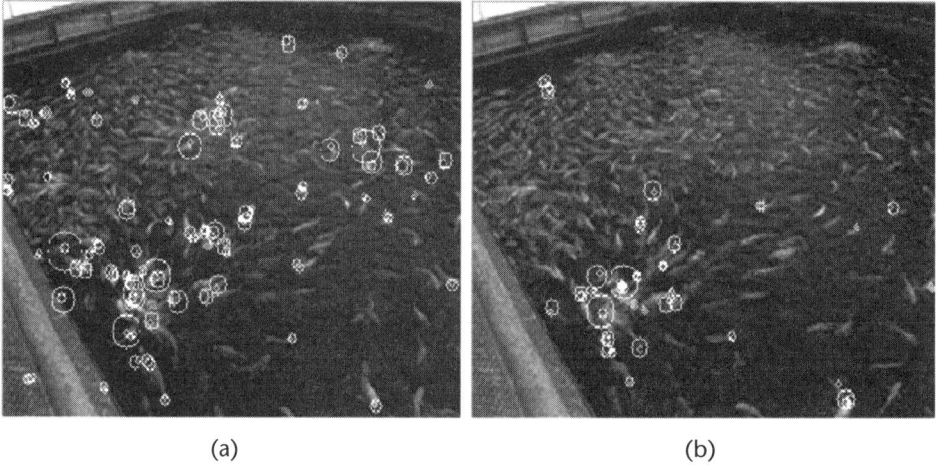

(a)   (b)

Figure 14.14 Complex scene: a massive number of fish moving towards fish feed. (a) Saliency result from the very first two frames only; (b) Saliency result at a later time.

tures as salient due to the lack of sufficient temporal information. Hence, there are salient responses all over the place.

Figure 14.14(b) shows saliency detection results at a later time. Although there are only two consecutive frames available to the saliency model, it has implicitly picked up earlier temporal information through parameter correction of the Kalman filters. Evidently, in Figure 14.14(b), the saliency model indeed emphasizes the region around the feeding spot with large saliency responses and smaller responses in other regions.

14.5 Conclusions

This chapter presented an unsupervised adaptive spatiotemporal saliency framework that combines low-level processing and statistical inference. Experimental results showed that the model using this framework has the capability of distinguishing between *predictable* and *unpredictable* motions, hence *spatiotemporal saliency* without the involvement of object detection and segmentation even in a complex scene. The framework was shown to be adaptive to the speed of two-dimensional features in order to increase the range of speed it can cope with.

Moreover, the described framework has the flexibility to be tuned to specific motion patterns depending on the requirements. Under a linear motion assumption, the Kalman filter is recommended as the predictor. The framework is totally flexible to target other motion models and consequently give rise to saliency accordingly.

References

[1] Arulampalam, M. S., et al., "A tutorial on particle filters for online nonlinear/non-gaussian bayesian tracking." *IEEE Transactions on Signal Processing*, Vol. 50(2), 2002, 174–188.

[2] Boiman, O. and Irani, M., "Detecting irregularities in images and in video." In *Proceedings. Tenth IEEE International Conference on Computer Vision*, Vol. 1, 2005, pp. 462–469, Beijing, China.

[3] Bouguet, J.-Y., "Pyramidal implementation of the lucas kanade feature tracker description of the algorithm." *Microprocessor Research Labs, Intel Corporation*, 2000.

[4] Harris, C. and Stephens, M., "A combined corner and edge detector." *Proceedings of the 4th Alvey Vision Conference*, 1988, pp. 147–151.

[5] Freeman, W. T., "The generic viewpoint assumption in a framework for visual perception." *Nature*, Vol. 368(6471), 1994, 542–5.

[6] Grewal, M. S. and Andrews, A. P., *Kalman filtering : theory and practice using MATLAB*. Wiley, New York; Chichester, 2nd ed., 2001.

[7] Itti, L. and Baldi, P., "A principled approach to detecting surprising events in video." *Proceedings. 2005 IEEE Computer Society Conference on Computer Vision and Pattern Recognition*, Vol. 1, 2005, pp. 631–637.

[8] Itti, L., Koch, C., and Niebur, E., "A model of saliency-based visual attention for rapid scene analysis." *IEEE Transactions on Pattern Analysis and Machine Intelligence*, Vol. 20(11), 1998, 1254–1259.

[9] Kadir, T. and Brady, M., "Saliency, scale and image description." *International Journal of Computer Vision*, Vol. 45(2), 2001, 83–105.

[10] Kailath, T., "The divergence and bhattacharyya distance measures in signal selection." *IEEE Transactions on Communication Technology*, Vol. 15, 1967, 52-60.

[11] Kalman, R. E., "A new approach to linear filtering and prediction problems." *Transactions of the ASME—Journal of Basic Engineering*, Vol. 82, 1960, 35-45.

[12] Kalman, R. E. and Bucy, R. S., "New results in linear filtering and prediction theory." *Transactions of the ASME—Journal of Basic Engineering*, Vol. 83, 1961, 95-107.

[13] Laptev, I. and Lindeberg, T., "Space-time interest points." *Proceedings Ninth IEEE International Conference on Computer Vision*, Vol. 1, 2003, pp. 432-439.

[14] Liu, Y., Bouganis, C. S., and Cheung, P. Y. K., "A spatiotemporal saliency framework." In *Proceedings of IEEE International Conference on Image Processing (ICIP 2006)*, Atlanta, GA, Oct. 2006, pp. 437-440.

[15] Oliva, A., et al., "Top-down control of visual attention in object detection." In *Proceedings. 2003 International Conference on Image Processing*, Vol. 1, 2003, pages I-253-6.

[16] Shi, J. and Tomasi, C., "Good features to track." *Proceedings. 1994 IEEE Computer Society Conference on Computer Vision and Pattern Recognition*, 1994, pp. 593-600.

[17] Zhai, Y. and Shah, M., "Visual attention detection in video sequences using spatiotemporal cues." In *Proceedings of the 14th annual ACM international conference on Multimedia*, 2006, pp. 815-824.

[18] Zhong, H., Shi, J., and Visontai, M., "Detecting unusual activity in video." *Proceedings. 2004 IEEE Computer Society Conference on Computer Vision and Pattern Recognition*, Vol. 2, 2004, pp. 819-826.

Acronyms and Abbreviations

ϕ	light intensity
γ	bulk threshold parameter
ϕ_0	semiconductor surface potential
ϕ_f	fermi potential
η_{PCE}	external quantum efficiency
ALU	arithmetic logic unit
AM	appearance model
ANSI	American National Standards Institute
ASIC	application specific integrated circuit
ATP	adenosine triphosphate
BA	behavior model
BIBO	bounded input, bounded output
biCMOS	bipolar complementary metal oxide semiconductor
C	cilium
Ca^{2+}	calcium ion
CCD	charged coupled device
CCTV	closed circuit television
cGMP	cyclic guanosine monophophosphate: leads to the opening of channels, allowing sodium ions to enter a photoreceptor cell.
CIE	Commission Internationale de l'Eclairage, or International Commission on Illumination
CIR	complete integration region
CLS	constrained least squares
CM	circular measure (keypoint class)
CMOS	complementary metal oxide semiconductor
CNN	cellular neural network
CNNUM	cellular neural network universal machine
CO	cytochrome-oxidase
C'ox	MOSFET gate oxide capacitance per unit area
CPU	central processing unit
CRF	classical receptive field
CT	computed tomography
CYCNN	Chua-Yang cellular neural network
DMA	direct memory access
DoG	difference of Gaussians
DOPC	difference of Poisson conjugate (kernel type)
DRAM	dynamic random access memory
DSP	digital signal processing
DTCWT	dual-tree complex wavelet transform

EKF	extended Kahlman filter
EM	expectation maximization
EQE	external quantum efficiency
eV	electron volt
FIFO	first-in, first-out
FIR	finite impulse response (filter)
FOV	field of view
FP	false positive: the event of a false object being classified as a true target object.
FPGA	field programmable gate array
FPR	false positive detection rate
FMP	filter magnitude products (keypoint class)
FN	false negative
GABA	γ-aminobutyric acid main inhibitory neurotransmitter in the central nervous system.
GC	guanylate cyclase
GDP	guanosine diphosphate
GMM	Gaussian mixture model
GMP	guanosine monophosphate
GPC receptor	G protein coupled receptor
GTP	guanosine triphosphate
GTPase	enzyme which hydrolyses GTP into GDP
HOMO	highest occupied molecular orbit
HR	high-resolution image
HVS	human visual system
I	intrinsic semiconductor
ICA	independent component analysis
I_{ds}	drain source current across a transistor
IIR	infinite impulse response (filter)
IPCE	internal photo conversion efficiency
ISE	integrated software environment
J_{cG}	rod/cone current mediated by cGMP
J_{SC}	short circuit current
k	Boltzmann's constant
K^+	potassium ion
K_{cG}	rod/cone channel activation constant
KCL	Kirchhoff's current law
L	MOSFET channel length
LED	light emitting diode
LGN	lateral geniculate neucleus
LoG	Laplacian of Gaussian function
LP	location prediction
LR	low resolution image
LUMO	lowest unoccupied molecular orbit
LUT	lookup table
MAP	maximum a posteriori

MAX	maximum
MD	motion detection
MIS	minimally invasive surgery
MLCNN	multi-layer CNN
MOS	metal oxide semiconductor
MOSFET	metal oxide semiconductor field effect transistor
MRF	Markov random field
MRI	magnetic resonance imaging
MT	middle temporal
n	diode factor; n-type semiconductor; the weak inversion slope factor; sub-threshold slope factor
Na^+	sodium ion
OLED	organic light emitting diode
OTA	operational transconductance amplifier
p	p-type semiconductor
PC	personal computer
PCBM	phenyl-C61-butyric acid methylester, a semiconducting small molecule
PCI	peripheral component interconnect
PDE	phosphodiesterase
PDF	probability density function
POCS	projection onto convex sets
PSF	point spread function
PR	perfect reconstruction
q	elementary charge
R	rhodopsin
RAM	random access memory
RF	receptive field
RGB	red, green, blue color scheme
RK	rhodopsin kinase
ROC	receiver operating characteristic
R_{sheet}	shunt resistance
RT	reaction time or response time
RTL	register transfer level
SDTP	spatial frequency dependent temporal poles
SF	summation field
SFG	signal flow graph
SIFT	scale-invariant feature transformation
SIMD	single instruction multiple data
SMAP	saliency map
SNR	signal to noise ratio
SRAM	static random access memory
STTF	spatiotemporal transfer function
SUM	summation
T	absolute temperature

TB	true blobs
TDCNN	temporal derivative CNN
TMF	temporal median filter
TMS	transcranial magnetic stimulation
TP	true positive. The event of a true target object being classified as a true target object.
TPR	true positive detection rate
TV	total variance
UKF	unscented Kahlman filter
USB	universal serial bus
V	voltage
V1	primary visual cortex
V2	second visual cortex area, the one receiving inputs from V1 and next stage from V1 in the anatomical pathway of the visual signal.
V_{db}	potential difference between the drain and the bulk.
V_{dd}	global supply voltage on the chip
V_{ds}	drain source voltage across a transistor
V_{eff}	effective gate source voltage across a transistor
VGA	video graphics array
V_{gs}	gate source voltage across a transistor
VHDL	VHSIC hardware description language
VHSIC	very high speed integrated circuit
via	interconnect between different horizontal planes on a chip
VLSI	very large scale integration
V_o	early voltage
V_{OC}	open circuit voltage
V_p	pinch off voltage
V_{sb}	potential difference between the source and the bulk
V_{to}	threshold voltage
YUV	Y (luminance) U, V (chromatic) color scheme
W	MOSFET channel width
ZBT	zero-bus turnaround

About the Editors

Anil Anthony Bharath is reader in image analysis at Imperial College London. He received his BEng in Electrical and Electronic Engineering from University College London, and his PhD from Imperial College London. He has published over 65 papers in the fields of image analysis, computer vision, imaging and ultrasound. Dr Bharath conceived the idea of the project *Reverse Engineering the Human Visual System*, establishing the links between research groups that resulted in this book.

Maria Petrou is the professor of signal processing and the head of the communications and signal processing group at Imperial College London. She has published more than 300 papers on topics in remote sensing, computer vision, machine learning, color analysis, medical signal and image processing, and other fields, and she is a coauthor of two books on image processing. Professor Petrou is a fellow of the Royal Academy of Engineering, IET, IoP and IAPR, a senior member of IEEE, and a distinguished fellow of the British Machine Vision Association. She earned her PhD at the Institute of Astronomy, Cambridge University, UK.

List of Contributors

Dylan Banks
Imperial College London
Research Associate
Institute of Biomedical Engineering
Bessemer Building
Imperial College London
South Kensington Campus
London SW7 2AZ, U.K.

Anil Anthony Bharath
Imperial College London
Reader in Image Analysis
Department of Bioengineering,
Room 3.13 RSM Building
Exhibition Road
London SW7 2AZ, U.K.

Christos-Savvas Bouganis
Imperial College London
Lecturer in Digital Systems
Dept of Electrical & Electronic Engineering
Exhibition Road
London SW7 2AZ, U.K.

Lichun Chen
Merck Chemicals Limited
Senior Physics Scientist
Merck Chemicals Ltd.
LC Chilworth Technical Centre
University Parkway
Southampton SO16 7QD, U.K.

Peter Y. K. Cheung
Imperial College London
Professor of Digital Systems
Room 912,
Dept of Electrical & Electronic Engineering
Exhibition Road
London SW7 2AZ, U.K.

Roy Davies
Royal Holloway University of London
Professor of Machine Vision
Department of Physics
Egham, Surrey, TW20 0EX, U.K.

Patrick Degenaar
Imperial College London
Lecturer in Neurobionics
Div. Neuroscience and the
Institute of Biomedical Engineering
B611 Bessemer Building
Exhibition Road
London SW7 2AZ, U.K.

Suhaib Fahmy
Trinity College Dublin
Research Fellow
Centre for Telecommunications
Value-Chain Research (CTVR)
Dublin, Ireland

H. M. D. Ip
Rutherford Appleton Laboratory
Microelectronics Design Engineer
ASIC Design Group
R76 First Floor
Science and Technology Facilities Council
Harwell Science and Innovation Campus
Didcot OX11 0QX, U.K.

Mirna Lertotic
Imperial College London
Research Associate
Institute of Biomedical Engineering
Bessemer Building
Exhibition Road
London SW7 2AZ, U.K.

Yang Liu
Formerly of Imperial College London
Dept of Electrical & Electronic Engineering
Imperial College
Exhibition Road
London SW7 2AZ, U.K.

Keith May
University of Bradford
Research Fellow
Vision Science Research Group
Richmond Road
Bradford, West Yorkshire
BD7 1DP, U.K.

Hiroshi Momiji
University of Nottingham
Research Associate
Department of Biomedical Science
University of Sheffield
Western Bank
Sheffield S10 2TN, U.K.

George P. Mylonas
Imperial College London
Research Associate
Institute of Biomedical Engineering
Exhibition Road
London SW7 2AZ, U.K.

Jeffrey Ng
Imperial College London
Research Fellow
Department of Bioengineering
Level 3 RSM Building
Exhibition Road
London SW7 2AZ, U.K.

Konstantin Nikolic
Imperial College London
Research Fellow (Corrigan Fellowship)
Institute of Biomedical Engineering
The Bessemer Building
London SW7 2AZ, U.K.

List of Contributors

Maria Petrou
Imperial College London
Professor of Signal Processing
Head of Signal Processing
and Communications Group
Department of Electrical and
Electronic Engineering
Exhibition Road
London SW7 2AZ, U.K.

Mark Sugrue
Research Associate
Dublin City University
School of Electronic Engineering
Faculty of Engineering and Computing
Dublin 9, Ireland

Guang-Zhong Yang
Imperial College London
Chair in Medical Image Computing
Royal Society/Wolfson MIC Laboratory
305/306 Huxley Building
Department of Computing
180 Queens Gate
London SW7 2BZ, U.K.

Li Zhaoping
University College London
Professor
Department of Computer Science
Gower Street
London WC1E 6BT, U.K.

Index

π-bonds, 270

A

Absorption, 265
  coefficient, 266
  light photon, 257
Acceleration methods, 382–83
Accumulator space, 134, 137
Active pixel sensor (APS), 304
Adaptive foveation, 324–30
  circuit implementation, 326–29
  defined, 324
  future, 329–30
  system algorithm, 325–26
  See also Foveation
Adaptive saliency responses, 411–12
Adaptive weights method, 152
Address event representation (AER), 317–24
  arbitration tree, 318–22
  bit scheme, 319, 320
  collision reduction, 323–24
  collisions, 322
  defined, 318
  readout, 329
  sparse coding, 322–23
  word system, 321
Adenosine triphosphate (ATP) molecules, 260
Admissibility condition, 106
Amacrine cells, 23
  defined, 4
  negative feedback mechanisms, 36
Analog retinomorphic circuitry, 289–330
  adaptive foveation, 324–30
  address event representation, 317–24
  digital methodologies versus, 296
  MOSFET, 292–95
  photo electric transduction, 296–300
  processing, 300–317
  processing principles, 290–96
Analog V1 platforms, 335–64
  characteristics, 335
  CNN, 340
  CNN and mixed domain spatiotemporal transfer functions, 342–45
  complex cell receptive field modeling, 362–63
  examples, 355–62
  linear CNN, 340–42
  networks with temporal derivative diffusion, 345–49
  signal flow graph-based implementation, 349–54
  vision chip comparison, 336–39
Angular frequency response, 116–18
Appearance model (AM), 231–32
  adaptive, 231
  algorithm, 232
  choice, 232
  feature tracking, 231
  module, 218
  optical flow, 231
  perfect, 240
Application-specific integrated circuits (ASICs), 340
A priori spline control, 205
ARAMA modeling, 207
Arbitration trees, 318–22
  defined, 318
  one-dimensional array of spiking addresses, 319
  two-dimension bit coded address, 319, 320
  word, 321
  See also Address event representation (AER)
Arithmetic logic unit (ALU), 382

Avalanche photodiodes, 261
Average reaction times (RTs), 74

B

Background modeling
   approach, 221
   pixel process calculations, 240
Bandpass analysis channels, 114
Bandpass filters
   fixed, 120
   kernels, 118
   output, 120
Behavioral reaction times (RTs), 75
Behavior analysis
   eye tracking in, 196–97
   module, 218
Behavior recognition, 237–39
   flagging, 237
   object classification, 239
Best linear unbiased estimator (BLUE), 146
Bhattacharyya distance, 406, 407
   Kalman filter and, 407
   for two-dimensional feature (linear motion), 408
   for two-dimensional feature (unpredictable motion), 409
Bidirectional subtraction circuit
   with current comparator switch, 309
   schematics, 308, 309
   with two channel current output, 308
Bilinear interpolation, 123
Binocular eye-tracking calibration, 204–6
   head shift, 206
   know intrinsic and extrinsic stereo endoscopic parameters, 205
   radial basis spline, 204
Biological gap junctions, 295
Bipolar cells, 4, 23
   circuitry, 305–6
   negative feedback mechanisms, 36
Bipolar current mode processing, 306–7
Bivariate Laplacian model, 129, 130
Blob sorting tracking, 233–36
   defined, 233
   qualitative results, 235
   rules, 234
   schematic, 235, 236
   sprite results, 236
Bottom-up saliencies
   behavioral studies, 69–70
   color-orientation asymmetry, 81–84
   color-orientation double feature, 84–87
   computation, 94
   discussion, 92–98
   interference by task-irrelevant features, 76–81
   map, 69–99
   materials and methods, 73–74
   orientation features grouping by spatial configurations, 87–91
   orientation-orientation double feature, 84–87
   procedure and data analysis, 74
   results, 75–91
   subjects, 74
   in visual selection, 94
"Bright-pupil" images, 195

C

Cable theory, 28
Camera-gain, 206
Canny edge detector, 161
CAVIAR database, 229
CCD, 290
   CMOS advantages over, 291
   mWatts consumption, 291
   photodetector arrays, 262–63
   sensors, 8
Cell interactions (V1), 59–61
Cell organization (V1), 53–58
   color selectivity, 56–57
   orientation selectivity, 55–56
   phase selectivity, 58
   scale selectivity, 57–58
Cellular neural networks. *See* CNNs
Center-off-surrender-on, 7
Center-on-surrounding-off, 6
Center-surround receptive fields, 27, 28
Charge coupled devices. *See* CCD
Chau-Yang CNNs (CYCNNs), 342, 345, 348
Circular measure (CM) keypoints, 140

Classical receptive fields (CRF), 60, 61, 72
Clutter, 240
CMOS, 290
   advantages, 291
   chips, 284
   circuitry, 291
   double junction, 267
   maturity, 291
   mWatts consumption, 291
   organic hybrid imaging chip, 285
   photodetector arrays, 263–65
   printing on, 284–85
   processes, 267
   readout, 291
   scaling down technology, 269
CNNs
   Chua-Yang (CYCNNs), 30, 342, 345, 348
   cloning templates, 341
   defined, 340
   implementations, 340
   linear, 340–42
   mixed domain STTFs and, 342–45
   multilayer (MLCNNs), 349, 352–54
   one-dimensional, 341
   templates, 344
   with temporal derivative diffusion, 346
   temporal-derived enhanced (TDCNN), 355
   universal machine (CNNUM), 340
Colinear facilitation, 73, 87, 90
Collisions, 322
   output matrix distortion, 322
   reduction, 323–24
Color characteristics, numerical simulation, 43–45
Color contrast, 82, 83
   irrelevant, 38, 82
   relevant, 83
Color filtering, 265–68
   in CMOS, 266
   techniques, 265
Color mechanisms, 34–36
Color orientation
   conjunctive cells, 84

double contrast border, 86
Color-orientation double feature, 84–87
   RT, 86
   saliency advantage, 85
Color selectivity (V1), 56–57
Compact Hough transform, 137
Compensatory eye movement, 191
Complete integration region (CIR), 54
Complex scene saliency analysis, 412–13
Complex wavelet pyramid on FPGA, 370–77
   corner detection, 377
   defined, 370
   FilterBank process, 371, 373
   FPGA design, 370–73
   host control, 373–74
   implementation analysis, 374–75
   kernels, 371
   ManageBuffer process, 371
   performance analysis, 375–77
   ProcessWindow process, 371
   See also Field-programmable gate arrays (FPGAs)
Computed tomography (CT), 187, 202
Condensation/particle filter, 232
Conditional-Gaussian statistical model, 125
Conductance-based models, 27
Cones, 4, 5
   function, 254
   L, 27
   light sensitivity, 254–55
   L/M ratio, 43, 45
   M, 27
   mimicking, 251–86
   modeling, 34
   mosaics, 32
   negative feedback mechanisms, 36
   S, 27
   total number of, 254
   types, 22
   See also Rods
Configuration superiority effect, 97
Conjunctive cells
   color orientation, 84

defined, 72
linking, 98
V1, 98
Constrained least squares (CLS), 173, 174–77
   defined, 174
   method comparison, 183
   reconstruction line profiles, 176
   super-resolution image reconstruction with, 174, 175
   See also Super-resolution
Contact printing, 282–84
Contextual influences (V1), 70
Continuous time signal flow graphs, 349–51
Continuous wavelet transform, 105–6
Contrast marginalized detection, 137
Contrast parameter, 130
Corner detection, 377, 378
Corner likelihood response, 123
Current comparator, 310–12
   bidirectional current subtraction circuit with, 309
   cascaded, 311
   illustrated, 307
   mirror, 311
   Monte Carlo evaluation, 311
   technique schematics, 310
Current mode approaches, 303–12
   bidirectional processing, 306–7
   improved horizontal cell circuitry, 305
   novel bipolar circuitry, 305–6
   See also Retinimorphic circuit processing
Current spreading
   foveal, 326
   function, 327
   network implementation, 327
Cyclic guanosine monophosphate (cGMP), 257
   molecule binding, 257
   replenishing, 258

D

"Dark-pupil" images, 195
Daubechies wavelet. *See* Haar wavelet

Decimation, 113
Depth recovery
   accuracy assessment, 202
   motion stabilization and, 206–9
   ocular vergence, 202–3
   on synthetic tissue, 207
Detection formulation, 126–27
Differential equations, 32–34
Direct memory access (DMA), 373
Discrete Fourier transform (DFT)
   from irregularly sampled data, 156–57
   for regularly sampled data, 157
Distribution distance, 406–10
Divisive normalization, 133–34, 141
Dual-channel tracking, 230–36
   appearance model, 231–32
   blob sorting, 233–36
   prediction approaches, 232–33
   schematic representation, 241
Dual-tree complex wavelet transform (DTCWT), 110, 138–40
Dyadic decompositions, 109

E

Edge detection, 158
   algorithm, 161
   Canny, 161
   postprocessing scheme, 160
   temporal, 221–24
   V1-inspired, 158–62
Edge-selective neurons, 51
Electro-oculography (EOG), 193
Endoscopy, 200
Equivalent circuit, 277–81
Erlang functions, 116
Expectation-maximization (EM) algorithm, 220
Expected fields, 126
   conditioned on shape position and rotation, 128
   notion of, 128–29
Extended Kalman filter (EKF), 232, 233
External quantum efficiency (EQE), 275
Eye-gain, 206
Eye movement, 191–92

abnormalities, 195
in fMRI, 199
types of, 191–92
Eye tracking
application of, 195–200
behavior analysis, 196–97
binocular, calibration, 204–6
human-computer interaction, 199–200
medicine, 197–99
P-CR-based, 206
psychology/psychiatry and cognitive sciences, 195
techniques, 192–95

F

Fabrication, organic semiconductor, 281–85
contact printing, 282–84
deposition techniques, 281
printing on CMOS, 284–85
False positive rate (FPR), 135
Feature trackers, 397, 399–404
goal, 399
Lucas-Kanade, 401–404
Feature tracking, 231
Feedback buffers, 298
Feed forward V1, 58–62
Field of view (FOV), in MIS, 185
Field-programmable gate arrays (FPGAs), 367–69
circuit design, 369
complex wavelet pyramid implementation, 370–77
defined, 367
design process, 369
fine-grain parallelism, 368
introduction, 367, 368
mapping two-dimensional filters onto, 369–70
programmable components, 386
reconfigurable properties, 368
trace transform implementation, 377–91
Xilinx, 390
Filter-bank paradigm, 112
Filter magnitude product (FMP) keypoints, 140

Fixational eye movements, 171
Focus-of-attention mechanisms, 138
Fourier transform, 118
discrete (DFT), 156–57
interpolation, calculating, 152
inverse, 152
Foveal current spreading technique, 326
Foveal image representation, 36–37
Foveal structure, 31–32
Foveation
adaptive, 324–25
defined, 324
effect, 324
utility versus scalability and, 328
Frame-by-frame detection, 230
Frames, 151
functions, 151
stacking, 221, 222
Frequency domain, 173
Frequency response
angular, 116–18
bandpass, 118
spatiotemporal cone filter, 356
Full trace transform system, 382–87
acceleration methods, 382–83
functional blocks, 386
initialization, 386–87
rotation block, 384–86
system overview, 383–84
target board, 383
top-level control, 384
See also Trace transform
Functional blocks, 386
lookup initialization data, 387
Type A, 388
Type B, 388–89
Type C, 389, 390
Functional MRI (fMRI), 198–99

G

Gabor envelopes, 108
Gabor functions, 107
complex, 113
cosine, 116
phase change, 109
sine, 116

Gabor patches
 defined, 107
 illustrated, 108, 109
Ganglion cells, 4, 6, 23–24
 defined, 23
 excitable membrane, 25
 firing rate, 28
 intelligent, 314–17
 midget, 24
 outputs, 43
 photosensitive, 24
 spikes, 34
Gaussian field estimators, 139–40
Gaussian gradient operators, 129
Gaussian kernels, 116
Gaussian mixture model (GMM), 220, 227
 computational cost, 230
 numerical results, 229
Gaussian scale parameter, 129
Gaussian smoothing, 129
Gaze-contingent displays, 199–200
 binocular eye-tracking calibration, 204–6
 conclusions, 209
 depth recovery and motion stabilization, 206–9
 ocular vergence for depth recovery, 202–3
 for robotic surgery, 200–209
GAZE Groupware system, 200
General surround suppression, 89
G-protein coupled (GPC) receptors, 256
Griding, 145

H

Haar wavelet
 3D, 230
 coefficients, 226
 defined, 225
 one-dimensional, 226
 spatiotemporal, 225–30
 s-t, 234
 threshold detection results, 228
Hardware
 mapping algorithms into, 367–92

 trace transform implementation, 377–91
Head-mounted eye tracker, 192
Head shift calibration, 206
Hidden variables, 131
Highest occupied molecular orbital (HOMO), 270, 272
High-resolution computed tomography (HRCT) images, 197, 198
Horizontal cells, 4, 23
 gap functions, 25–26
 improved circuitry, 305, 306
 negative feedback mechanisms, 36
Horizontal RESistor (HRES), 301
Human-computer interaction, 199–200
Human eye, 3–10
 cells, 4
 cones, 4, 5
 cross section, 4
 issues, 8–10
 photograph, 3
 photoreceptors, 4
 phototransduction, 253–60
 physiology, 253–55
 retina, 5, 6
 rods, 5
Human visual system, 1–17
 color, 36
 human eye, 3–10
 LGN, 10–12
 motion analysis and V5, 15
 motion detection, 217
 motion processing, 218–19
 overview, 2–15
 reverse engineering, 167–68, 171–88
 V1 region and visual cortex, 12–15
 wiring diagram, 242
Hybrid Monte Carlo particle filter, 233
Hyperacuity, 172–73
 defined, 171
 fixational eye movements and, 171–72
Hyperpolarization, 257
Hypothesis testing, 71

I

Ice-cube model, 54
Image difference histogram, 223

Image processing, 145–67
  interpolation errors, 145
  linear, 145–57
  nonlinear, 157–67
  subconscious, 158
Image registration, 183–84
Independent component analysis (ICA), 59
Indium tin oxide (ITO), 273
In-phase component, 112–13
Integrated circuits, 290
Integration-based photodetection circuits, 298–99
Intelligent ganglion cells, 314–17
  ON-OFF, 315–16
  pulse width encoding, 316–17
Intel SIMD Integrated Performance Primitives libraries, 376
Intercellular dynamics, 34
Interference
  color-orientation asymmetry in, 81–84
  Snowden's, 83
  by task-irrelevant features, 76–81
Internal photocurrent conversion efficiency (IPCE), 275
Interpolation
  errors, avoiding, 145
  Fourier transform of, 152
  of irregularly sampled data, 146–56
  iterative error correction, 151–53
  Kriging, 146–51
  normalized convolution, 153–56
Intracortical interactions (V1), 61–62
Inverse Fourier transform, 152
Ionic current models, 27–28
Irrelevant color contrast, 82, 83
Iterative error correction, 151–53

J

Just-noticeable difference (JND), 80
J-V curve, 275

K

Kalman filter, 404
  Bhattacharyya distance and, 407
  defined, 232
  extended (EKF), 232, 233
  measurement update stage, 405
  noise covariance matrices, 406
  prediction stage, 405
  probability density and, 405
  stages, 405
  unscented (UKF), 232, 233
  variables for two-dimensional feature (linear motion), 408
  variables for two-dimensional feature (unpredictable motion), 409
Keypoints
  CM, 140
  detection with DTCWT, 138–40
  FMP, 140
Kingsbury's DTCWT, 110
Kirchhoff's current law (KCL), 343
Koniocellular, 12
Kriging, 146–51
  application to image interpolation, 151
  defined, 146
  estimation, 146
  estimation of covariance matrix of data, 147–48
  variogram models, 148–49

L

Lagrange multiplier, 147
Lateral geniculate nucleus (LGN), 2, 10–12
  cross section, 11
  incoming connections, 54
  interlayer connections, 55
  koniocellular, 12
  layers, 11
  magnocellular pathway, 10–12
  midget cell, 10
  neurons, 12
  parasol cell, 10
  parvocellular, 12
L-cones, 27
Leaky integrate-and-fire model, 27
Lie-algebra perspective, 119
Linear CNNs, 340–42
  diffusion in, 344

general STTF, 348
implementation by means of
 transconductors, 343
temporal derivative diffusion, 347
See also CNNs
Linear image processing, 145-57
 basis, 145
 DFT from irregularly sampled data,
  156-57
 interpolation of irregularly sampled
  data, 146-56
 iterative error correction, 151-53
 Kriging, 146-51
 normalized convolution, 153-56
 See also Image processing
Linear motions, 406, 408, 409
Line tracer, 385
L/M-cone
 populations, 44
 ratio, 43, 45
Location prediction (LP) module, 218
Logarithmic sensors, 297-98
Lookup tables (LUTs), 368, 390
Lowest unoccupied molecular orbital
 (LUMO), 270, 272
Lucas-Kanade optical flow algorithm,
 401-404

M
Macula region, 191
Magnetic resonance imaging (MRI), 162,
 187
 functional (fMRI), 198
 original/enhanced, 161, 162
Magnocellular pathway, 10-12, 24
Markov chain Monte Carlo, 232
Markov random field (MRF)
 with Gibbs PDF, 180
 prior, 180-83
Mathematical modeling, 27-30, 45
 retina and it's functions, 28-30
 single cells of retina, 27-28
Maximum a posteriori (MAP), 173
 defined, 180
 line profiles, 182
 method comparison, 183

MRF method, 180-83
 reconstruction, 181
MAX rule, 80, 81, 84, 92
M-cones, 27
Measurement matrix, 404
Medicine, eye tracking in, 197-99
Membrane potentials, 40
Metal oxide semiconductor field effect
 transistor (MOSFET), 292-95
 biological gap junctions, 295
 complementary forms, 292
 defined, 292
 inversion characteristics, 294-95
 moderate inversion, 294
 nMOS devices, 292, 293
 pMOS devices, 292, 293
 strong inversion, 295
 threshold voltage, 293
 transconductance characteristics,
  293-94
 transistor operation, 293
 weak inversion, 294, 295
Metal oxide semiconductor (MOS), 262
Mexican Hat function, 224
Microsaccades, 8
Middle temporal (MT). *See* V5
Midget cells, 10, 24
Miniature eye movement, 191
Minimally invasive surgery (MIS), 184-87
 defined, 184
 endoscopy, 200
 example image, 186
 FOV, 185
 reconstructions, 186, 187
 recovery times and, 185
 robotic-assisted, 201
 setup diagram, 185
 soft tissue defamation, 201
 super-resolution application, 184-87
Moderate inversion, 294
Momiji's model, 41
Moore-Penrose pseudoinverse, 119
Motion channel, 238
Motion detection (MD), 219-30
 categories, 219-20
 computational cost, 230

evolution, 240
HVS, 217
module, 218
motion distillation, 220
perfect, 240
pixel/feature optical flow approach, 219
pixelwise statistical background modeling and subtraction, 220
principle, 239
robust system, 232
spatiotemporal Haar wavelet, 225–29
temporal edge detection, 221–24
wavelet decomposition, 224–25
Motion distillation, 220
Motion processing, 218–19
Motion stabilization, 206–210
Multilayer CNNs (MLCNNs), 349, 352–54
feedforward interlayer, 354
linearized, 352
SFG relations with, 352–54
spatiotemporal filtering capability, 352
STTF implementable by, 354
See also CNNs
Multirate filter bank structures, 108
Multirate schemes, 113
Multiscale coding framework, 59
Mutual suppression, 90

N

Negative impedance converters (NICs), 303
Neurological dorsal/ventral pathway mimicking, 217–42
Neurons
color selectivity, 57
color-tuned, 57
edge-selective, 51
highest bidding, 95
intracortical interactions between, 72
V1, 14, 51, 52, 62
V5, 219
NMOS devices, 292, 293
Nonconjunctive cells, 98
Nonlinear image processing, 157–67
edge detection, 158

features from trace transform, 165–67
thresholding, 157
trace transform, 162–64
V1-inspired edge detection, 158–62
See also Image processing
Nonlinear motions, 406
Normalized convolution, 153–56
Nugget, 148
Numerical simulation examples, 38–46
color characteristics, 43–45
parameters and visual stimuli, 38–39
spatial characteristics, 41–43
temporal characteristics, 39–41
nVidia GeForce Graphics platform, 141

O

Observation data, 124
Ocular vergence
for depth recovery, 202–3
dynamics of, 210
recovered deformation from, 208
ON-OFF ganglion cells, 315–16
Operational transconductance amplifiers (OTAs), 342
Optical flow, 231
Optokinetic eye movement, 192
Organic photodetection, 271–73
Organic semiconductors, 269–85
fabrication, 281–85
forms, 271
photodetection, 271–73
photodiode electronic characteristics, 274–81
photodiode structure, 273–74
phototransduction with, 269–85
polymers, 271
principles, 269, 270–71
small molecules, 271
Orientation
complex dominance field, 122
contrast, 89
dominance, 121
estimation, 119–20
features, emergent grouping, 87–91
irrelevant, 83
maps, generating, 121–23

segmentation, 81, 82
selectivity (V1), 55–56
Orientation-orientation double features, 92
Oriented bandpass frequency response, 118

P

Parasol cell, 10
Particle filters, 239, 240
Parvocellular pathway, 12, 24
PCBM, 271
    as electron-conducting layer, 274
    as electron transporter, 280
    image, 271
PDMS stamps, 282, 283
Phase correlation methods, 184
Phase estimation, 123–24
Phase-invariant feature maps, 138–40
Phase selectivity (V1), 58
Phosphodiesterase (PDE)
    activation, 257
    activation energy barrier, 260
    deactivation, 258
Photoconductors, 261
Photocurrent, 274–77
    current-mode readout, 300
    nonlinear variations, 278
Photodiodes, 261
    bias, 278
    effective response, 279
    efficiency, 274–77
    electronic characteristics, 274–81
    equivalent circuit, 277–81
    operation as voltage and current sources, 296
    photocurrent, 274–77
    sensitivity, 278
    shunt resistance, 277–81
    spectral response characteristics, 281
    structure, 273–74
    technology, 273
    working structure, 273
    See also Organic semiconductors
Photo electric transduction, 296–300
    feedback buffers, 298
    integration-based photodetection circuits, 298–99
    logarithmic sensors, 297–98
    photocurrent current-mode readout, 300
Photogates, 262
Photoreceptor groups, 325
Photoreceptors, 4, 22–23
    classes, 261–62
    cones, 22
    engineering aspects, 259–60
    light adaptation of, 258–59
    process, 255
    rods, 23
    semiconductor, 251–86
    spiking circuit, 314
Phototransduction
    human eye, 253–60
    with organic semiconductor devices, 269–85
    in silicon, 260–69
Phototransduction cascade, 255–58
    absorption, 257
    amplification, 260
    deactivation, 258
    G-protein activation, 257
    ion channel closing, 257
    light activation, 257–58
    PDE activation, 257
Phototransistors, 261
Piecewise-linear method, 152
Pipelining, 383
Pixelwise statistical background modeling/subtraction, 220
Plausibility
    by demonstration, 131–32
    by real image data, 132–33
    vector model, 129–30
PMOS devices, 292, 293, 299
Poisson distributions, 396
Poisson kernels, 116
Polaris optical tracker, 207
Polar separable wavelet design, 113–20
    angular frequency response, 116–18
    filter, 114–16
    filter kernels, 118–19

overview, 113
radial frequency, 114–16
steering and orientation estimation, 119–20
See also V1 wavelet models
Polymers, 271
film thickness, 283
image, 271
Pomerantz's configuration superiority effect, 97
Predict and match tracking, 231
Printing
on CMOS, 284–85
contact, 282–84
Projection onto convex sets (POCS), 152, 173, 177–79
defined, 177
HR images, 178
line profiles, 179
method comparison, 183
reconstruction calculation with, 178
as superior CLS reconstruction method, 178
See also Super-resolution
Pulse width encoding, 316–17
Pupil minus corneal reflection (P-CR), 193
Purkinje images, 194
Pursuit eye movement, 191

Q

Quadrature component, 113
Queuing systems, 323

R

Radial frequency, 114–16
Radon transform, 162, 163
Random access memories (RAMs), 374, 375
available block, 375
external, 375, 377, 384
image word, 386
lookup, 386, 387
on-board, 384
static (SRAMs), 383, 385
Range, 148
Reaction times (RTs)

average, 74
behavioral, 75
color-orientation double feature, 86
shortening by feature doubling, 86
Real-time spatiotemporal saliency, 395–413
adaptive saliency responses, 411–12
complex scene saliency analysis, 412–13
distribution distance, 406–10
feature trackers, 397, 399–404
framework illustration, 397
framework overview, 396–98
framework realization, 398–411
introduction, 395–96
performance evaluation, 411–13
prediction, 404–6
suppression, 410–11
two-dimensional feature detection, 398–99
Receiver operating characteristics (ROC)
contrast marginalized detection, 137
curve, 135
estimation, 136
performance comparison, 136
Receptive fields, 6, 26
center-surround, 27, 28
classical (CRF), 60, 61, 72
complex, modeling, 362–63
eccentricities estimation, 41
Gaussian-spreading, time-decaying, 29–30
tremor and, 40
V1, 51
visual cortical, modeling, 360–62
Recombinant adenoviruses, 60
Reconfigurable fields, 312–13
global control, 313
simulated output, 313
Recursive filter, 404
Register transfer level (RTL), 369
Relative absorbance, wavelength relations, 35
Retina, 5
behavior emulation, 9
defined, 254
eccentricity, 25

hexagonal cells, 7
sensors, 6
single cells, mathematical modeling, 27–28
Retinal anatomy, 21–25
  amacrine cells, 23
  bipolar cells, 23
  cell layers, 22
  ganglion cells, 23–24
  horizontal cells, 23
  pathways, 24
  photoreceptors, 22–23
Retinal function model, 30–38
  color characteristics, 43–45
  color mechanisms, 34–36
  differential equations, 32–34
  foveal image representation, 36–37
  foveal structure, 31–32
  motion modeling, 37–38
  numerical simulation examples, 38–46
  parameters and visual stimuli, 38–39
  spatial characteristics, 41–43
  temporal characteristics, 39–41
Retinal motion, 37–38
Retinal physiology, 25–27
Retinal signal processing, 289
Retinimorphic circuit processing, 300–317
  current mode approaches, 303–12
  image arrays, 300
  intelligent ganglion cells, 314–17
  reconfigurable fields, 312–13
  voltage mode resistive networks, 301–3
Reverse correlation technique, 60
Reverse engineering, 167–68, 171–88
  conclusions/challenges, 188
  hyperacuity and super-resolution, 172–73
  introduction, 171–72
  super-resolution applications, 184–88
  super-resolution image reconstruction methods, 173–84
Rhodopsin, 255–56
  deactivation, 258
  lifetime, 258
Richardson constant, 279
Robotic surgery
  gaze-contingent control for, 200–209
  MIS, 201
Robotic vision, 29
Rods, 5, 23
  function, 254
  light sensitivity, 254–55
  mimicking, 251–86
  See also Cones
Rotation block, 384–86

S

Saccades, 8, 192
Saccadic eye movement, 191
Saliency computation, 94, 96
Saliency hypothesis (V1), 71–73
  color-orientation asymmetry, 81–84
  color-orientation double feature, 84–87
  cortical areas and, 92
  discussion, 92–98
  intracortical interactions, 72
  irrelevant color contrast, 82, 83
  isofeature suppression, 73
  materials and methods, 73–74
  orientation features grouping by spatial configurations, 87–91
  orientation-orientation double feature, 84–87
  procedure and data analysis, 74
  relevant mechanisms, 72
  results, 75–91
  subjects, 74
  task irrelevant features, 73, 76–81
  See also V1
Scale-invariant feature transform (SIFT), 107
Scale selectivity (V1), 57–58
Scale-space image representation, 111
S-cones, 27
SDTP placement technique, 362
Semiconductor photoreceptors, 260–69
  bandwidth, 261
  CCD arrays, 262–63
  chromatic response, 261
  classes, 261–62
  CMOS arrays, 263–65
  color filtering, 265–68

dynamic range, 261
as intrinsic systems, 280
introduction, 251–53
linearity, 261
noise, 261
organic, 269–85
performance characteristics, 261
quantum efficiency, 261
scaling considerations, 268–69
See also Photoreceptors
Sequential Monte Carlo, 232
Shi and Tomasi algorithm, 398–400
Shiftable multiscale transforms, 109
Shift invariance, 109
Shunt resistance, 277–81
Signal flow graphs (SFGs)
cone filtering network implementation, 357
continuous time, 349–51
implementation schematic, 351
relations with MLCNN, 352–54
Signal-to-noise ratio (SNR), 252, 253
Significance test, 76
Silicon
color absorption within, 265
phantom heart, 203
phototransduction in, 260–69
Sill, 148
Single-neuron responses, 52–53
Sinograms, 162
Small molecules, 271
SMAP (x), 70–72, 92
Snowden's interference, 83
Sobel filter, 312
Sparsity, 138
Spatial-acuity tasks, 172
Spatial characteristics, numerical simulation, 41–43
Spatial configurations
orientation feature grouping by, 87–91
testing and predictions, 88
Spatial image Gaussian filter, 303
Spatial saliency, 395
Spatiotemporal cone filter, 355–60
approximation network, 357
defined, 355

frequency response, 356
ideal response, 355
magnitude response, 356
preferred input, 360
SFG for implementation, 357
Spatiotemporal Haar wavelet, 225–30
convolution output, 226
decomposition, 226
See also Wavelets
Spatiotemporal saliency framework, 395–413
adaptive saliency responses, 411–12
complex scene saliency analysis, 412–13
defined, 395
distribution distance, 406–10
efficiency, 398
feature trackers, 397, 399–404
illustrated, 397
overview, 396–98
performance evaluation, 411–13
prediction, 404–6
realization, 398–410
suppression, 410
two-dimensional feature detection, 398–99
Spatiotemporal transfer functions (STTFs), 342, 345
general, 348
mixed-domain, 347
in mixed temporal-Laplace/spatial Z-domain, 346
MLCNN and, 353
rational, nonseparable, 345
stability, 348–49
Spectral response characteristics, 281
Stacking the frames, 221, 222
Static random access memories (SRAMs), 383, 385
Steerable filtering framework, 119
Steering
estimation, 119–20
functions, 120
Strong inversion, 295
Subconscious image processing, 158
SUM rule, 80, 81, 84, 87, 91, 93
Super-resolution, 10, 172–73

algorithm goals, 177
CLS approach, 174–77
defined, 171
high-resolution image recovery, 174
image edges, 188
image reconstruction methods, 173–84
image registration, 183–84
low-resolution images in, 173
MAP, 180
method comparison, 183
in minimally invasive surgery, 184–87
MRF, 180–83
POCS, 177–79
Suppression
isofeature, 73
mutual, 90
spatiotemporal saliency framework, 410
surround, 89
Surround suppression, 89
Surveillance systems, 237

T

Target board, 383
Task irrelevant features, 73
interference, 76–81
interference illustrations, 79
invisible for saliency, 76
noise added by, 96
prediction, 77
Temporal characteristics, numerical simulation, 39–41
Temporal-derived CNN (TDCNN), 355
Temporal edge detection, 221–24
Temporal median filter (TMF), 220, 227
computational cost, 230
numerical results, 229
Tessellating pixel-group layout, 328
Texture features, 96
Thresholding, 157
Totally endoscopic coronary artery bypass grafts (TECAB), 201
recovered epicardial surface deformation for, 209
in vivo results, 202
Trace functionals, 380

Trace transform, 162–64
accelerating, 382–83
area results, 389–91
column processing, 166
computation, 164
computational complexity, 381–82
conclusions, 391
features from, 165–67
flexible functions for exploration, 387–89
full system, 382–87
functional blocks, 386
functional coverage, 389
functions used to produce, 165
hardware implementation, 377–91
initialization, 386–87
introduction to, 377–81
parameters, 382
performance, 389–91
rotation block, 384–86
system overview, 383–84
target board, 383
top-level control, 384
Tracking
blob sorting, 233–36
dual-channel, 230–36
feature, 231
paradigms, 217
predict and match, 231
theory of, 239–41
visual systems, 240
See also Motion detection
Transcranial magnetic stimulation (TMS), 14
Trichromacy, 22
True positive rate (TPR), 135
Type A functional block, 388
Type B functional block, 388–89
Type C functional block, 389, 390

U

Uniform intensity circle, 129
Universal serial bus (USB), 384, 387
Unmatched blobs, 235
Unscented Kalman filter (UKF), 232

V

V1, 12–15, 51–63
  analog platforms, 335–64
  bottom-up saliency map, 69–99
  cell interactions, 59–61
  cell organization, 53–58
  color selectivity, 56–57
  columns, 13
  conjunctive cells, 98
  contextual influences, 70
  defined, 2
  enhanced responses, 52
  feed forward, 58–62
  fingerprints, 98
  for first stage of vision, 13
  as frequency analyzer, 14
  global computation, 59–61
  ice-cube model, 54
  intracortical interactions, 61–62
  issues, 14–15
  low-level decomposition, 51
  neuron arrangement, 51
  neurons, 14, 52, 62
  organization and functions, 52–58
  orientation selectivity, 55–56
  phase selectivity, 58
  receptive fields, 51
  retinotopic map, 53
  role, 3
  saliency hypothesis, 71–72
  scale selectivity, 57–58
  single-neuron responses, 52–53
  spatial phase pooling, 54
  visual stimuli representation, 53
V1-inspired edge detection, 158–62
V1 wavelet models, 105–41
  in computer vision, 120–24
  corner likelihood response, 123
  inference, 124–34
  orientation map generation, 121–23
  overview, 120–21
  phase estimation, 123–24
  polar separable design, 113–20
V5, 218–19
  defined, 15

  neurons, 219
Variogram models, 148–49
  characteristic parameters, 148
  exponential, 149
  fractal, 148
  Gaussian, 149
  linear, 149
  spherical, 149
Vector fields
  conditional random deviation, 131
  mean, 131
  variable constant model, 130–31
Vector image fields, 125
Vergence eye movement, 191
Vernier-type tasks, 172
Very large scale integrated (VLSI) analog circuits, 14–15
Video-oculography (VOG), 193, 194
Virtual ground and scaling circuit, 304
Vision
  chip comparison, 336–39
  computer, V1-like wavelet models in, 120–24
  first stage, 13
  robotic, 29
  wavelets in, 107
Visual cortex, V1 region, 12–15
Visual cortical receptive field modeling, 360–62
Visual inference, 124–34
  divisive normalization, 133–34
  expected fields, 126, 128–29
  formulation of detection, 126–27
  observation data, 124
  plausibility by demonstration, 131–32
  plausibility from real image data, 132–33
  sampling of (B, X), 127–28
  uniform intensity circle, 129
  variable contrast model, 130–31
  vector fields, 125, 130–31
  vector model plausibility and extension, 129–30
Visual stimuli, V1, 53
Voltage mode resistive networks, 301–3

HRES, 301-2
  limitations, 303
  Mead and Mahowald retina chip, 302, 303
  spatial image Gaussian filter, 303
  two-dimensional chips, 302-3
Voronoi method, 152, 153

W

Wavelets, 105-7
  admissibility condition, 106
  choices, 107-11
  color component decompositions, 111
  complex decomposition, 120
  continuous transform, 105-6
  Daubechies, 225
  decomposition, 224-25
  defined, 105
  in image analysis, 107
  linear versus nonlinear mappings, 112-13
  one-dimensional, 105
  polar separable design, 113-20
  properties, 106
  shift invariance, 109
  spatiotemporal Haar, 225-30
  steerable, 116
  two-dimensional complex dual-tree, 110
  in vision, 107
Weak inversion, 294, 295
Weber-Fechner's law, 258-59
Weight vector components, 119
Wiley-Marvasti method, 152

X

Xilinx FPGAs, 390

Z

Zatebradine, 46
Z-transform, 342, 343, 346